Sergey Makarov • Marc Horner
Gregory Noetscher
Editors

Brain and Human Body Modeling

Computational Human Modeling at EMBC 2018

Editors
Sergey Makarov
Massachusetts General Hospital
Boston, MA, USA

Worcester Polytechnic Institute
Worcester, MA, USA

Gregory Noetscher
Worcester Polytechnic Institute
Worcester, MA, USA

Marc Horner
ANSYS, Inc.
Evanston, IL, USA

https://doi.org/10.1007/978-3-030-21293-3

Preface to *Computation Human Models and Brain Modeling: EMBC 2018*

Numerical modeling has become an essential enabling technology in a variety of engineering fields, including mechanics, chemistry, fluid dynamics, electromagnetics, and acoustics. Modeling accelerates the product development cycle, giving scientists and engineers the opportunity to explore design iterations and scenarios in virtual space, and allowing optimization over a host of external conditions that would be time and cost prohibitive to experimentally characterize and quantify. Furthermore, simulations permit the examination of resulting field values, such as internal current distributions or energy absorption in tissues that would typically not be available to an experimentalist due to safety or ethics concerns.

However, regardless of the physics under consideration or the method utilized, every practitioner in the field of numerical modeling knows one fundamental rule: the simulation is only as good as the underlying model being employed. This is a more articulate or eloquent way of saying "garbage in equals garbage out," but regardless of how it is phrased, the message is the same. If there are fundamental flaws or inaccuracies in the model that mask or modify the physics under examination, even if the simulation itself runs flawlessly, results might be erroneous and predictions based on that simulation will not accurately embody the intended aspects of the physical world. It is with this motivation in mind that developers of phantoms characterizing the human body and its corresponding physiological processes have continuously advanced the state of the art and pursued ever more accurate representations of human anatomy at a variety of geometric scales.

Advancements in human phantoms are a product of many converging disciplines, ranging from the basic sciences of chemistry, biology, and physics to more applied areas such as electrical and computer engineering, material science, medical data acquisition and segmentation, surface and volumetric mesh manipulation, and large-scale data processing. The memory and computational processing limitations encountered in previous model generations, where human bodies were represented with basic, homogeneous geometric primitives or highly de-featured faceted models, no longer apply to modern simulation platforms. The rapid advance in computing hardware permits a new generation of ever more detailed models with substantially enhanced levels of anatomical accuracy. Similarly, the incorporation of sophisticated material properties to support coupled multi-physics simulations is also now possible. Furthermore, advances in our understanding of the anatomy and physiology of the human body continue to provide ever-growing insights into the tissue properties and their detailed organization at the macro- and microscopic levels, thus enabling models that increasingly capture the most relevant properties.

As human models have improved, the scope of applications examined via simulation has also grown, providing researchers and engineers with powerful tools to explore new and exciting hypotheses regarding human physiology, pathophysiology, and biomedical engineering. The application of electromagnetic fields in biomedical engineering has produced promising diagnostic and therapeutic methodologies and protocols that may now be competently and thoroughly studied to generate detailed analyses on estimated efficacy and patient safety. Topics of recent interest to the research and medical communities are broadly distributed across the electromagnetic frequency spectrum and include: cancer ablation via radio frequency (RF) heating; safety and efficacy assessments of patients with and without implanted medical devices during procedures such as magnetic resonance imaging (MRI); new and varied coil designs for optimal MRI protocols; treatment of brain disorders, such as depression, via noninvasive brain stimulation techniques like transcranial magnetic stimulation (TMS) and transcranial direct current stimulation (tDCS); optimal design, configuration, and placement of single or multiple coils or electrodes for focused and deep internal electromagnetic field generation; pain management therapies that rely on noninvasive nerve stimulation rather than potentially addictive pharmaceuticals; and many others. While seemingly disparate, these applications are united in their need for high-quality computational human phantoms and optimized simulation methods that enable fast and accurate approximations of the underlying physics that govern responses of the body to externally applied electromagnetic stimuli. This is the motivation that drives the research contained in this work and has provided inspiration to the researchers and engineers laboring in this field.

This work is a collection of selected papers presented during the third Annual Invited Session on Computational Human Models. The session was conducted from July 17 to 21, 2018, in Honolulu, HI, as part of the 40th Annual International Conference of the IEEE Engineering in Medicine and Biology Society (EMBS), which provided a collaborative platform showcasing academic and commercial research representing the current state of the art in computational human models and applications for which they are employed. The various session tracks brought together subject matter experts in diverse fields representing academia, government institutions, and industry partners. A clear outcome of this effort was a comprehensive, multidisciplinary review of each area, and the promotion of a rich dialogue on promising future paths in human phantom development, numerical methods, and simulation applications. The chapters presented here provide an overview of the invited session and highlight a myriad of potential avenues of development and exploration during future EMBS conferences.

The first section presents chapters devoted to models specifically tailored for noninvasive stimulation. A collection of techniques that employ the physics of electromagnetism to stimulate specific regions of human anatomy are reviewed. The research is aimed at treating various pathologies, including neurological disorders treated with noninvasive brain stimulation and chronic pain treated via peripheral nerve stimulation. Several brain stimulation modalities are presented along with custom models that have been generated to best represent the anatomic features most affected by these treatments.

The second section is devoted to tumor-treating fields (TTFields), which is a new and promising treatment for glioblastoma that was recently approved by the US Food and Drug Administration. The simulations employed in these chapters include human models that inform practitioners on the impact of electrode placements on the surface of the body, leading to optimization of electrode configurations and knowledge-based estimates of the resulting field strengths within the body. This enables practitioners of TTFields protocols to optimally align the direction of the fields produced by the electrodes, examine field penetration, and conduct studies investigating the effects of numerous parameters (including field frequency and intensity) on estimated tumor and glioblastoma treatment.

Section three is a collection of investigations into how computational human models may be used to evaluate safety concerns for a variety of applications. These investigations include patient-specific models generated from medical imaging data to customize treatments as well as modified models adapted to integrate implanted medical devices for assessing safety during MRI. The section also includes an examination of bioelectricity at the cellular level and a study on techniques related to microwave ablation. Industrial radiography accidents and models employed to examine brain hemorrhage characteristics are also considered.

The final section details efforts related to customized human models tailored to specific applications. These include incorporating a dynamic breathing sequence into a normally static model to simulate human respiration, integration of highly resolved and detailed ear canal structures for simulation of wearable devices, conversion of voxel-based models to polygon surface models, and a new technique for measuring material conductivity.

While the exciting work presented here is indeed impressive, there is much yet to accomplish to enhance current modeling and simulation capabilities. Several sessions at the upcoming 41st Annual International Conference of the IEEE Engineering in Medicine and Biology Society to be held in Berlin, Germany, on July 23–27, 2019, will be devoted to model generation and related applications. These sessions will offer both extensions to the results given in 2018 and new research that will expand the field of computational human phantom generation.

Berenson-Allen Center for Noninvasive Brain Stimulation and Division for Cognitive Neurology, Beth Israel Deaconess Medical Center, and Harvard Medical School, Boston, MA, USA

Alvaro Pascual-Leone

Institut Guttman de Neurorehabilitación, Universitat Autónoma de Barcelona, Barcelona, Spain

Contents

Part I
Human Body Models for Non-invasive Stimulation

Chapter 1
SimNIBS 2.1: A Comprehensive Pipeline for Individualized Electric Field Modelling for Transcranial Brain Stimulation

Guilherme B. Saturnino, Oula Puonti, Jesper D. Nielsen, Daria Antonenko, Kristoffer H. Madsen, and Axel Thielscher

1.1 Introduction

Non-invasive brain stimulation (NIBS) aims at modulating brain activity by inducing electric fields in the brain [1]. The electric fields are generated either by a magnetic coil, in the case of transcranial magnetic stimulation (TMS), or by a current source and electrodes placed directly on the scalp, in the case of transcranial electric stimulation (TES). In both cases, the induced electric fields in the brain have a complex and often counter-intuitive spatial distribution, which is dependent on the individual anatomy of a target subject. In recent years, there has been a growing interest in moving away from a one-size-fits-all stimulation approach in NIBS to more individually informed protocols [2]. The driving force behind this shift is the

---

Guilherme B. Saturnino and Oula Puonti contributed equally to this chapter.

G. B. Saturnino · A. Thielscher (✉)
Danish Research Centre for Magnetic Resonance, Centre for Functional and Diagnostic Imaging and Research, Copenhagen University Hospital Hvidovre, Hvidovre, Denmark

Department of Health Technology, Technical University of Denmark, Kongens, Lyngby, Denmark
e-mail: axelt@drcmr.dk

O. Puonti
Danish Research Centre for Magnetic Resonance, Centre for Functional and Diagnostic Imaging and Research, Copenhagen University Hospital Hvidovre, Hvidovre, Denmark

J. D. Nielsen · K. H. Madsen
Danish Research Centre for Magnetic Resonance, Centre for Functional and Diagnostic Imaging and Research, Copenhagen University Hospital Hvidovre, Hvidovre, Denmark

Department of Applied Mathematics and Computer Science, Technical University of Denmark, Kongens, Lyngby, Denmark

D. Antonenko
Department of Neurology, Universitätsmedizin Greifswald, Greifswald, Germany

S. Makarov et al. (eds.), *Brain and Human Body Modeling*,
https://doi.org/10.1007/978-3-030-21293-3_1

widely reported variation of NIBS effects within and between individuals [3], which could be explained in part by the interplay of the individual anatomy and the electric field propagation [4]. Although software tools have become available that generate realistic anatomical models of the head based on magnetic resonance imaging (MRI) scans and use those models to numerically estimate the electric field induced in the brain, they are still not predominantly used in NIBS studies. This is likely due to the lack of robustness and usability of the previous generation of tools, in turn hampering the individualized application of NIBS in both mapping the human brain function and as a rehabilitation tool in various neuropathologies [5, 6].

The aim of SimNIBS is to facilitate the use of individualized stimulation modelling by providing easy-to-use software tools for creating head models, setting up electric field simulations, and visualizing and post-processing the results both at individual and group levels. SimNIBS was first released in 2013 [7], had a major update in 2015, with the release of version 2 [2], and more recently another major update with the release of version 2.1, described in the current work. SimNIBS 2.1 is a free software, distributed under a GPL 3 license, and runs on all major operating systems (Windows, Linux and MacOS). In this tutorial, we will concentrate on **what** SimNIBS 2.1 can be used for and **how** the analyses are performed in practice with step-by-step examples. The chapter is structured as follows: First, we give a general overview of the simulation pipeline and of its building blocks. Next, we provide a step-by-step example of how to run a simulation in a single subject, and then we demonstrate a set of MATLAB tools developed for easy processing of multiple subjects. Finally, we conclude with an analysis of the accuracy of automated electrode positioning approaches. More information, as well as detailed tutorials and documentation can be found from the website www.simnibs.org.

1.2 Overview of the SimNIBS Workflow

Figure 1.1 shows an overview of the SimNIBS workflow for an individualized electric field simulation. The workflow starts with the subject's anatomical MRI images, and optionally diffusion-weighted MRI images. These images are segmented into major head tissues (white and grey matter, cerebrospinal fluid, skull and scalp). From the segmentations, a volume conductor model is created, and used for performing the electric field simulations. The simulations can be set up in a graphical user interface (GUI) or by scripting. Finally, the results can be mapped into standard spaces, such as the Montreal Neurological Institute (MNI) space or FreeSurfer's FsAverage.

1.2.1 Structural Magnetic Resonance Imaging Scans

The minimum requirement for running an individualized SimNIBS simulation is a T1-weighted structural scan of a subject's head anatomy. Although SimNIBS will run on almost all types of T1-weighted scans, we have found that setting the readout

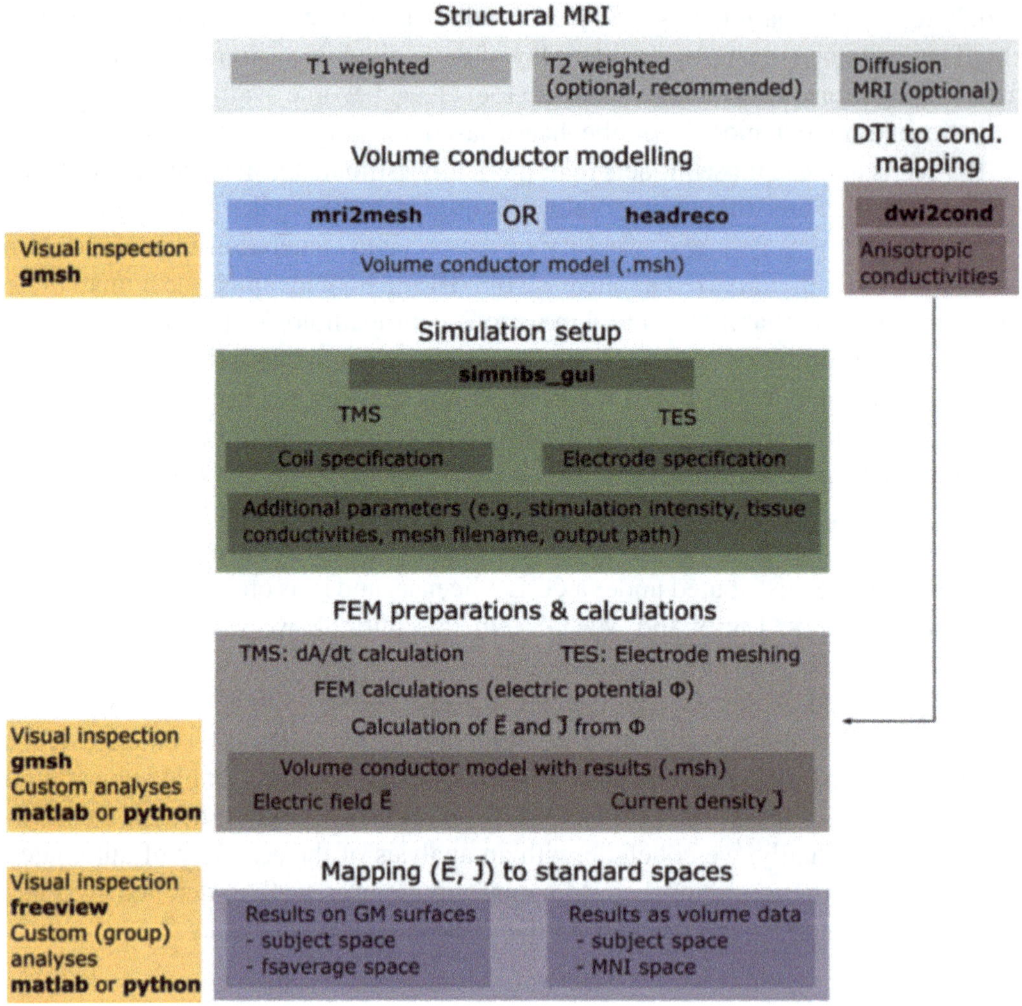

Fig. 1.1 Overview of the SimNIBS workflow

bandwidth low to ensure a good signal-to-noise ratio in the brain region and using a fat suppression method, such as selective water excitation, to minimize the signal from spongy bone, typically ensure a high quality of the resulting head models. See Fig. 1.2 for an example of good quality scans we found to work well with SimNIBS and [8] for the details of the sequences.

Including a T2-weighted scan is optional, but highly recommended as it facilitates accurate segmentation of the border between skull and cerebrospinal fluid (CSF). Both skull and CSF appear dark in T1-weighted scans, whereas in T2-weighted scans the CSF lights up, thus guiding the separation between the tissues. Skull has a low electric conductivity, while CSF is highly conducting, meaning that any segmentation errors in these two compartments can have a large effect on the resulting electric field distribution inside the head, especially when TES is applied [8]. If you are interested in modelling the neck region in detail, we recommend using neck coils if these are available at the imaging site.

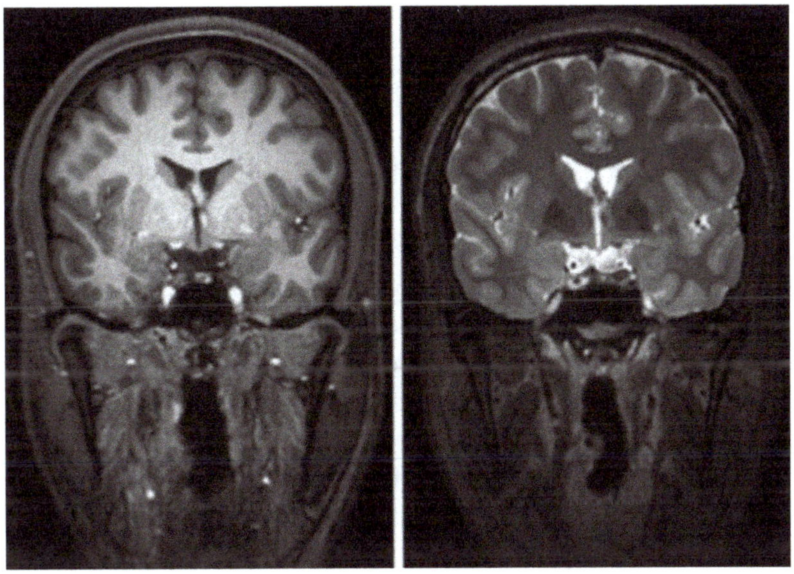

Fig. 1.2 Example of high-quality T1- and T2-weighted scans likely to work well with SimNIBS. Note that in the T1-weighted scan, the skull appears dark due to the fat suppression

Optionally, SimNIBS also supports modelling of anisotropic conductivities for grey (GM) and white matter (WM), which requires a diffusion-weighted MRI scan (dMRI). Only single shell data (i.e. with a single b-value in addition to some b = 0 images) with a single phase encoding direction for the echo planar imaging (EPI) readout is supported.

1.2.2 Volume Conductor Modelling

The first step in the pipeline is the generation of a volume conductor model of the head, which is needed for simulating the induced electric fields. In order to create this finite element (FEM) mesh, we need to assign each voxel in the MRI scan(s) to a specific tissue class, i.e. to segment the scan into the different head tissues. Currently, SimNIBS offers two options for segmentation: **mri2mesh** [7] and **headreco** [8].

mri2mesh combines FSL [9] (version 5.0.5 or newer) and FreeSurfer [10] (version 5.3.0 or newer) to segment the head tissues. FSL is used to segment the extra-cerebral tissues, while FreeSurfer is used to segment the brain and to generate accurate surface reconstructions of the grey matter sheet. Note that **mri2mesh** is restricted only to the head and does not create models of the neck region.

headreco uses the SPM12 [11] toolbox for segmenting the MRI scan, and is now the recommended option in SimNIBS. It has been shown to be more accurate in segmenting the extra-cerebral structures, especially the skull, compared to **mri2mesh** [8], while also providing accurate segmentations of the brain tissues. The computational anatomy toolbox (CAT12, recommended) [12] provided with

SPM can be used to create surface reconstructions of the grey matter sheet which are on par with the accuracy of those generated by FreeSurfer [12]. In addition, **headreco** has an extended field of view, also modelling the neck region. For ease of use, both SPM12 and CAT12 are distributed together with SimNIBS.

Once the segmentation by either method has finished successfully, the tissue maps are cleaned by applying simple morphological operations, and used to create surface reconstructions. As a final step, the FEM mesh is generated by filling in tetrahedrons between the tissue surfaces using Gmsh [13].

Neither **mri2mesh** nor **headreco** have off-the-shelf support for pathologies such as tumours or lesions. These can however be included into the head models by manually editing the segmentation masks generated by the methods. When using **mri2mesh**, please consult the FreeSurfer website (https://surfer.nmr.mgh.harvard.edu/fswiki/FsTutorial/WhiteMatterEdits_freeview) on how to handle scans with pathologies. Manual edits using **headreco** should be done on the output segmentation masks in the *mask_prep* folder located within the *m2m_{subID}* folder. Once corrections have been made, the surface meshing step (**"headreco surfacemesh subID"**) and volume meshing step (**"headreco volumemesh subID"**) should be re-run to generate the edited head model. Note that when creating head models from scans with pathologies, the CAT12 toolbox should *not* be used.

dwi2cond (optional) uses FSL (version 5.0.5 or newer) to prepare diffusion tensors for GM and WM from dMRI data. The tensors are used by SimNIBS to estimate anisotropic conductivities in WM and GM during the FEM calculations.

1.2.3 Simulation Setup

Simulations can be set up using the graphical user interface (GUI), which provides an interactive view of the head model. This allows users to easily select parameters such as coil positions, electrode positions and shapes, as well as more advanced settings such as tissue conductivities and post-processing options.

It might also be of interest to do simulations of one or a few different setups across a group of subjects. With this in mind, version 2.1.1 introduced a new interface for setting up simulations using MATLAB or Python scripts.

The GUI as well as the scripts will be described in more detail in Sect. 1.3, as well as on the website www.simnibs.org.

1.2.4 Finite Element Method Calculations

Transcranial direct current stimulation (tDCS) simulations begin by adding electrodes to the head model. In this step, nodes in the skin surface are shifted to form the shape of the electrode, while keeping good quality elements. Afterwards, the body of the electrodes is constructed by filling in tetrahedra. As this step does not require re-meshing the entire head, it can be done much more efficiently compared to other methods that require re-meshing, especially when only a few electrodes are used.

TMS simulations start by calculating the change in the magnetic vector potential $\boldsymbol{A}$, that is the $d\boldsymbol{A}/dt$ field in the elements of the volume conductor mesh for the appropriate coil model, position and current. There are currently two types of coil models:

.ccd files: Created from geometric models of the coil and represented as a set of magnetic dipoles from which we can calculate the $d\boldsymbol{A}/dt$ field using a simple formula [14].

.nii files: Created either from geometric models of the coils or direct measurement of the magnetic field [15]. Here, the $d\boldsymbol{A}/dt$ field is defined over a large volume, and the calculation of the $d\boldsymbol{A}/dt$ at the mesh elements is done via interpolation. This allows for faster simulation setup at little to no cost in simulation accuracy.

Both simulation problems are solved using the FEM with linear basis functions. This consists of constructing and solving a linear system of the type $\boldsymbol{Mu} = \boldsymbol{b}$, where $\boldsymbol{M}$ is a large (in SimNIBS typically ~$10^6 \times 10^6$) but sparse matrix, called the "stiffness matrix", $\boldsymbol{u}$ are the electric potentials at the nodes and the right-hand side $\boldsymbol{b}$ contains information about boundary conditions (such as potentials in electrode surfaces in tDCS simulations), and source terms (such as the $d\boldsymbol{A}/dt$ field in TMS simulations). SimNIBS solves the linear system using an iterative preconditioned conjugate gradient method [16]. SimNIBS 2.1 uses GetDP [17] to form the linear system, which in turn calls PETSc [18] to solve it.

TDCS simulations can also be easily extended to simulations of transcranial alternating current stimulation (tACS). In the frequency ranges used in tACS, a quasi-static approximation holds [19]. In the quasi-static approximation, the relationship between input currents $I(t)$ and the electric field at the positions $\boldsymbol{x}$, $E(\boldsymbol{x})$ is linear:

$$E(\boldsymbol{x},t) = \alpha(\boldsymbol{x}) I(t)$$

where $\alpha(\boldsymbol{x})$ is a proportionality constant, meaning that it does not vary during the oscillation. This constant can be obtained simply by running a simulation where we set the input current to unity. $I(t)$ is the input current. For example, a sinusoidal current input can be written as

$$I(t) = I_o \sin\left(2\pi \, t/f + \phi\right)$$

where f is the stimulator frequency, ϕ the stimulator phase and I_o the stimulator amplitude, which corresponds to half of the peak-to-peak current. Usually, we would visualize the electric field at the maximum or minimum of $I(t)$, which corresponds to $\pm I_o$. In case several stimulators are used at different frequencies of phases, we have several pairs $(\alpha_i(\boldsymbol{x})I_i(t))$, one for each stimulator, and the total electric field at a given time point is given by the sum of their individual contributions

$$E(\boldsymbol{x},t) = \sum_{i=1}^{n} \alpha_i(\boldsymbol{x}) I_i(t)$$

1.2.5 *Mapping Fields*

The result of the FEM calculation is the electric field at each tetrahedral element of the subject's head mesh. However, visualization is often easier using cortical surfaces or NifTI volumes. Therefore, SimNIBS 2.1 can transform fields from the native mesh format to these formats via interpolation. Our interpolation algorithm is based on the superconvergent patch recovery method [20], which ensures interpolated electric field values that are consistent with tissue boundaries.

When performing simulations on multiple subjects, we often want to be able to directly compare the electric field across subjects to, for example, correlate the electric field with behavioural or physiological data on the stimulation effects [21]. For this purpose, SimNIBS can also transform simulation results to the MNI template, using linear and non-linear co-registrations, as well as to the FreeSurfer's FsAverage surface.

1.3 Practical Examples and Use Cases

1.3.1 *Hello SimNIBS: How to Process a Single Subject*

Here we describe how to run a TMS and a tDCS simulation on a single example subject. The example subject "Ernie" can be downloaded from the SimNIBS website, and the steps below can be reproduced step by step to get familiar with SimNIBS.

Generating the Volume Conductor Model

Open a terminal and go to the directory "ernie" to access the example data set. Copy the content of the "org"-subfolder to another location in order to not overwrite the files of the original example dataset. Next, go to the folder where you copied the data, and call headreco to generate the volume conductor model:

headreco all --cat ernie ernie_T1.nii.gz ernie_T2.nii.gz

In the command, the first argument, "**all**", tells headreco to run all reconstruction steps including: segmentation, clean-up of tissue maps, surface meshing, and volume meshing. The second argument, "**--cat**" is a flag for using the CAT12 toolbox for accurate reconstruction of the cortical surface. The third argument, "**ernie**", is a subject identifier (subID), which is used to name generated folders, e.g. m2m_ernie, and output files, e.g. ernie.msh. The two final arguments are the paths to the T1- and T2-weighted structural scans.

A few extra input options are useful to know:

-d no-conform Adding this option will prevent headreco from modifying, i.e. transforming and resampling, the original MRI scan. This might be desirable when a one-to-one correspondence between the head model coordinates and the neural navigation system coordinates is required.

-v < density > This option allows you to set the resolution, or vertex density (nodes per mm^2), of the FEM mesh surfaces. By default, SimNIBS uses 0.5 nodes/mm^2 as the **<density >** value.

In general, we recommend using the **--cat** option; however, the execution time will be longer compared to omitting the option. In addition, if you want to process scans with pathologies, you should not use CAT12, as the cortical reconstruction is not designed to work with pathologies.

After **headreco** has finished, please check the quality of the head model by calling:

headreco check ernie

If needed, open a new terminal for this operation and go into the folder in which you started headreco the first time. For our example case, the subject identifier is "ernie", but please replace this one with whichever subID was used in the first call to **headreco**. Note that we recommend that you have installed freeview (provided by FreeSurfer, available on Linux and Mac OS X platforms) to visualize the results. The **check** function displays two windows for inspecting the output. The first window shows the T1-weighted scan with the segmentation and structure borders overlaid (Fig. 1.3, left). We recommend de-selecting the segmentation (**ernie_final_contr.nii**) in freeview, and checking that the segmentation borders follow the intensity gradients of different tissues (Fig. 1.3, middle). Fig. 1.4 shows the second freeview window, which displays the T1-weighted scan co-registered to the MNI template. We recommend checking if the T1-weighted scan overlaps well with the MNI template by de-selecting the T1-weighted scan (**T1fs_nu_nonlin_MNI.nii**) in freeview (Fig. 1.4, right). Figure 1.5 shows an example of a segmentation error where the skull is erroneously labelled as skin. This can be seen in the front of the head, where the skin label protrudes into the skull. This example emphasizes the need for fat-suppressed data when only a T1-weighted scan is used. In the scan shown in Fig. 1.5, spongy bone is bright with intensities comparable to those of scalp, causing the segmentation method to mis-classify it as extra-cerebral soft tissue. Small segmen-

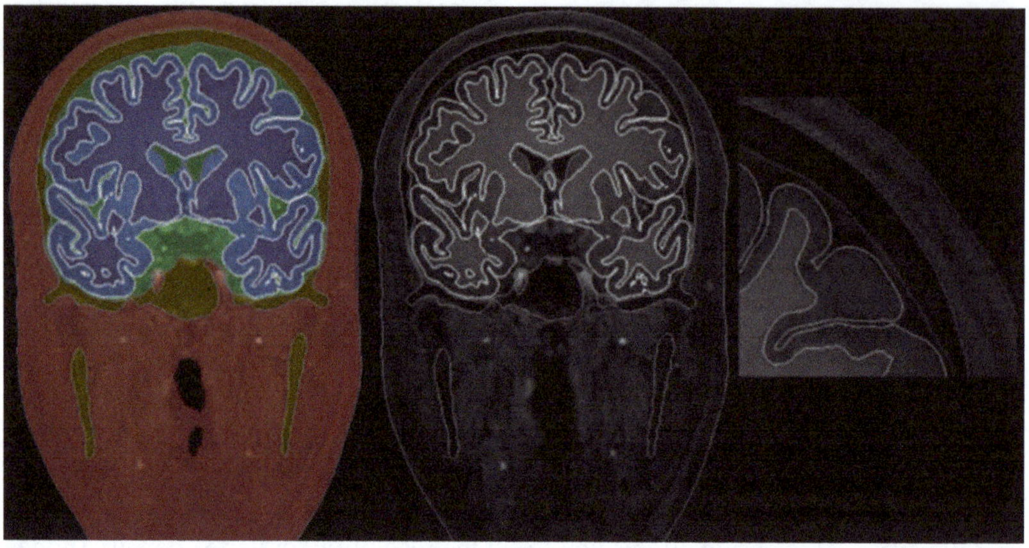

Fig. 1.3 Data displayed after calling the **check** option. Left: T1-weighted scan with the segmentation and structure borders overlaid. Middle: structure borders overlaid on the T1-weighted scan after de-selecting the segmentation in freeview. Right: zoom-in of the cortex. Note that the segmentation borders nicely follow the intensity borders between the tissues

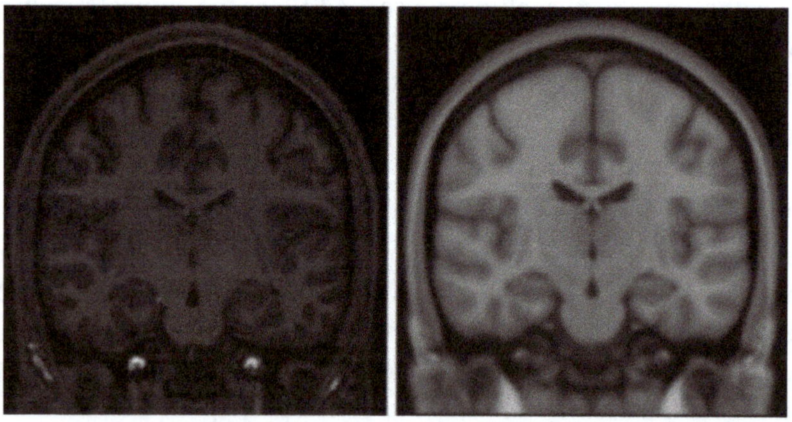

Fig. 1.4 Data displayed after calling the **check** option. Left: T1-weighted scan co-registered to the MNI template. Right: MNI template shown after de-selecting the T1-weighted scan in freeview. Note that the scans seem to be well registered

tation errors like this can be corrected by manually re-labelling the segmentation masks in the "*mask_prep*" folder located in the *m2m_{subID}* folder, and re-running the surface and volume meshing steps. If you are not familiar with using freeview, please refer to the tutorial on the SimNIBS website (*http://www.simnibs.org/_media/docu2.1.1/tutorial_2.1.pdf*). If you do not have access to freeview, the visualizations will be displayed using SPM. However, these are very primitive and are not recommended for checking the output from **headreco**.

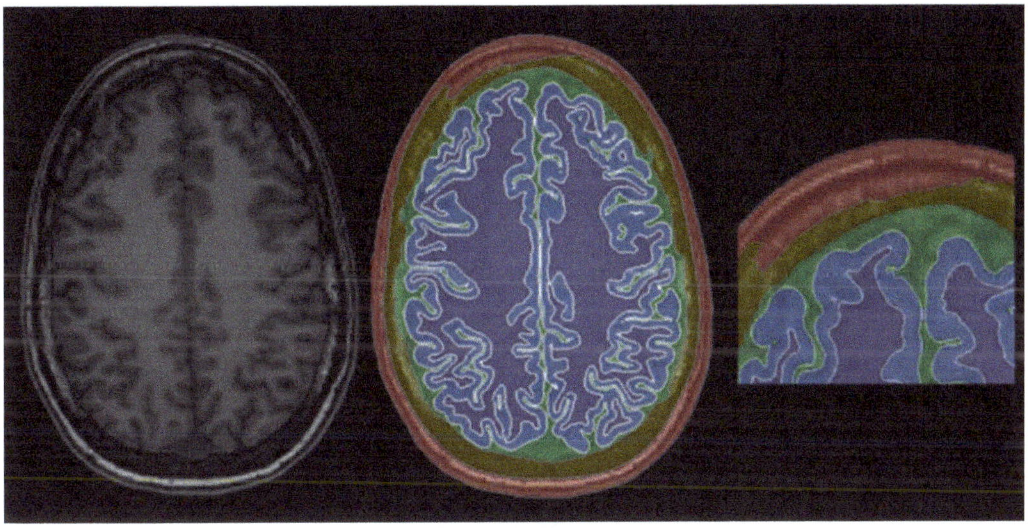

Fig. 1.5 Example of a segmentation error after headreco processing. The spongy bone is erroneously labelled as skin. This example emphasizes the need for fat-suppression when using only a T1-weighted scan

Finally, you should inspect the volume conductor mesh for any obvious errors. This can be done by calling:

gmsh ernie.msh

in the subject folder. This call opens a gmsh window displaying the generated head model; please see the tutorial on the website if you are not familiar with gmsh (*http://www.simnibs.org/_media/docu2.1.1/tutorial_2.1.pdf*).

The folder structure and most important files are shown in Table 1.1.

- *eeg_positions/* Folder containing the 10-10 electrode positions for the subject both as a "*.csv*", used for acquiring electrode positions, and a "*.geo*" file, used for visualization of the positions in Gmsh. If you have custom electrode positions, they should be added here as a .csv file.
- *mask_prep/* Folder containing the cleaned tissue maps along with the white matter and pial surface files if CAT12 was used. In case there are errors in the segmentation, the masks can be manually corrected and a new head model can subsequently be generated. Note that the CAT12 WM and GM surfaces can currently not be modified.
- *headreco_log.html*, a log-file with output from the **headreco** run. If something goes wrong, the log-file helps with troubleshooting, and should be sent as an attachment when contacting the SimNIBS support email list (support@simnibs.org).
- *ernie.msh*, the FEM head model used for the simulations.
- *ernie_T1fs_conform.nii.gz*, the input scan in the conform space defined by the –d option. This scan has the same millimetre space as the head model, and can be used to annotate landmarks which can then be directly transformed onto the head model.

Table 1.1 The folder structure after **headreco** has finished. In this table, only the most important folders and files are listed

```
ernie/
├── m2m_ernie/
│   ├── eeg_positions/
│   ├── mask_prep/
│   ├── ...
│   └── headreco_log.html
├── ernie.msh
└── ernie_T1fs_conform.nii.gz
```

Setting Up a Simulation

Once the head model is ready, we can set up tDCS and TMS simulations interactively using the GUI. The GUI can be started on the command line by calling:

simnibs_gui

In the GUI, the user can:

- Visualize and interact with head models.
- Define electrode and coil positions by clicking in the model or selecting a position from the EEG 10-10 system.
- Visually define electrode shapes and sizes.
- Select from the available coil models.
- Change tissue conductivity parameters and set up simulations with anisotropic conductivity distributions.
- Run simulations.

In the GUI, there are two types of tabs, one for tDCS simulations, and another for TMS simulations, shown respectively in the top and bottom of Fig. 1.6. The tDCS tabs define a single tDCS field simulation with an arbitrary number of electrodes. On the other hand, TMS tabs can define several TMS field simulations using the same coil. For this example, we will set up a tDCS simulation with a 5 × 5 cm anode placed over C3 and a 7 × 5 cm cathode placed over AF4, and a TMS simulation with the coil placed over the motor cortex, pointing posteriorly. Details on how to use the graphical interface can be found on the website (*http://www.simnibs.org/_media/docu2.1.1/tutorial_2.1.pdf*).

After the simulation setup, click on the **Run** button to start the simulations. Running both simulations takes 10–15 minutes, depending on the computer, and uses around 6 GB of memory. As a note, before starting the simulations, you can set additional options (in the menu Edit➔Simulation Options) to let SimNIBS write out the results as surface data or NifTI volume data. This is not further covered in this basic example, but the output files created in these cases are described in the next example. The results of the simulation will be written in the output folder specified in the GUI, in this case "*simnibs_simulation/*". The folder has the files shown below in Table 1.2.

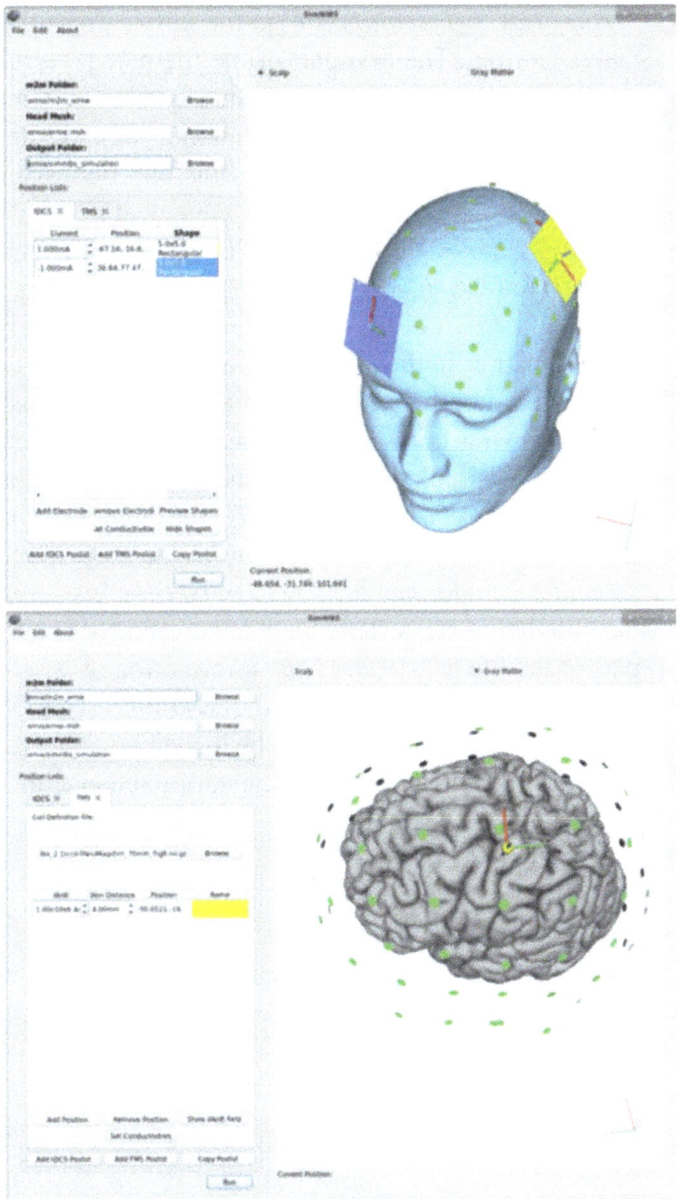

Fig. 1.6 Set-up of a tDCS (top) and a TMS (bottom) simulation in the graphical user interface

Table 1.2 The output folder of a simple tDCS and TMS simulation

```
simnibs_simulation/
├── ernie_TDCS_1_scalar.msh
├── ernie_TMS_2-0001_Magstim_70mm_Fig8_nii_scalar.msh
├── ernie_TMS_2-0001_Magstim_70mm_Fig8_nii_coil_pos.geo
├── simnibs_simulation_20180920-130401.log
└── simnibs_simulation_20180920-130401.mat
```

- "*ernie_TDCS_1_scalar.msh*" is the output from the tDCS simulation, in Gmsh "*.msh*" format. The first part of the file name, "*ernie*", is the subID. The second part, "*TDCS*", informs us that this is a tDCS simulation. The third part, "*1*", denotes that this was the first simulation we have defined in the GUI, and finally, "*scalar*" tells us have used scalar (as opposed to anisotropic) conductivities for the simulations.
- "*ernie_TMS_2-0001_Magstim_70mm_Fig8_nii_scalar.msh*" is the output of the second simulation, also in gmsh "*.msh*" format. As is the case for the tDCS output, the first part of the file name is the subID, and the second is the number of the simulation in the simulation list. We next see the number of the TMS position, as it might happen that several TMS positions are defined in a single TMS list. Following this, "*Magstim_40mm_Fig8_nii*" gives us the name of the coil used for the simulation, and "*scalar*" the type of conductivity.
- "*ernie_TMS_2-0001_Magstim_70mm_Fig8_nii_coil_pos.geo*" is a Gmsh ".geo" file which shows the coil position for the corresponding simulation.
- "*simnibs_simulation_20180920-13041.log*" is a text file with a detailed log of the simulation steps. This file can be used for troubleshooting. Here, the second part of the file is date and time information of when the simulation started.
- "*simnibs_simulation_20180920-13041.mat*" is a MATLAB data file with the simulation setups. This file can be loaded into the GUI or MATLAB at a later time to check the simulation parameters, or to change them and re-run the simulation.

Visualizing Fields

The electric field ***E*** is a vector field meaning that the electric field has both a norm (i.e. vector length or magnitude) and a direction in space, as shown in Fig. 1.7. As visualizations of the entire vector are challenging and often unclear, in SimNIBS we

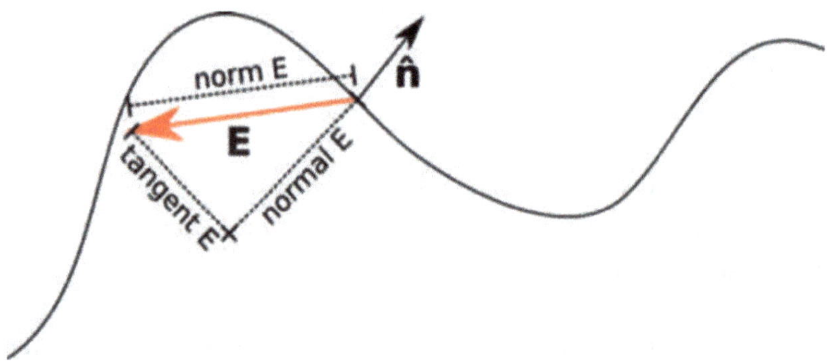

Fig. 1.7 Decomposition of a vector **E** in relation to a surface. The norm corresponds to the length of the vector. At each point, the surface defines a normal vector $\hat{\mathbf{n}}$, and this vector is perpendicular to the tangent plane to the surface at that point. Given the normal vector, we can decompose the vector **E** into normal and tangent components. The normal component is the part of **E** in the same line as the normal vector, and the tangent component is perpendicular to it. The normal component also has a sign, indicating if the field is entering or leaving the surface. In SimNIBS, a positive normal indicates that the field is entering the surface, and a negative normal indicate the field is leaving the surface

usually visualize the **norm** (or strength) of the electric field instead. The norm of the electric field corresponds to the size of the electric field vector, and therefore is always positive and does not contain any information about the direction of the electric field.

One way we can quickly visualize the simulation results is to use the **mesh_show_results** MATLAB function. This function comes as a part of SimNIBS version 2.1.2, and provides visualizations of the output fields using MATLAB plotting tools, as well as some summary values for the field strength and focality. For example, when running the function on the output tDCS mesh, we obtain the plot shown in Fig. 1.8a, and the values below in Table 1.3.

The first lines in Table 1.3 show that the displayed data is the field "norm E", that is the norm or strength of the electric field, calculated in the region number 2, which corresponds to the GM volume. Afterwards, we have information on the peak electric fields. We see that the value of 0.161 V/m corresponds to the 95th percentile of the norm of the electric field, the value of 0.201 V/m to the 99th percentile and 0.249 to the 99.9th percentile. We also have information about the focality of the

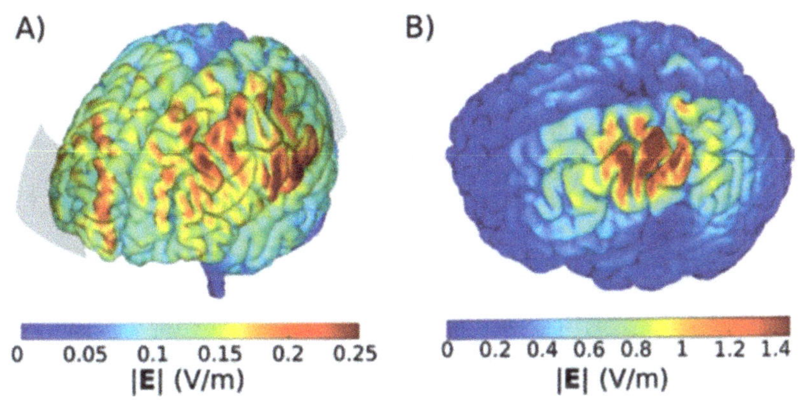

Fig. 1.8 Visualization of (**a**) tDCS and (**b**) TMS electric field norms in MATLAB

Table 1.3 Output of **mesh_show_results** for the tDCS simulation

```
---------------------------------------------
SUMMARY
field name: normE
region indices: 2

peak fields
percentiles:   95      99      99.9
values:        0.161   0.201   0.249 (in [V/m])

focality
cutoffs:       50        75 (in % of 99.9 percentile)
values:        1.4e+05   1.29e+04 (in cubic mm)
---------------------------------------------
```

electric field. Here, focality is measured as the GM volume with an electric field greater or equal to 50% or 75% of the peak value. To avoid the effect of outliers, the peak value is defined as the 99.9th percentile.

Running the same function on the TMS result file, we obtain the plot shown in Fig. 1.8b, as well as the peak fields and focality measures shown below in Table 1.4.

We can see that the peak fields for TMS are much higher than for tDCS, even though we simulated with a current of 10^6 A/s, very low for TMS. In the focality measures, we see that the TMS electric fields are much more focal than the tDCS electric fields, with around five times less GM volume exceeding 75% of the peak value than tDCS.

Additionally, the "*.msh*" files can be opened with the Gmsh viewer, producing 3D visualizations as shown in Fig. 1.9. Gmsh has a vast range of functionalities, such as clipping planes, but can be harder to use than **mesh_show_results**.

Table 1.4 Output of **mesh_show_results** for the TMS simulation

```
----------------------------------------------
SUMMARY
field name: normE
region indices: 2

peak fields
percentiles:   95       99      99.9
values:        0.446    0.849   1.41 (in [V/m])

focality
cutoffs:       50       75 (in % of 99.9 percentile)
values:        1.28e+04  3.32e+03 (in cubic mm)
----------------------------------------------
```

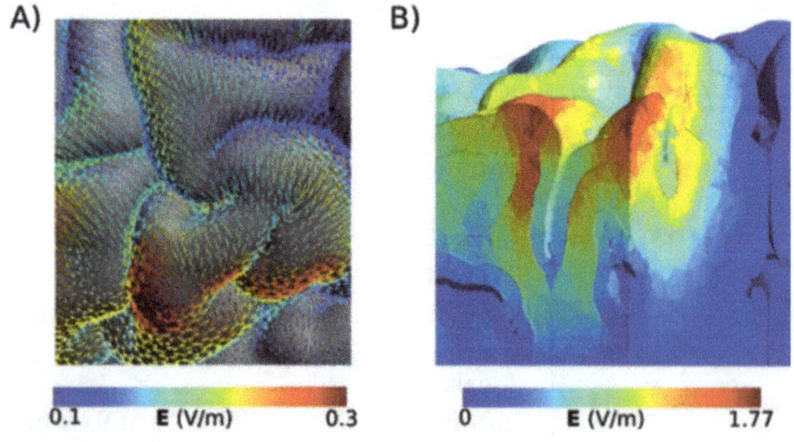

Fig. 1.9 Visualization in Gmsh of (**a**) electric field vectors around central gyrus for the tDCS simulation and (**b**) TMS electric field depth profile in the hotspot

1.3.2 Advanced Usage: Group Analysis

Now, we want to simulate one tDCS montage, with a 5 × 5 cm electrode over C3 and a 5 × 7 cm electrode over AF4 in five subjects, called "sub01", "sub09", "sub10", "sub12", "sub15" and visualize the results in a common space, namely the FsAverage surface. The subjects and example scripts can be downloaded from: https://osf.io/ah5eu/

Head Meshing

For each subject, follow the steps in section "Generating the Volume Conductor Model".

Write a Python or MATLAB Script

We can set up the simulation of each subject using the GUI, as described in the first example. However, when working with multiple subjects, it can be advantageous to script the simulations for efficiency. SimNIBS provides both MATLAB and Python interfaces to set up simulations. Script 1.1 shows how to set up and run a simulation with a 5 × 5cm anode placed over C3 and a 7 × 5cm cathode placed over AF4 for all subjects. The output of Script 1.1 for sub01 is shown in Table 1.5.

To define the rectangular electrodes, we need two coordinates. The "*centre*" defines where the electrode will be centred, and "*pos_ydir*" how the electrode will be rotated. More precisely, the electrode's "y" axis is defined as a unit vector starting at "centre" and pointing towards "*pos_ydir*". Fig. 1.10 shows one of the cathodes (return electrode) defined using the script above, with the coordinate system and EEG positions overlaid. We can see that the electrode is centred in AF4, and its Y axis points towards F6. "*pos_ydir*" does not need to be set when the electrodes are round.

When the *map_to_fsavg* option is set to *true*, SimNIBS computes the electric fields in a surface located in the middle of the GM layer. This cortical surface, along with the norm, normal and tangent components of the electric field at the cortical surface and the angle between the electric field and the cortical surface can found in the *subject_overlays* folder, for both the left hemisphere (*lh*) and for the right hemisphere (*rh*) as shown in Table 1.5. Afterwards, these quantities are transformed into the FsAverage space. The transformed quantities can be found in the *fsavg_overlays* folder, as shown in Table 1.5. Additionally, we have the electric field and its norm in MNI space in the *mni_volumes* folder.

Table 1.5 Output files and folders of Script 1 for sub01. The ".angle", ".norm",.. files are FreeSurfer overlay files and the ".central" files are FreeSurfer surface files

```
bipolar/
├── fsavg_overlays
│   ├── lh.sub01_TDCS_scalar.fsavg.E.angle
│   ├── lh.sub01_TDCS_scalar.fsavg.E.norm
│   ├── lh.sub01_TDCS_scalar.fsavg.E.normal
│   ├── lh.sub01_TDCS_scalar.fsavg.E.tangent
│   ├── rh.sub01_TDCS_scalar.fsavg.E.angle
│   └── ...
├── mni_volumes
│   ├── sub01_TDCS_scalar_MNI_E.nii.gz
│   └── sub01_TDCS_scalar_MNI_normE.nii.gz
├── simnibs_simulation_20181019-100530.log
├── simnibs_simulation_20181019-100530.mat
└── subject_overlays
    ├── lh.central
    ├── lh.sub01_TDCS_scalar.central.E.angle
    ├── lh.sub01_TDCS_scalar.central.E.norm
    ├── lh.sub01_TDCS_scalar.central.E.normal
    ├── ...
    ├── rh.central
    ├── rh.sub01_TDCS_scalar.fsavg.E.angle
    └── ...
```

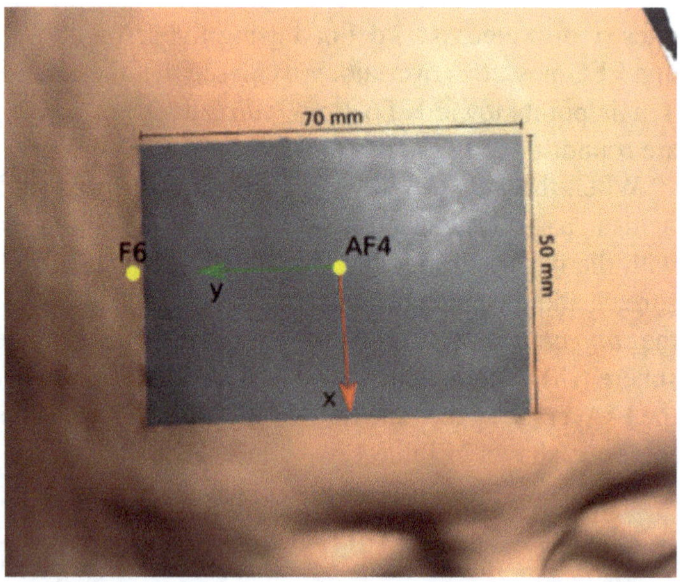

Fig. 1.10 50 × 70 mm electrode defined with a "centre" in AF4 and a "pos_ydir" in F6

```
path_to_headmodels = "/path/to/head/models/";
subjects = ["sub01", "sub09", "sub10", "sub12", "sub15"];
results_folder = "bipolar/fsavg_overlays";

normals = {};
for i = 1:length(subjects)
    sub = subjects(i);
    % Load normal field data
    normal_surf = sprintf('lh.%s_TDCS_1_scalar.fsavg.E.normal', sub);
    m = mesh_load_fsresults(char(...
        fullfile(path_to_headmodels, sub, results_folder, normal_surf)));
    % Add to cell
    normals{i} = m.node_data{1}.data;
end
% Calculate average and standard deviation of the normal at each node
normals = cell2mat(normals);
avg_normal = mean(normals, 2);
std_normal = std(normals, 0, 2);
% Place the fields in the mesh structure
m.node_data{1}.data = avg_normal;
m.node_data{1}.name = 'E.normal.avg';
m.node_data{2}.data = std_normal;
m.node_data{2}.name = 'E.normal.std';
% Plot the fields
mesh_show_surface(m, 'field_idx', 'E.normal.avg')
mesh_show_surface(m, 'field_idx', 'E.normal.std')
```

Script 1.1 Script for running a tDCS simulations with an anode over C3 and a cathode over AF4 in five subjects and transforming the results to FSAverage and MNI spaces.

```
path_to_headmodels = "/path/to/head/models/" ;
subjects = ["sub01", "sub09", "sub10", "sub12", "sub15"];
results_folder = "bipolar/fsavg_overlays" ;

normals = {};
for i = 1:length(subjects)
    sub = subjects(i);
    % Load normal field data
    normal_surf = sprintf( 'lh.%s_TDCS_1_scalar.fsavg.E.normal' , sub);
    m = mesh_load_fsresults(char( ...
        fullfile(path_to_headmodels, sub, results_folder, normal_surf)));
    % Add to cell
    normals{i} = m.node_data{1}.data;
end
% Calculate average and standard deviation of the normal at each node
normals = cell2mat(normals);
avg_normal = mean(normals, 2);
std_normal = std(normals, 0, 2);
% Place the fields in the mesh structure
m.node_data{1}.data = avg_normal;
m.node_data{1}.name = 'E.normal.avg' ;
m.node_data{2}.data = std_normal;
m.node_data{2}.name = 'E.normal.std' ;
% Plot the fields
mesh_show_surface(m, 'field_idx' , 'E.normal.avg' )
mesh_show_surface(m, 'field_idx' , 'E.normal.std' )
```

Script 1.2 Analysis of simulation results in FSAverage space.

Visualizing Results

We can also make use of the MATLAB library within SimNIBS to analyze the results from the simulations. Here, we are interested in the average and standard deviation of the normal component of the electric field in the cortex. The normal component, as shown in Fig. 1.7, is the part of the electric field which is either entering or leaving the cortex.

Script 1.2 loads the normal field component data for each subject and calculates the mean and the standard deviation across subjects at each position of the FsAverage template. The fields are then visualized using MATLAB visualization tools. The results are shown in Fig. 1.11. We can, for example, see strong current in-flow in the central gyrus, and large variations in the normal component in frontal regions.

1.4 The Accuracy of Automatic EEG Positioning

Here, we compare EEG 10-10 positions obtained either from:

A. Transforming EEG 10-10 electrode positions defined in MNI space to the subject space using a non-linear transform, and then projecting the positions to the scalp. This is done for both **mri2mesh** and **headreco** head models.
B. Manually locating the fiducials: left pre-auricular point (LPA), right pre-auricular point (RPA), nasion (Nz) and inion (Iz) on MRI images, and afterwards calculating the EEG positions using the definitions in [22].

Calculations using method A require no user input and are automatically performed in both **mri2mesh** and **headreco** head modelling pipelines, while calculations using method B require the user to manually select the fiducial positions.

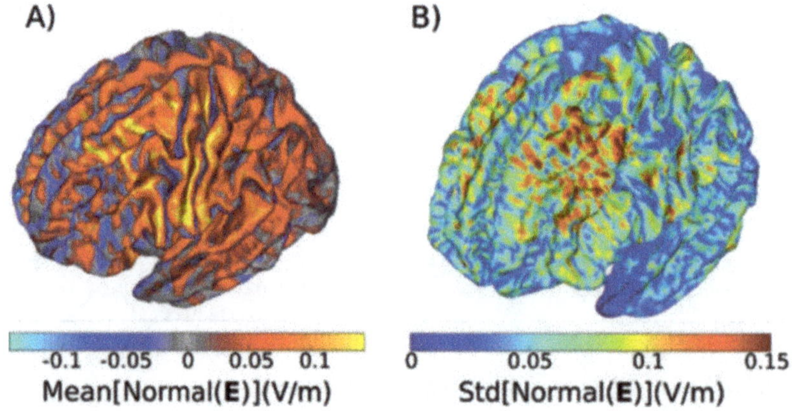

Fig. 1.11 (**a**) Mean and (**b**) Standard deviation of the normal field component across 5 subjects. The fields were caused by tDCS with an anode over C3 and a cathode over AF4. Positive values in (**a**) denote inflowing currents, and negative values outflowing currents

To compare the methods A and B to position the electrodes, we calculated the EEG 10-10 positions using both ways for MR data of 17 subjects. The data was acquired as part of a larger study. The subjects gave written informed consent before the scan, and the study was approved by the local ethics committee of the University of Greifswald (Germany). The 17 datasets were acquired on a 3-Tesla Siemens Verio scanner (Siemens Healthcare, Erlangen, Germany) using a 32-channel head coil (T1: 1 × 1 × 1 mm^3, TR 2300 ms, TE 900 ms, flip angle 9°, with selective water excitation for fat suppression; T2: 1 × 1 × 1 mm^3, TR 12770 ms, TE 86 ms, flip angle 111°). For method B, the fiducials were manually located for each subject by a trained investigator on the T1- and T2-weighted images. The later had no knowledge of the automatically determined positions. The fiducials Nz, Iz, LPA and RPA were set in freeview, following the procedure described in [22] and additionally verified using the SimNIBS GUI. The subject-specific coordinates of the fiducials were extracted, and these manually set positions were then compared with those calculated by the automatic algorithm in each individual.

Table 1.6 shows the maximal distance across all subjects between the fiducials obtained using method A and manually selected fiducials (B). We see that Nz is the most consistent fiducial, where we have the least deviation, whereas Iz is where we have the highest deviation. Also, the maximal difference in position across the two methods is ~1 cm, indicating that method A works well to approximate the positions of the fiducials.

Furthermore, in Fig. 1.12, we compare the two methods for all electrode positions in the EEG 10-10 system. The deviation in positioning each electrode was calculated as the mean of the distance between the positions obtained with either headreco or mri2mesh to the manually located fiducial positions, across all 17 subjects and for each electrode.

The errors for all electrodes are below 1 cm, indicating that the two algorithms for placing EEG electrodes are in agreement. We can also see that the errors in the EEG positions obtained from **headreco** are on average lower than the ones obtained from **mri2mesh**. It also seems that the anterior electrodes have less errors than the posterior electrodes. Interestingly, the location of the errors is different across the two pipelines, with **mri2mesh** being more inaccurate in superior regions and **headreco** more inaccurate in posterior regions. This might be caused by differences in the way FSL (**mri2mesh**) and SPM (**headreco**) calculate non-linear MNI transfor-

Table 1.6 Maximum and mean distance between the fiducial positions selected by hand and obtained from the MNI transformations across 17 subjects, for the two head modelling pipelines

	mri2mesh		headreco	
Fiducial	Max distance (mm)	Mean distance ± standard deviation (mm)	Max distance (mm)	Mean distance ± standard deviation (mm)
LPA	6.4	3.2 ± 1.5	8.7	5.4 ± 2.0
RPA	8.9	3.0 ± 1.6	10.6	5.9 ± 1.7
Nz	3.9	2.1 ± 1.0	6.0	3.9 ± 1.6
Iz	14.3	4.0 ± 3.5	13.2	5.2 ± 3.3

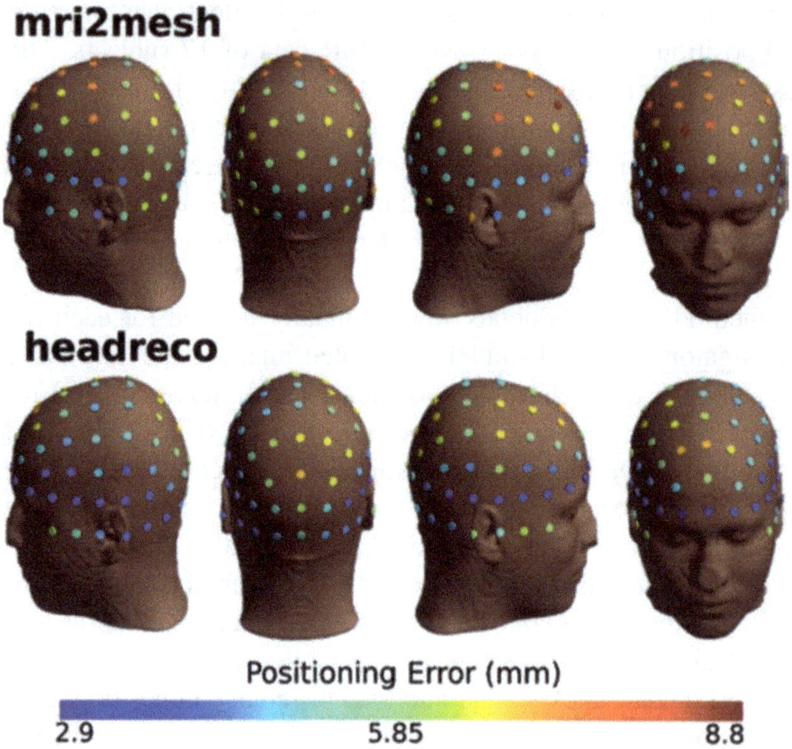

Fig. 1.12 Positioning error for electrodes in the EEG 10-10 system. The error is calculated by comparing the positions calculated based on manually selected fiducials to positions calculated based on non-linear MNI transformations

mations is different. The average error across all positions was 5.6 mm for **mri2mesh** head models and 4.9 mm for **headreco** head models indicating good accuracy.

1.5 Conclusion

We presented SimNIBS 2.1 (www.simnibs.org), a software for individualized modelling of electric fields caused by non-invasive brain stimulation. SimNIBS is free software and avaliable for all major platforms. SimNIBS does not require the installation of any additional software in order to run simulations on the example dataset. To construct head models, SimNIBS relies either on MATLAB, SPM12 and CAT12 (**headreco**) or on FSL and FreeSurfer (**mri2mesh**).

We also presented two examples of workflows in SimNIBS. In the first example, we started by using **headreco** to construct a head model. Following this, we used the GUI to set up a tDCS and a TMS simulation in an interactive way, and finally visualized the results. In the second example, we constructed several head models and used a MATLAB script to run simulations for each subject. We then calculated the mean and the stardard deviation of the electric field norm across all subjects, using the FreeSurfer's FsAverage brain template. Finally, we show results validating our automatic procedure to obtain electrode positions for the EEG 10-10 system.

SimNIBS is still being actively developed, and we expect further updates to be implemented in the future.

Acknowledgements Lundbeckfonden (grant Nr. R118-A11308), and NovoNordisk fonden (grant Nr. NNF14OC0011413). This project has received funding from the European Union's Horizon 2020 research and innovation programme under grant agreement no. 731827 "STIPED". The results and conclusions in this chapter present the authors' own views, and do not reflect those of the EU Commission.

References

1. Fregni, F., & Pascual-Leone, A. (2007). Technology insight: Noninvasive brain stimulation in neurology – Perspectives on the therapeutic potential of rTMS and tDCS. *Nature Clinical Practice Neurology, 3*(7): 282–393
2. Thielscher, A., Antunes, A., and Saturnino, G. B.. (2015). Field modeling for transcranial magnetic stimulation: A useful tool to understand the physiological effects of TMS? In *Proc. Annu. Int. Conf. IEEE Eng. Med. Biol. Soc. EMBS* (pp. 222–225).
3. Guerra, A., López-Alonso, V., Cheeran, B., & Suppa, A. (2018). Solutions for managing variability in non-invasive brain stimulation studies. *Neuroscience Letters.*
4. Li, L. M., Uehara, K., & Hanakawa, T. (2015). The contribution of interindividual factors to variability of response in transcranial direct current stimulation studies. *Frontiers in Cellular Neuroscience, 9*, 181.
5. Kubis, N. (2016). Non-invasive brain stimulation to enhance post-stroke recovery. *Front. Neural Circuits, 10*, 56.
6. Morishita, T., & Hummel, F. C. (2017). Non-invasive Brain Stimulation (NIBS) in motor recovery after stroke: Concepts to increase efficacy. *Curr. Behav. Neurosci. Reports, 4*(3), 280–289.
7. Windhoff, M., Opitz, A., & Thielscher, A. (2013). Electric field calculations in brain stimulation based on finite elements: An optimized processing pipeline for the generation and usage of accurate individual head models. *Human Brain Mapping, 34*(4), 923–935.
8. Nielsen, J. D., Madsen, K. H., Puonti, O., Siebner, H. R., Bauer, C., Madsen, C. G., Saturnino, G. B., & Thielscher, A. (2018). Automatic skull segmentation from MR images for realistic volume conductor models of the head: Assessment of the state-of-the-art. *NeuroImage, 174*, 587–598.
9. Smith, S. M., Jenkinson, M., Woolrich, M. W., Beckmann, C. F., Behrens, T. E. J., Johansen-Berg, H., Bannister, P. R., De Luca, M., Drobnjak, I., Flitney, D. E., Niazy, R. K., Saunders, J., Vickers, J., Zhang, Y., De Stefano, N., Brady, J. M., & Matthews, P. M. (2004). Advances in functional and structural MR image analysis and implementation as FSL. *NeuroImage, 23*, S208–S219.
10. Fischl, B. (2012). FreeSurfer. *NeuroImage, 62*(2), 774–781.
11. W. Penny, K. Friston, J. Ashburner, S. Kiebel, and T. Nichols, Statistical parametric mapping: The analysis of functional brain images. 2007. Elsevier; London, UK.
12. Dahnke, R., Yotter, R. A., & Gaser, C. (2013). Cortical thickness and central surface estimation. *NeuroImage, 65*, 336–348.
13. Geuzaine, C., & Remacle, J.-F. (2009). Gmsh: A three-dimensional finite element mesh generator with built-in pre-and post-processing facilities. *International Journal for Numerical Methods in Engineering, 79*(11), 1309–1331.
14. Thielscher, A., & Kammer, T. (2004). Electric field properties of two commercial figure-8 coils in TMS: Calculation of focality and efficiency. *Clinical Neurophysiology, 115*(7), 1697–1708.

15. Madsen, K. H., Ewald, L., Siebner, H. R., & Thielscher, A. (2015). Transcranial magnetic stimulation: An automated procedure to obtain coil-specific models for field calculations. *Brain Stimulation 8*(6), 1205–1208.
16. Saad, Y. (2003). *Iterative methods for sparse linear systems* (2nd ed.). SIAM. Philadelphia, PA, USA.
17. Geuzaine, C. (2007). GetDP: A general finite-element solver for the de Rham complex. *PAMM, 7*(1), 1010603–1010604.
18. Balay, S., Abhyankar, S., Adams, M., Brown, J., Brune, P., Buschelman, K., Dalcin, L., Eijkhout, V., Gropp, W., Kaushik, D., Knepley, M., May, D., McInnes, L. C., Mills, R. T., Munson, T., Rupp, K., Sanan, P., Smith, B., Zampini, S., Zhang, H., and Zhang, H. (2018). {PETS}c {W}eb page.
19. Opitz, A., Falchier, A., Yan, C. G., Yeagle, E. M., Linn, G. S., Megevand, P., Thielscher, A., Deborah, R. A., Milham, M. P., Mehta, A. D., & Schroeder, C. E. (2016). Spatiotemporal structure of intracranial electric fields induced by transcranial electric stimulation in humans and nonhuman primates. *Scientific Reports, 6*, 1–11.
20. Zienkiewicz, O. C., & Zhu, J. Z. (1992). The superconvergent patch recovery and a posteriori error estimates. Part 1: The recovery technique. *International Journal for Numerical Methods in Engineering, 33*(7), 1331–1364.
21. Bungert, A., Antunes, A., Espenhahn, S., & Thielscher, A. (2017). Where does TMS stimulate the motor cortex? Combining electrophysiological measurements and realistic field estimates to reveal the affected cortex position. *Cerebral Cortex, 27*(11), 5083–5094.
22. Jurcak, V., Tsuzuki, D., & Dan, I. (2007). 10/20, 10/10, and 10/5 systems revisited: Their validity as relative head-surface-based positioning systems. *NeuroImage, 34*(4), 1600–1611.

Chapter 2
Finite Element Modelling Framework for Electroconvulsive Therapy and Other Transcranial Stimulations

Azam Ahmad Bakir, Siwei Bai, Nigel H. Lovell, Donel Martin, Colleen Loo, and Socrates Dokos

2.1 Introduction

Electroconvulsive therapy (ECT) has been used to ameliorate major depressive disorder for patients who are resistant to drug therapy. The treatment involves applying a train of alternating pulses across two electrodes placed on the scalp. ECT is an effective treatment [1], but also carries a risk of cognitive side effects, such as disorientation and memory loss [2]. Treatment efficacy has been noted to rely on multiple factors, such as electrode placement and stimulus dose [3]. In addition, there is currently also great interest in other brain stimulation techniques for therapeutic neuromodulation or neurostimulation, including transcranial direct current stimulation.

Due to electrical conductivity variation across different tissues, the current pathways induced by electrical stimulation are not straightforward to identify. The presence of highly resistive skull and air-filled paranasal sinuses impedes the passage of electrical currents, forcing the majority of currents to travel through the less resistive

---

A. Ahmad Bakir · N. H. Lovell · S. Dokos
Graduate School of Biomedical Engineering, Faculty of Engineering, University of New South Wales (UNSW Sydney), Sydney, NSW, Australia

S. Bai (✉)
Graduate School of Biomedical Engineering, Faculty of Engineering, University of New South Wales (UNSW Sydney), Sydney, NSW, Australia

Department of Electrical and Computer Engineering, Technical University of Munich (TUM), Munich, Germany

Munich School of BioEngineering, TUM, Garching, Germany
e-mail: siwei.bai@tum.de

D. Martin · C. Loo
School of Psychiatry, UNSW Sydney, Sydney, NSW, Australia

Black Dog Institute, Sydney, NSW, Australia

S. Makarov et al. (eds.), *Brain and Human Body Modeling*,
https://doi.org/10.1007/978-3-030-21293-3_2

regions [4]. Furthermore, the white matter exhibits a strongly anisotropic conductivity due to its myelinated structure, thus determining the prefential current pathway within the brain [5, 6]. As such, the electrical current distribution in the brain resulting from brain stimulation is complex and cannot be readily imaged, and is impractical to be measured empirically. Alternatively, the electrical current and electric field (E field) in the brain can be simulated via computational modelling. The finite element (FE) method is one of the most popular numerical approaches for solving models expressed as partial differential and integral equations.

The main goal of computational modelling for ECT and other brain stimulation techniques is to determine the region(s) modulated by the electrical stimulus. It is believed that non-invasive brain stimulation shifts the tissue's membrane potential, subsequently affecting neuronal firing [7]. The use of computational modelling to examine differences in regional E fields as the ECT stimulation approach is altered allows for a better understanding of the relationship between brain stimulation and clinical effects with current forms of ECT, as well as offering the potential for futher improvements in ECT stimulation techniques. In this chapter, we will discuss approaches and steps necessary to implement computational modelling of the human head to determine the voltage and E field distribution during the application of ECT and other transcranial electric stimulation techniques.

2.2 Methods

Figure 2.1 describes the steps needed to undertake a computational study of electrical brain stimulation. Similar steps have been performed as part of previous finite element studies [3, 5, 8, 9], with minimal variation to suit the need for each study.

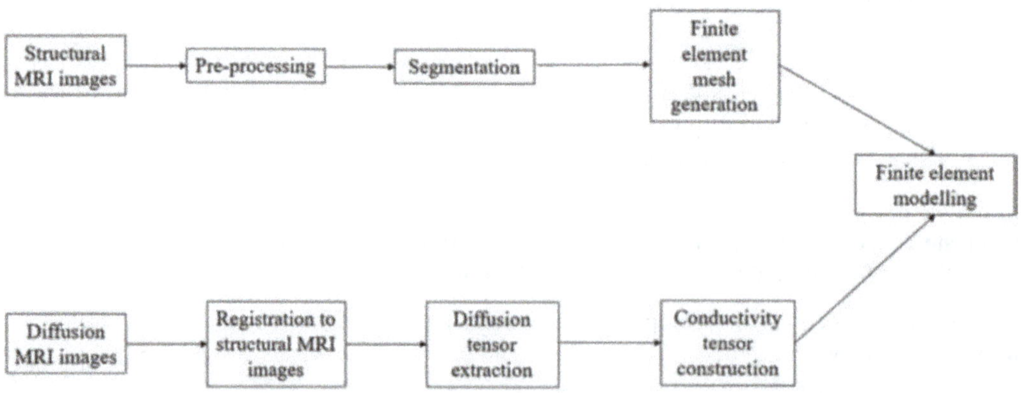

Fig. 2.1 Flowchart describing the workflow needed to implement a finite element model of the head, mainly involving segmenting the head structures (top row) and extracting white matter anisotropy (bottom row)

2.2.1 Image Pre-processing

In order to simulate the properties of different head structures, these structures need to be individually reconstructed from the acquired images. The process of partitioning the image into different domains or "masks" is known as image segmentation.

In order to increase the accuracy and reduce the effort of image segmentation, certain pre-processing procedures are performed prior to segmentation. These may include resampling (to reduce the resolution), cropping (to restrict the image set to the region of interest, i.e. ROI), artefact correction (such as motion, metal and bias field artefacts), edge and contrast enhancement and image registration. These operations can be performed using a selection of open-source image-processing software, such as ImageJ (https://imagej.nih.gov/ij/), 3D Slicer (https://www.slicer.org/) and ITK-SNAP (http://www.itksnap.org/). Among these, bias field correction and image registration are highly common pre-processing steps in the segmentation of MR head scans.

Bias Field Correction

Bias field noises are caused by low intensity and smooth signals that distort the MRI images, and are present especially in older MR devices [10]. This type of noise causes regional differences in signal intensity in the images, leading to non-uniform intensities in the same head structure, as shown in Fig. 2.2. If left uncorrected, segmentation quality may be affected. Bias field correction can be performed prior to segmentation using open-source tools designed specifically for head segmentation, as listed in Table 2.1.

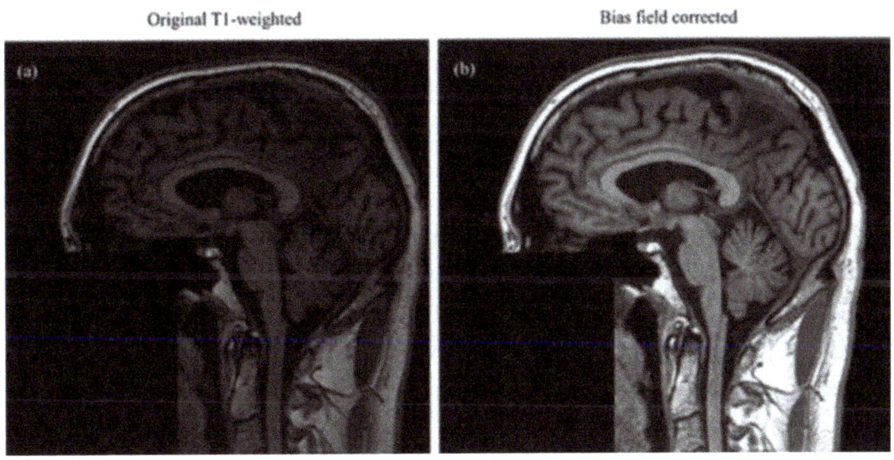

Fig. 2.2 The effect of bias field correction (**a**) before and (**b**) after a correction performed in 3D Slicer. The patient's face was hidden for privacy

Table 2.1 List of open-source software packages for brain segmentation

Software package	Developer
BrainSuite (http://brainsuite.org/)	University of California, Los Angeles and University of Southern California
FSL (https://fsl.fmrib.ox.ac.uk/fsl/fslwiki/)	Oxford University
SPM (https://www.fil.ion.ucl.ac.uk/spm/)	University College London
FreeSurfer (https://surfer.nmr.mgh.harvard.edu/)	Harvard University
SimNIBS[a] (http://simnibs.de/)	Copenhagen University Hospital Hvidovre and multiple institutions

Further details on each software tool are available from the listed websites
[a]SimNIBS combines other software such as FreeSurfer and SPM to build a pipeline for brain stimulation modelling

Image Registration

It is not uncommon to obtain multimodal MRI scans, such as T1-weighted scans together with T2-weighted, proton density (PD)-weighted or diffusion-weighted scans, to provide complementary information regarding tissue structures in the brain. As these scans may be acquired in different coordinate systems, it is essential to perform image registration to transform these into the same coordinate system prior to segmentation. A common registration method is affine transformation, which is a linear transformation aligning two sets of images together through translation, scaling, shear mapping and rotation [11]. When a linear registration method is not able to provide a satisfactory outcome, such as when registering scans from different subjects, a non-linear transformation method should be applied. These transformations are performed, automatically or manually guided, through identification of anatomical landmarks, such as the corners of the ventricles, which are easily distinguishable from the images. Image registration is typically available in image-processing software packages.

Image Segmentation

The structural domains are separated through their identifiable landmarks and/or edges. T1-weighted MRI is typically used as it provides a good contrast between whole head structures, especially between grey and white matter. Skull extraction can be challenging, especially at the ethmoid sinus region, but this can be rectified by combining the T1-scan data with CT, T2-weighted or PD-weighted MRI scans [12].

Several open-source software packages designed for brain segmentation have been developed by different research groups. These can automatically extract major head structures, such as grey and white matter, skull and cerebrospinal fluid (CSF), from MRI images. Several of these can also further partition the grey matter into various cortices based on pre-defined atlases. A list of automatic brain segmentation software packages is provided in Table 2.1.

It is good practice to perform additional checks following automatic segmentation to ensure there are no segmentation errors. This is usually performed in image processing software that allows manual segmentation, e.g. open-source 3D Slicer and ITK-SNAP, as well as commercially available tools such as Materialise Mimics Innovation Suite (https://www.materialise.com/), Amira (https://www.fei.com/software/amira-for-life-sciences/) and Simpleware (https://www.synopsys.com/simpleware). Other processing, such as smoothing, can also be performed to improve segmentation quality. Several software packages provide training datasets and online tutorials to assist learning.

Manual Segmentation

Thresholding is a critical step in manual segmentation. A thresholding filter can be applied to select particular brain regions. For example, grey and white matter can be easily discerned from T1-weighted MRI images since they exhibit different image intensities. As such, they can be readily segmented into individual masks. In addition to grey and white matter, the CSF space can also be easily recognised from its high intensity in T2-weighted images. This facilitates masking of CSF in between the pial surface (outer grey matter surface) and the dura surface (inner skull surface) as well as the interior brain ventricular system.

Segmentation can also be performed with other image processing techniques such as “seeding and growing”, Boolean operations and mask growing/shrinking. These options are available as standard features of most image segmentation software packages [13]. “Seeding and growing” begins by manually placing seed points in a particular region. The seed points are then expanded to adjacent pixels based on certain region membership criteria, such as pixel intensity and connectivity, until all the connected pixels cover the structure of interest. This prevents the inclusion of other regions of similar pixel intensities into the same mask: for example, segmenting the brain by thresholding alone, ignoring connectivity, may inadvertently include the bone marrow of the skull.

Boolean operation techniques work directly on segmented masks, creating a union, intersection or difference between two masks. These can be used to obtain regional domains encapsulated between two domains. For example, the CSF is encapsulated between the dura and pia structures. Rather than directly segmenting the CSF, it can instead be obtained by performing Boolean subtraction of the encapsulating domains enclosing the brain and skull. This ensures continuity of the surfaces between domains in addition to segmentation efficiency. Boolean operations can also be used to detect boundary intersections between masks resulting from segmenting errors, as detailed in Section “Challenges and Tips in Segmentation”.

Mask growing/shrinking is another technique that operates directly on segmented masks. It is similar to performing a mask scaling. Depending on the algorithm, this may be performed in 2D or 3D, or using a uniform or non-uniform approach based on pixel intensity. The combination of these two operations may be used to remove islands, close holes, or interpolate between every two or three image slices.

Surface Smoothing

Following manual or automatic segmentation, output masks are often rough and contain sharp edges. Small islands, i.e. disconnected shells, may also be formed during segmentation and should be removed. These issues can be addressed using a smoothing process as shown in Fig. 2.3.

Surface smoothing can be performed on the masks within the image processing software, using either Gaussian, median or Laplace smoothing. It can also be performed after the masks have been exported as surface triangulated objects, usually in *.stl* (stereolithography) format. Operating platforms that can perform smoothing include Blender (https://www.blender.org/), Geomagic Wrap (https://www.3dsystems.com/software/geomagic-wrap) and Materialise 3-matic (https://www.materialise.com/). The smoothing strength must be tuned so that the accuracy of the structure is not compromised. Any sharp edges in the form of spikes need to be removed, since this may prevent efficient meshing in later stages of the modelling effort. Furthermore, such a structure is unlikely to be correct, particularly if located between the brain gyri.

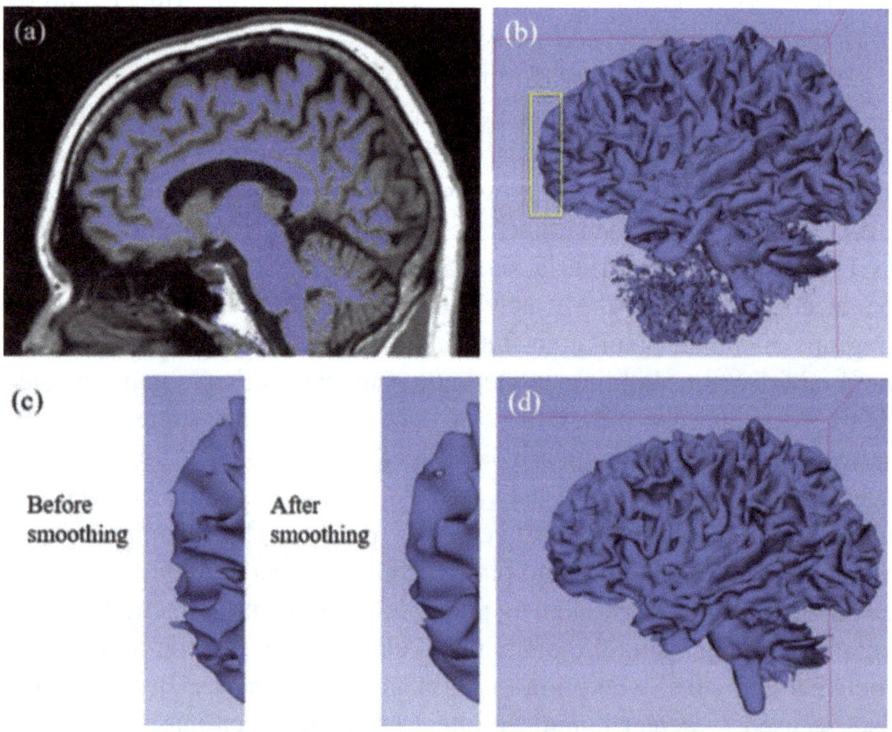

Fig. 2.3 (**a**) Thresholding of white matter, where the mask was initially generated automatically by FSL and imported into 3D Slicer for further processing. (**b**) The initial surface output from (**a**). (**c**) Gaussian smoothing applied to (**b**), with zoomed in view in the yellow box region in (**b**). (**d**) The final smoothed structure with segmentation errors in the form of small shells removed and remaining holes patched

Cortical Structure Labelling

Different brain regions are responsible for different physical and mental functions. For this reason, it is often of interest to observe the effect of ECT or other transcranial electric stimulation techniques on specific brain substructures. Nonetheless, extracting these structures can be challenging since it requires knowledge of anatomical landmarks, which are not readily discernible. In addition, manual segmentation can risk losing consistency among multiple subjects, especially if the segmentation is performed by different people.

Some brain segmentation software provide automatic labelling of brain regions, as shown in Fig. 2.4. This labelling is based on established brain atlases where each region has been meticulously mapped. Examples of brain atlases are BrainSuite's BCI-DNI_brain [14] and USCBrain [15], which are available in the BrainSuite software.

The first step in brain region labelling involves registering the subject's skull-stripped brain images to the atlas. This utilises linear and nonlinear warping to align the subject's brain with the atlas. Subsequently, the regions are automatically labelled based on anatomical landmarks [14, 16].

Challenges and Tips in Segmentation

Several challenges can be encountered during the segmentation process. In this section, several tips and precautions are presented.

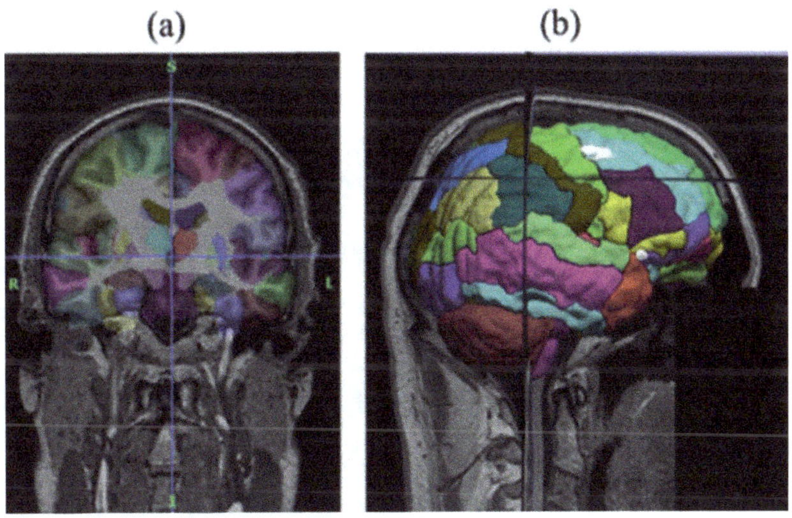

Fig. 2.4 Brain labelling tool in BrainSuite. (**a**) Masks overlaid on T1-weighted images where different colours signify distinct brain regions. (**b**) 3D surfaces generated from the masks. Smoothed individual cortical regions can then be produced by intersecting these labelled masks with the smoothed grey matter mask created previously

Unwanted intersections between surfaces can present difficulties during the volumetric meshing stage. As these intersecting surfaces are usually small, as shown in Fig. 2.5, very fine mesh elements will be created around these surfaces. The intersections could also lead to errors during simulation since these structures are merely segmentation artefacts. For example, if the inner skull surface protrudes into the outer skull, this will create a hole in the skull domain, connecting the CSF directly to the scalp. Since the skull is highly resistive, a preferential current pathway is unintentionally created, producing a simulation error.

To prevent such intersections, segmentation can be performed from the outermost surface first, gradually moving towards the inner surface. Inner surface segmentation can then be performed using the outer surface boundary as a guide. Small intersections can also be highlighted using Boolean operators and subsequently corrected manually. It should be noted that intersections can exist even for automated segmentation tools such as those of FSL and BrainSuite. As such, the segmentation output must always be visually inspected.

Other challenges can also arise when segmenting the skull in T1-weighted images, since both compact/cortical bone and CSF appear dark. This may be resolved by directly obtaining the skull mask from CT scans. However, it is often difficult to acquire CT scans due to concerns over unnecessary radiation exposure. The skull is thus extracted by defining the outer skull surface using T1- or PD-weighted images, and the inner skull surface using T2- or PD-weighted images. The spongy bone of the skull is often identifiable as a bright region between the two thin dark regions (compact bone) outside the brain. Segmenting the air-filled paranasal sinuses within the skull may also be difficult as they appear dark. They are often recognised as frontal bone regions where spongy bone is missing.

It is not uncommon to generate a mutually exclusive mask for each domain. This may however create surface continuity problems, especially after surface smooth-

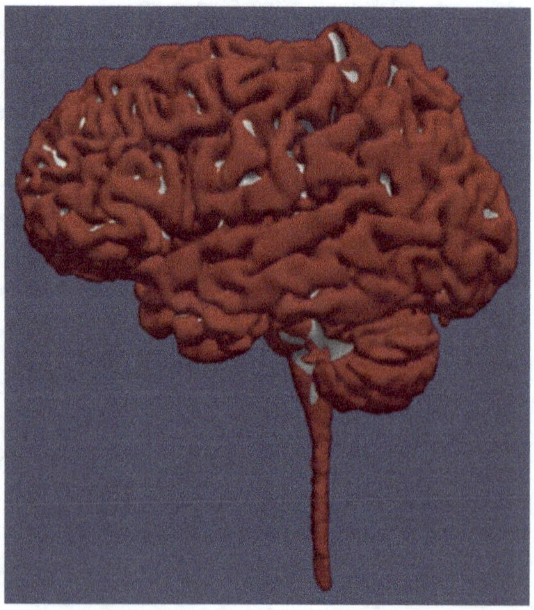

Fig. 2.5 Intersections between grey and white matter due to segmentation error. This can be overcome if segmentation of the grey matter is performed first, followed by white matter segmentation

ing, during which both inner and outer surfaces of each mask are exclusively modified. A better practice would be creating inclusive masks, i.e. an outer domain mask within which all other domains of interest are contained. For instance, instead of creating a grey matter-only mask, a brain mask that also contains CSF in the ventricles as well as white matter can be generated. Separation can be accomplished using Boolean operations after all necessary modifications are performed. In some FE meshing tools, separation is not only unnecessary but may also cause contact surface problems between domains.

2.2.2 *White Matter Anisotropy*

White matter consists of myelinated neuronal tracts, which contribute to its highly anisotropic behaviour. This microanatomical characteristic influences the electrical conductivity such that it is more conductive along the tract than in the transverse direction [6]. Consequently, this affects the spread of the ECT electric field. Simulation studies by Lee et al. [5] and Bai et al. [6] show that disregarding this anisotropy can result in errors in deeper brain structures such as the corpus collosum and hippocampus. As such, the anisotropy helps direct current towards deeper brain regions, where significant effects have been noted following ECT [17].

The linear relationship between electrical conductivity and water diffusion tensor has been experimentally validated [18, 19], suggesting that the conductivity tensor shares the same eigenvectors as the diffusion tensor. The water diffusion tensor can be extracted using the diffusion tensor model for diffusion-weighted MRI (DW-MRI) [20], which can be performed in FSL using the probabilistic tracking algorithm from the FDT diffusion toolbox [21–23].

Two separate files containing *b*-values and *b*-matrices for all gradient directions are required as input for the diffusion tensor calculation. The former summarises the sensitivity to diffusion for each gradient direction, whereas the latter reflects the attenuation effect in *x*, *y* and *z* for each gradient direction [20]. In addition, the input also requires a 3D NIfTI image file of the brain region of interest (ROI), and a 4D NIfTI image file combining all gradient direction scans. Eigenvectors and fractional anisotropy (FA) are then calculated.

FA determines the anisotropic characteristics of a tensor. In brief, an FA value of 0 indicates complete isotropy whilst a value of 1 indicates complete anisotropy. The FA value is determined from the eigenvalues (λ_1, λ_2, λ_3) of the diffusion tensor as follows [24]:

$$FA = \sqrt{\frac{3}{2}}\left(\frac{\sqrt{\left(\lambda_1-\hat{\lambda}\right)^2+\left(\lambda_2-\hat{\lambda}\right)^2+\left(\lambda_3-\hat{\lambda}\right)^2}}{\sqrt{\lambda_1^{\,2}+\lambda_2^{\,2}+\lambda_3^{\,2}}}\right), \tag{2.1}$$

where $\hat{\lambda}$ is the average of the three eigenvalues. Regions with a low FA value (typically FA < 0.45), which suggests a low local anisotropy, are removed from the tensor analysis [25] (Fig. 2.6).

The conductivity tensor of white matter, **σ**, is calculated from:

$$\sigma = \mathbf{S}\,\mathbf{diag}\left(\sigma_l,\sigma_t,\sigma_t\right)\mathbf{S}^{\mathrm{T}}, \tag{2.2}$$

where **S** is the orthogonal matrix of unit eigenvectors obtained from the white matter diffusion tensor, and σ_l and σ_t are the conductivities in the longitudinal and transverse fibre directions, respectively, which may be calculated using various methods [26, 27]. Following the diffusion tensor calculation, conductivity tensors of data points in the DW-MRI scans can then be linked to their individual coordinates. Only conductivity data with a strong anisotropy signal (FA ≥ 0.45) should be exported. Afterwards, the conductivity at the undefined region can be linearly interpolated using the neighbouring strong anisotropy signals.

2.2.3 FE Meshing

After segmentation, the masks are exported as triangulated surface objects, usually in the form of an *.stl* file. The masks need to be polyhedralised (tetrahedralised in most scenarios), before they are ready to be used in FE analysis. Polyhedralisation (or FE meshing) is the process of generating polyhedral mesh elements to approximate a geometric domain, and these elements are the basis of the FE method. FE meshing is performed in dedicated meshing software, such as the open-source SALOME (https://www.salome-platform.org/), Materialise 3-matics, Simpleware, as well as ICEM CFD and Fluent which are both from ANSYS, Inc. (https://www.ansys.com/). The volumetric mesh will then be imported into FE simulation software such as COMSOL Multiphysics (https://www.comsol.com/) or ANSYS Workbench.

A tight contact between masks is essential to ensure continuity between meshed domains. It is thus often advisable in many of these meshing software packages to not import perfectly mutually exclusive masks. One practice is to generate inclusive masks, as detailed in Section "Challenges and Tips in Segmentation". Another is to generate an open intersecting surface in one mask; for example, an open intersecting surface at the end of the spinal cord extending beyond the closed surface of the skull. Afterwards, an intersecting curve must be formed at the intersection so that both surfaces can be meshed, followed by the remaining volumes.

In addition, the shape of these triangulated surface objects may be approximated with non-uniform rational basis spline (NURBS) surfaces, whose shapes are determined from a series of control points. The NURBS-approximated objects, exported in IGES format, can be imported as geometry into FE software and readily meshed into a cluster of polyhedral mesh elements. This provides flexibility in modifying

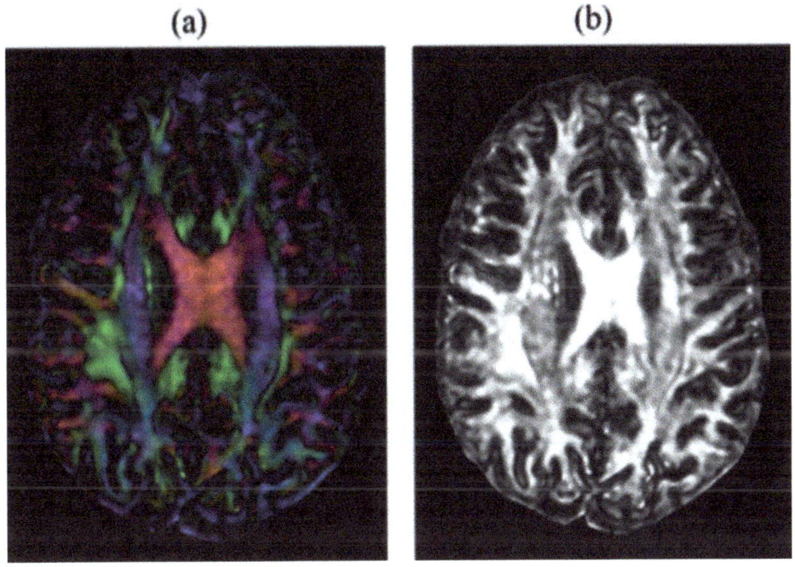

Fig. 2.6 (**a**) Original diffusion tensor images with colour denoting the principal direction (largest eigenvalues) of the diffusion tensor (red indicates left-right, green indicates anterior-posterior, and blue is inferior-superior). (**b**) Fractional anisotropy with lighter colour indicating higher anisotropy. These images were generated using 3D Slicer software

the model geometry directly within the FE analysis platform. The NURBS-conversion process is available in 3-matics, Geomagic, Blender etc.; however, tedious manual operations are inevitable if the object has a complex structure.

Whilst meshing is mostly performed automatically by such software, care should still be taken in setting up the meshing method, including maximum and minimum element size, element growth factor, mesh smoothing parameters and if necessary, mesh coarsening paremeters. It is also recommended to check mesh quality to identify poor quality elements, duplicate elements or uncovered faces. These errors may need to be repaired manually.

2.2.4 *Physics and Property Settings*

Bioelectromagnetism is the study of electric, magnetic and electromagnetic phenomena arising from living cells, tissues or organisms. In the field of bioelectromagnetism, biological tissues are generally considered as "volume conductors", in which the inductive component of the impedance is neglected, and resistances, capacitances and voltage sources are distributed throughout a three-dimensional (3D) region [28].

In the low-frequency band, where the frequency of internal bioelectric events lies, capacitive and electromagnetic propagation can be neglected [29, 30], thus treating bioelectric currents and voltages in living tissues as stationary [31]. This is known as the *quasi-static* approximation. A recent modelling study by Bossetti et al. [32]

investigated the difference in neural activation between solving the quasi-static field approximation and solving the full inhomogeneous Helmholtz equation using square-pulse current stimuli. They found that for commonly used stimulus parameters, the exact solution for the potential (including capacitive tissue effects) can be well approximated by the quasi-static case. Given the relatively low values of permittivity and magnetic permeability in living tissues, the quasi-static approximation can therefore be employed in computational head models of transcranial stimulation.

The electrical potential φ resulting from ECT can be obtained by solving the Laplace equation:

$$\nabla \cdot \left(-\sigma \nabla \varphi\right) = 0, \tag{2.3}$$

where φ is the electric potential and $\boldsymbol{\sigma}$ is the electrical conductivity tensor. In order to solve Eq. 2.3, boundary conditions have to be defined at all domain boundaries/surfaces. Based on the "quasi-uniform" assumption, the degree of activation in a target region is proportional to the local electric field magnitude $\left|\vec{E}\right|$ [33]. The electric field $\vec{E}$ is determined from Maxwell's equations under quasi-static conditions using

$$\vec{E} = -\nabla \varphi. \tag{2.4}$$

Simulation results can be analysed by comparing the average electric field magnitude $\overline{E}$ in several ROIs of the brain, which is determined using:

$$\overline{E} = \frac{\iiint \left|\vec{E}\right| dV}{\iiint dV}, \tag{2.5}$$

where $\left|\vec{E}\right|$ is the local electric field magnitude at every spatial point in the ROI, and the denominator is simply the volume of the ROI.

Tissue Conductivity

It is typical to assume homogeneity and isotropy in most head tissues, except for white matter. Tissue conductivities for each domain, presented in Table 2.2, were determined from previous experimental studies, as described in Bai et al. [25].

Electrode Placement

Conventional ECT electrode placements including bifrontal (BF), bitemporal (BT) and right unilateral (RUL) placements have been substantially investigated using computational modelling [3, 5, 26]. Several variations of electrode placement have also been investigated with estimates of electric field strengths in key brain regions, with an aim to improve existing ECT protocols [34, 35].

Table 2.2 List of tissue conductivities employed by Bai et al. [3, 25]

Head tissue	Conductivity (S/m)
Scalp	0.41
Eyes	0.41
Sinus	0[a]
CSF	1.79
Grey matter	0.31
White matter – fibre	0.65
White matter – transverse	0.065
Skull – compact	0.006
Skull – spongy	0.028

[a]Note that setting the conductivity to zero in any tissue region may present a numerical issue in the simulations. The common practice is to either set this domain as inactive in the simulations or to assign an extremely low value such as 1e-8 S/m

There are several ways to define the location of an electrode on the scalp:

1. Creating a geometric object representing the physical ECT electrode on the scalp, e.g. a short cylinder with a diameter of 5 cm and a conductivity of 9.8×10^5 S/m [5]. A normal inward current density $J|_{electrode}$ can be defined at the top boundary of one electrode,

$$J = \frac{I}{A}, \tag{2.6}$$

where I is the stimulus current, e.g. 800 mA for ECT, and A is the top boundary electrode area. A ground condition ($\varphi = 0$) can be defined at the top boundary of the other electrode. Current continuity or other conditions representing the skin-electrode interface can be defined at the scalp surface of both electrodes.

2. An isolated geometric boundary (e.g. a circular boundary of diameter 5 cm) defined on the scalp surface. The boundary condition for the ECT electrodes can thus be defined, with other conditions representing the skin-electrode interface over this isolated boundary.
3. A mathematically defined boundary created by intersecting the scalp surface with a geometry defined by analytic functions (e.g. an analytically defined sphere with a diameter of 5 cm for creating ECT electrodes) [25, 36]. An evenly distributed normal current density J is then applied over the analytically defined geometry. A normal inward current boundary condition is defined over the entire scalp such that everywhere, except at the ECG electrodes, the normal current density is zero. The other electrode is defined to have a normal outward current density, $-J|_{electrode}$.

Other Boundary Conditions

The neck region on the lower boundary of the head can be defined using a ground [25] or distributed impedance boundary condition [3]. The latter sets the neck as being connected to a resistive compartment, which in turn is connected to ground. This permits some flow of current through the neck boundary, albeit negligible in magnitude (0.001% of delivered ECT current). The boundary condition of this distributed impedance is defined as follows:

$$\vec{n} \cdot J \big|_{\text{neck}} = \frac{\sigma_{\text{neck}}}{d_s} \left(\varphi - \varphi_{\text{ref}} \right), \tag{2.7}$$

where $\vec{n} \cdot J \big|_{\text{neck}}$ is the normal outward current density at the neck boundary, σ_{neck} is the conductivity assigned to the boundary, d_s is the thickness of the boundary and φ_{ref} is the reference voltage, which is set to zero for ground.

The rest of the scalp is set as an insulated boundary (i.e. zero normal component of current density). If the sinuses are inactive in the simulation, their boundaries should be set to insulated as well. All other internal boundaries must be set as continuous current density interfaces to ensure electric continuity between domains.

Numerical Solver Settings

Under quasi-static assumptions, the stimulus amplitude and the resulting voltage, electric field and current density are all in a linear relationship. Therefore, it is sufficient to employ a steady-state solver. A detailed head model usually involves computing over a large number of mesh elements (>5 million). In COMSOL, there are two classes of linear solvers for computation: direct solvers, such as PARDISO, which are time-efficient but require large computational memory, and iterative solvers, such as conjugate gradient, which approach the solution gradually, and thus are memory-efficient but may require substantial computational time. An iterative solver is a better choice for standard desktop workstations with 24 to 64 Gb RAM. The absolute tolerance of the error in previous works was set to 10^{-5} [36] or even at 10^{-8} [5, 9]. In general, a lower absolute tolerance yields a more accurate result, provided it is greater than the numerical precision of the computer processor.

2.3 Simulation Results

2.3.1 Electric Feld for Three ECT Electrode Configurations

The MRI scan of a patient (39-year-old male) with bipolar disorder was acquired at Neuroscience Research Australia following an ECT session. The patient provided an informed consent for study participation, which also received ethics approval by the University of New South Wales. A T1-weighted 3 T head scan was obtained along with diffusion tensor imaging in 32 directions. The head scan was truncated at the chin, and the voxel size was 1 mm in all directions. The head was segmented into multiple domains corresponding to the tissues listed in Table 2.2. White matter anisotropy was obtained by the methods described in Sect. 2.2.2.

Three common ECT electrode configurations were simulated: bifrontal (BF), bitemporal (BT) and right unilateral (RUL) as depicted in Fig. 2.7. An electrical ECT stimulus was applied using an isolated circular boundary defined to be 5 cm diameter. This boundary was supplied with a current density J at the anode and $-J$ at the cathode. The stimulus current was set at 800 mA in all electrode configurations. The lower neck boundary was set to a distributed impedance as in Eq. (2.7), with $\boldsymbol{d_s}$ set to 5 cm from the ground reference, whilst $\boldsymbol{\sigma_{neck}}$ was set to the scalp conductivity.

The analysis was performed on regions of the brain associated with emotional responses, directing attention, memory, verbal and learning skills, among others [37], namely the cingulate gyri, parahippocampal gyri, subcallosal gyri, amygdala, inferior frontal gyri, hippocampus and middle frontal gyri. These ROIs were extracted automatically using BrainSuite's labelling tool and BCI-DNI atlas. This approach allows specific comparison in the particular brain region rather than qualitative observation in the whole brain.

Average $\left|\vec{E}\right|$ was calculated for each ROI using Eq. (2.5) and the results for each electrode configuration were compared. The results in Fig. 2.8 indicate that the right middle frontal gyrus displayed the largest average E-field for all three electrode configurations. RUL produced a maximum E-field in all right sides of the analysed locations as well as the left cingulate gyrus. Otherwise, the maximum average $\left|\vec{E}\right|$ of the other left-brain regions was achieved via the BT configuration.

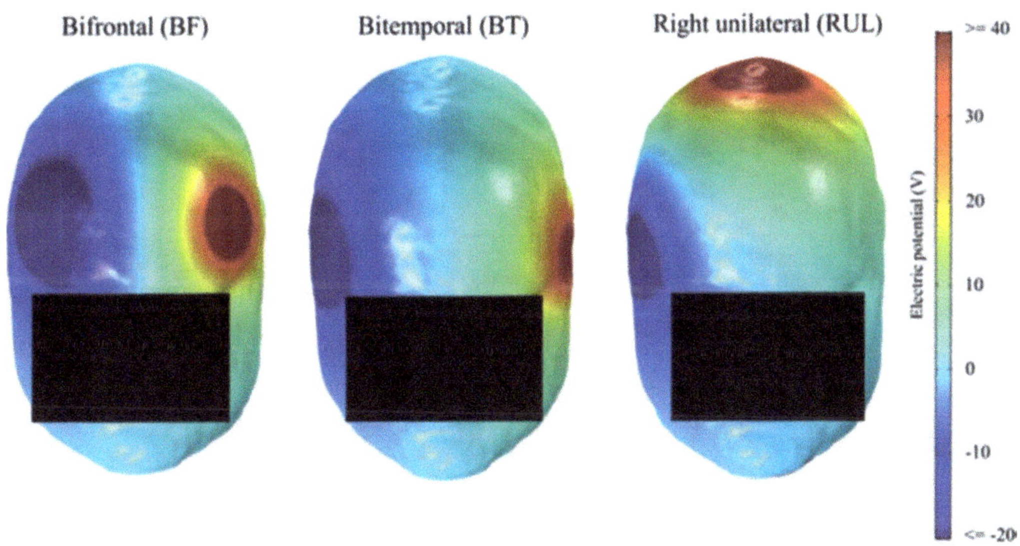

Fig. 2.7 Simulated electric potential on the scalp surface under three ECT electrode configurations. The sites of maximum and minimum electric potential are the electrode placement sites. The face of the patient is hidden for privacy. Results are shown for a single ECT pulse

The resulting electric field magnitude, $\left|\vec{E}\right|$, is displayed in Fig. 2.9. The electric field in the BF and BT configurations showed a more symmetric electric field profile, whilst RUL displayed a larger $\left|\vec{E}\right|$ over the right brain hemisphere relative to the left. The BF configuration resulted in higher $\left|\vec{E}\right|$ at the frontal lobe regions of the brain only. On the other hand, the $\left|\vec{E}\right|$ distribution in the BT configuration encompassed both the frontal and temporal lobe regions.

The BT configuration has been noted to be the most efficacious among the three electrode placements using the least amount of stimulus dose; however, it is also associated with a larger rate of cognitive side effects [38]. These effects are possibly due to the larger brain area affected by the electrical field as shown in Figs. 2.8 and 2.9. The RUL configuration is known to result in less verbal impairment [39], possibly due to the lesser impact on the speech area at the left inferior frontal region. Nonetheless, the RUL configuration typically requires a suprathreshold stimulus (exceeding seizure threshold) to be effective [39]. The BF configuration, on the other hand, is also known to result in less cognitive impairment than BT, but nonetheless using lesser stimulus dosage than RUL [40]. As such, electrode placement sites and stimulus dosage can affect the outcome of ECT, which can be investigated further using the modelling framework described in this chapter.

Modelling results such as those shown here can provide additional understanding of clinical ECT findings in a given patient. The electric field, voltage and current data obtained from the simulation can be used to infer ECT impact on the brain.

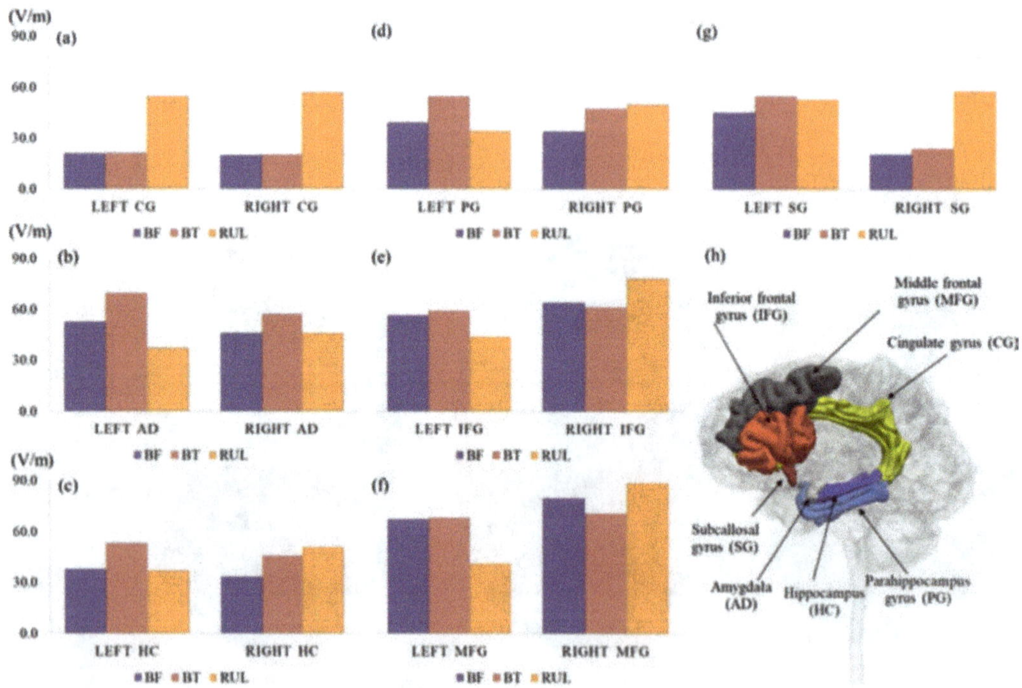

Fig. 2.8 Average electric field magnitude in specific ROIs: (**a**) cingulate gyri (CG), (**b**) amygdala (AD), (**c**) hippocampus (HC), (**d**) parahippocampal gyri (PG), (**e**) inferior frontal gyri (IFG), (**f**) middle frontal gyri (MFG) and (**g**) subcallosal gyri (SG). The position of these ROIs within the brain is described in (**h**)

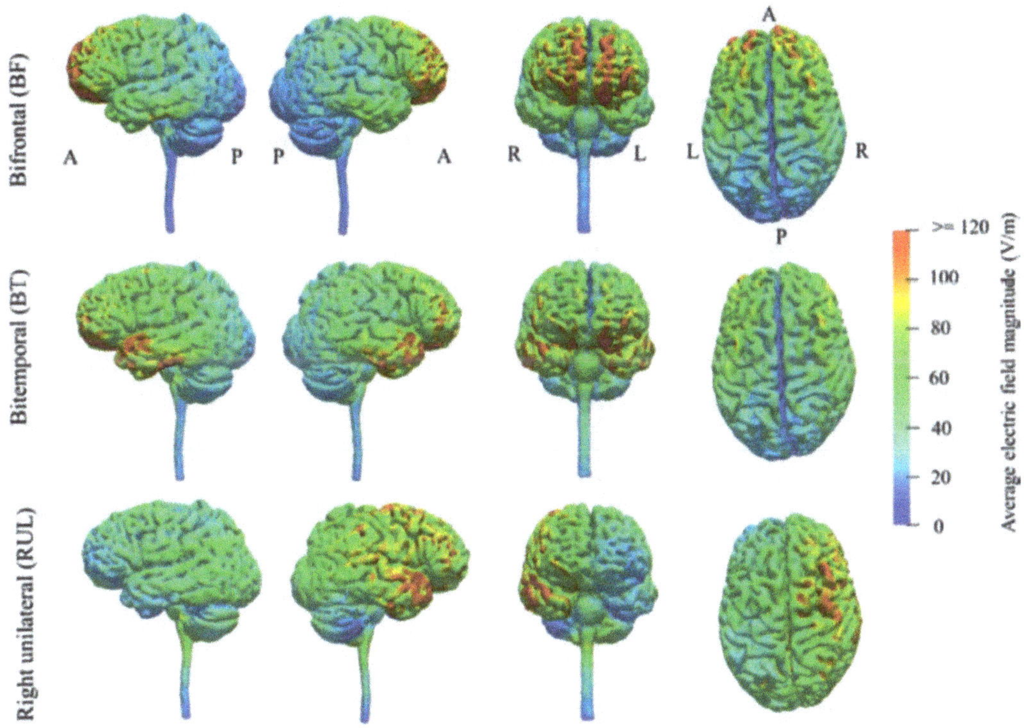

Fig. 2.9 Distribution of electric field magnitude in V/m on the grey matter surface. The maximum electric field is capped to 200 V/m for image clarity. The labels on the first row indicate the image orientation as follows: A anterior, P posterior, R right and L left

Nevertheless, these results are preliminary and should be treated with caution since results may differ from subject to subject. However, such results can be used to associate brain regions affected by ECT with patient responses following the treatment, providing for the future development of safer and more effective ECT protocols.

2.4 Discussion

2.4.1 Model Extensions

The modelling framework described in this chapter only addresses the computation of average electric field within different brain regions. Other possibilities for data analysis include: focality analysis by masking regions with electric field magnitude below a specified threshold [41], reconstruction of binarised subtraction maps for direct comparison of stimulus effects among different electrode placements [3], and analysis of heating during brain stimulation by incorporating a bioheat equation [42].

Moreover, the framework described here only simulates the head as a passive volume conductor under electrical stimulus, disregarding the complex excitable tissue properties of brain neurons. Nevertheless, the underlying brain activity is still poorly understood. The neuronal tissue itself acts as an internal source of current, which may disrupt the externally applied stimulus. A review by Ye and Steiger summarizes the evidence for such phenomena from various experimental and simulation studies [43]. Brain neurons are also interconnected to more distal regions due to the existence of neural tracts, which connect different regions of the brain. Thus, excitation of one region will propagate along the axon [37].

Several modelling studies have addressed the issue of brain activation. A recent study by Riel et al. [44] performed a preliminary activation analysis along the white matter fibre tracts using a volume conductor model. The fibre tracts were extracted from a DW-MRI image using constrained spherical deconvolution, and electrical potentials were subsequently interpolated along the fibre tracts for calculating the activation function [44]. Bai et al. [36] presented a finite element (FE) whole-head model incorporating Hodgkin-Huxley-based continuum excitable neural descriptions in the brain, which was able to simulate the dynamic changes of brain activation directly elicited by ECT, allowing investigation of parameters such as pulse duration [36]. Nevertheless, the computation was rather lengthy. In addition, the intracellular potential in the model was assumed to be resistively tied to a remote fixed potential, whose physiological meaning was difficult to interpret. This constraint did not allow for the spread of excitation through neural networks in the brain.

McIntyre et al. have, over the years, introduced a representation of white matter (WM) fibres in the vicinity of the subthalamic nucleus (STN), combined with a volume conductor model of deep brain stimulation (DBS) [45, 46]. After the electric potential induced by a DBS device was calculated by the FE solver, the time-dependent transmembrane potential was solved in NEURON software using a Hodgkin-Huxley-type model based on the interpolated potential distribution along the length of each axon. The model was able to predict activation in STN neurons and internal capsule fibres, and the degree of activation matched well with animal experimental data [45].

Subject-Specific Tissue Conductivity

The electrical conductivities of head tissues in most modelling studies are based on in vitro measurements. It is anticipated that some variation exists due to experimental conditions and sample preparation, let alone any inter-subject differences. However, it would be highly invasive to perform in vivo measurements of electrical conductivity, especially within brain structures.

Recent work by Fernández-Corazza et al. used electrical impedance tomography (EIT) to noninvasively obtain the conductivity of head tissues [4]. They injected small amounts of electrical current through multiple electrode pair configurations. A Laplace equation, as in Eq. (2.3), was then solved for every electrode pair. Afterwards, an inverse problem was solved to optimise the electrical conductivity using Newton's optimisation method to fit the Laplace equations to EIT experimental data. It was observed that the accuracy of this method also depends on the accuracy of the skull segmentation. Their study showed that an over-smoothed and compact skull geometry, with closed foramen, can overestimate the conductivity by almost 30% relative to a more accurate skull segmentation.

2.5 Conclusion

Computational head models and FE simulation provide additional insights into understanding regions of the brain affected by ECT and other transcranial stimulation techniques by using metrics such as the electric field distribution, which are difficult to obtain by direct experimental measurement. Furthermore, these head models can serve as a tool for testing novel ECT protocols prior to animal and clinical studies.

Acknowledgements Siwei Bai is a postdoctoral researcher sponsored by the European Commission under the Individual Fellowship scheme within the Marie Skłodowska-Curie Actions. Azam Ahmad Bakir is partly supported by NHMRC Project Grant 1126742.

References

1. Carney, S., et al. (2003). Efficacy and safety of electroconvulsive therapy in depressive disorders: A systematic review and meta-analysis. *Lancet, 361*(9360), 799–808.
2. Elias, A., Phutane, V. H., Clarke, S., & Prudic, J. (2018). Electroconvulsive therapy in the continuation and maintenance treatment of depression: Systematic review and meta-analyses. *Australian & New Zealand Journal of Psychiatry, 52*(5), 415–424.
3. Bai, S., Galvez, V., Dokos, S., Martin, D., Bikson, M., & Loo, C. (2017). Computational models of bitemporal, bifrontal and right unilateral ECT predict differential stimulation of brain regions associated with efficacy and cognitive side effects. *European Psychiatry, 41*, 21–29.
4. Fernández-Corazza, M., Turovets, S., Luu, P., Price, N., Muravchik, C. H., & Tucker, D. (2018). Skull modeling effects in conductivity estimates using parametric electrical impedance tomography. *IEEE Transactions on Biomedical Engineering, 65*(8), 1785–1797.
5. Lee, W. H., Deng, Z.-D., Kim, T.-S., Laine, A. F., Lisanby, S. H., & Peterchev, A. V. (2012). Regional electric field induced by electroconvulsive therapy in a realistic finite element head model: Influence of white matter anisotropic conductivity. *NeuroImage, 59*(3), 2110–2123.
6. Bai, S., Loo, C., Geng, G., & Dokos, S. (2011). Effect of white matter anisotropy in modeling electroconvulsive therapy. In *Engineering in Medicine and Biology Society, EMBC, 2011 annual international conference of the IEEE* (pp. 5492–5495). IEEE.
7. Bergmann, T. O., Karabanov, A., Hartwigsen, G., Thielscher, A., & Siebner, H. R. (2016). Combining non-invasive transcranial brain stimulation with neuroimaging and electrophysiology: Current approaches and future perspectives. *NeuroImage, 140*, 4–19.
8. Kim, D., Jeong, J., Jeong, S., Kim, S., Jun, S. C., & Chung, E. (2015). Validation of computational studies for electrical brain stimulation with phantom head experiments. *Brain Stimulation, 8*(5), 914–925.
9. Laakso, I., Tanaka, S., Mikkonen, M., Koyama, S., Sadato, N., & Hirata, A. (2016). Electric fields of motor and frontal tDCS in a standard brain space: A computer simulation study. *NeuroImage, 137*, 140–151.
10. Juntu, J., Sijbers, J., Van Dyck, D., & Gielen, J. (2005). *Bias field correction for MRI images* (pp. 543–551). Berlin, Heidelberg: Springer Berlin Heidelberg.
11. Johnson, H., Harris, G., & Williams, K. (2007). BRAINSFit: Mutual information rigid registrations of whole-brain 3D images, using the insight toolkit. *Insight J, 57*(1).

12. Nielsen, J. D., et al. (2018). Automatic skull segmentation from MR images for realistic volume conductor models of the head: Assessment of the state-of-the-art. *NeuroImage, 174*, 587–598.
13. Gao, Y., Kikinis, R., Bouix, S., Shenton, M., & Tannenbaum, A. (2012). A 3D interactive multi-object segmentation tool using local robust statistics driven active contours. *Medical Image Analysis, 16*(6), 1216–1227. /08/01/ 2012.
14. Pantazis, D., et al. (2010). Comparison of landmark-based and automatic methods for cortical surface registration. *NeuroImage, 49*(3), 2479.
15. Joshi, A. A., et al. (2017). A whole brain atlas with sub-parcellation of cortical gyri using resting fMRI. In *Medical imaging 2017: Image processing* (Vol. 10133, p. 101330O). International Society for Optics and Photonics.
16. Joshi, A. A., Shattuck, D. W., Thompson, P. M., & Leahy, R. M. (2007). Surface-constrained volumetric brain registration using harmonic mappings. *IEEE Transactions on Medical Imaging, 26*(12), 1657–1669.
17. Oltedal, L., et al. (2018). Volume of the human hippocampus and clinical response following electroconvulsive therapy. *Biological Psychiatry, 84*, 574.
18. Oh, S., Lee, S., Cho, M., Kim, T., & Kim, I. (2006). Electrical conductivity estimation from diffusion tensor and T2: A silk yarn phantom study. *Proc Intl Soc Mag Reson Med, 14*, 3034.
19. Tuch, D. S., Wedeen, V. J., Dale, A. M., George, J. S., & Belliveau, J. W. (2001). Conductivity tensor mapping of the human brain using diffusion tensor MRI. *Proceedings of the National Academy of Sciences, 98*(20), 11697–11701.
20. Basser, P. J., & Jones, D. K. (2002). Diffusion-tensor MRI: Theory, experimental design and data analysis–a technical review. *NMR in Biomedicine: An International Journal Devoted to the Development and Application of Magnetic Resonance In Vivo, 15*(7–8), 456–467.
21. Behrens, T. E., Berg, H. J., Jbabdi, S., Rushworth, M. F., & Woolrich, M. W. (2007). Probabilistic diffusion tractography with multiple fibre orientations: What can we gain? *NeuroImage, 34*(1), 144–155.
22. Behrens, T. E., et al. (2003). Non-invasive mapping of connections between human thalamus and cortex using diffusion imaging. *Nature Neuroscience, 6*(7), 750.
23. Behrens, T. E., et al. (2003). Characterization and propagation of uncertainty in diffusion-weighted MR imaging. *Magnetic Resonance in Medicine: An Official Journal of the International Society for Magnetic Resonance in Medicine, 50*(5), 1077–1088.
24. Basser, P. J. (1995). Inferring microstructural features and the physiological state of tissues from diffusion-weighted images. *NMR in Biomedicine, 8*(7), 333–344.
25. Bai, S., Dokos, S., Ho, K.-A., & Loo, C. (2014). A computational modelling study of transcranial direct current stimulation montages used in depression. *NeuroImage, 87*, 332–344.
26. Lee, W. H., Lisanby, S. H., Laine, A. F., & Peterchev, A. V. (2016). Comparison of electric field strength and spatial distribution of electroconvulsive therapy and magnetic seizure therapy in a realistic human head model. *European Psychiatry, 36*, 55–64.
27. Wolters, C. H., Anwander, A., Tricoche, X., Weinstein, D., Koch, M. A., & Macleod, R. S. (2006). Influence of tissue conductivity anisotropy on EEG/MEG field and return current computation in a realistic head model: A simulation and visualization study using high-resolution finite element modeling. *NeuroImage, 30*(3), 813–826.
28. Malmivuo, J., & Plonsey, R. (1995). *Bioelectromagnetism: Principles and applications of bioelectric and biomagnetic fields*. USA: Oxford University Press.
29. Geselowitz, D. B. (1963). The concept of an equivalent cardiac generator. *Biomedical Sciences Instrumentation, 1*, 325–330.
30. Schwan, H. P., & Kay, C. F. (1957). Capacitive properties of body tissues. *Circulation Research, 5*(4), 439–443.
31. Plonsey, R., & Heppner, D. B. (1967). Considerations of quasi-stationarity in electrophysiological systems. *The Bulletin of Mathematical Biophysics, 29*(4), 657–664.
32. Bossetti, C. A., Birdno, M. J., & Grill, W. M. (2007). Analysis of the quasi-static approximation for calculating potentials generated by neural stimulation. *Journal of Neural Engineering, 5*(1), 44.

33. Bikson, M., Dmochowski, J., & Rahman, A. (2013). The "quasi-uniform" assumption in animal and computational models of non-invasive electrical stimulation. *Brain Stimulation, 6*(4), 704.
34. Loo, C. K., Bai, S., Donel Martin, M., Gálvez, V., & Dokos, S. (2015). Revisiting frontoparietal montage in electroconvulsive therapy: Clinical observations and computer modeling a future treatment option for unilateral electroconvulsive therapy. *The Journal of ECT, 31*(1), e7–e13.
35. Bai, S., Loo, C., Lovell, N. H., & Dokos, S. (2013). Comparison of three right-unilateral electroconvulsive therapy montages. In *Engineering in medicine and biology society (EMBC), 2013 35th annual international conference of the IEEE* (pp. 819–822). IEEE.
36. Bai, S., Loo, C., Al Abed, A., & Dokos, S. (2012). A computational model of direct brain excitation induced by electroconvulsive therapy: Comparison among three conventional electrode placements. *Brain Stimulation, 5*(3), 408–421.
37. Guyton, A. C., & Hall, J. E. (2006). *Textbook of medical physiology*. Elsevier Saunders.
38. Kellner, C. H., Tobias, K. G., & Wiegand, J. (2010). Electrode placement in electroconvulsive therapy (ECT): A review of the literature. *The Journal of ECT, 26*(3), 175–180.
39. Kolshus, E., Jelovac, A., & McLoughlin, D. M. (2017). Bitemporal v. high-dose right unilateral electroconvulsive therapy for depression: A systematic review and meta-analysis of randomized controlled trials. *Psychological Medicine, 47*(3), 518–530.
40. Bailine, S. H., et al. (2000). Comparison of bifrontal and bitemporal ECT for major depression. *American Journal of Psychiatry, 157*(1), 121–123.
41. Deng, Z.-D., Lisanby, S. H., & Peterchev, A. V. (2011). Electric field strength and focality in electroconvulsive therapy and magnetic seizure therapy: A finite element simulation study. *Journal of Neural Engineering, 8*(1), 016007.
42. Khadka, N., Zannou, A. L., Zunara, F., Truong, D. Q., Dmochowski, J., & Bikson, M. (2018). Minimal heating at the skin surface during transcranial direct current stimulation. *Neuromodulation: Technology at the Neural Interface, 21*(4), 334–339.
43. Ye, H., & Steiger, A. (2015). Neuron matters: Electric activation of neuronal tissue is dependent on the interaction between the neuron and the electric field. *Journal of Neuroengineering and Rehabilitation, 12*(1), 65.
44. Riel, S., Bashiri, M., Hemmert, W., & Bai, S., A tractography analysis for electroconvulsive therapy in IEEE Eng Med Biol Soc, Honolulu, HI, 2018: IEEE.
45. Butson, C. R., & McIntyre, C. C. (2005). Role of electrode design on the volume of tissue activated during deep brain stimulation. *Journal of Neural Engineering, 3*(1), 1.
46. Gunalan, K., et al. (2017). Creating and parameterizing patient-specific deep brain stimulation pathway-activation models using the hyperdirect pathway as an example. *PLoS One, 12*(4), e0176132.

Chapter 3
Estimates of Peak Electric Fields Induced by Transcranial Magnetic Stimulation in Pregnant Women as Patients or Operators Using an FEM Full-Body Model

Janakinadh Yanamadala, Raunak Borwankar, Sergey Makarov, and Alvaro Pascual-Leone

3.1 Introduction

Recent studies confirm the efficacy of transcranial magnetic stimulation (TMS) as a noninvasive treatment of medication-resistant depression [1, 2]. Four different devices, the Neuronetics Neurostar Stimulator, Brainsway H-Coil system, Magstim Magnetic Stimulator, and MagVenture Stimulator, have been cleared by the U.S. Food and Drug Administration (FDA) for the treatment of medication-resistant depression [3, 4].

Even though TMS coil holders, and even robots, have been developed that might make the application of TMS more spatially precise and efficient, to date, TMS is often applied by an operator who manually positions and retains the TMS coil over the subject's head. A potential safety concern is thus generated when the operator is a woman and is pregnant. There are no studies to date that assess the safety of TMS

J. Yanamadala (✉)
ECE Department, Worcester Polytechnic Institute, Worcester, MA, USA

MathWorks, Natick, MA, USA
e-mail: jyanmadala@wpi.edu

R. Borwankar
ECE Department, Worcester Polytechnic Institute, Worcester, MA, USA

S. Makarov
Massachusetts General Hospital, Boston, MA, USA

Worcester Polytechnic Institute, Worcester, MA, USA

A. Pascual-Leone
Berenson-Allen Center for Noninvasive Brain Stimulation and Division for Cognitive Neurology, Beth Israel Deaconess Medical Center, Harvard Medical School, Boston, MA, USA

S. Makarov et al. (eds.), *Brain and Human Body Modeling*,
https://doi.org/10.1007/978-3-030-21293-3_3

for a fetus. In the case of a pregnant woman as a TMS operator, we must consider two possibilities:

- Standard operation with the TMS coil held at distances of approximately 1–2 ft from the belly
- Accidental TMS coil discharge when the coil is in direct contact with the belly or in its immediate vicinity

In addition to the scenario of a pregnant woman as a TMS operator, the possibility of a pregnant woman as a TMS patient is also important to consider. TMS can cause a generalized tonic seizure, which can pose a significant risk for the integrity of a pregnancy. Therefore, in most instances, pregnancy will be an exclusion criterion for TMS. However, a considerable percentage of women experience symptoms of depression during pregnancy and develop clinical depression requiring medical intervention. TMS has been proposed as a method to treat maternal depression while avoiding fetal exposure to drugs [5, 6]. So while the risk-benefit profile is argued to be better for TMS than for medications, one must consider that TMS may cause fetal exposure to high induced currents.

In estimating acceptable levels of induced currents, we refer to guidelines from the International Commission on Non-Ionizing Radiation Protection (ICNIRP) [7, 8]. The 2010 ICNIRP basic restrictions for occupational exposure to time-varying electric and magnetic fields for frequencies in the band 1 Hz–100 kHz [8] recommend that the exposure should be limited to electric fields in the head and body of less than 800 mV/m in order to avoid peripheral and central myelinated nerve stimulation. ICNIRP also recommends that the restrictions on electric or magnetic fields including transient or very short-term peak fields (which are encountered during TMS) be regarded as instantaneous values which should not be time averaged.

We assume that the estimate of 0.8 V/m maximum peak field should also apply to the fetal brain, body, and trunk.

3.2 Methods and Materials

3.2.1 Existing Computational Models of a Pregnant Woman

Induction currents in the entire human body (or bodies) of a pregnant subject caused by a TMS coil can be established in every particular case via numerical electromagnetic modeling. One of the primary investigated concerns has been a *significant electric current density*, which may develop in the highly conducting amniotic fluid surrounding the fetus and subject to an external time-varying magnetic field [9, 10]. Table 3.1 lists computational models of a pregnant woman and/or a fetus currently available for electromagnetic and radiological simulations.

These models (except for Refs. [10, 11], which are highlighted in the table) are based on insertion of a fetus into an existing nonpregnant female model. Meanwhile, the models developed from scans of pregnant females include the abdominal region only [10, 11].

Table 3.1 Computational pregnant-woman models (after ~2004)

	Model name	A/H/W	Da	TYPE	RES, mm[c]	FV	D	References
IT'IS Found., Switzerland	PREGNANT WOMAN[a]	3, 7, 9 months fetus	N	V	0.5 x 0.5 x 1.0 h 0.9 x 0.9 x 2b	Y	N	[15, 16]
Natl. Inst. of Inform. and Comm. Technol., Japan	PREGNANT WOMAN (based on non-pregnant model [21])	22/160/53 12, 20, 23, 26, 29, 32 and 33 weeks fetus	N	V	2 x 2 x 2	Y	N	[14, 16, 18]
Imperial College, UK	PREGNANT WOMAN[b]	28 weeks fetus	N	V	1 x 1 x 5	N	N	[11]
Health Protection Agency, UK	PREGNANT WOMAN[c]	23/163/60 8, 13, 26, 38 weeks fetus	N	V	2 x 2 x 2	N	N	[9]
Graz University of Technology, Austria	SILVY	89 kg 30 weeks fetus	N	V	2 x 2 x 7	N	N	[10]
Helmholtz Zentrum Munchen, Germany CST AG, Germany	KATJA	43/163/62 24 weeks fetus	N	V	1.8 x 1.8 x 4.8	N	N	[19]
Rensselaer Poly. Institute, NY, USA	PREGNANT WOMAN[d]	30 weeks fetus	N	V	6 x 6 x 7	N	N	[10]
Rensselaer Poly. Institute, NY, USA	RPI P-3, P-6, P-9[e]	First, second, third trimesters fetus	N	NURBS, V	6 x 6 x 7 for fetus	N	N	[11]

Notes: A/H/W – age(years)/height (cm)/weight (kg); Da – original image dataset made available for independent evaluation of true model accuracy (Y/N); TYPE (V – voxel; NURBS – CAD model); RES – lowest image resolution (before or after post-processing) of the model declared by the provider (h = head, b = body); FV – free version for research available (Y/N); D – deformable/posable (Y/N)

[a]Based on ELLA [15, 16, 20]

[b]Abdominal region only

[c]Based on NAOMI [20]

[d]Abdominal region only (body from above liver to below pubic symphysis)

[e]Anatomical data for the pregnant female and the fetus are gathered from several origins

All models in Table 3.1 are *voxel* models, except for Ref. [11] which used B-splines or NURBS, while some parts of the body are adapted from the Visible Human Project [12, 13]. Figure 3.1 shows a NURBS model from Ref. [11], while Fig. 3.2 shows the voxel model family from Ref. [14].

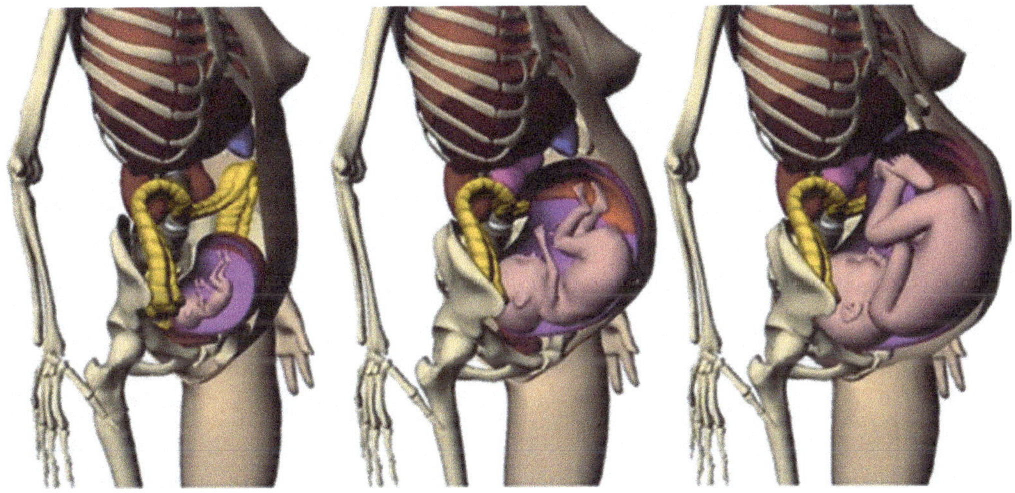

Fig. 3.1 NURBS model of a pregnant woman [11]

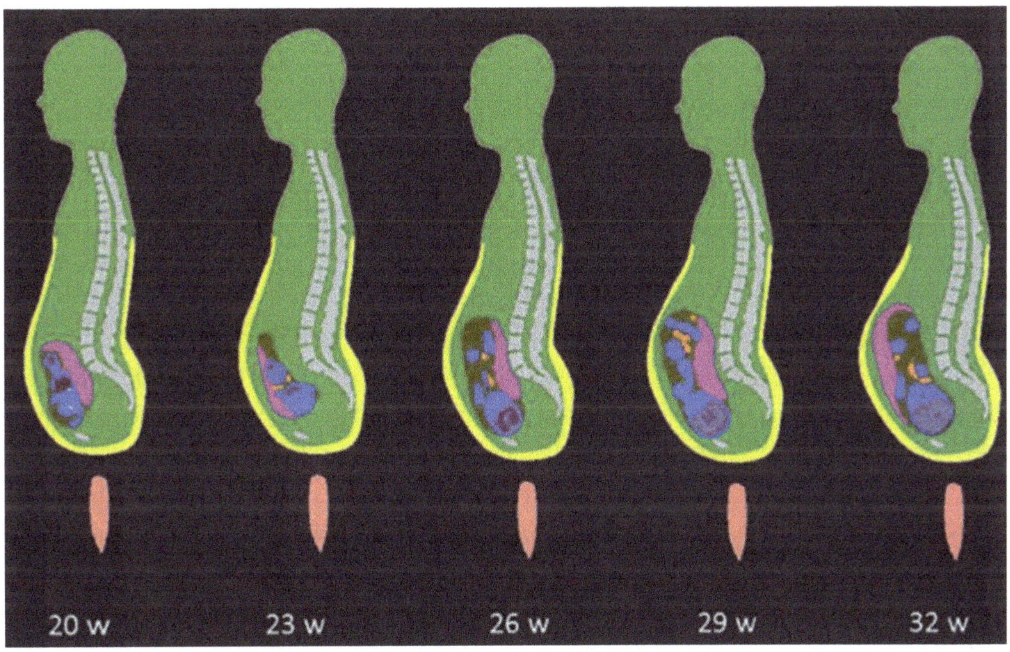

Fig. 3.2 Voxel model of a pregnant woman [16]

3.2.2 Construction of FEM (CAD) Full-Body Pregnant Woman Model and Model Topology

The voxel models listed in Table 3.1 are perfectly fine for radiation dosimetry studies [10, 11, 18] and for high-frequency and RF simulation studies of specific absorption rates [14–16, 18] based on the finite-difference time-domain (FDTD) approach. However, they are not suitable for the finite element method (FEM), which is generally employed by the TMS community [23–33]. This method more accurately captures complicated coil geometry(s) and curved boundaries between tissues. The NURBS surfaces [11] also have limited value for an FEM solver, which internally operates with geometry primitives: triangular facets and tetrahedra. A conversion from NURBS surfaces to triangular surfaces may require (very) significant additional meshing time.

To enable FEM analysis, a full-body CAD model of a pregnant female in the form of triangular surface meshes has been developed. As an initial dataset, we have chosen the detailed voxel model of a pregnant female by Nagaoka [14, 17, 18]; see second row of Table 3.1. We received the voxel model after signing a licensing agreement with the National Institute of Information and Communications Technology, Japan. This model is based on a 22-year-old pregnant Japanese female (26th week or second trimester) [17]. The original pregnant female voxel model was developed from MRI data collected on a nonpregnant Japanese woman who was 160 cm tall and weighed 53 kg. Further, abdominal MR images of a 26-week pregnant woman were segmented and inserted into this full-body model.

We converted this voxel model into an FEM CAD model using isosurface extraction in ITK-SNAP [35] and MATLAB. Mesh decimation, healing, and smoothing were performed using custom MATLAB scripts and ANSYS SpaceClaim. Standard mesh intersection approaches [36–41] typically result in a large number of triangles close to intersection chains and loops. Furthermore, they leave coincident faces, which might create compatibility problems. Resulting object intersections (which are usually "shallow" intersections) were resolved by locally moving intersecting surfaces in their respective normal directions with a step size of 0.2 mm or so until the intersection was no longer present [42, 43].

A well-known problem with FEM models is object matching in a contact region. Usually, the contact region is not explicitly defined in an imported CAD model and has to be discovered separately by testing for face-to-face overlaps and matching CAD faces/edges [44]. This circumstance may lead to problems for certain CAD kernels such as ACIS. To prevent CAD import errors, a thin gap was introduced between all tissue objects and was filled with "average body properties" of an outer enclosing shell. In some sense, this gap represents membranes separating different tissues. If the gap is reasonably small, it provides a close approximation to reality for different physical processes.

In order to construct the fetus model representing a pregnant female during the first and third trimester, we used the base data for the second trimester and the deformation approach described in Refs. [14, 18].

Figure 3.3 shows three variations of the CAD model constructed for the present study. The corresponding tissue mesh inventory is summarized in Supplement I. To our knowledge, this is the only detailed FEM-compatible model of a pregnant female currently available. The current model contains approximately 100 individual parts. Its distinct feature is a continuous CSF shell around the gray matter for both the mother and the fetus. Creation and testing required about 12 man months and continues as a work in progress. Figure 3.4 illustrates the corresponding fetal volume (second trimester) on a larger scale.

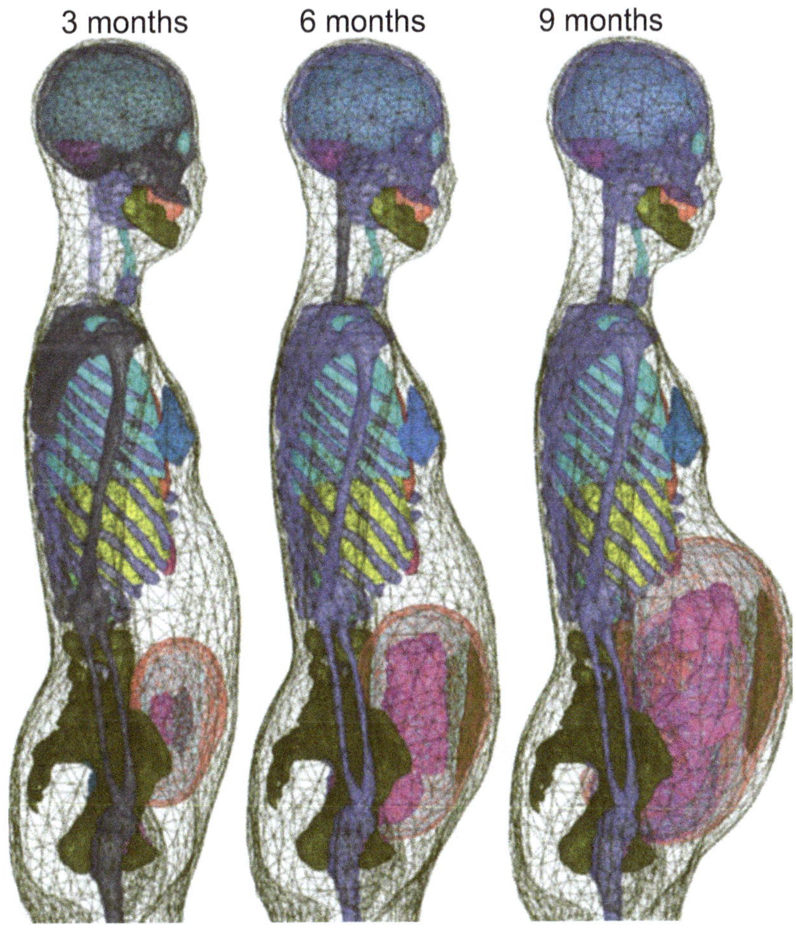

Fig. 3.3 CAD models of a pregnant woman during the first, second, and third trimesters

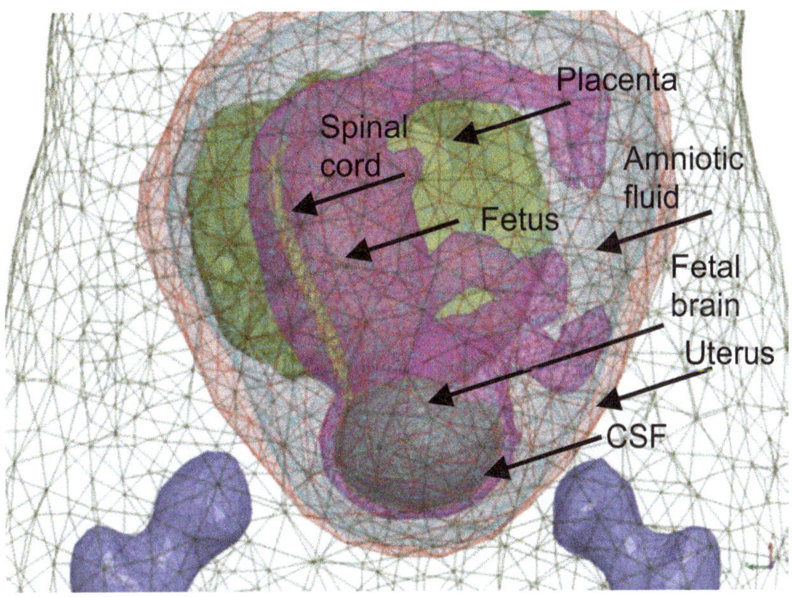

Fig. 3.4 Detailed view of the fetus for the second trimester model – posterior view with pelvic bones and other nearby tissues removed

Table 3.2 Material properties used in mother/fetus models [11, 51]

Tissue	σ (S/m)/ε_r
Amniotic fluid	Cerebrospinal fluid
Fetus	Mean of muscle, uterus, and blood
Fetal brain	$\left[\left(\frac{\sigma\_\text{fetalbrain}(64\,\text{MHz})}{\sigma\_\text{fetus}(64\,\text{MHz})}\right)+\left(\frac{\sigma\_\text{fetalbrain}(127\,\text{MHz})}{\sigma\_\text{fetus}(127\,\text{MHz})}\right)/2\right]\times\sigma\_\text{fetus}$
	$\left[\left(\frac{\varepsilon\_\text{fetalbrain}(64\,\text{MHz})}{\varepsilon\_\text{fetus}(64\,\text{MHz})}\right)+\left(\frac{\varepsilon\_\text{fetalbrain}(127\,\text{MHz})}{\varepsilon\_\text{fetus}(127\,\text{MHz})}\right)/2\right]\times\varepsilon\_\text{fetus}$
Placenta	Average muscle

Other biomechanical CAD models of pregnant women having different degrees of approximation have also been constructed [45–48]. However, these models do not include detailed geometry of the fetus suitable for EM simulation [47–50]. Figure 3.4 demonstrates the corresponding fetal volume (second trimester) on a larger scale.

3.2.3 *Tissue Properties*

Most tissues were assigned material properties (conductivity and dielectric constant) following the Gabriel & Gabriel database [49], which is further replicated in the IT'IS database [50]. Fetal properties follow Refs. [11, 51] and are outlined in Table 3.2. The conductivity and permittivity of fetal brain are comparable to that of the fetus and also behave similarly. Hence, the fetal brain is assigned scaled fetus material properties; the scaling factor is obtained from the available dataset. The material property values assigned to all tissues are also provided in the supplement.

3.3 Study Design

3.3.1 *TMS Coil*

Similar to Ref. [26], the base coil is a figure-eight straight coil with a loop radius of 35 mm. However, instead of a stranded conductor, a solid conductor (copper) with a diameter of 8 mm was used.

3.3.2 *Pulse Form and Duration*

TMS pulse forms vary widely in shape and duration [52–55]. Table 3.3 summarizes the data for four common FDA-approved TMS machines.

In order to take the majority of cases into consideration, we have chosen a simple biphasic harmonic coil pulse current:

$$I(t) = I_0 \sin(2\pi t/\tau), \quad \frac{dI}{dt} = \frac{2\pi}{\tau} I_0 \cos(2\pi t/\tau), \tag{3.1}$$

$$0 \le t \le \tau, \; I(t) = 0 \text{ otherwise}$$

The derivative of the coil current, I, is proportional to the induced electric field/induced electric current in the body. In order to include the majority of cases from Table 3.3, the total pulse duration or length τ was evaluated for two limiting values:

$$\tau = 1 \text{ ms}, \quad \tau = 0.1 \text{ ms} \tag{3.2}$$

Please note that the equivalent frequency of the biphasic harmonic pulse given by (3.1) is

$$f = 1/\tau \tag{3.3}$$

Other more elaborate pulse forms have also been studied [62].

3.3.3 Coil Current

For every pulse duration and coil position, the coil current amplitude I_0 in (3.1) has been found from the condition of one standard motor threshold (SMT) unit [61, 62]. One SMT means that the electric field at a point 2 cm from the surface of the head beneath the coil center reaches the motor threshold value of approximately 130 V/m [61, 62]. Motor threshold, a measure of the TMS intensity necessary to evoke a peripheral motor response, is variable across individuals but is also remarkably constant in a given individual [63]. For example, the peak coil current for a 0.1-ms-long pulse was found to be approximately 9000 A·turns magnetomotive force (mmf) at 1 SMT unit.

Table 3.3 Pulse widths and SMT values for different TMS setups

#, Pulse form	TMS system	Pulse duration (ms)	Std. motor threshold (SMT)	References
1. Biphasic	Brainsway	0.370	0.60–1.4	[56]
2. Monophasic	Magstim 200[2]	1.000	NA	[57]
3. Biphasic	MagVenture	0.290	0–1.7	[58]
4. Biphasic	Neuronetics	0.185	0.22–1.6	[58, 59]

3.3.4 Coil Positions

Two coil positions for a *pregnant patient* have been considered (see Fig. 3.5). In the first case (Fig. 3.5a), the straight coil is located 10 mm above the top of the head. In the second case (Fig. 3.5b), the straight coil is translated and then tilted by 60 degrees. The first case might represent a standard TMS coil placement for studies aimed at evaluating central motor conduction, though a circular TMS coil would be generally used in such instances. The second case aims to approximate the position of the TMS for the treatment of depression.

A coil positioning map for the *pregnant operator* is shown in Fig. 3.6 for the second trimester. The closest distance from the coil center to the body is 115 mm. We consider three representative polarizations of the major current dipole of the coil:

A. In the coronal plane (z-polarization in Fig. 3.6, Config. A, labeled as A1-A6)
B. In the sagittal plane (y-polarization in Fig. 3.6, Config. B, labeled as B1-B6)
C. In the transverse plane (x-polarization in Fig. 3.6, Config. C, labeled as C1-C4)

For every polarization type, four to six representative coil locations have been tested in the sagittal plane as shown in Fig. 3.6. The operator could achieve any of these positions by moving their right arm along with the coil holder. This results in a total of 16 test cases. For each test case, the eddy current density and the corresponding electric field everywhere in the body were computed. Each test case is conducted for the first, second, and third trimesters. For the first trimester, the closest distance from the coil center to the body is 45 mm (B) or 80 mm (C), 115 mm (B) and (C) for the second, and for the third – 40 mm (B) and 80 mm (C). The remaining topology is the same.

3.3.5 Accidental Coil Discharge

Two extreme cases have also been considered, which are not shown in Fig. 3.6. These are when the coil is moved in the xy-plane until it is as close to the body as possible. These cases will be labeled as B7, B8 and C5, C6, respectively.

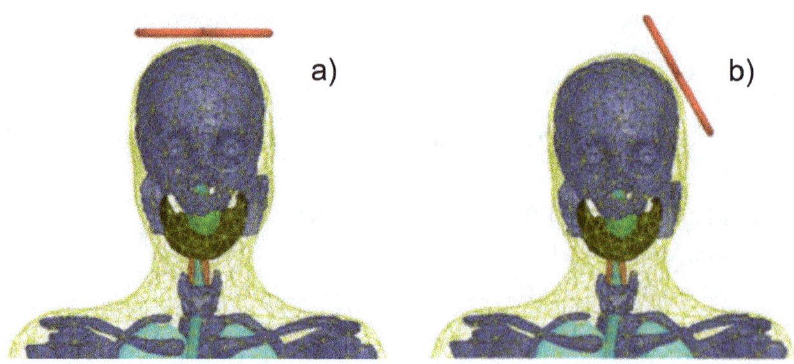

Fig. 3.5 Two coil positions used for the pregnant patient study

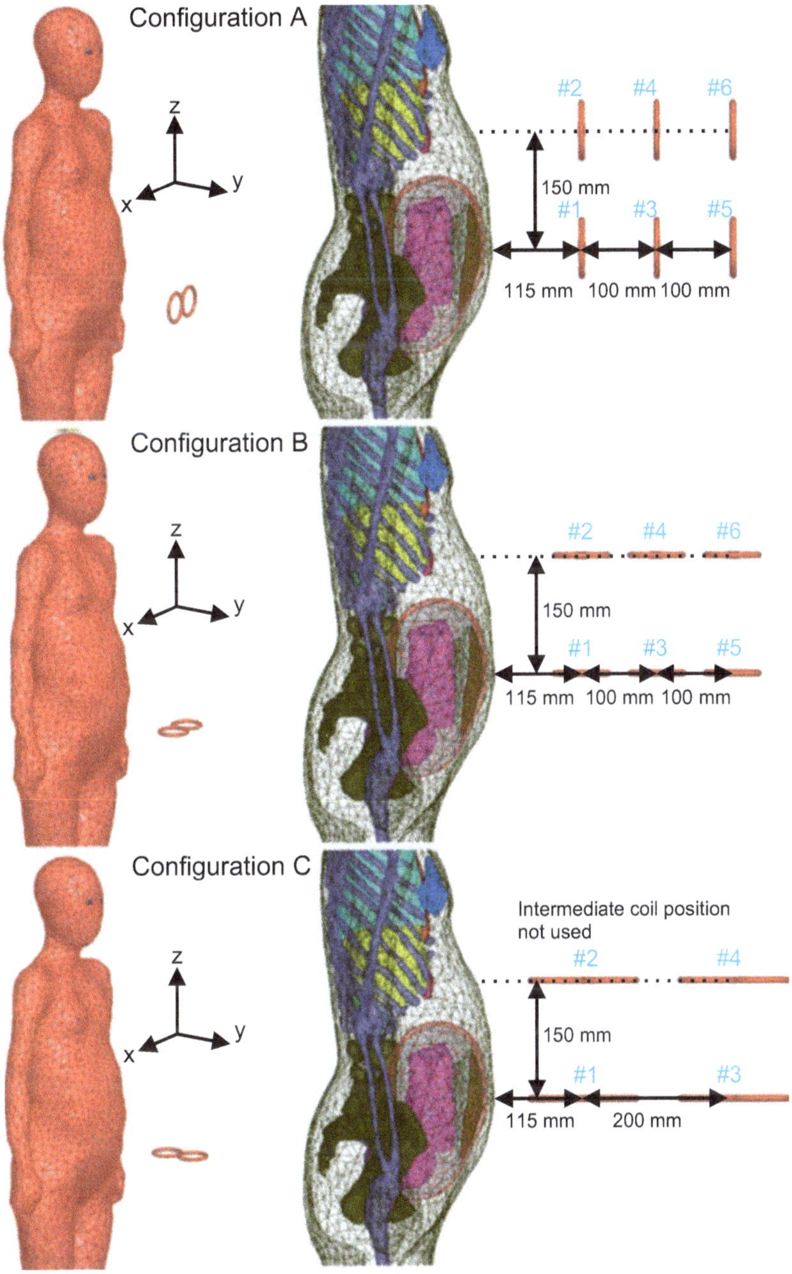

Fig. 3.6 Coil positioning map for a pregnant operator (second trimester) study. Some tissues are hidden for clarity

3.3.6 Frequency-Domain Computations

All simulations were performed in the frequency domain using ANSYS Maxwell 3D FEM software (v. 16). ANSYS Maxwell 3D is a commercial FEM software package with adaptive mesh refinement and has been extensively used for eddy current computations, similar to the earlier studies [24–26]. The software takes into account both conduction and displacement currents (as well as free and polarization charges), and solves the full-wave Maxwell equation for the magnetic field, ***H***, in the frequency domain

$$\nabla\times\frac{1}{\sigma+j\omega\varepsilon}\times\mathbf{H}=-j\omega\mu\mathbf{H} \tag{3.4}$$

where σ is the local medium conductivity, and ε and μ are the local permittivity and permeability, respectively. The major difference from the full-wave case is that the phase is assumed to be constant over the volume of interest. Although Maxwell 3D also has a transient FEM solver, this solver does not take into account the displacement currents and was therefore not used.

All simulations made use of the automated adaptive meshing technology available in Maxwell to iteratively refine the mesh. Five adaptive meshing passes were performed during the calculation, with the final meshes approaching about 2 M tetrahedra. Details of the adaptive mesh refinement procedure have been discussed previously [60].

3.3.7 *Time-Domain Computations*

Frequency-domain results (coil excitation with a sinusoidal waveform) for fields and currents have been collected for multiple frequencies (a logarithmic frequency sweep) over the band from 300 Hz to 3 MHz in order to generate the required pulse forms via the fast Fourier transform (FFT) and inverse FFT (IFFT) as described in Ref. [60]. The corresponding method has been described in the same reference; it is time-consuming but accurate. The time-domain solution is required for any pulse form including the harmonic pulses given by (3.1) since they are distorted quite differently from the harmonic wave of the same frequency. This solution is also important for other (nonharmonic) pulse forms [60].

3.3.8 *Finding Maximum Peak Current Density/Electric Field Strength in Individual Tissues*

A uniform 5 × 5 × 5 mm grid of observation points was introduced within a rectangular box, which covers the abdominal area only. This resulted in approximately 150,000 observation points within the body, where the induced current and the electric field are evaluated. For *every* such point, the pulse form has been restored via IFFT. Then, interpolation of peak pulse values onto a finer 2 × 2 × 2 mm grid was performed, followed by averaging over each small tissue volume as recommended in Ref. [8]. Finally, the absolute maximum peak current/field has been evaluated for every all tissues.

3.4 Results: Pregnant Patient

3.4.1 Qualitative Behavior of Induced Currents in the Body of a Pregnant Patient at Different Frequencies (Pulse Durations)

Figure 3.7 shows eddy current amplitude distribution in a coronal plane of a pregnant patient for three representative frequencies: 3 kHz, 30 kHz, and 300 kHz. The coil current amplitude is 10,000 A (10,000 A·turns mmf). Note that the color scale has been multiplied by the factor of 10 for every subsequent figure.

We observe that the peak current in the fetal area does not exceed 0.1 mA/m^2, 1 mA/m^2, and 40 mA/m^2. Comparing Fig. 3.7a, b, the induced current initially appears to behave as a linear function of frequency. However, the behavior becomes nonlinear after 30 kHz or so as seen in Fig. 3.7c.

We also observe that the bulk of the induced current at any frequency is primarily excited in the amniotic fluid, but not in the fetus. This is to be expected due the very high conductivity of the amniotic fluid.

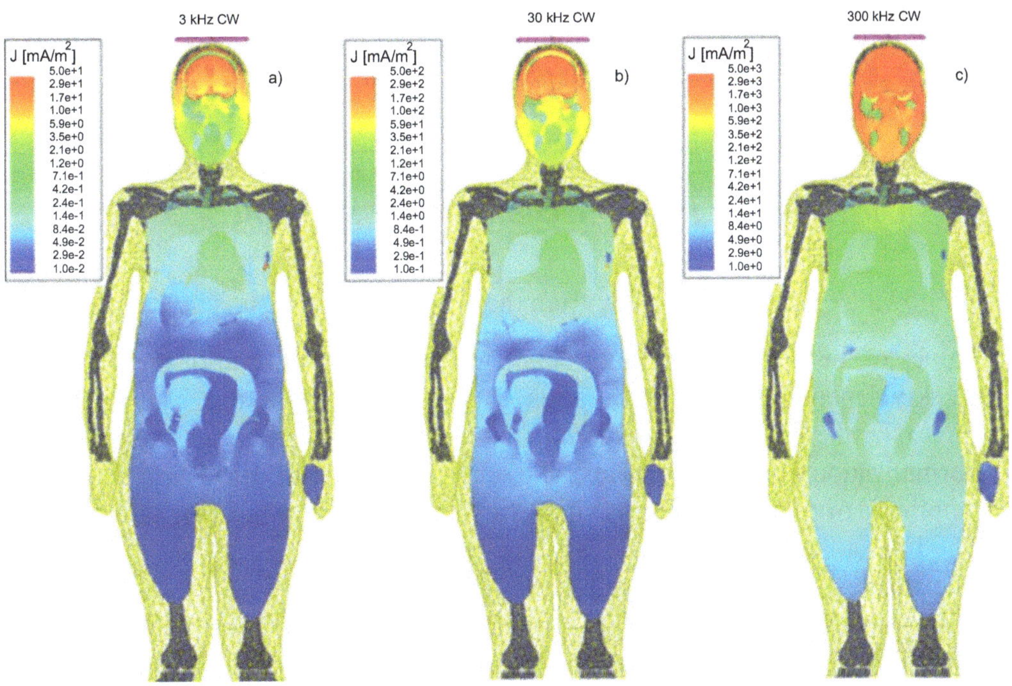

Fig. 3.7 Eddy current amplitude distribution in a coronal plane for three representative frequencies: 3 kHz, 30 kHz, and 300 kHz (second trimester model). The coil current amplitude is 10,000 A (10,000 A·turns mmf). Note that the color scale was increased by a factor of 10 and 100 for the middle and right-hand figures, respectively

3.4.2 *Quantitative Results for Maximum Peak Electric Field at One SMT Unit*

In the subsequent study, the excitation is always given by a biphasic pulse from Eqs. (3.1)–(3.3) and the TMS intensity is always equal to one SMT unit. Figure 3.8 presents the results for the maximum peak electric field (maximum magnitude of the electric field vector, $\mathbf{E}(t)$) for the two coil configurations in Fig. 3.5, respectively, and for every involved tissue. The first, second, and third trimester models were used in this study.

3.4.3 *Comparison with the Recommended Safe Value of Electric Field*

According to the safety requirements discussed in the Introduction, the peak electric field throughout the *fetal volume* (including fetus, placenta, uterus, and amniotic fluid) shall not exceed 800 mV/m. This condition is *certainly met* for all cases given in Fig. 3.8, even using a reduction factor of 10. One obvious reason is that the magnetic field from the coil decays very rapidly far from the head, being approximately proportional to the inverse third power of the distance [64].

3.4.4 *Observations from the Quantitative Solution*

The following observations follow from the analysis of the results given in Fig. 3.8:

- Values of the peak electric field obtained using the condition of one STM unit weakly depend on the pulse duration. This is in contrast to the results shown in Fig. 3.7, where the dependence on frequency is paramount. The reason is the normalization condition of one SMT unit, which means, for example, that the amplitude of the coil current is significantly increased for the 1.0 ms pulse.
- The largest fields are observed in the placenta/uterus.
- The smallest fields are observed in the fetal brain.
- Peak values for two different coil orientations are quite similar.
- The third trimester is characterized by somewhat larger values of the maximum peak electrical field as compared to the first and second trimesters.

These observations suggest that the results given in Fig. 3.8 are rather general and should be valid for a wide variety of coil orientations and pulse durations.

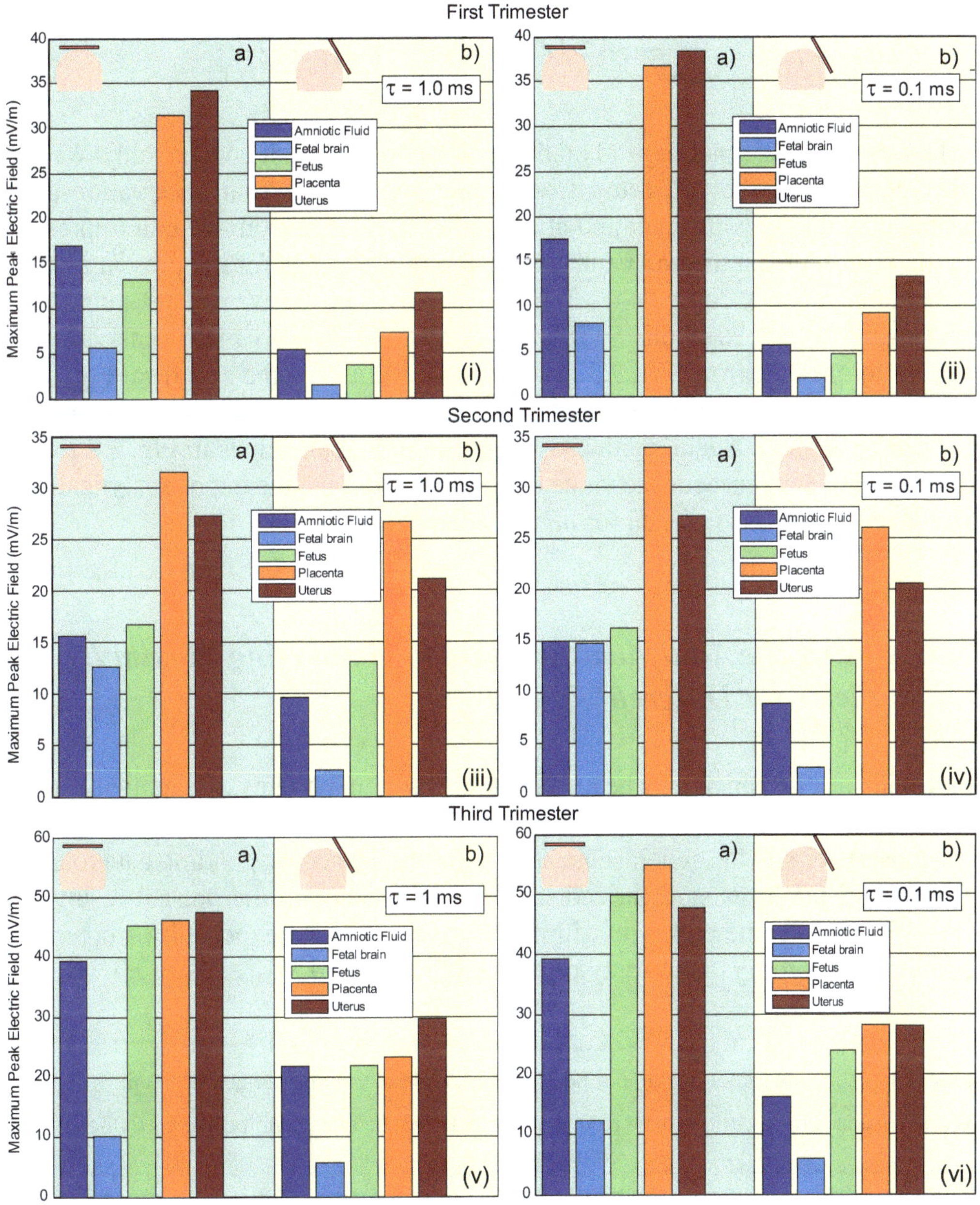

Fig. 3.8 Maximum peak electric field values at two different pulse durations, two different coil positions, and three different stages of pregnancy computed separately for each fetal tissue

3.4.5 Comparison with Upper Analytical Estimate for Electric Fields/Eddy Currents

Using a simplified upper analytical estimate for eddy current/induced electric field in the human body [60], the local electric field anywhere within the body is expressed directly through a time-varying lumped coil current, $I_0f(t)$, in the following form

$$\boldsymbol{E} = -\frac{\partial \boldsymbol{A}^{\text{inc}}}{\partial t}, \; \boldsymbol{A}^{\text{inc}}(\boldsymbol{r},t) = \frac{\mu_0 I_0 f(t)}{4\pi} \oint_C \frac{d\boldsymbol{l}}{\left|\boldsymbol{r} - \boldsymbol{r}'(l)\right|} \tag{3.5}$$

This estimate does not depend on the specific human model under study. We first apply eq. (3.5) to the coil setup from Fig. 3.5a, assuming an observation point located beneath the coil center and at a distance of 62 mm from the coil (representing the distance from the coil center to the top of the uterus for the present model). The resulting upper electric field estimate is obtained as 90 mV/m at any pulse duration (when normalized to one SMT unit). Neither of the maximum peak values in Fig. 3.8 for the coil from Fig. 3.5a exceeds this value. For the coil from Fig. 3.5b, the same or a more elaborate estimate (with spatial averaging to undo a loci effect) [60] can be applied. Again, neither of the maximum peak values in Fig. 3.8 for the coil from Fig. 3.5b exceeds the value of 90 mV/m. Hence, the upper analytical estimate given by (3.5) is justified for *all* considered cases.

3.4.6 *Using the Analytical Estimate for Predicting Maximum Fields for Different Patients*

To provide results which may be expected for different patients, we apply the upper analytical estimate of Eq. (3.5) to different distances from the coil center to the top (or a closest point in the general case) of the uterus. The corresponding data rounded to within ±3 mV/m is summarized in Table 3.4. Although the present results are given for one specific coil type, similar estimates may be expected for other coil geometries according to the study performed in Ref. [60].

Table 3.4 Estimates for the maximum peak field in the fetal volume at one SMT unit

Nearest distance from coil center to uterus	Rounded estimate for the maximum peak electric field in the entire fetal area
50 cm	<170 mV/m
60 cm	<100 mV/m
70 cm	<60 mV/m
80 cm	<40 mV/m
90 cm	<30 mV/m

3.5 Results: Pregnant Operator and Accidental Coil Discharge

3.5.1 Quantitative Results for Maximum Peak Electric Field at One SMT Unit

All coil configurations shown in Fig. 3.6 have been studied for three stages of pregnancy and for different pulse durations. Figure 3.9 presents typical data for the second trimester and for the biphasic pulse of 0.1 ms duration. The following observations can be made from these and other relevant computations:

- For coil positions in close proximity to the belly (A-1, A-2, C-1), the peak electric field in the fetal volume may exceed the safe limit of 800 mV.
- When the distance from the coil center to the nearest point of the uterus is less than 60 cm, the maximum peak values in excess of 100 mV/m may be observed (this number is adopted from Table 3.4).
- When the distance from the coil center to the nearest point of the uterus is greater than 60 cm, the upper estimate from Table 3.4 can be applied.

Coil polarization B in Fig. 3.6 creates the smallest values of the peak electric field. This result is to be expected since the equivalent dipole of the figure-eight coil is essentially perpendicular to the abdominal surface.

3.5.2 Accidental Coil Discharge

In the two extreme cases (polarization B and C in Fig. 3.6), the coil is placed as close to the body as possible by moving it in the *xy*-plane. The corresponding cases for polarization B and C in Fig. 3.6 have been labeled as B7, B8 and C5, C6, respectively. The corresponding maximum peak field values for the entire fetal volume are shown in Fig. 3.10. For these cases, the suggested limit of 800 mV/m may be exceeded by a factor of ten or higher. Similar results were obtained for polarization A for similar extreme coil placements.

3.6 Conclusion

At present, safe limits of fetal exposure to TMS electric and magnetic fields are an open subject. This study aimed to perform both numerical and analytical analyses of this important issue.

As a limit of the maximum peak electric field observed in the fetal volume, we have chosen the value of 800 mV/m, which allows us to avoid peripheral and central myelinated nerve stimulation [8].

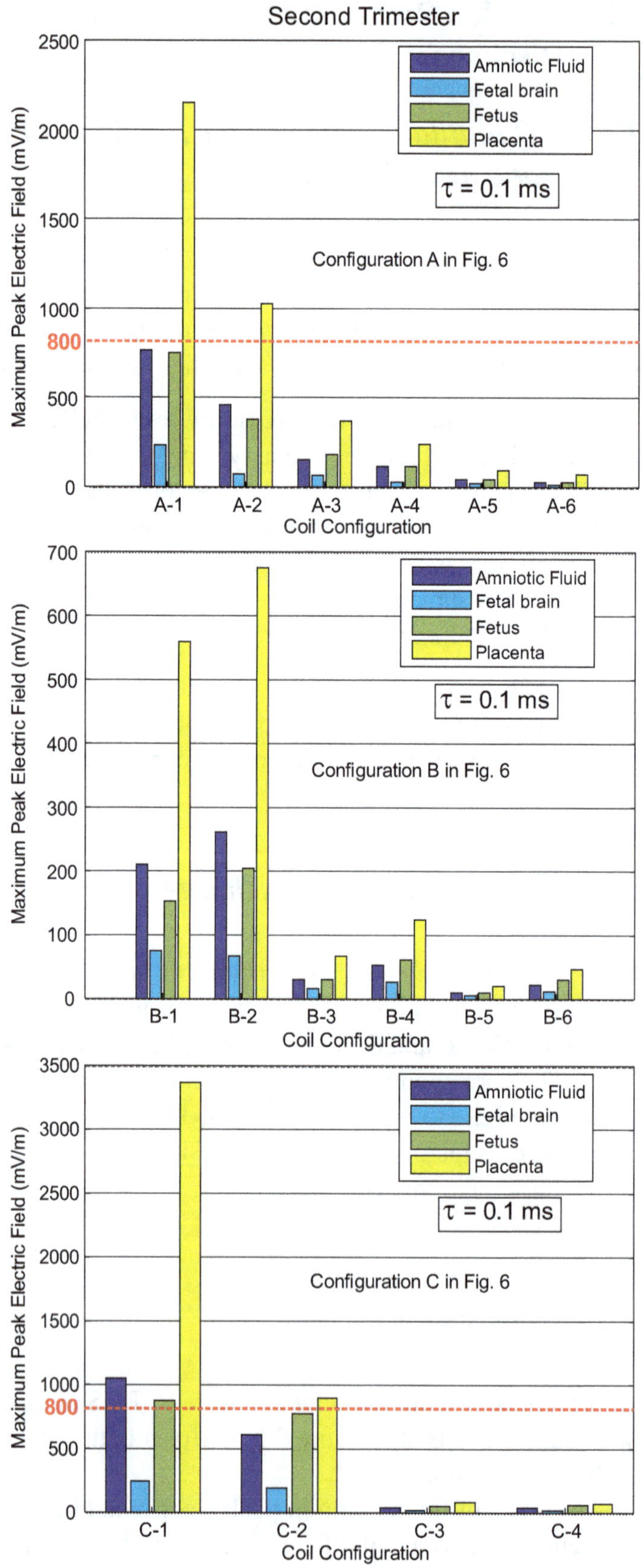

Fig. 3.9 Maximum peak electric field values for the second trimester for all coil configurations/positions depicted in Fig. 3.6

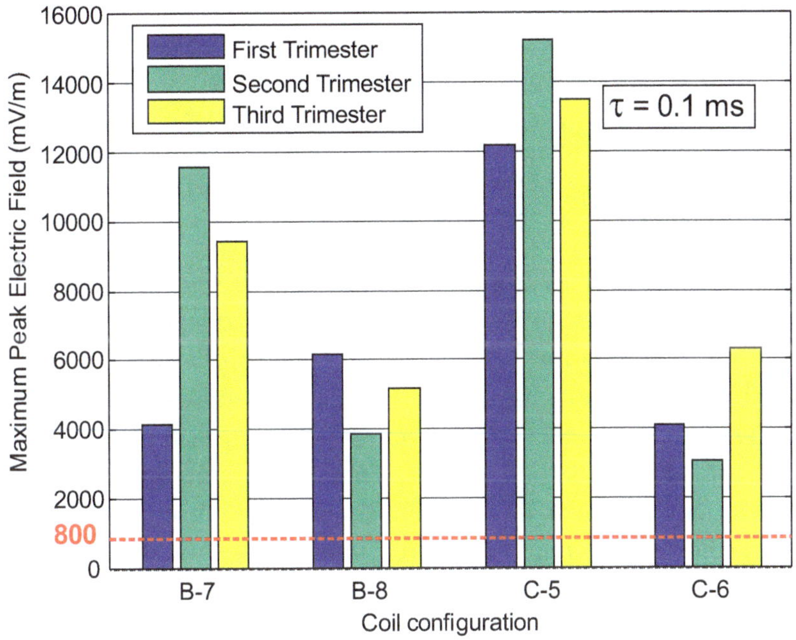

Fig. 3.10 Maximum peak electric field values for all trimesters and for all coil positions listed in Fig. 3.6

Our numerical and analytical estimates for biphasic TMS pulses of different durations provide similar safety estimates. They reveal that:

1. For the TMS intensity of one SMT unit and when the closest distance between the center of the coil and the uterus is greater than or equal to 60 cm (2 ft.), the maximum peak electric field in the fetal volume (including fetus, placenta, uterus, amniotic fluid) is expected to be less than or equal to 100 mV/m. This value is significantly lower than the recommended safe limit of 800 mV/m.
2. The estimate given above was shown for any stage of pregnancy, for two realistic pulse durations, and for pregnant woman either as a patient or an operator.
3. This estimate appears to scale linearly with TMS intensity. For example, at the TMS intensity of 1.5 SMT unit, the peak field in the fetal volume is less than or equal to $100 \times 1.5 = 150$ mV/m when the closest distance between the coil center and the uterus is still 60 cm or greater.
4. This estimate is scaled approximately proportional to the inverse third power of the distance. For example, at the TMS intensity of one SMT unit, the peak field in the fetal volume is less than $100 \times (6/5)^3 \sim 170$ mV/m when the closest distance between the coil center and the uterus is 50 cm or greater.
5. The following approximate equation for the maximum peak electric field E in the fetal volume is suggested

$$E \leq 100\left(\frac{60}{d}\right)^3 I\left[\mathrm{mV/m}\right] \quad (3.6)$$

where d is the closest distance between the coil center and the uterus in cm and I is the TMS intensity in SMT units. We expect (3.6) to hold at the distances d gretater than 30 cm.

6. FDA-cleared TMS devices employ a TMS coil holder. A pregnant operator can (and should) maintain a larger – and thus safer – distance when using a coil holder during TMS treatment. However, the possibility of accidental TMS coil discharge close to the belly has to be considered. In this case, the suggested limit of 800 mV/m may be exceeded by a factor of ten or greater.
7. Given the unknown biological consequences of a large number of pulses in a typical treatment sequence, the decision of whether to use TMS for treatment of depression (the only currently approved indication) should be based on a risk-benefit analysis. In considering the risk-benefit balance, it is important to contemplate the fetal risks posed by pharmacologic treatments for depression in pregnant patients [65, 66]. For more experimental and less evidence-supported indications, a prudent course of action would be to avoid the use of TMS in pregnant women. In any case, appropriate informed consent is critical.

The content of this chapter is solely the responsibility of the authors and does not necessarily represent the official views of Harvard Catalyst, Harvard University, and its affiliated academic health care centers, the National Institutes of Health, or the Sidney R. Baer Jr. Foundation.

Acknowledgements The authors thank Dr. Gregory Noetscher, Mr. Matthew W. Piazza, Ms. Tsering Dolma, Ms. Mariya Zagalskaya, Mr. Edward Burnham, Mr. Harshal Tankaria, Mr. Goutham Kodumudi Srichandhru, Mr. Anh Le Tran, and Mr. David Kelly, all of Worcester Polytechnic Inst., MA for their help in development and improvement of the pregnant female CAD computational human model and in performing multiple simulations in ANSYS Maxwell 3D. The authors also thank Dr. Swarnalatha Gummadi of St. Vincent De Paul Hospital, Thrissur, India, for her feedback in the development of the pregnant models.

Japanese Virtual Model (JVM) Finite-Element Model Version 1.1 (6 months)

Table 3.5 List of triangular surface meshes – version 1.1

Hard tissues		Soft tissues		Individual tissues	
Mesh #	Tissue name	Number of triangles	Mesh quality	Min. Edge length, mm	Tissue type
1	JVM amniotic fluid	1286	0.007942	1.850163	Cerebrospinal fluid
2	JVM bile	122	0.272007	2	Bile
3	JVM bladder	420	0.054992	0.273501	Bladder
4	JVM breast fat left	836	0.068894	0.614704	Fat
5	JVM breast fat right	782	0.037711	0.632222	Fat
6	JVM cerebellum	366	0.052604	1.568904	Cerebellum
7	JVM clavicle left	944	0.061578	1.087026	Bone
8	JVM clavicle right	828	0.024572	0.594466	Bone
9	JVM cornea left	76	0.787694	2.000000	Cornea
10	JVM cornea right	76	0.787694	2.000000	Cornea
11	JVM CSF	6514	0.054137	0.446563	Cerebrospinal fluid
12	JVM duodenum	1030	0.004117	0.44492	Duodenum
13	JVM esophagus	756	0.014063	0.429978	Esophagus
14	JVM femur left	2914	0.123556	0.912948	Bone
15	JVM femur right	3216	0.133124	0.408209	Bone
16	JVM fetal brain	854	0.172681	1.861426	Fetal brain
17	JVM fetus	5000	0.007519	1.107346	Fetus
18	JVM fetus CSF	1282	0.039712	0.293805	Cerebrospinal fluid
19	JVM gallbladder	686	0.088411	2	Gallbladder
20	JVM gray matter	5454	0.004558	0.167652	Gray matter
21	JVM heart	1000	0.032791	1.660989	Heart muscle
22	JVM Humerus ulna left	3368	0.007025	0.393429	Bone
23	JVM Humerus ulna right	3558	0.018193	0.093016	Bone
24	JVM jaw	424	0.007359	0.359137	Bone
25	JVM kidney left	568	0.024197	1.163018	Kidney
26	JVM kidney right	636	0.082014	0.314779	Kidney
27	JVM large intestine	3914	0.002854	0.03227	Large intestine
28	JVM large intestine content	5886	1.23E-08	3.05E-05	Large intestine
29	JVM lens left	18	0.787694	2.00000	Lens
30	JVM lens right	18	0.787694	2.00000	Lens
31	JVM liver	2512	0.001358	0.786026	Liver
32	JVM lungs left	1138	0.001854	0.47502	Lung

Hard tissues		Soft tissues		Individual tissues	
Mesh #	Tissue name	Number of triangles	Mesh quality	Min. Edge length, mm	Tissue type
33	JVM lungs right	1146	0.038088	2.120805	Lung
34	JVM ovary left	798	0.207408	0.942092	Ovary
35	JVM ovary right	230	0.170639	0.640317	Ovary
36	JVM pancreas	600	0.015017	0.522665	Pancreas
37	JVM Patella left	304	0.379008	1.276178	Bone
38	JVM Patella right	424	0.005271	1.340134	Bone
39	JVM pelvic	10012	1.6E-05	0.02115	Bone
40	JVM placenta	902	0.008601	2.09859	Placenta
41	JVM rib left 1	194	0.037071	1.501538	Bone
42	JVM rib left 2	246	0.019692	0.874205	Bone
43	JVM rib left 3	298	0.033745	1.198342	Bone
44	JVM rib left 4	376	0.017842	0.792734	Bone
45	JVM rib left 5	552	0.008819	0.223525	Bone
46	JVM rib left 6	482	0.058624	0.928572	Bone
47	JVM rib left 7	516	0.044063	1.117289	Bone
48	JVM rib left 8	498	0.019398	0.611004	Bone
49	JVM rib left 9	438	0.032979	0.525822	Bone
50	JVM rib left 10	594	0.013294	0.127242	Bone
51	JVM rib left 11	282	0.01784	1.493664	Bone
52	JVM rib left 12	158	0.020202	1.013902	Bone
53	JVM rib right 1	194	0.121647	0.908528	Bone
54	JVM rib right 2	252	0.019442	0.85563	Bone
55	JVM rib right 3	298	0.034585	0.900052	Bone
56	JVM rib right 4	374	0.017841	0.917949	Bone
57	JVM rib right 5	624	0.010604	0.159166	Bone
58	JVM rib right 6	474	0.109128	0.928528	Bone
59	JVM rib right 7	516	0.026893	1.025215	Bone
60	JVM rib right 8	502	0.007985	0.541626	Bone
61	JVM rib right 9	564	0.015767	0.125747	Bone
62	JVM rib right 10	596	0.013823	0.193589	Bone
63	JVM rib right 11	282	0.012179	1.474308	Bone
64	JVM rib right 12	156	0.017841	0.320081	Bone
65	JVM salivary gland left	1208	0.036962	2.000000	Salivary gland
66	JVM salivary gland right	1006	0.088411	0.48356	Salivary gland
67	JVM scapula left	1618	0.006059	0.77685	Bone
68	JVM scapula right	1594	0.051858	0.330764	Bone
69	JVM sclera left	556	0.642238	2.000000	Eye (Vitrous humor)
70	JVM sclera right	556	0.642238	2.000000	Eye (Vitrous humor)

(continued)

Hard tissues		Soft tissues		Individual tissues	
Mesh #	Tissue name	Number of triangles	Mesh quality	Min. Edge length, mm	Tissue type
71	JVM skin	8042	0.005529	4.349768	Skin
72	JVM skull	8196	0.005778	0.024727	Bone
73	JVM small intestine	4574	0.000462857	0.121570	Small intestine
74	JVM spine	4694	0.025728	0.503403	Bone
75	JVM stomach	1996	0.00633	0.528757	Stomach
76	JVM stomach contents 1	1388	0.088411	2.000000	Stomach
77	JVM stomach contents 2	968	0.076652	2.000000	Stomach
78	JVM thalamus	574	0.103805	2.000000	Thalamus
79	JVM thyroid	806	0.026187	1.999985	Thyroid
80	JVM tibia fibia left	2686	0.15568	0.965745	Bone
81	JVM tibia fibia right	3332	0.019527	0.407727	Bone
82	JVM tongue	260	0.092234	2.353283	Tongue
83	JVM trachea	870	0.123615	0.597242	Trachea
84	JVM urine	308	0.018274	0.531584	Urine
85	JVM uterus	596	0.300344	8.490075	Uterus
86	JVM white matter	9088	0.000959	0.148079	White matter

References

1. Rossi, S., et al.; Safety of TMS Consensus Group. (2009). Safety ethical considerations, and application guidelines for the use of transcranial magnetic stimulation in clinical practice and research. *Clinical Neurophysiology, 120*(12), 2008–2039.
2. Fox, M. D., et al. (2013). Identification of reproducible individualized targets for treatment of depression with TMS based on intrinsic connectivity. *NeuroImage, 66*, 151–160.
3. DeNoon, D. J. (2008, October). Brain-stimulating device cleared for depression treatment after 1 drug failure. WebMD. [Online]. Available: http://www.webmd.com/depression/news/20081008/fda-oks-tms-depression-device
4. Brainsway. (2013, Jan.) gets FDA approval for anti-depression device. Reuters. [Online]. Available: http://www.reuters.com/article/2013/01/09/brainsway-fda-idUSL5E9C99OU20130109
5. Kim, D. R., et al. (2011). An open label pilot study of transcranial magnetic stimulation for pregnant women with major depressive disorder. *Journal of Women's Health, 20*(2), 255–261.
6. Kim, D. R., et al. (2011). A survey of patient acceptability of repetitive transcranial magnetic stimulation (TMS) during pregnancy. *Journal of Affective Disorders, 129*(1–3), 385–390.
7. ICNIRP. (1998). Guidelines for limiting exposure to time-varying electric, magnetic and electromagnetic fields (up to 300 GHz). *Health Physics, 74*(4), 494–522.
8. ICNIRP. (2010). Guidelines for limiting exposure to time-varying electric and magnetic fields (1 Hz – 100 kHz). *Health Physics, 99*(6), 818–836.
9. Dimbylow, P. (2006). Development of pregnant female, hybrid voxel-mathematical models and their application to the dosimetry of applied magnetic and electric fields at 50 Hz. *Physics in Medicine and Biology, 51*(10), 2383–2394.

10. Cech, R., et al. (2007). Fetal exposure to low frequency electric and magnetic fields. *Physics in Medicine and Biology, 52*(4), 879–888; Shi, C. Y., & Xu, X. G. (2004). Development of a 30-week-pregnant female tomographic model from computed tomography (CT) images for Monte Carlo organ dose calculations. Medical Physics, *31*(9), 2491–2497.
11. Hand, J. W., et al. (2006). Prediction of specific absorption rate in mother and fetus associated with MRI examinations during pregnancy. *Magnetic Resonance in Medicine, 55*(4), 883–893; Xu, X. G., et al. (2007). A boundary-representation method for designing whole-body radiation dosimetry models: pregnant females at the ends of three gestational periods--RPI-P3, -P6 and -P9. *Physics in Medicine and Biology, 52*(23), 7023–7044.
12. Ackerman, M. J. (1998). The visible human project. *Proceedings of the IEEE, 86*(3), 504–511.
13. Ackerman, M. J. (2016). *The visible human project®: From body to bits*. Orlando: EMBC 2016.
14. Nagaoka, T., et al. (2015). *SAR calculation in semi-homogeneous human models of pregnancy for RF exposure*. Asia-Pacific International Symposium on Electromagnetic Compatibility. Taipei, pp. 444–447.
15. The Virtual Population. (2016, April). High-resolution anatomical models for computational life sciences. IT'IS Foundation/FDA Flyer, European Conference on Antennas and Propagation.
16. Gosselin, M. C., et al. (2014). Development of a new generation of high-resolution anatomical models for medical device evaluation: The Virtual Population 3.0. *Physics in Medicine and Biology, 59*(18), 5287–5303.
17. Nagaoka, T., et al. (2007). An anatomically realistic whole-body pregnant-woman model and specific absorption rates for pregnant-woman exposure to electromagnetic plane waves from 10 MHz to 2 GHz. *Physics in Medicine and Biology, 52*(22), 6731–6745.
18. Nagaoka, T., et al. (2008). Estimating specific absorption rates in pregnant women by using models at 12-, 20-, and 26-weeks' gestation for plane wave exposures. International Symposium on Electromagnetic Compatibility – EMC Europe, Hamburg, pp. 1–4.
19. Becker, J., et al. (2008). Katja – the 24th week of virtual pregnancy for dosimetric calculations. *Polish Journal of Medical Physics and Engineering, 14*(1), 13–19.
20. Christ, A., et al. (2001). The Virtual Family – development of surface-based anatomical models of two adults and two children for dosimetric simulations. *Physics in Medicine and Biology, 55*(2), 23–38.
21. Dimbylow, P. J. (2005). Development of the female voxel phantom, NAOMI, and its application to calculations of induced current densities and electric fields from applied low frequency magnetic and electric fields. *Physics in Medicine and Biology, 50*(6), 1047–1070.
22. Nagaoka, T., et al. (2004). Development of realistic high resolution whole-body voxel models of Japanese adult male and female of average height and weight, and application of models to radio-frequency electromagnetic-field dosimetry. *Physics in Medicine and Biology, 49*(1), 1–15.
23. Starzynski, J., et al. (2002). Simulation of magnetic stimulation of the brain. *IEEE Transactions on Magnetics, 38*(2), 1237–1240.
24. Wagner, T. A., et al. (2004). Three-dimensional head model simulation of transcranial magnetic stimulation. *IEEE Transactions on Biomedical Engineering, 51*(9), 1586–1598.
25. Wagner, T. A., et al. (2006). Transcranial magnetic stimulation and stroke: A computer-based human model study. *NeuroImage, 30*(3), 857–870.
26. Wagner, T. A., et al. (2014). Impact of brain tissue filtering on neurostimulation fields: A modeling study. *NeuroImage, 85*(3), 1048–1057.
27. Miranda, P. C., et al. (2003). The electric field induced in the brain by magnetic stimulation: A 3-D finite-element analysis of the effect of tissue heterogeneity and anisotropy. *IEEE Transactions on Biomedical Engineering, 50*(9), 1074–1085.
28. Miranda, P. C., et al. (2016). Computational models of non-invasive brain and spinal cord stimulation. EMBC 2016, Orlando.

29. Wenger, C., et al. (2015). The electric field distribution in the brain during TTFields therapy and its dependence on tissue dielectric properties and anatomy: A computational study. *Physics in Medicine and Biology, 60*, 7339–7357.
30. Chen, M., & Mogul, D. J. (2009). A structurally detailed finite element human head model for simulation of transcranial magnetic stimulation. *Journal of Neuroscience Methods, 179*(1), 111–120.
31. Opitz, A., et al. (2013). Physiological observations validate finite element models for estimating subject-specific electric field distributions induced by transcranial magnetic stimulation of the human motor cortex. *NeuroImage, 81*, 253–264.
32. Bottauscio, O., et al. (2014). Evaluation of electromagnetic phenomena induced by transcranial magnetic stimulation. *IEEE Transactions on Magnetics, 50*(2), 1033–1036.
33. Deng, Z. D. (2013). *Electromagnetic field modeling of transcranial electric and magnetic stimulation: Targeting, individualization, and safety of convulsive and subconvulsive applications*. Ph.D. dissertation, Department of Electrical Engineering, Columbia University, Ithaca, New York.
34. Lee, W. H. (2014). Noninvasive neuromodulation: Modeling and analysis of transcranial brain stimulation with applications to electric and magnetic seizure therapy. Ph.D. dissertation, Department of Biomedical Engineering, Columbia University, Ithaca, New York.
35. Yushkevich, P. A., et al. (2016). *ITK-SNAP: An interactive tool for semi-automatic segmentation of multi-modality biomedical images*. Orlando: EMBC.
36. Lo, S. H. (1995). Automatic mesh generation over intersecting surfaces. *International Journal for Numerical Methods in Engineering, 38*, 943–954.
37. Lo, S. H., & Wang, W. X. (2004). A fast robust algorithm for the intersection of triangulated surfaces. *Engineering with Computers, 20*, 11–21.
38. Elsheikh, A. H., & Elsheikh, M. (2014). A reliable triangular mesh intersection algorithm and its application in geological modelling. *Engineering with Computers, 30*, 143–157.
39. Coelho, L. C., et al. (2000). Intersecting and trimming parametric meshes on finite-element shells. *International Journal for Numerical Methods in Engineering, 47*, 777–800.
40. Lira, W.M., et al. (2002). Multiple intersections of finite-element surface meshes. 11th International Meshing Roundtable, Ithaca, New York.
41. Lindenbeck, C. H., et al. (2002). TRICUT: A program to clip triangle meshes using the rapid and triangle libraries and the visualization toolkit. *Computers & Geosciences, 28*, 841–850.
42. Yanamadala, J., et al. (2014, October 1–5). Segmentation of the visible human project® (VHP) female cryosection images within MATLAB® environment. 23rd International Meshing Roundtable (IMR23), London.
43. Noetscher, G.M., et al. (2015, October 12–14). VHP-Female v3.0 FEM/BEM computational human phantom. 24th International Meshing Roundtable (IMR24), Austin.
44. Gammon, M. (2014, October 12–15). CAD clean-up for meshing. What could possibly go wrong?" Short Course. 23rd International Meshing Roundtable, London, pp. 1–70.
45. Moorcroft, M. D., et al. (2003). Computational model of the pregnant occupant: Predicting the risk of injury in automobile crashes. *American Journal of Obstetrics and Gynecology, 189*(2), 540–544.
46. Duma, S.M., et al. (2006). *Analysis of pregnant occupant crash exposure and the potential effectiveness of four-point seatbelts in far side crashes*. The Proceedings of the 50th Association for the Advancement of Automotive Medicine Conference, 50.
47. Duma, S.M., et al. (2005). A computational model of the pregnant occupant: Effects of restraint usage and occupant position on fetal injury risk. Proceedings. of the 19th International Conference for the Enhanced Safety Vehicles, Washington, D.C.
48. Kitagawa, Y., & Yasuki, T. (2010). Development of pregnant FE model and prediction of kinematics in frontal impact. CiNii. [Online]. Available: http://ci.nii.ac.jp/naid/130004515562/.
49. Gabriel, C., & Gabriel, S. (1997). Compilation of the dielectric properties of body tissues at RF and microwave frequencies. [Online]. Available: http://niremf.ifac.cnr.it/docs/DIELECTRIC/Report.html.

50. Hasgall, P. A., et al. (2015). IT'IS Database for thermal and electromagnetic parameters of biological tissues. In *Version 2.6, January 13th*. www.itis.ethz.ch/database.
51. Schepps, J. L., & Foster, K. R. (1980). The UHF and microwave dielectric properties of normal and tumour tissues: Variation in dielectric properties with tissue water content. *Physics in Medicine and Biology, 25*, 1149–1159.
52. Peterchev, A. V., et al. (2008). A transcranial magnetic stimulator inducing near-rectangular pulses with controllable pulse width (cTMS). *IEEE Transactions on Biomedical Engineering, 55*(1), 257–266.
53. Thielscher, A. Transcranial magnetic stimulation [Online], Available: ftp://ftp.kyb.mpg.de/kyb/chaimow/For%20Me/TMS/Thielscher_Lecture_Session1.pdf
54. Peterchev, A. V., et al. (2011). Repetitive transcranial magnetic stimulator with controllable pulse parameters. *Journal of Neural Engineering, 8*(3), 1–24.
55. Peterchev, A. V., et al. (2014). Controllable pulse parameter transcranial magnetic stimulator with enhanced circuit topology and pulse shaping. *Journal of Neural Engineering, 11*(5), 1–12.
56. Brainsway Deep TMS System, Brainsway Ltd., Jerusalem, Israel, 2013, pp. 5–8.
57. MAGSTIM 2002 P/N 3001-23-04, The Magstim Company Ltd., Whitland SA34 0HR, United Kingdom, 2005, pp. 27–28.
58. MagVita TMS Therapy System, Tonica Elektronik A/S, Farum, Denmark, 2015, pp. 5–4.
59. NeuroStar TMS Therapy System, Neuronetics, Inc., Malvern, PA, 2014, pp. 1–30.
60. Makarov, S. N., et al. (2016). Preliminary upper estimate of peak currents in transcranial magnetic stimulation at distant locations from a TMS coil. *IEEE Transactions on Biomedical Engineering, 63*(9), 1944–1955.
61. Epstein, C. M., et al. (2008). *The Oxford handbook of transcranial stimulation*. New York: Oxford University Press.
62. Guidance for Industry and Food and Drug Administration Staff. (2011, July 26). Class II Special Controls Guidance Document: Repetitive Transcranial Magnetic Stimulation (rTMS) Systems. Center for Devices and Radiological Health, FDA.
63. Herbsman, T., et al. (2009). Motor threshold in transcranial magnetic stimulation: the impact of white matter Fiber orientation and skull-to-cortex distance. *Human Brain Mapping, 30*(7), 2044–2055.
64. Makarov, S. N., et al. (2015). *Modeling of low frequency electromagnetic fields in electrical and biological systems*. New York: Wiley.
65. Pearlstein, T. (2008). Perinatal depression: Treatment options and dilemmas. *Journal of Psychiatry & Neuroscience, 33*(4), 302–318.
66. Divya, M. P., et al. (2016). Depression in pregnancy-consequences and treatment modalities. *International Journal of Pharmacy and Pharmaceutical Sciences, 8*(3).

Chapter 4
Electric Field Modeling for Transcranial Magnetic Stimulation and Electroconvulsive Therapy

Zhi-De Deng, Conor Liston, Faith M. Gunning, Marc J. Dubin, Egill Axfjörð Fridgeirsson, Joseph Lilien, Guido van Wingen, and Jeroen van Waarde

4.1 Introduction

Major depressive disorder (MDD) is a highly prevalent condition with a lifetime prevalence of nearly 20% [1]. MDD is currently the second leading cause of disability worldwide, and the World Health Organization (WHO) has predicted that, by

Z.-D. Deng (✉)
Noninvasive Neuromodulation Unit, Experimental Therapeutics and Pathophysiology Branch, National Institute of Mental Health, National Institutes of Health, Bethesda, MD, USA

Department of Psychiatry and Behavioral Sciences, Duke University School of Medicine, Durham, NC, USA
e-mail: zhi-de.deng@nih.gov

C. Liston
Department of Psychiatry, Weill Cornell Medical College–New York Presbyterian Hospital, New York, NY, USA

Feil Family Brain and Mind Research Institute, Weill Cornell Medical College–New York Presbyterian Hospital, New York, NY, USA

Sackler Institute for Developmental Psychobiology, Weill Cornell Medical College–New York Presbyterian Hospital, New York, NY, USA

F. M. Gunning
Department of Psychiatry, Weill Cornell Medical College–New York Presbyterian Hospital, New York, NY, USA

Institute of Geriatric Psychiatry, Weill Cornell Medical College–New York Presbyterian Hospital, New York, NY, USA

M. J. Dubin
Department of Psychiatry, Weill Cornell Medical College–New York Presbyterian Hospital, New York, NY, USA

Feil Family Brain and Mind Research Institute, Weill Cornell Medical College–New York Presbyterian Hospital, New York, NY, USA

S. Makarov et al. (eds.), *Brain and Human Body Modeling*,
https://doi.org/10.1007/978-3-030-21293-3_4

2020, it will be the leading cause of disability. In the Diagnostic and Statistical Manual of Mental Disorders, MDD is also the diagnosis that is most strongly associated with suicide attempts, a phenomenon whose rates have sharply increased over the past two decades in the USA [2]. Present first-line treatment options for MDD include antidepressant medications and cognitive-based therapies. However, a large proportion of patients remain unresponsive to these treatment options [3]. This underscores the urgent need for more personalized approaches to treatments as well as alternative antidepressant therapies, such as noninvasive brain stimulation.

Several noninvasive brain stimulation techniques are now available for the treatment of MDD. Electroconvulsive therapy (ECT) is a highly effective treatment for patients with severe and medication-resistant depression. ECT delivers a series of electrical pulse trains to the brain via scalp electrodes that induce a generalized tonic–clonic seizure in anesthetized patients. For the treatment of MDD in adults, ECT has a sustained response rate of approximately 80% and a remission rate of 75% [4]. Despite this superior clinical efficacy, little is known about the interindividual variability in the electric field (E-field) strength and distribution induced by ECT. In this work, we aimed to quantify E-field variability in a depressed patient population and to explore correlates with antidepressant treatment outcome.

Another FDA-cleared treatment for depression is repetitive transcranial magnetic stimulation (rTMS). In depressed patients receiving rTMS, interindividual variability in the induced E-field strength and distribution has not been well characterized. It is not known, for example, what aspect of the E-field is related to improvements in depression symptoms. Such information would be useful for patient selection and/or guide treatment target and dosing.

Conventional magnetic neurostimulation systems use a current-carrying coil to generate a time-varying magnetic field pulse, which in turn produces a spatially varying electric field – via electromagnetic induction – in the central or peripheral nervous system. An alternative approach to generating the time-varying magnetic field is by means of moving permanent magnets. Several systems have been proposed [5–7], involving rotation of high-strength neodymium magnets. One of these systems, termed synchronized transcranial magnetic stimulation (sTMS), was explored as a treatment of depression [8].

The sTMS device is comprised of a configuration of three cylindrical neodymium magnets mounted over the midline frontal polar region, the superior frontal gyrus, and the parietal cortex. The speed of rotation for the magnets was set to the

E. A. Fridgeirsson · G. van Wingen
Department of Psychiatry, University of Amsterdam, Amsterdam, Netherlands

J. Lilien
Department of Psychiatry and Behavioral Sciences, Duke University School of Medicine, Durham, NC, USA

J. van Waarde
Department of Psychiatry, Rijnstate Hospital, Arnhem, Netherlands

patient's individualized peak alpha frequency of neural oscillations, as obtained by pretreatment electroencephalo-graphy recorded from the prefrontal and occipital regions while the patient remained in an eyes-closed, resting state [9]. The hypothesized mechanism of action is that entrainment of alpha oscillations, via exogenous subthreshold sinusoidal stimulation produced by sTMS, could reset neural oscillators, enhance cortical plasticity, normalize cerebral blood flow, and altogether ameliorate depressive symptoms [10]. In a multi-center, double-blind, sham-controlled trial of sTMS treatment of MDD, there was no difference in efficacy between active and sham in the intent-to-treat sample [8]. No direct electrophysiological evidence of the hypothesized mechanism of sTMS was reported, nor was the stimulation intensity and distribution well characterized. In this work, we evaluate the electric field characteristics of sTMS using the finite element method.

4.2 Modeling Methods

4.2.1 ECT Modeling

This study included 67 patients who received ECT at Rijnstate Hospital in Arnhem, the Netherlands [11]. ECT was administered using a Thymatron System IV (pulse amplitude = 900 mA; Somatics LLC, Lake Bluff, IL, USA), with bilateral and/or right unilateral electrode placements (see Fig. 4.1). Depression severity was assessed using the Montgomery–Åsberg Depression Rating Scale (MADRS). T1-weighted

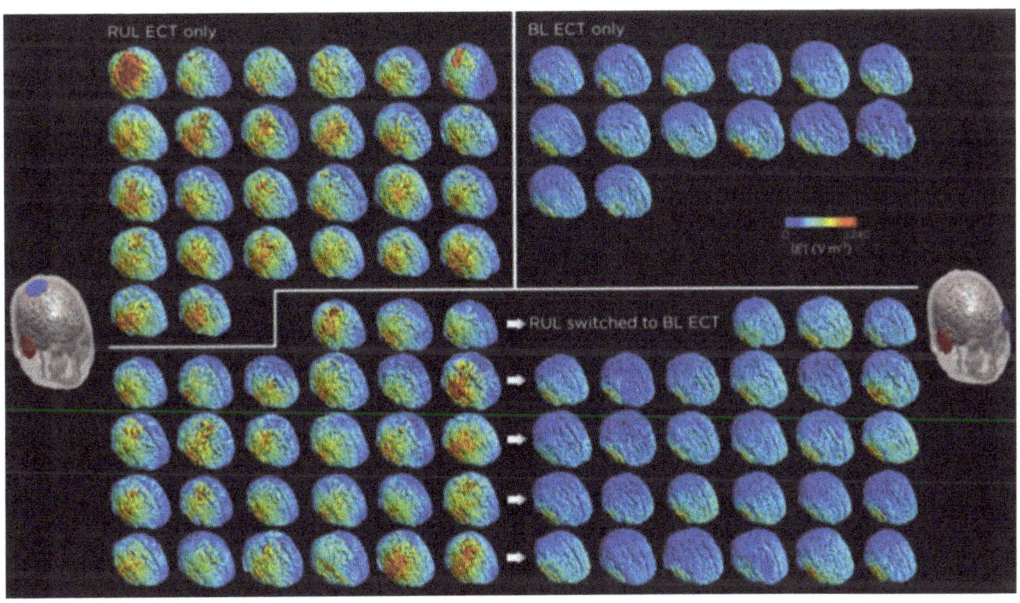

Fig. 4.1 E-field induced in 26 patients receiving right unilateral (RUL) ECT only, 14 patients receiving bilateral (BL) ECT only, and 27 patients who started with RUL ECT and switched to BL ECT

MRI (1.1 mm isotropic voxel) was acquired at baseline. We examined the relationship between E-field strength and post-treatment MADRS score using a general linear model, controlling for age, sex, baseline MADRS score, and the number of ECT sessions.

4.2.2 rTMS Modeling

The Institutional Review Board of Weill Cornell Medical College approved this rTMS study. Twenty-six treatment-resistant depressed patients (age 21–68) participated in the study. Patients received daily 10 Hz rTMS over the left dorsolateral prefrontal cortex (DLPFC) using the NeuroStar system 5 days per week for 5 weeks [12]. Treatment response was assessed using the 24-item Hamilton Rating Scale for Depression (HAMD-24) at baseline and after the course of rTMS. T1-weighted MRIs were acquired within 7 days prior to starting rTMS and within 3 days of completing the treatment. Diffusion tensor images were acquired using a single-shot spin echo imaging sequence. Motor threshold was determined by visualization of movement technique at baseline. Anatomically realistic finite element head models were constructed from individual MRIs using SimNIBS 2.0.1 [13]. The rTMS coil was centered on the F3 site according to the International 10–20 system [14], oriented 45 degrees toward midline. We evaluated the E-field strength at the DLPFC gray matter (middle frontal gyrus (MFG), inferior frontal sulcus (IFS), and superior frontal sulcus (SFS)).

4.2.3 sTMS Modeling

Patient MRI data was not available to model sTMS. The finite element model was implemented in COMSOL Multiphysics (COMSOL, Burlington, MA) using its version of the IEEE Specific Anthropomorphic Mannequin (SAM) phantom as a basis for the geometry (Fig. 4.2). The head model (stator) has uniform, isotropic electrical conductivity of 0.33 S m^{-1} and a relative permeability of 1. Three cylindrical magnets (rotors) were positioned along the midline: magnet #1 was located over the frontal pole just above the eyebrows, magnet #2 was 7.1 cm from Magnet #1, approximately overlying the superior frontal gyrus, and magnet #3 was 9.2 cm from magnet #2, approximately overlying the parietal cortex. Each magnet was 2.54 cm in diameter and height, diametrically magnetized, with a residual flux density of 0.64 T. The axes of rotations were perpendicular to the sagittal plane and the rotation velocity was set to 10 Hz, corresponding to approximately peak alpha frequency. The resulting adaptive mesh consisted of 56,825 tetrahedral elements.

Under the vector potential formulation, Ampère's law was first applied to all domains:

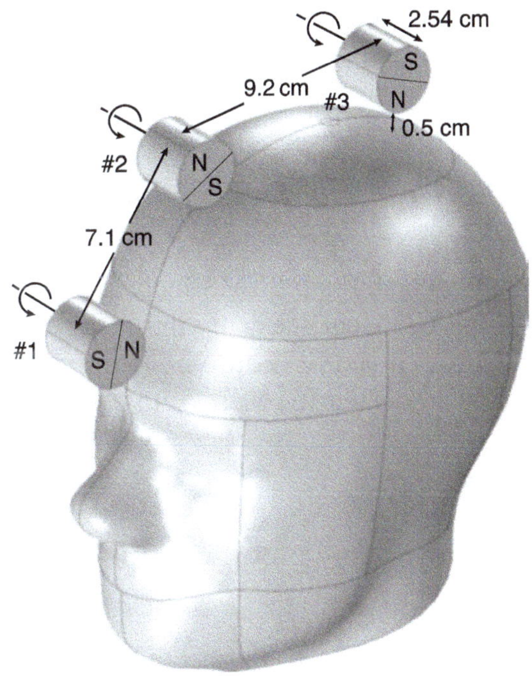

Fig. 4.2 Dimensions and placement of the three cylindrical magnets in the sTMS system

$$\sigma \frac{\partial \mathbf{A}}{\partial t} + \nabla \times \left(\frac{1}{\mu} \nabla \times \mathbf{A} \right) = 0,$$

and a magnetic flux conservation equation for the scalar magnetic potential was applied to current-free parts of both the rotor and stator:

$$-\nabla \cdot \left(\mu \nabla V_{\mathrm{m}} - \mathbf{B}_{\mathrm{r}} \right) = 0.$$

Continuity in the scalar magnetic potential was enforced at the interface between the rotor and stator. A stationary solution was first obtained using a direct solver (MUMPS), and then the time-dependent problem (in 10 degrees rotation steps) was solved. This assumes that the transient effects of initiating the rotating magnets have decayed, and the final solution reflects steady-state behavior.

4.3 Results

4.3.1 Electric Field Induced by ECT

Figure 4.1 shows the E-field distribution induced by bilateral and unilateral ECT in the study patients. For right unilateral ECT, the maximum E-field strength induced in the brain is 513.0 ± 113.2 V m^{-1}. For bilateral ECT, the largest cluster of white matter voxels where the E-field strength is significantly correlated with the post-treatment MADRS score includes parts of the right inferior fronto-occipital fasciculus, inferior longitudinal fasciculus, uncinate fasciculus, anterior thalamic radiation, and the corticospinal tract, where there is high E-field strength.

4.3.2 *Electric Field Induced by rTMS*

Figure 4.3 shows the TMS-induced E-field distribution in a representative patient. At the treatment intensity, the mean maximum induced E-field strengths at the MFG, IFG, and SFG are 92.2 V m^{-1}, 56.5 V m^{-1}, and 79.6 V m^{-1}, respectively. Stimulator intensity was positively correlated with E-field strength at the MFG

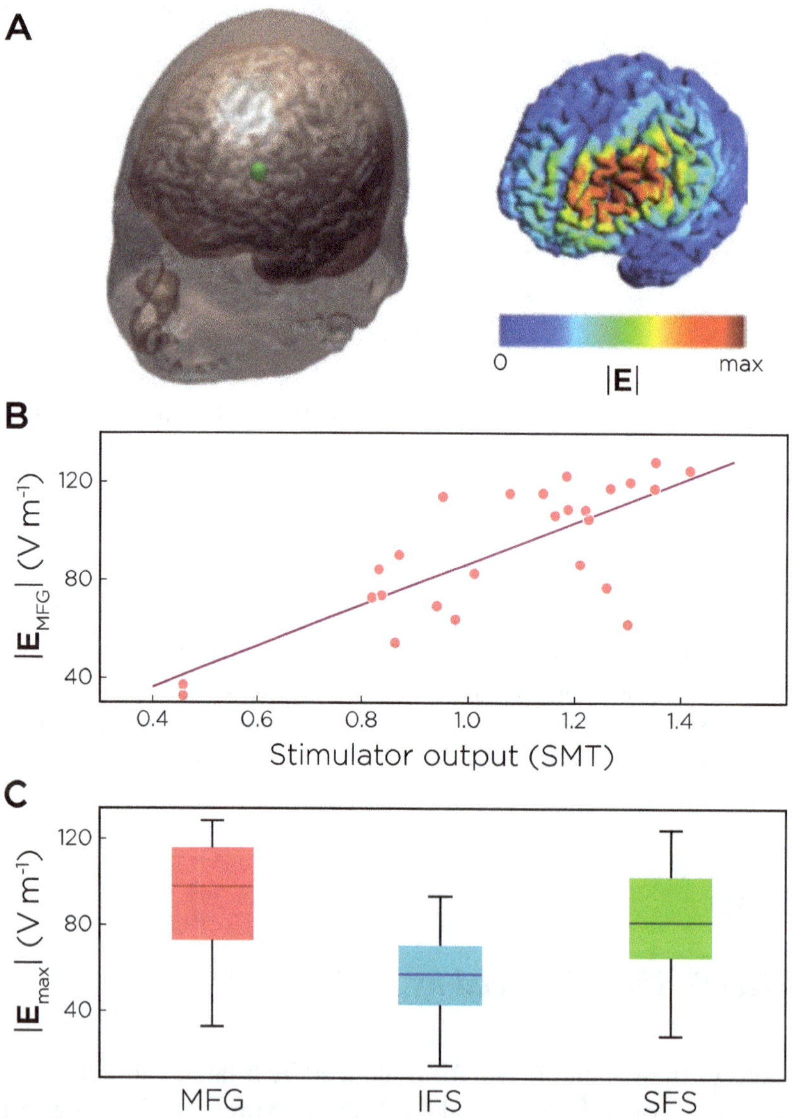

Fig. 4.3 rTMS-induced E-field. (**a**) Head model and E-field distribution in a representative patient. The green dot on the head model indicates location of the TMS target. (**b**) Correlation between stimulation intensity (in standardized motor threshold (SMT) units) and maximum E-field strength at MFG. (**c**) Distribution of E-field strengths at the MFG, IFS, and SFS, for the 26 patients

($r = 0.77$, $p < 0.001$). However, E-field strengths at the MFG, IFG, and SFG were not correlated with changes in HAMD-24.

4.3.3 Electric Field Induced by sTMS

Figurc 4.4 shows thc clcctric ficld distribution of thc full sTMS configuration in the SAM head model. The stimulation is broadly distributed over midline frontal polar, medial frontal, and parietal regions. The peak-induced electric field strength at the surface of the head is approximately 0.06 V m^{-1}. At a depth of 1.5 cm from the head surface, corresponding to the depth of the cortex, the electric field strength attenuates to approximately 0.02 V m^{-1}.

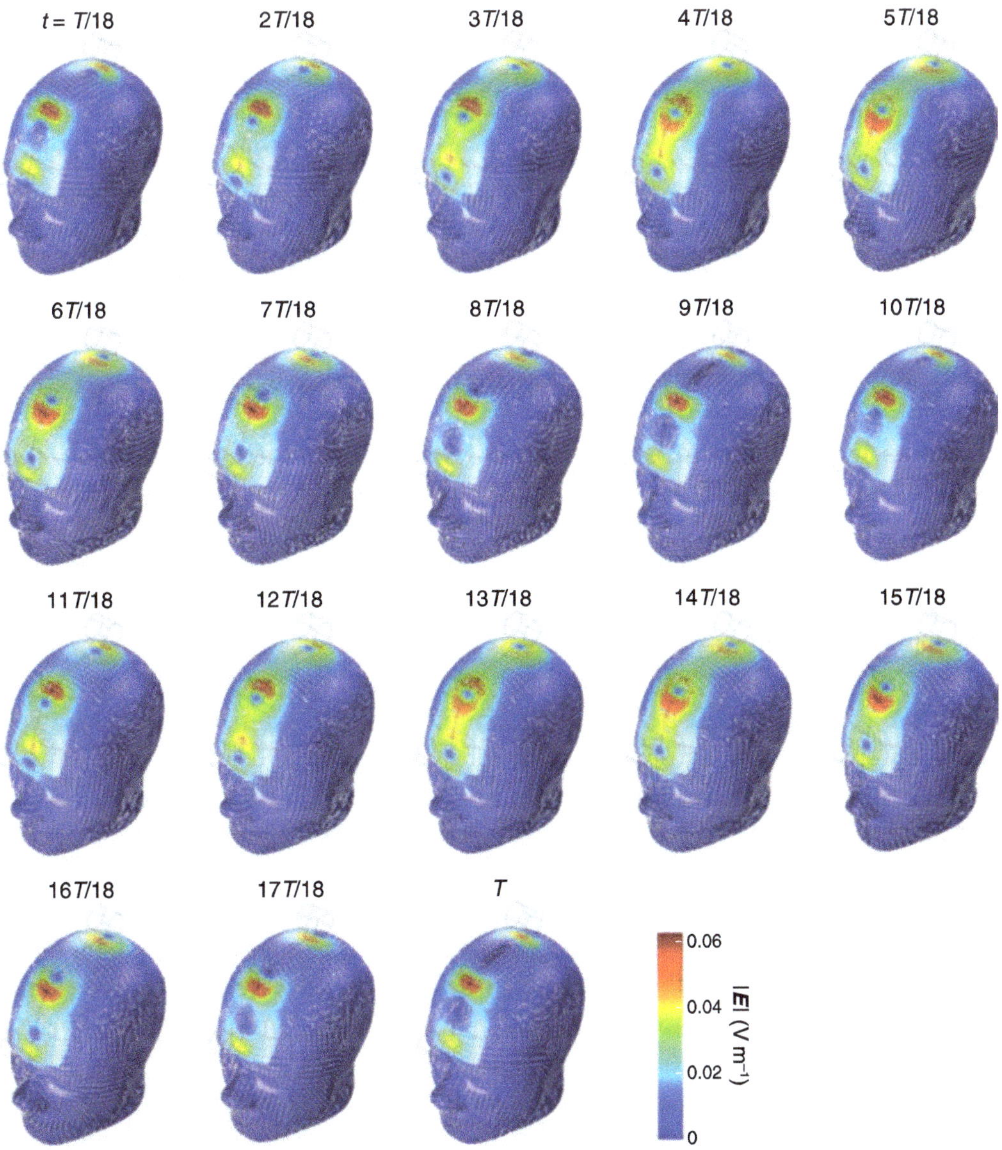

Fig. 4.4 One full revolution (period = T) of the full sTMS configuration in steady-state

4.4 Discussion

There is marked variability in the distribution of E-field induced by ECT across individuals, with approximately 22% variation in the maximum E-field strength attributed to anatomical differences. Stimulation of anterior–posterior oriented white matter tracts on the right hemisphere, such as the inferior fronto-occipital fasciculus and inferior longitudinal fasciculus, appears to be related to clinical outcome.

There is also marked variability in the induced E-field strength at the DLPFC in patients receiving rTMS. Region of interest analysis of the E-field distribution in combination with clinical outcome could inform targeting and dosing strategies.

Jin and Phillips estimated the intensity of sTMS stimulation to be approximately 0.1% that of standard TMS [9]. However, this estimate was based on comparison of maximum surface fields and does not account for boundary conditions of the head. Our simulation with a head model suggests that the peak electric field strength at the level of the cortex is approximately 0.02 V m^{-1}. This field strength is an order of magnitude lower compared to those induced by transcranial current stimulation (tCS) [15] and low-field magnetic stimulation (LFMS) [16, 17]. The sTMS field strength is comparable to that of low-intensity repetitive magnetic stimulation (LI-rMS) in an in vitro model, which has been shown to alter cellular activation and gene expression in an organotypic hindbrain explant and in a stimulation frequency-specific manner [18]. Thus, the low field strength of sTMS could be biologically active.

Helekar and Voss proposed a device comprised of an assembly of high-speed rotating cylindrical magnets [7]. These N52 grade magnets are smaller (3/8 inch in height and 1/4 inch in diameter) and have stronger surface field (B_r = 1.48 T) compared to the sTMS magnets. The magnets are axially magnetized, but the axis of rotation is perpendicular to the axis of the cylinder. The motor provides a no-load speed of 24,000 rpm (400 Hz). Since the induced electric field strength is proportional to the angular frequency of rotation, higher rotational speed can increase the electric field strength. Helekar and Voss estimated the intensity of their high-speed rotating magnet device to be approximately 6% that of TMS, based on voltage measurements made with an inductor search coil [19, 20]. However, measurements made in air and without the conductivity boundaries of the head would likely overestimate the electric field strength. Furthermore, smaller magnets have faster field attenuation with distance compared to larger magnets.

Watterson proposed and tested a similar high-speed rotating magnet device for stimulation of muscle nerves [6]. In a series of in vitro experiments on the cane toad sciatic nerve and attached gastrocnemius muscle, Watterson and Nicholson observed that nerve activation was achievable with a rotational frequency of 230 Hz [21]. The activation of peripheral nerves is thought to be more sensitive to the gradient of the electric field. To maximize the field gradient, Watterson's device employs a "bipole" configuration, comprising two diametrically magnetized cylindrical magnets next to one another with opposite magnetization directions [21].

In this work, we simulated the sTMS system at a fixed rotational frequency of 10 Hz. The frequency of peak alpha oscillation across individuals can vary between 8 and 13 Hz. As mentioned above, the induced electric field strength is proportional to the frequency of rotation of the magnets. Therefore, individualizing the rotational frequency could introduce variability in the induced electric field strength across individuals. Higher field strength can be achieved by increasing the rotational speed. However, neuronal activation becomes inefficient at very high frequencies. Finally, the interaction between field strength and excitation frequency could be nonlinear. For example, it has been demonstrated that when 140 Hz transcranial alternating current stimulation is applied to the motor cortex, low current amplitude of 0.4 mA results in reduction of motor evoked potential (MEP) amplitudes, intermediate amplitudes of 0.6 and 0.8 mA showed no effect on MEP, and high amplitude of 1 mA results in enhancement of MEP amplitudes [22].

4.5 Conclusion

We evaluated the electric field characteristics of ECT, rTMS, and the sTMS system of rotating magnets using the finite element method. We found substantial variability in E-field strength across patients receiving ECT and rTMS, possibly contributing to variability in clinical outcome. For the experimental sTMS treatment, we found that the maximum induced electric field strength at the level of the cortex is approximately 0.02 V m^{-1}, which is an order of magnitude lower compared to those delivered by transcranial current stimulation and low-field magnetic stimulation. Future work will include simulation of sTMS in anatomically-accurate head models derived from individual brain scans and treatment parameters. Direct electrophysiological data should also be collected to validate the proposed mechanism of action.

Acknowledgments Zhi-De Deng is supported by the NIMH Intramural Research Program and NARSAD Young Investigator Award from the Brain & Behavior Research Foundation.

References

1. Ferrari, A. J., et al. (2013). Burden of depressive disorders by country, sex, age, and year: Findings from the global burden of disease study 2010. *PLoS Medicine, 10*(11), e1001547.
2. Stone, D. M., et al. (2018). Vital signs: Trends in state suicide rates — United States, 1999–2016 and circumstances contributing to suicide — 27 states, 2015. *MMWR. Morbidity and Mortality Weekly Report, 67*(22), 617–624.
3. Rush, A. J., et al. (2006). Acute and longer-term outcomes in depressed outpatients requiring one or several treatment steps: A STAR∗D report, (in eng). *The American Journal of Psychiatry, 163*(11), 1905–1917.
4. Husain, M. M., et al. (2004). Speed of response and remission in major depressive disorder with acute electroconvulsive therapy (ECT): A Consortium for Research in ECT (CORE) report. *The Journal of Clinical Psychiatry, 65*(4), 485–491.
5. Phillips, J. W., & Jin, Y. (2013). *Devices and methods of low frequency magnetic stimulation therapy*. USA.
6. Watterson, P. A. (2014). Device including moving magnet configurations.

7. Helekar, S. A., & Voss, H. U. (2014). Method and apparatus for providing trancranial magnetic stimulation (TMS) to a patient.
8. Leuchter, A. F., et al. (2015). Efficacy and safety of low-field synchronized transcranial magnetic stimulation (sTMS) for treatment of major depression. *Brain Stimulation, 8*(4), 787–794.
9. Jin, Y., & Phillips, B. (2014). A pilot study of the use of EEG-based synchronized transcranial magnetic stimulation (sTMS) for treatment of major depression. *BMC Psychiatry, 14*(1), 13.
10. Leuchter, A. F., Cook, I. A., Jin, Y., & Phillips, B. (2013). The relationship between brain oscillatory activity and therapeutic effectiveness of transcranial magnetic stimulation in the treatment of major depressive disorder. *Frontiers in Human Neuroscience, 7*(37).
11. van Waarde, J. A., van Oudheusden, L. J., Verwey, B., Giltay, E. J., & van der Mast, R. C. (2013). Clinical predictors of seizure threshold in electroconvulsive therapy: A prospective study. *European Archives of Psychiatry and Clinical Neuroscience, 263*(2), 167–175.
12. Liston, C., et al. (2014). Default mode network mechanisms of transcranial magnetic stimulation in depression. *Biological Psychiatry, 76*(7), 517–526.
13. Thielscher, A., Antunes, A., & Saturnino, G. (2015). Field modeling for transcranial magnetic stimulation: A useful tool to understand the physiological effects of TMS? *Conference Proceedings IEEE Engineering in Medicine and Biology Society*, 222–225.
14. Beam, W., Borckardt, J. J., Reeves, S. T., & George, M. S. (2009). An efficient and accurate new method for locating the F3 position for prefrontal TMS applications. *Brain Stimulation, 2*(1), 50–54.
15. Miranda, P. C., Mekonnen, A., Salvador, R., & Ruffini, G. (2013). The electric field in the cortex during transcranial current stimulation. *NeuroImage, 70*, 48–58.
16. Rohan, M. L., et al. (2014). Rapid mood-elevating effects of low field magnetic stimulation in depression. *Biological Psychiatry, 76*(3), 186–193.
17. Wang, B., et al. (2018). Redesigning existing transcranial magnetic stimulation coils to reduce energy: Application to low field magnetic stimulation. *Journal of Neural Engineering, 15*(3), 036022.
18. Grehl, S., Martina, D., Goyenvalle, C., Deng, Z.-D., Rodger, J., & Sherrard, R. M. (2016). In vitro magnetic stimulation: A simple stimulation device to deliver defined low intensity electromagnetic fields. *Front Neural Circuits, 10*(85).
19. Helekar, S. A., & Voss, H. U. (2016). Transcranial brain stimulation with rapidly spinning high-field permanent magnets. *IEEE Access, 4*, 2520–2528.
20. Helekar, S. A., et al. (2018). The strength and spread of the electric field induced by transcranial rotating permanent magnet stimulation in comparison with conventional transcranial magnetic stimulation. *Journal of Neuroscience Methods, 309*, 153–160.
21. Watterson, P. A., & Nicholson, G. M. (2016). Nerve-muscle activation by rotating permanent magnet configurations. *The Journal of Physiology, 594*(7), 1799–1819.
22. Moliadze, V., Atalay, D., Antal, A., & Paulus, W. (2012). Close to threshold transcranial electrical stimulation preferentially activates inhibitory networks before switching to excitation with higher intensities. *Brain Stimulation, 5*(4), 505–511.

Chapter 5
Design and Analysis of a Whole-Body Noncontact Electromagnetic Subthreshold Stimulation Device with Field Modulation Targeting Nonspecific Neuropathic Pain

Check for updates

Sergey Makarov, Gene Bogdanov, Gregory Noetscher, William Appleyard, Reinhold Ludwig, Juho Joutsa, and Zhi-De Deng

5.1 Introduction

Pain is distinguished by duration as acute (less than 6 weeks), subacute (6–12 weeks), and chronic (12 weeks or more) pain. Approximately 100 million US adults suffer from common chronic pain conditions, more than the number affected by heart disease, diabetes, and cancer combined [1]. The economic cost of chronic pain in adults, including health care expenses and lost productivity, is $560–630 billion annually

S. Makarov (✉)
Massachusetts General Hospital, Boston, MA, USA

Worcester Polytechnic Institute, Worcester, MA, USA
e-mail: makarov@wpi.edu

G. Bogdanov · W. Appleyard · R. Ludwig
ECE Department, Worcester Polytechnic Inst, Worcester, MA, USA

G. Noetscher
Worcester Polytechnic Institute, Worcester, MA, USA

J. Joutsa
Athinoula A. Martinos Center for Biomedical Imaging, Massachusetts General Hospital, Harvard Medical School, Charlestown, MA, USA

Department of Neurology, University of Turku, Turku, Finland

Division of Clinical Neurosciences, Turku University Hospital, Turku, Finland

Z.-D. Deng
Noninvasive Neuromodulation Unit, Experimental Therapeutics & Pathophysiology Branch, National Institutes of Mental Health, NIH, Bethesda, MD, USA

S. Makarov et al. (eds.), *Brain and Human Body Modeling*,
https://doi.org/10.1007/978-3-030-21293-3_5

[1]. Seven in ten Americans feel that pain research and management should be one of the medical community's top few priorities (16%) or a high priority (55%) [2].

One form of chronic pain is nociceptive pain, which is the normal response to injury of tissues such as muscles, visceral organs, joints, or bones. Another form is neuropathic pain, which involves dysfunction of (i) the peripheral nervous system (PNS) or (ii) the central nervous system (CNS). The latter case is amplification and generation of pain within the CNS itself due to distorted sensory processing, malfunctioning of pain-inhibitory mechanisms, and enhancing pain-facilitatory mechanisms [3–6]. An example is psychogenic pain, which does not usually have a physical origin [7, 8]. Highly prevalent symptoms in chronic pain are depression and anxiety [7–11], which are reported by more than 50% of patients with chronic pain [9]. Pain and depression may create a vicious cycle in which pain worsens depression and vice versa [10].

Low back pain, either acute or chronic, dominates other pain types [7] and affects about 80–84% of the population at least once at some point in life [12–14]. In the US Armed Services alone, low back pain was the primary diagnosis for more than seven million ambulatory care visits between 2000 and 2009 [15]. Current estimates are that approximately 25% of people with acute low back pain experience recurrent episodes, while 7–10% progress to a chronic state [15] and can experience significant physical, psychological, and social sequelae that affect their long-term functioning and quality of life [16]. According to [11], 70% of subjects with chronic low back pain report fatigue and 18% report depression. According to [17], 59% of the patients with chronic low back pain report poor sleep.

Chronic low back pain accounts for 22% of all chronic pain cases and for 35% of most persistent pain sites [7]. The classification of low back pain is complicated by the varying presentation and complex nature of pain [14]. The most common diffuse neuropathic pain without radiating beyond the buttocks is classified as *nonspecific* low back pain [14], which makes up 60% of individuals suffering from chronic low back pain [18].

The initial treatment for acute nonspecific low back pain is conservative, including nonopioid analgesics (acetaminophen, aspirin), nonsteroidal anti-inflammatory drugs (ibuprofen, ketoprofen), physiotherapy, dynamic strengthening exercises, thermotherapy, and, if necessary, a short course of muscle relaxants [13, 14]. Further, conservative methods include traction treatment, manual therapy, and transcutaneous electrical nerve stimulation (TENS) [14]. A commonly prescribed treatment for chronic pain is opioids (codeine, oxycodone, hydrocodone, and morphine) [13, 19]. The use of opioids is controversial due to severe addiction and misuse [13, 15, 16, 20, 21]. It is argued that chronic use of opioids is detrimental to people with back pain because they can aggravate depression, leading to a worsening of the pain [13]. Whenever possible, opioid medications in chronic noncancer pain should be avoided [20].

TENS is a common drug-free alternative treatment technique that stimulates selected sensory nerves and muscles via electrodes placed on the skin over the painful area [22–28]. The electrodes inject electric currents and, most importantly, electric fields into body. The theory is that the local electric field stimulation can modify both cause and perception of chronic pain. A number of systematic reviews of the effect of TENS on various painful conditions, such as labor pain, rheumatoid arthritis, phantom limb pain, and chronic lower back pain, are available [22, 24–28]. However, these reviews indicate that most controlled randomized clinical trials failed to show significant effects of the existing small-scale TENS systems with a strongly localized electric field distribution.

This design-based study is driven by the limitations of TENS. We introduce a conceptually different electromagnetic stimulation device. Instead of local high-intensity and suprathreshold TENS, we suggest to stimulate the PNS and muscular system of the *entire* lower body in a noncontact, patient-friendly way. At the same time, we suggest to use low or subthreshold power levels. In other words, we propose mild yet more broad electromagnetic treatment, potentially beneficial for nonspecific chronic pain. The proposed device [29, 30] would primarily affect peripheral nerves, the spinal cord, muscles, joints, and bone. Simultaneously, it could influence the somatosensory cortex via many affected pathways, in line with the modern concept of central control of pain [8]. Based on the numbers cited above [7, 14, 18], we can estimate that some 14% of all chronic pain cases might be subject to the proposed alternative treatment.

The text is organized as follows. Section 5.2 describes a theoretical device model, specifies the field distribution within the resonator, and describes hardware design, test, and functionality, including semiautomatic operation/tuning and representative continuous run times. Section 5.3 provides computational results for the electric field distribution within the body obtained via two independent numerical methods. Section 5.4 discusses possible device modifications, as well as potential application scenarios. Section 5.5 concludes the chapter.

5.2 Materials and Methods

5.2.1 Suprathreshold Versus Subthreshold Stimulation

The vast majority of transcutaneous electrical nerve stimulation (TENS) devices [14, 22–28] today are suprathreshold. They excite local currents/electric fields that are strong enough to produce an *action potential* in peripheral axons assembled in bundles of sensory nerve fibers. A case in point is a small pulsed-current InTENSity™ Twin Stim® III device [31] similar to that shown in Fig. 5.1a [32], intended for home use. It creates a prominent zapping sensation when the electrodes are connected to the lower back. The majority of PEMF (pulsed electromagnetic field) devices operate in a similar fashion.

Alternatively, a continuous current injection with a more powerful stationary device may be employed by chiropractors as shown in Fig. 5.1b [33]. When connected to the lower back, it creates a rather strong burning sensation. The electric field may also be injected via a noncontact induction coil [34, 35] as in Fig. 5.1c.

Fig. 5.1 (**a**) Portable pulsed-current TENS (or PEMF) device from National Health Service (NHS) England website [32]. (**b**) Stationary continuous-current TENS device with sponge electrodes for low back pain treatment in Hannover, Germany [33]. (**c**) Induction coil-based peripheral neuropathic pain stimulation for low back pain treatment [34, 35]

Although a broader coverage with many electrodes is now possible, local suprathreshold effects still dominate.

With respect to treatment-persistent depression, only three nonpharmacological therapies have been approved by the FDA to date; all of them are electromagnetic and suprathreshold: transcranial magnetic stimulation [36], vagus nerve stimulation [37], and electroconvulsive therapy [38].

Subthreshold refers to a low-power electromagnetic stimulation that is too small to elicit action potentials. However, it still alters the axonal membrane potential [39]. This effect accumulates and maximizes toward axon terminals, i.e., synapses [39]. It is *synaptic efficacy* (or natural neurotransmission efficacy) that is altered and enhanced by the subthreshold stimulation [39]. In a chain of neurons, this stimulation could cause an incremental relay effect, which may further enhance neuronal network activity [39]. The theory of subthreshold stimulation [39–42] has been developed in application to transcranial current stimulation [39–47]. Low-field magnetic stimulation of depression is an active area of research [48–54]. The subthreshold technique is also a modern research direction in spinal cord stimulation [55–59], as it eliminates noxious and off-target paresthesia, while being efficient, as well as in vagus [60], occipital [61], proximal [62], and parasympathetic [63] nerve stimulation, and in neuromuscular stimulation [64]. Electrocutaneous subthreshold stimulation has been found to improve sleep and reduce reactive anxiety/depression [65].

5.2.2 Concept of the Magnetic Stimulator. Two-Dimensional Analytical Solution for Solenoidal E-Field

Figure 5.2 shows the anticipated device concept. An external, uniform, *rotating* (or circularly-polarized) magnetic flux density **B** with amplitude B_0 is created in the transverse plane around a tissue volume, depicted in Fig. 5.2a. By Faraday's law of induction, this field excites an *axial,* rotating solenoidal electric field **E** in free space or in a tissue volume, which is expressed in terms of the magnetic vector potential **A**,

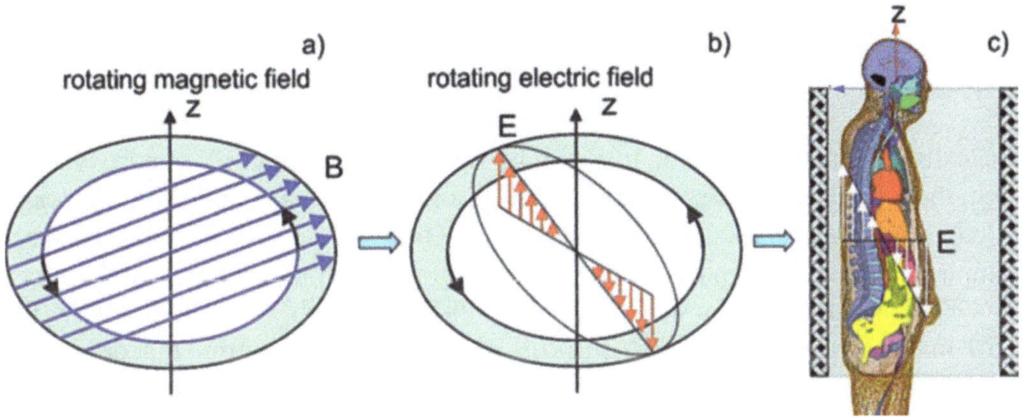

Fig. 5.2 Concept of non-invasive electric field excitation via the induction mechanism. A rotating magnetic field shown in (**a**) excites the rotating electric field within the conductor (**b**) and within the body (**c**), respectively

$$E = -\partial A / \partial t \tag{5.1}$$

as shown in Fig. 5.2b; we set **B** = ∇ × **A**. Thus, when a biological body is placed into this volume, a significant noninvasively excited electric field in the *axial* direction will appear parallel to the major peripheral nerves, spinal cord, long bones, major arteries, veins, and other structures. This is in contrast to a solenoidal coil wound around the body creating electric fields and currents in the less desirable transverse plane.

The rotational character of the field also assures that not only one body cross section (e.g., coronal or sagittal) will be subject to the electric field excitation, but the entire body volume.

In the ideal, two-dimensional case and for any conducting target with a strict cylindrical symmetry placed into the device, either homogeneous or not, the corresponding two-dimensional problem, shown in Fig. 5.2a, b, will have an exact analytical solution in the quasi-static (or eddy current) approximation. The electric field within the target is given by [66].

$$E_x = 0, \quad E_y = 0, \quad E_z = \omega B_0 r \cos(\omega t - \phi) \tag{5.2}$$

$$J = \sigma E \tag{5.3}$$

where r is the radial distance from the coil axis in cylindrical coordinates, ω is the angular frequency, ϕ is an arbitrary phase, J is induced current density, and σ is the (local) medium conductivity which is either constant or obeys cylindrical symmetry. Although Eqs. (5.2) and (5.3) might be used to roughly estimate the electric field in the body based on a cylindrically symmetric assumption, its actual value will deviate as shown below.

5.2.3 Three-Dimensional Coil Resonator Design. Solenoidal E-Field

The external rotating magnetic flux density **B** is created using a volumetric resonator in the form of a low-pass birdcage coil. Resonators of this type (called "birdcage coils") are routinely used as MRI radio frequency (RF) coils [67, 68], but for a completely different purpose, namely atomic spin excitation and RF signal acquisition. The resonant frequency in this application is very high, typically 64 MHz (for 1.5 T magnets) or higher. For our purpose, we decided to reconstruct that design for a much lower frequency band of 100 kHz or less. In particular, the band of 10 kHz has recently demonstrated great promise for spinal cord stimulation for back and leg chronic pain management [55–59, 69–77] with and without previous back surgeries [70, 72], and is utilized by TENS [78]. In addition to superior pain relief, the 10 kHz band may provide long-term improvements in quality of life and functionality for subjects with chronic low back and leg pain [77]. On the other hand, a wider band of 4–30 kHz has been used for polyneuropathy (a general degeneration of peripheral nerves that spreads toward the center of the body) electrostimulation treatment [79–82].

We chose the birdcage coil because it can produce a very homogeneous B-field in the transverse plane, and because it can produce a circularly polarized B-field. These features relax the requirements for accurate patient positioning relative to the coil. The patient has significant freedom of movement transversely within the coil, including freedom of rotation (thanks to circular polarization). This should enhance patient comfort and permit long treatment sessions.

The resonator concept will allow for an arbitrary "tonic" modulation [30] of the carrier frequency, which was found to beneficially address the variable nature of chronic pain across different patients [76]. This modulation can be either open- or closed-loop (e.g., a single-channel EEG signal fed back to the modulator).

When the resonant frequency becomes low as in the present case, the standard RF birdcage coil will possess very low inductance L. Tuning such a coil toward resonance at low frequencies would require large capacitance C. This, however, means a low Q-factor (quality factor $Q = \sqrt{L / C} / R$ is the "gain" of the series resonator) and higher costs, as well as higher fabrication complexity [83, 84], and will restrict the use of the conventional birdcage coil to frequencies above at least a few megahertz. Different methods to overcome this difficulty have been suggested [83–88], but they are all limited to *small-size* coils.

Our design described in the subsequent patent application [29] utilizes a unique large-scale low-pass birdcage coil topology with an intentionally very large number (144) of long rungs (boosting inductance) and bridging capacitors seen in Figs. 5.6 and 5.7. After this point, the improvement in adding more rungs is small. Several other reasons for the number of rungs include:

- To distribute the required capacitors uniformly around the coil circumference
- To improve the mechanical stiffness of this self-supporting coil structure

- To reduce rung tube diameter, simplifying assembly
- To facilitate direct drive of the coil at a single rung (for each mode). The equivalent parallel resistance at resonance was significantly above 50 ohms for the 144 rung configuration, permitting the use of capacitive-only matching networks. Later in the development process, we decided not to use direct drive, but the option remains

However, we are not claiming that the particular number of rungs we used is optimal. All the coil geometrical parameters (number of rungs, rung and end ring tube diameters, coil length and diameter) are subject to optimization during design and construction of the next prototype. We may even consider switching to a fundamentally different coil winding design, such a saddle coil. That said, any optimization is not expected to dramatically improve the coil's B-field. Improvements up to 20% may be possible (slightly more if the coil is made smaller).

Along with this, we employ carefully designed inductive power coupling. The inductive coupling blocks DC and acts as a balun (balanced-unbalanced transformer). This is useful in terms of both circuit design (we avoid having to install a transformer) and safety concerns (no direct path from the AC to the coil). It also allows adjusting the load resistance at resonance. Inductive coupling perturbs the current distribution in the coil significantly less than directly driving an individual rung. As a result, the resonator possesses a superior quality factor of approximately 300 [30]. Therefore, we may achieve any desired electric field levels of up to 50–100 V/m within the lower body due to the resonance effect and still use standard power electronics equipment.

A computational model of a particular resonator constructed in this study is shown in Fig. 5.3a with the electric current distribution to scale. The coil has a diameter of 0.94 m and a length of 1.10 m; the coil resonates at 100 kHz or at 145 kHz depending on the values of the bridging capacitors. The coil consists of two rings (top and bottom) joined via multiple straight rungs, each bridged with a lumped capacitor at its center. The capacitors control the coil's resonant frequency. The resonating coil is fed via two lumped ports in quadrature, or using inductive coupling with two loops in quadrature as explained below.

From the modeling point of view, the resonant electric current in both rings at any fixed time instant behaves like a full period of a sine function of polar angle φ. This ring current distribution is shown in Fig. 5.3a. As time progresses, the ring current distribution shown in Fig. 5.3a rotates with angular frequency ω. As a result, the time-domain ring current $i(t,\varphi)$ in the top and bottom rings can be expressed in the form

$$i(t,\varphi) = \pm I_0 \cos(\omega t - \varphi + \pi/2), \quad \varphi \in 0, 2\pi \tag{5.4}$$

where I_0 is the current amplitude determined by the excitation power and by the quality factor (or the "gain") of the resonator.

The AC current in each rung shown in Fig. 5.3a does not change along its length. Simultaneously, at any fixed time instant, it also varies from rung to rung as a harmonic function of the polar angle φ with the full period corresponding to the ring circumference. This rung current distribution is shown in Fig. 5.3a. As time progresses, the rung current distribution shown in Fig. 5.3a also rotates with angular frequency ω. Each individual time-domain rung current density $j(t, \varphi)$ can be expressed in the form

$$j(t,\varphi)=\frac{I_0}{F}\cos(\varphi+\omega t),\quad F=\frac{1}{2}\sum_{m=0}^{\frac{N}{2}-1}\sin\left(\frac{2\pi m}{N}\right),\quad \varphi\in 0, 2\pi \tag{5.5}$$

where N is the total number of rungs. This form obeys the current conservation law, or Kirchhoff's Current Law (KCL), at every ring-rung junction.

The useful current, which creates a nearly constant horizontal rotating magnetic field B_r with amplitude B_0 and axial rotating electric field E_z according to Eq. (5.2), is the rung current density $j(t, \varphi)$. Contributions of each rung add up in a constructive manner. The ring current, on the other hand, does not contribute to the axial (or vertical) electric field, E_z. However, it may create a strong transverse electric field very close to the rings.

It should be pointed out that Eqs. (5.4) and (5.5) describe the rotating current behavior, which is a combination of *two* elementary resonant modes. Each elementary mode does not rotate and appears as depicted in Fig. 5.3a. However, when excited in quadrature (with a 90 degree phase shift and a 90 degree excitation offset along the coil circumference), both modes combine to create the current distribution given by Eqs. (5.4) and (5.5) and the associated rotating electric field. The rotation phenomenon enables us to treat the entire body and not merely a singular component or region.

5.2.4 *Solenoidal Electric Field Distribution with and without a Simple Conducting Object*

Figure 5.3b–d shows the resulting electric field distribution in the coil (coronal plane) when the current amplitude $I_0 = 1$ ampere in either ring given by Eq. (5.4). The results are given for one resonant mode, as shown in Fig. 5.3a. Due to linearity, this result can simply be scaled for other excitation levels. Accurate field computations have been performed with the fast multipole method described in [89]. The magnitude of the axial component E_z in V/m for an empty coil is shown in Fig. 5.3b. The electric field is indeed zero at the coil's center.

When a conducting object representing a load is inserted into the coil, the field distribution changes. Figure 5.3c shows the distribution when a conducting cylinder with a diameter of 0.4 m and a length of 1 m is inserted into the coil along its axis. The particular conductivity value σ does not matter since only the conductivity contrast, $(\sigma - \sigma_{air})/(\sigma + \sigma_{air})$, is present in the solution [66]. This value is always unity since $\sigma_{air} = 0$.

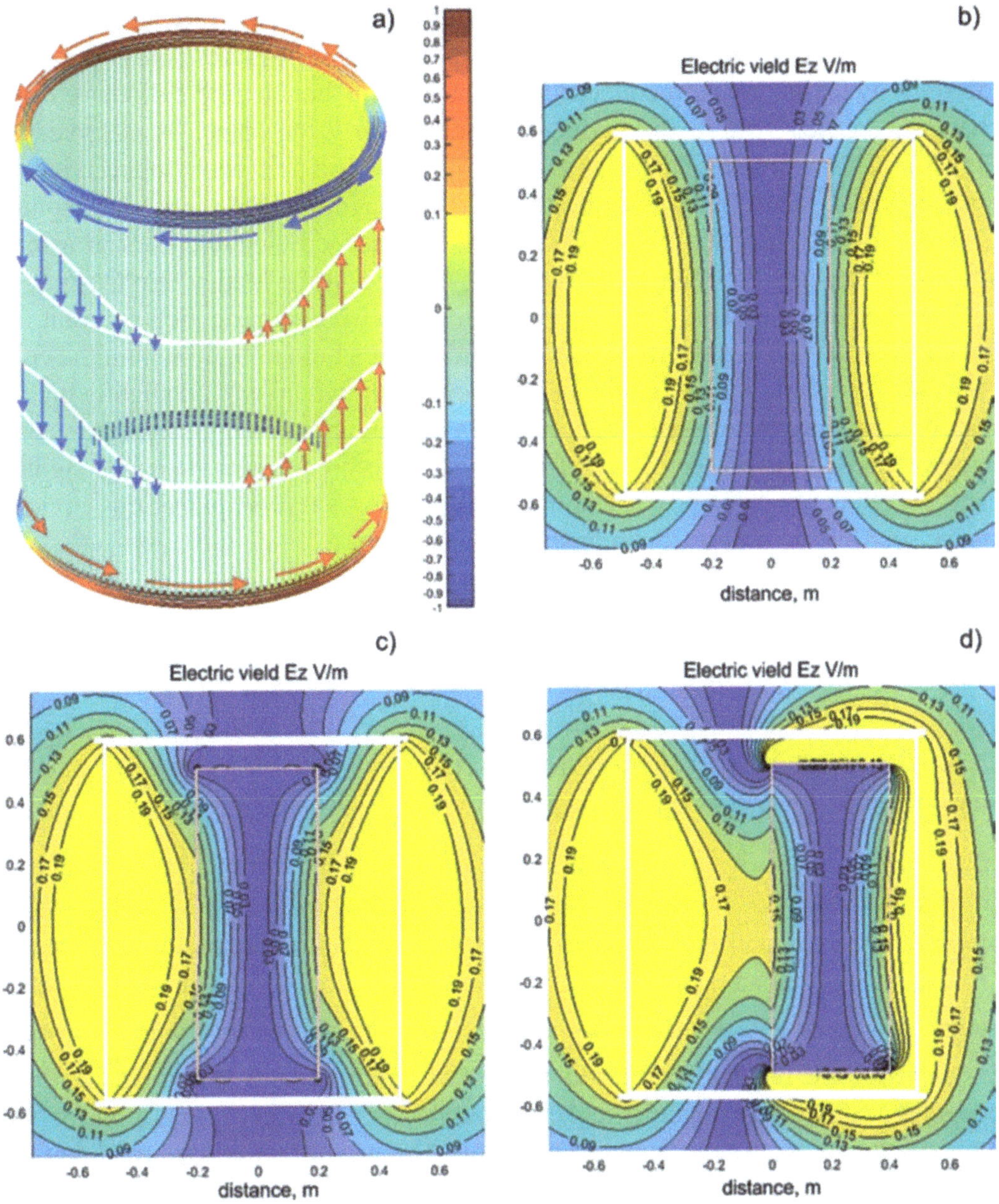

Fig. 5.3 Current distribution in the coil resonator and the associated *solenoidal* electric field created by the coil when the ring current amplitude is 1 A for one resonant mode. (**a**) Electric current distribution in the coil along with the current color bar to scale; (**b**) Magnitude of the vertical electric field E_z in V/m for an empty coil in the coronal plane; (**c**) Magnitude of the vertical electric field E_z in V/m for the coil with a coaxial conducting cylinder 0.4 × 1 m inside; (**d**) Magnitude of the vertical electric field E_z in V/m for the coil with a conducting cylinder 0.4 × 1 m shifted in the transverse plane inside

An interesting and useful effect is observed in Fig. 5.3c: we see a "pulling" of the electric field into the cylinder close to the coil center. This is due to surface charges that appear at and near the cylinder tips. As a result, the electric field close to the cylinder surface at the center plane of the coil increases by nearly 36%.

Another remarkable observation (this effect is common in MRI RF coils) follows from Fig. 5.3d where the conducting cylinder has been shifted from the coil axis to the right by 0.2 m. While the electric field outside the conducting cylinder clearly changes, the field within the cylinder remains nearly the same, as observed in Fig. 5.3c. This may be explained as a result of the electric field being induced by the magnetic field, similar to eddy currents. Since the magnetic field is relatively homogeneous in the transverse plane of the coil, the induced electric fields in a load should not strongly depend on the transverse position of the load within the coil.

These rudimentary simulations allow us to establish two basic facts relevant for the analysis of realistic electric field distributions in a human body within the coil. First, we expect that the average transcutaneous electric field will be slightly higher than predicted by the air-filled coil model in dorsal, abdominal, and lumbar body regions. Second, we expect that the field within the body will not change significantly when the body is moved within the coil in the transverse plane; this circumstance seems to be useful from a practical point of view.

5.2.5 *Contribution of Unpaired Electric Charges*

Generally, the total electric field within the coil is expressed in terms of two auxiliary potentials. Instead of Eq. (5.1), one has

$$\boldsymbol{E} = -\partial \boldsymbol{A} / \partial t - \nabla \varphi \tag{5.6}$$

where φ is the scalar electric potential and $\boldsymbol{A}$ is the magnetic vector potential. In the quasistatic approximation to Maxwell's equations, the time derivative in the Lorentz gauge $\frac{1}{c^2}\partial\varphi / \partial t + \nabla \cdot \boldsymbol{A} = 0$ is neglected (which gives us the Coulomb gauge, $\nabla \cdot \boldsymbol{A} = 0$), while it is still kept in Eq. (5.6). As a result, the $-\nabla\varphi$ term in Eq. (5.6) becomes a conservative electric field contribution due to charge density alone, while the $-\partial \boldsymbol{A}/\partial t$ term is a true solenoidal electric field contribution due to current density alone.

In accordance with the (quasi)electrostatic theory [90], the conservative electric field is blocked by charges induced on a surface of a conducting object and does not penetrate into the object. Therefore, its contribution is ignored in the present study, similar to the theoretical models of transcranial magnetic stimulation or TMS. Note that this charge contribution may be quite large in the present problem, close to the bridging capacitors.

We have also performed full-wave ANSYS ED simulations of this coil and found that the capacitor voltage drop E-field does not significantly affect the E-field *within* the patient. The externally applied conservative E-field is expelled from the patient by the high conductivity of tissues. Only the E-field induced from the B-field is important.

5.2.6 *Power Amplifier/Driver*

In order to create the rotating magnetic and electric fields, as seen in Figs. 5.2a, b, two resonant modes are excited in the coil resonator. These modes display the same current distribution as shown in Fig. 5.3a, but rotated by 90 degrees about the coil axis with respect to each other as well as having an additional temporal phase shift of 90 degrees.

To accomplish this, a custom designed class-D, high-efficiency, single-frequency power amplifier (PA), whose circuit schematic is presented in Fig. 5.4a, was constructed and prototyped, as shown in Fig. 5.4b, c. The upper block in Fig. 5.4a is a class-S modulator, which is followed by two class-D output stages in quadrature, exciting the two resonant modes. The PA has two outputs, one for each resonant mode, and generates a harmonic power RF signal at a fixed carrier frequency at each output. At present, this frequency is typically around 100 kHz. At the same time, the same PA may be tuned to operate at any carrier frequency from 30 kHz to 300 kHz in the LF band. The PA operation, including variable power levels, an optional variable modulation or tonic frequency, and a semiautomatic patient-specific RF frequency tuning procedure, which is automated via a microcomputer board, can be seen in Figs. 5.4c and 5.5d.

The PA output stage is powered by a 3 kW Sorensen DCS 150–20 Variable Regulated DC (direct current) power supply seen at the bottom of Fig. 5.4c. When connected to a standard three-phase 208 VAC outlet, the max RF output power is about 2.9 kW, based on 3 kW DC power. Alternatively, when connected into a single-phase 240 VAC outlet, the max RF power reduces to about 2.3 kW based on 2.4 kW DC power.

Arbitrary *modulation* (pulse or CW) of the carrier signal with a maximum modulation frequency component of 1 kHz is available via the modulator. The modulation bandwidth is limited mainly by the coil envelope time constant of about 1 ms. Typical modulation is sinusoidal in the 0.5–100 Hz range, generated by the PA firmware.

The PA also monitors its output power and load impedance. It uses this information to automatically adjust the carrier frequency in a narrow band such that the output power remains on target. The amplifier cost, including the DC power supply, is under $10,000. The prototype 100 kHz PA was assembled in a rackmount case shown in Figs. 5.4b, c.

The reason for designing a custom, fixed-frequency PA is the lack of an affordable and appropriate commercial model. Industrial low-frequency RF power supplies, e.g., Comdel's CLB3000, are costly and require a matched 50 Ω load. Because our load impedance varies widely with frequency, keeping the load matched is a challenge. It would require load impedance monitoring and fine frequency control (potentially difficult with a commercial unit), and/or a software-controlled matching network (costly). Additionally, generating two outputs in quadrature would require either a 90° hybrid (another costly component), or phase-locking two commercial PAs at a 90° phase difference, which can be difficult. Finally, the majority

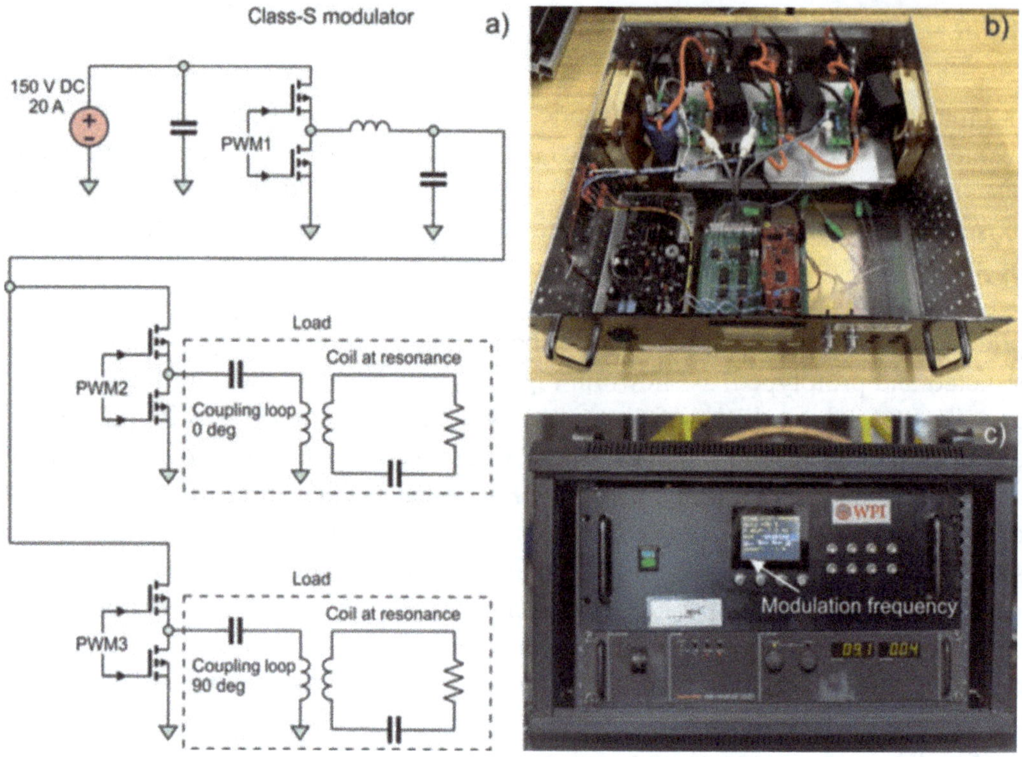

Fig. 5.4 (**a**) High-level circuit schematic of the two-channel power amplifier. (**b**) Rackmount air-cooled assembly of the electronic hardware. (**c**) Amplifier display controlling output power and modulation frequency (if used)

of commercial PAs require water cooling, whereas our PA relies on air cooling. One disadvantage of our custom design is its unknown reliability, a factor that will be proven over time.

5.2.7 Coupling and Matching the Power Amplifier to the Resonating Coil

The amplifier is coupled to the resonating coil inductively via two proximate loops. One such loop is shown in Fig. 5.5c. Apart from certain technical advantages of the inductive coupling, this methodology assures that there is no direct current path from the AC power outlet to the coil. This design enhances overall device safety at any power level, including high-power operation.

The matching network for a single coil port is shown in Fig. 5.5a. Two ports with identical matching networks are located 90° apart around the coil structure, as shown in Fig. 5.5b. The port matching network consists of a series capacitance C_1, series inductance L_1, and the fixed inductance L_2 of the inductive loop seen in Fig. 5.5c.

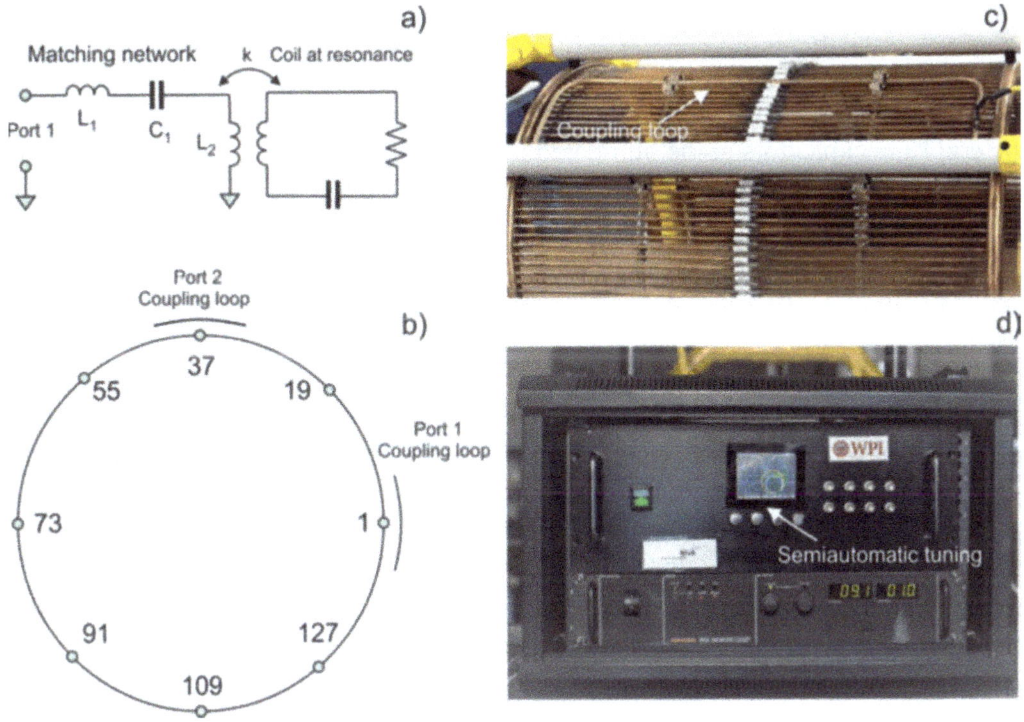

Fig. 5.5 (**a**) Matching and tuning network of the power amplifier. (**b**) Assembly of two coupling-loop feed around the coil circumference with 144 rungs. (**c**) Noncontact inductive coupling of the power amplifier to the coil resonator at one port. (**d**) Smith chart/reflection coefficient display of the power amplifier controller used for semiautomatic tuning at any desired time instant

The matching network is tuned such that the load looks mostly resistive over a small frequency band around the coil's resonance. For example, the load reactance stays quite low from 99.85 kHz to 100.15 kHz, while the resistance varies from 1.3 Ω to 6 Ω. Because the coil resonance shifts as the coil heats up, the operating frequency must be actively adjusted to compensate for this change, or the output power will vary.

We used two Cornell Dubilier Electronics 940C20S47K-F per rung ($C = 0.094$ μF per rung). These are 0.047 μF, 2 kV DC, 500 V AC-rated polypropylene film capacitors. They have a typical ESR of 12 mΩ at 100 kHz ($Q = 2800$), and a max RMS current rating of 5.2 A at 70 °C. We exceed this current rating by about 50% at full power. However, we have measured capacitor temperatures using an IR camera. They are below 70 °C, well within the operating range.

Another important safety feature of the matching network is its benign power envelope step response. The matching network avoids large spikes in PA output current while energy is building up in the resonating coil.

Finally, the matching network presents a sufficiently inductive impedance to higher harmonics of the PA output voltage. This protects the output stage, and ensures that voltage transitions occur when the output current is low, thereby improving efficiency. The efficiency of the PA with the expected load is estimated to be greater than 90% over a wide output power range.

5.2.8 Tuning Procedure

The primary adjustable components are the series capacitance C_1 in Fig. 5.5a and the coil rung capacitors at the numbered locations in Fig. 5.5b. First, the coil needs to be manually tuned by installing smaller capacitors in parallel with the primary coil capacitors at strategic locations. Coil adjustments include mode decoupling, tuning of each mode to the same frequency, and impedance matching for each mode (at the coupling loop). Since capacitance is normally only added, the coil only tunes down in frequency.

The semiautomatic tuning procedure implies adjusting PA frequency in a very narrow frequency range. It ensures that the reflection coefficient of both modes stays below −25 dB when matched to the maximum-power coil impedance of $Z_0 = 2.5\ \Omega$ and that both resonances are within a 20 Hz band. The tuning procedure is controlled and guided by the Smith chart/reflection coefficient display of the PA controller seen in Fig. 5.5d. It includes a number of well-defined steps, and is applied to the coil at its designated operating location in an effort to account for the presence of large nearby metal objects. The tuning procedure is simple to perform.

5.2.9 Coil Assembly, Device Setup, and Operation

A resonator coil prototype made of thin-walled, light copper tubing was constructed; it is shown in Figs. 5.6a, b and 5.7. Tubing thickness was kept at ¾ mm or greater to avoid excessive eddy current losses in the copper. The capacitor size in Fig. 5.6a is relatively large since those components must operate at significant currents levels, up to 18 A RMS per rung for a maximum power of 3 kW, and at large voltages, up to 300 V RMS across the capacitor. The total coil weight without the frame is approximately 120 lbs. (54 kg).

This durable coil prototype was then framed, augmented with a horizontal bed, and placed horizontally to enable a subject to rest in the coil, as shown in Fig. 5.6a, which simultaneously shows the complete device setup. The entire coil frame is portable. The distance between the PA, which is connected to the inductive coupling loops of the coil via two isolated cables, can vary from 1 to 3 m, although larger distances may be possible. As mentioned above, there is no direct ohmic current path from the AC power outlet to the coil, which is an important safety feature.

An operator sets the power level, the modulation frequency, and performs RF tuning at the beginning of the resonator operation and for a particular coil load. At the maximum power level, the ring conductors of the coil heat up to approximately 70–75 °C at continuous operation, as illustrated in Fig. 5.8. Continuous coil operation at the maximum input power of 3 kW was tested multiple times with an uninterrupted operation time of up to 2 h and with a cumulative operating time in excess of 100 h.

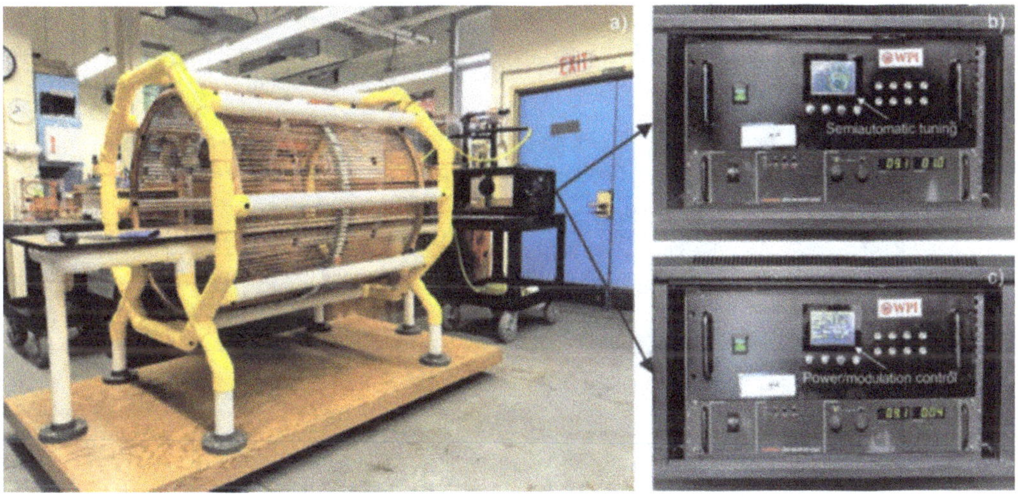

Fig. 5.6 (**a**) Complete framed resonator coil unit with the PA. (**b**) Smith chart/reflection coefficient display of the power amplifier controller to be used for semiautomatic tuning for an individual subject/patient. (**c**) Amplifier display controlling output power and modulation frequency for harmonic modulation

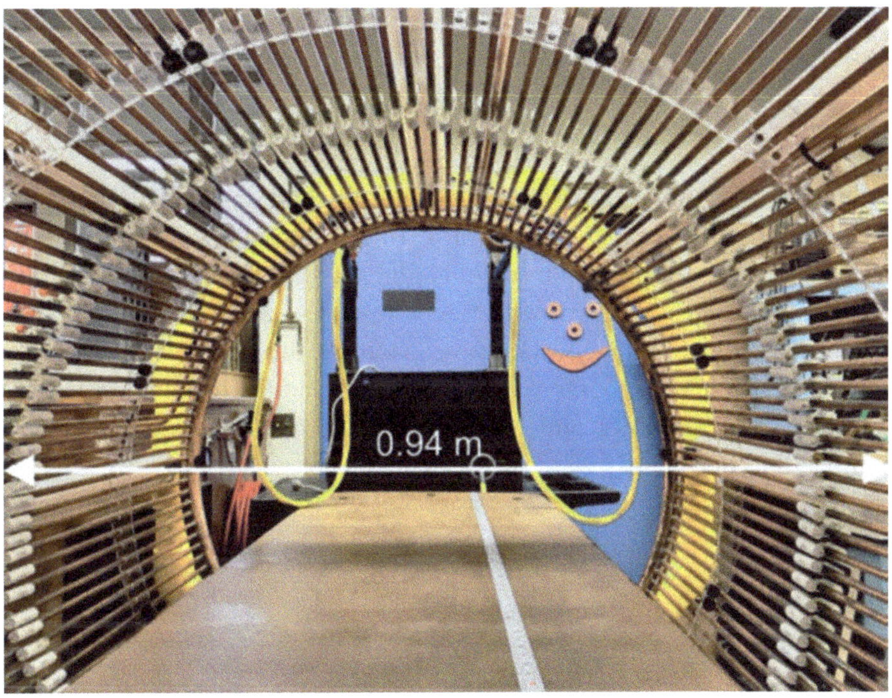

Fig. 5.7 Active area of the subthreshold resonator device

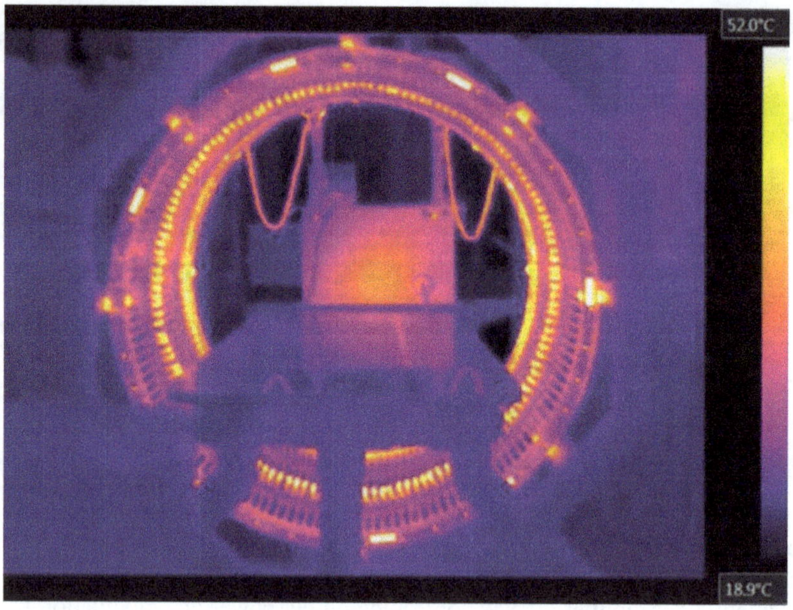

Fig. 5.8 Infrared map of the coil and PA(rear) temperature distribution after 30 min of operation at maximum power obtained with FLIR A325sc IR camera

5.2.10 Quality Factor of the Resonator and the Magnetic Field Strength

The achievable field strength in the coil is determined by three factors: the strength of the PA, the quality factor Q or the "gain" of the resonator, and the coil volume. When the quality factor is high, large field values within the coil can be achieved at a modest input power.

When measured across one of its rung capacitors, the birdcage coil behaves like a parallel resonator in a narrow frequency range around the resonant mode. Using a setup with a signal generator and oscilloscope, the resonator's quality factor has been estimated in the form:

$$Q = \frac{f_0}{\left(f_U - f_L\right)\left(1 - \frac{V_1}{V_0}\right)} \tag{5.7}$$

where f_0 is the resonance frequency and voltage V_0 is the open-circuited generator voltage. Derivation of Eq. (5.7) is given in Appendix A. Voltage V_1 is measured at resonance (where it is maximized). f_L and f_U are the lower and upper frequencies, respectively, where voltage V_1 drops by 3 dB from its peak at resonance. This method is accurate in the high Q limit. The experimental data for 100 kHz and 145 kHz are given in Table 5.1. Table 5.1 reports a Q-factor value of about 300 and

Table 5.1 Measured Q-factors for the coil resonator at 100 kHz and 145 kHz, respectively

Coil	f_0, kHz	f_L, kHz	f_U, kHz	Q
Unloaded	145.30	144.567	145.875	295.8
Loaded	145.28	144.566	145.879	292.2
Unloaded	101.42	101.008	101.827	277.3
Loaded	101.43	101.009	101.835	275.0

The load is a 200 lb. subject

Table 5.2 Characteristics of existing low-frequency RF coils given for comparison with the present resonator prototype

Ref.#	Type of the coil	Frequency, kHz	Q (unloaded)
[83]	Wound birdcage coil (84 mm long and has a diameter of 73 mm)	386	180
[84]	Wound birdcage coils and a solenoid. The diameter and the length of the coils are 70 mm	238/425	100–280
[85]	27 tTurn saddle coil made of Litz wire with 8 cm diameter	373	105
[86]	4-Coil Whiting-Lee configuration, 33 cm long	83.6	100
[87]	Solenoidal coils; 6–46 cm in length and 4–52 cm in diameter	210/275	60–30, reduced Q
[88]	Cylindrical saddle-shaped loops (5 saddle pairs of 10 turns each), coil diameter is 26 mm	87	NA

a minimum difference between loaded (with a human body) and unloaded coil, which is to be expected at this low frequency. These values agree with the theoretical/simulation predictions to within 10%. With decreasing frequency, the Q-factor will decrease approximately proportional to the square root of the resonant frequency.

The established quality factor values are superior to the values reported in the literature for known low-frequency resonator coils (used for low-field MRI) in Table 5.2. Note that all listed competitors have a much smaller coil size/volume and typically a lower quality factor.

It is important to point out again that the quality factor in Table 5.1 is weakly affected by body loading, in contrast to conventional high-frequency MRI RF coils. This observation, also mentioned in Ref. [84] and other sources, is a limitation of the present electromagnetic stimulator. The RF power losses are mostly in the coil itself, and not in the human body.

B-field measurements have been performed via a calibrated single-axis coil probe located at the coil axis. The B-field magnitude was 1.01 mT at the coil center and at full power (3 kW DC) at 100 kHz. The measured and theoretical results differ by no more than 10%. At the full input power level of 3 kW, the amplitude of the resonant ring current I_0 at 100 kHz in two coil rings reaches 603 A, while the amplitude of the rung current reaches 26 A.

5.3 Device Safety Estimates

5.3.1 Peripheral Nervous System (PNS) Stimulation Threshold

The present low-frequency subthreshold electrostimulation device must not exceed the PNS simulation threshold to operate safely and without unpleasant sensation. Guidelines from the International Commission on Non-Ionizing Radiation Protection (Table 5.2 of [91]) require the occupational exposure to an electric field to be limited to a value of approximately 27 V/m RMS at 100 kHz and by a value of 39 V/m RMS at 145 kHz (the so-called basic restrictions [91]). These restrictions are mainly due to limits on peripheral nerve stimulation [91] and should therefore be respected. Other relevant results on the PNS stimulation thresholds at lower frequencies are presented in Refs. [92–94].

5.3.2 Specific Absorption Rate (SAR)

Safety estimates also rely upon the levels of the specific absorption rate (SAR) within the body. The SAR is the energy absorption rate that causes body temperature to rise due to an imposed electromagnetic field. The maximum value of SAR_{1g} in the body must be below 10 W/kg required by the FDA-accepted safety standard [95, 96]. The global-body *SAR* must be below the 2 W/kg limit [95, 96].

5.3.3 Method of Analysis

SAR and electric field measurements cannot be performed easily for human subjects in vivo. SAR and device performance estimates are typically derived and accepted today from computational electromagnetics (CEM) simulations performed with detailed virtual humans [97]. In this study, we use the multi-tissue CEM phantom VHP-Female v. 5.0 (female/60 year/162 cm/88 kg, obese) [97–103] derived from the cryosection dataset archived within the Visible Human Project® of the US National Library of Medicine [104]. The phantom includes about 250 individual tissues and is augmented with material property values from the IT'IS database [105]. The average-body conductivity is assigned as 0.25 S/m, which reflects a mixture of muscle and fat.

The primary CEM software used in this study is the accurate commercial FEM solver ANSYS® Electromagnetic Suite 18.2.0 with rigorous adaptive mesh refinement. In addition, and for verification/validation purposes, we employ an in-house boundary element fast multipole method (BEM-FMM) described in Ref. [89]. In the latter case, a higher near-surface resolution can be achieved and the original surface phantom model can be refined and smoothed from approximately 0.5 M triangles to 3.5 M triangles. In this particular study, the human model is placed in the coil at the shoulder landmark, as shown in Fig. 5.9, so that the top of the shoulder coincides with the ring plane. Other configurations have also been considered.

Results obtained with both software packages differ by no more than 2% in the unloaded coil (field at the coil center) and by no more than 25–50% in the coil loaded with the multi-tissue human body. The latter deviation may be explained by somewhat different surface meshes.

Below we report simulations at two power levels: 1.5 kW input power and 3 kW input power. The first power level is the half power level of the amplifier driver; the second power level corresponds to full power. At full power level, the amplitude of the resonant ring current I_0 in Eqs. (5.4 and 5.5) reaches 603 A, while the amplitude of the rung current reaches 26 A.

5.3.4 *Electric Field Levels*

Figure 5.10 shows the simulated RMS levels of the electric field in the body at 100 kHz and at the input power level of 1.5 kW obtained via the BEM-FMM simulations. We observe that, at half power level of the amplifier driver, the fields everywhere in the body do not generally exceed 30 V/m RMS and are thus within the

Fig. 5.9 Multi-tissue CEM phantom VHP-Female v. 5.0 within the resonant coil (ANSYS 18.2.0)

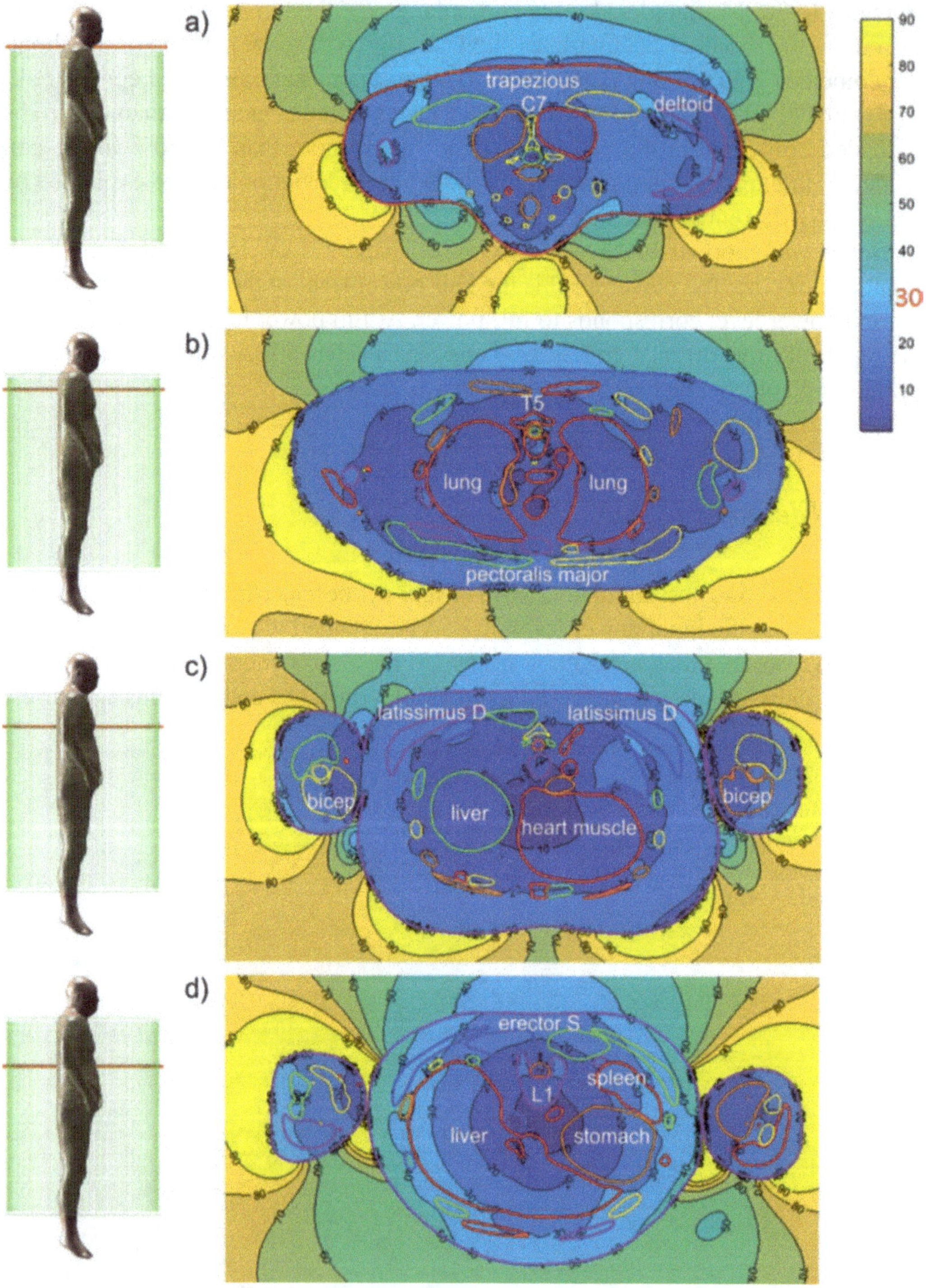

Fig. 5.10 Complex RMS magnitude of the electric field (V/m) at 1500 W input power

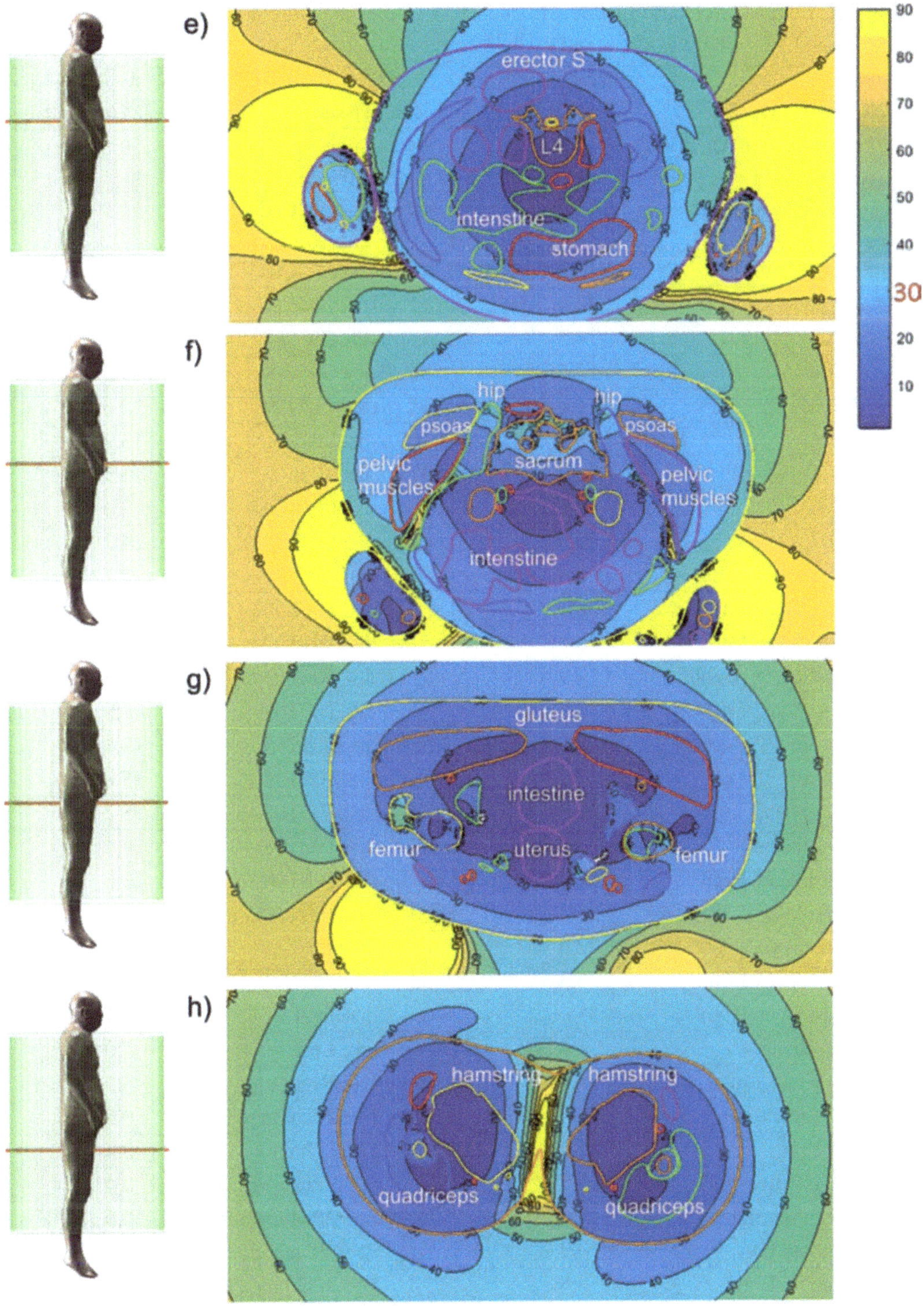

Fig. 5.10 (continued)

Table 5.3 Computed electric field levels (V/m RMS) in every individual tissue at 1.5 kW input power (ANSYS® Electromagnetic Suite 18.2.0)

Mesh	Tissue	Avg. E field (V/m RMS)	Mesh	Tissue	Avg. E field (V/m RMS)
1	Air Internal Maxillary Sinus Left	7.7	39	Cuneiform Medial right	0.6
2	Air Internal Maxillary Sinus Right	6.9	40	discC02C03	10.3
3	Arteries	10.5	41	discC03C04	11.6
4	Bladder	28.0	42	discC04C05	14.3
5	C01	14.5	43	discC05C06	16.9
6	C02	13.6	44	discC06C07	19.0
7	C03	14.8	45	discC07T01	20.6
8	C04	18.5	46	discL01L02	13.9
9	C05	21.7	47	discL02L03	11.5
10	C06	26.1	48	discL03L04	8.4
11	C07	29.6	49	discL04L05	4.7
12	Calcaneous left	0.6	50	discL05L06	7.1
13	Calcaneous right	1.1	51	discL06S00	13.8
14	Cartilage1 Left	18.5	52	discT01T02	20.2
15	Cartilage1 Right	19.9	53	discT02T03	17.6
16	Cartilage2 Left	19.7	54	discT03T04	17.9
17	Cartilage2 Right	20.3	55	discT04T05	17.4
18	Cartilage3 Left	21.2	56	discT05T06	15.8
19	Cartilage3 Right	20.9	57	discT06T07	15.2
20	Cartilage4 Left	22.5	58	discT07T08	14.4
21	Cartilage4 Right	21.8	59	discT08T09	13.6
22	Cartilage5 Left	24.3	60	discT09T10	13.2
23	Cartilage5 Right	22.6	61	discT10T11	12.5
24	Cartilage6 Left	36.3	62	discT11T12	13.2
25	Cartilage6 Right	35.9	63	discT12L01	13.4
26	Cerebellum	1.4	64	Eye Left	5.0
27	Clavicle left	55.6	65	Eye Right	5.3
28	Clavicle right	34.9	66	Feet1Phalange left	0.5
29	Coccyx	42.6	67	Feet1Phalange right	0.4
30	CSF OuterShell	3.5	68	Feet2Phalange left	0.4
31	CSF Ventricles	0.4	69	Feet2Phalange right	0.4
32	Cuboid Left	0.9	70	Feet3Phalange left	0.3
33	Cuboid Right	0.6	71	Feet3Phalange right	0.4
34	Cuneiform Intermediate left	1.3	72	Feet4Phalange left	0.4
35	Cuneiform Intermediate right	0.5	73	Feet4Phalange right	0.6
36	Cuneiform Lateral left	1.1	74	Feet5Phalange left	0.4
37	Cuneiform Lateral right	0.4	75	Feet5Phalange right	0.7

Mesh	Tissue	Avg. E field (V/m RMS)	Mesh	Tissue	Avg. E field (V/m RMS)
38	Cuneiform Medial left	1.3	76	Femur Bone Marrow Left	7.1
77	Femur Bone Marrow Right	8.6	117	Humerus right	23.1
78	Femur left	69.3	118	Intestine	20.6
79	Femur right	83.7	119	Jaw lower	10.0
80	Fibula left	5.9	120	Kidney left	29.3
81	Fibula right	5.4	121	Kidney right	27.2
82	Gray Matter Spinal Cord	1.5	122	L01	27.9
83	Hands1 1Phalange left	10.2	123	L02	25.3
84	Hands1 1Phalange right	9.5	124	L03	22.1
85	Hands1 2Phalange left	9.5	125	L04	19.0
86	Hands1 2Phalange right	12.1	126	L05	14.4
87	Hands1 3Phalange left	12.2	127	L06	17.9
88	Hands1 3Phalange right	13.4	128	Liver	29.8
89	Hands2 1Phalange left	7.6	129	Lungs	19.4
90	Hands2 1Phalange right	7.1	130	Median Nerve left	11.6
91	Hands2 2Phalange left	8.1	131	Median Nerve right	13.0
92	Hands2 2Phalange right	8.5	132	Muscle Bicep left	11.9
93	Hands2 3Phalange left	7.1	133	Muscle Bicep right	12.9
94	Hands2 3Phalange right	9.5	134	Muscle Calf left	5.0
95	Hands3 1Phalange left	6.4	135	Muscle Calf right	5.2
96	Hands3 1Phalange right	6.2	136	Muscle Deltoid left	18.7
97	Hands3 2Phalange left	8.1	137	Muscle Deltoid right	19.3
98	Hands3 2Phalange right	8.6	138	Muscle Erector spinae left	26.8
99	Hands3 3Phalange left	9.3	139	Muscle Erector spinae right	26.9
100	Hands3 3Phalange right	11.0	140	Muscle Forearm Extensors left	6.9
101	Hands4 1Phalange left	7.2	141	Muscle Forearm Extensors right	8.6
102	Hands4 1Phalange right	6.9	142	Muscle Forearm Flexors left	6.9
103	Hands4 2Phalange left	10.4	143	Muscle Forearm Flexors right	7.2
104	Hands4 2Phalange right	10.0	144	Muscle Gluteus left	27.9
105	Hands4 3Phalange left	11.1	145	Muscle Gluteus right	27.2
106	Hands4 3Phalange right	10.7	146	Muscle Hamstring left	18.9
107	Hands5 1Phalange left	9.0	147	Muscle Hamstring right	19.1
108	Hands5 1Phalange right	10.3	148	Muscle Latissimus Dorsi left	36.6

(continued)

Table 5.3 (continued)

Mesh	Tissue	Avg. E field (V/m RMS)	Mesh	Tissue	Avg. E field (V/m RMS)
109	Hands5 2Phalange left	11.0	149	Muscle Latissimus Dorsi right	38.5
110	Hands5 2Phalange right	12.6	150	Muscle Neck Combined left	13.6
111	Hands5 3Phalange left	10.3	151	Muscle Neck Combined right	13.4
112	Hands5 3Phalange right	12.1	152	Muscle Obliques left	39.5
113	Heart Muscle	14.2	153	Muscle Obliques right	40.1
114	Hip left	60.0	154	Muscle Pectoralis major left	21.7
115	Hip right	61.4	155	Muscle Pectoralis major right	20.9
116	Humerus left	20.7	156	Muscle Pectoralis minor left	19.2
157	Muscle Pectoralis minor right	18.6	194	Ribs left8	47.3
158	Muscle Pelvic Combined left	25.7	195	Ribs left9	46.1
159	Muscle Pelvic Combined right	25.0	196	Ribs left10	48.5
160	Muscle Psoas left	13.9	197	Ribs left11	51.7
161	Muscle Psoas right	13.9	198	Ribs left12	39.0
162	Muscle Quadriceps left	20.2	199	Ribs right1	29.6
163	Muscle Quadriceps right	19.9	200	Ribs right2	26.2
164	Muscle Rectus Abdominis left bottom	32.3	201	Ribs right3	25.2
165	Muscle Rectus Abdominis left middle	34.9	202	Ribs right4	26.1
166	Muscle Rectus Abdominis left top	39.1	203	Ribs right5	27.2
167	Muscle Rectus Abdominis right bottom	32.5	204	Ribs right6	29.9
168	Muscle Rectus Abdominis right middle	35.4	205	Ribs right7	35.6
169	Muscle Rectus Abdominis right top	38.1	206	Ribs right8	43.2
170	Muscle Sartorius left	18.6	207	Ribs right9	53.9
171	Muscle Sartorius right	17.5	208	Ribs right10	58.9
172	Muscle Tibialis Anterior left	6.2	209	Ribs right11	56.9
173	Muscle Tibialis Anterior right	5.8	210	Ribs right12	40.9
174	Muscle Trapezius left	23.6	211	Sacrum	45.7
175	Muscle Trapezius right	24.0	212	Scapula left	38.2

Mesh	Tissue	Avg. E field (V/m RMS)	Mesh	Tissue	Avg. E field (V/m RMS)
176	Muscle Tricep left	12.0	213	Scapula right	38.8
177	Muscle Tricep right	14.0	214	Skin Shell	27.8
178	Navicular left	1.8	215	Skull	22.8
179	Navicular right	0.7	216	Sphenoid	8.9
180	Patella left	24.3	217	Spleen	33.9
181	Patella right	22.6	218	Sternum	25.2
182	Peripheral Nerve left	17.1	219	Stomach	22.6
183	Peripheral Nerve Right	14.1	220	T01	28.4
184	Pubic Symphysis	32.1	221	T02	27.2
185	Radial Nerve left	14.6	222	T03	27.2
186	Radial Nerve right	12.4	223	T04	26.9
187	Ribs left1	26.4	224	T05	25.7
188	Ribs left2	30.1	225	T06	25.5
189	Ribs left3	26.4	226	T07	26.2
190	Ribs left4	26.7	227	T08	26.4
191	Ribs left5	28.3	228	T09	26.9
192	Ribs left6	31.5	229	T10	26.1
193	Ribs left7	37.3	230	T11	26.4
231	T12	27.5	240	Trabecular upper right	0.9
232	Talus left	1.3	241	Trachea Sinus	12.4
233	Talus right	0.6	242	Ulna Radius left	8.1
234	Tibia left	8.3	243	Ulna Radius right	7.8
235	Tibia right	7.9	244	Uterus	17.3
236	Tongue	5.2	245	Veins lower	12.5
237	Trabecular lower left	0.5	246	Veins upper	12.4
238	Trabecular lower right	0.8	247	White Matter	1.0
239	Trabecular upper left	0.7			

ICNIRP guidelines. As expected, higher field levels are observed closer to the surface; the field gradually decreases toward the center of the body.

Quantitative estimates of the average electric field for every particular tissue obtained via ANSYS Electromagnetic Suite 18.2.0 are given in Table. 5.3. Note the lower electric fields in the intracranial volume. Additionally, we observe higher electric fields in the individual body muscles. It is also interesting to observe that the fields in bone may be quite high, in particular in the femur and pelvic bones.

However, the computed local electric fields may considerably exceed the values reported in Table 5.3, in particular by 1.5–6 times. These peak values are less accurate. One potential source of the numerical error is insufficient resolution of lengthy and time-consuming full-body computations very close to the interfaces where higher fields are usually observed.

5.3.5 SAR Levels

The body-averaged or whole-body (global-body) SAR_{body} is given by averaging the local *SAR* over the entire body volume. In terms of the complex field phasor $\boldsymbol{E}(\boldsymbol{r})$, one has

$$SAR_{\text{body}} = \frac{1}{V_{\text{body}}} \int_{V_{\text{body}}} \frac{\sigma(\boldsymbol{r})}{2\rho(\boldsymbol{r})} \boldsymbol{E}(\boldsymbol{r}) \cdot \boldsymbol{E}(\boldsymbol{r})^{*} \, dV \tag{5.8}$$

Here, $\sigma(\boldsymbol{r})$ is the local tissue conductivity and $\rho(\boldsymbol{r})$ is the local mass density. At full power of 3 kW and positioned at the shoulder landmark, the global-body SAR computed via ANSYS Electromagnetic Suite 18.2.0 is 0.25 W/kg. Thus, the total power dissipation in the body does not exceed 30 W, i.e., 1% of the total power. The same percentage ratio is valid at half input power.

The second critical estimate is SAR_{1g}, which is given by averaging over a contiguous volume with the weight of 1 g,

$$SAR_{1g}(\boldsymbol{r}) = \frac{1}{V_{1g}} \int_{V_{1g}} \frac{\sigma(\boldsymbol{r})}{2\rho(\boldsymbol{r})} \boldsymbol{E}(\boldsymbol{r}) \cdot \boldsymbol{E}(\boldsymbol{r})^{*} \, dV \tag{5.9}$$

The maximum value of SAR_{1g} in the body computed via ANSYS Electromagnetic Suite 18.2.0 at the full power of 3 kW and located at the shoulder landmark is 4.55 W/kg.

Although this last value might appear to be relatively high, it is still within the corresponding SAR limits in MRI machines [95, 96]. In particular, the major applicable MRI safety standard, issued by the International Electrotechnical Commission (IEC) and also accepted by the U.S. Food and Drug Administration, in the normal mode (mode of operation that causes no physiological stress to patients) limits global-body SAR to 2 W/kg, global-head SAR to 3.2 W/kg, local head and torso SAR to 10 W/kg, and local extremity SAR to 20 W/kg [96]. The global SAR limits are intended to ensure a body core temperature of 39 °C or less [95, 96].

5.4 Discussion

5.4.1 Efficacy of Stimulation

The present study establishes safety and potential feasibility of the resonant neurostimulation device. However, its efficacy for treatment of chronic back pain remains largely unknown. Only clinical trials, which would ideally thoroughly investigate both short-term and cumulative effects of the suggested lower-body electromagnetic treatment, could probably answer this question. Our aim is to provide a doctor with the possibility to vary power, resonant frequency, tonic frequency, and electromagnetic pulse envelope to enable the best possible outcome during the anticipated clinical trial.

5.4.2 *Integrated Effect of Stimulation*

The present subthreshold stimulation device will not only affect the PNS of the lower back but also muscles, bones, tendons, and cartilage. Evidence suggests that subthreshold pulsed electromagnetic fields may stimulate osteogenesis in vitro and in vivo [106, 107], improve bone quality in osteoporotic and nonosteoporotic cell-based studies [108, 109], human studies [110–114], animal studies [115–121], and augment bone fracture healing [107, 122–124]. Further evidence suggests that TENS therapy stimulates a change in the biochemical and physiological muscle conditions that may lead to muscle relaxation [125, 126]. Some evidence also suggests that the kHz stimulation of the lower body will increase the vascular endothelial growth factor receptor on circulating hematopoietic stem cells [81], whose local niche (the bone marrow of the pelvis, femur, and sternum [127]) might be well affected by the present stimulation device.

Another extremely interesting effect of the kHz peripheral nerve stimulation observed previously in [65, 80] and implicitly in the present device is a potential for sleep improvement. It is not clear how to describe and account these integrated effects of the stimulation. We will attempt to carefully document and report prior relevant literature findings and the corresponding stimulation conditions for nerve/muscle/bone/marrow, and link them to the present stimulation conditions.

5.4.3 *Operation as an EMAT*

The present electrostimulation device may also operate as an electromagnetic acoustic transducer (EMAT) when a DC current is injected into the tissue via surface electrodes at a specified location. The Lorentz force will excite an ultrasonic field whose frequency is the resonant frequency.

5.4.4 *Variation of Resonant Frequency*

While fine tuning with a low-loss ferromagnetic load is straightforward, it is quite challenging, however, to vary the resonant frequency of a power resonator, which is usually cast in stone, allowing only a narrow tuning range. In order to do this, we have studied three different methods: a bank of electronically controlled switched power capacitors, a bank of fixed capacitors with low-resistance power relay switches, and a mechanically replaceable bolted joints-based fixed-capacitor bank. Such banks need be constructed for each of the 144 rungs of the coil in Fig. 5.6 or Fig. 5.7 in order to vary the resonant or carrier frequency over the band of, say, 10–100 kHz.

Although the first two approaches are fast and elegant, they are unfeasible. The key is the equivalent series resistance (ESR) of switched capacitors and power relay switches. Existing series switches increase ESR by about 10 times or even more. This dramatically lowers the resonator quality factor Q, resulting in about three times lower field values for a given input power. The switched capacitor solution has other serious drawbacks. Each switched capacitor block is much larger than a fixed capacitor, which will result in issues related to physically accommodating all components. On the other hand, for a relay with an exceptionally small contact resistance of 5 mΩ, Q will change by a factor of 0.6 at 100 kHz and 0.3 at 10 kHz. As the relays cycle, their contact resistance may go up significantly, especially if we do not follow the guidelines for minimum switched current (to create an arc that cleans the contacts). Thus, Q could continue dropping with cycling. Therefore, we plan to implement low-ESR mechanically replaceable bolted-joints based fixed-capacitor banks. The frequency-switching operation will take approximately 3 h.

5.5 Conclusion

In this technical study, we described a whole-body noncontact subthreshold electromagnetic stimulation device based on the concept of a familiar MRI RF resonating coil, but at a much lower resonant frequency (100–150 kHz and potentially down to 10 kHz), with a field modulation option (0.5–100 Hz), and with an input power level of up to 3 kW. Its unique features include a relatively high electric field level within the subject's biological tissue due to the resonant effect but at low power dissipation, or SAR level, in the body itself.

We emphasize that in the low-frequency limit and at moderate field levels, SAR rather weakly correlates with the deposited electric field. One reason for this is that SAR is proportional to the field squared, and is thus quite small at moderate and low field levels. A second reason is that the tissue conductivity itself is lower (at 100 kHz, it is twice as low as at 100 MHz for muscle and five times lower for fat [105]).

Due to the large resonator volume and its noncontact nature, the subject may be conveniently located anywhere within the resonating coil over a prolonged period of time at moderate and safe electric field levels. The electric field effect does not depend on a particular body position within the resonator. The field penetration is deep everywhere in the body, including the extremities; muscles, bones, and peripheral tissues are mostly affected. Over a shorter period of time, the electric field levels could be increased to relatively large values with an amplitude of about 1 V/cm.

We envision treatment of chronic pain, and particularly neuropathic pain, as the primary potential clinical application for the device. The device enables whole-body coverage, which could be useful in the treatment of widespread pain conditions, such as painful polyneuropathy or fibromyalgia. In addition, a deeper tissue penetration can be achieved without side-effects caused by high current density in the skin associated with the traditional contact electrodes of TENS. It should be noted that these potential clinical applications are speculative and warrant empirical testing in the future.

Considerable attention has been paid to device safety including both the AC power safety and human exposure to electromagnetic fields. In the former case, we have used inductive coupling, which assures that there is no direct current path from the AC power outlet to the coil. This design enhances overall device safety at any power level, including high-power operation. As with more traditional MRI devices, no large metal objects should be located in the immediate vicinity of the coil.

Human exposure to the electromagnetic field within the coil has been evaluated by performing extensive modeling with two independent numerical methods and with an anatomically realistic multi-tissue human phantom. We have shown that the SAR levels within the body correspond to the safety standards of the International Electrotechnical Commission when the input power level of the amplifier driver does not exceed 3 kW. We have also shown that the electric field levels generally comply with the safety standards of the International Commission on Non-Ionizing Radiation Protection when the input power level of the amplifier driver does not exceed 1.5 kW.

Acknowledgments The authors are thankful to Dr. James O'Rourke, Dr. John McNeill, Ms. Leah Morales, Mr. Brandon Weyant (all from Worcester Polytechnic Institute), MD Irina V. Zhdanova (ClockCoach), and Dr. Aapo Nummenmaa (Massachusetts General Hospital) for useful discussions. Dr. Deng is supported by the Intramural Research Program of the National Institute of Mental Health, NIH.

Appendix A: Derivation of Eq. (5.7) and Coil Q

The corresponding measurement circuit is given in Fig. 5.11.

The derivation of Eq. (5.7) is as follows. The coil, when measured at a rung capacitor, looks like a parallel resonator (assuming the two degenerate modes are decoupled). In a narrow band around the high-Q resonance, the parallel resonator impedance can be approximated as:

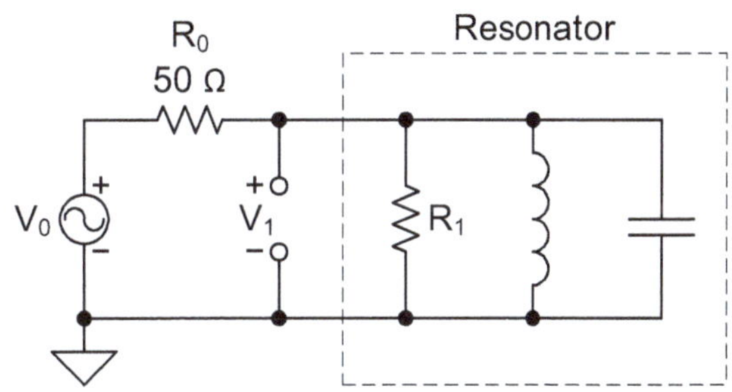

Fig. 5.11 Measurement circuit to evaluate coil Q

$$Z_{\text{res}} = \frac{R_1}{1 + j2Q\dfrac{\Delta f}{f_0}} \tag{5.A1}$$

where f_0 is the resonance frequency and Δf is the deviation from it. Then,

$$\frac{V_1}{V_0} = \frac{Z_{\text{res}}}{R_0 + Z_{\text{res}}} \tag{5.A2}$$

Taking the absolute value and simplifying, one has

$$\left|\frac{V_1}{V_0}\right| = \frac{R_1}{\sqrt{(R_0 + R_1)^2 + 4R_0^2 Q^2 \dfrac{\Delta f^2}{f_0^2}}} \tag{5.A3}$$

At resonance,

$$\frac{V_1}{V_0} = \frac{R_1}{R_0 + R_1} \tag{5.A4}$$

At the V_1 3 dB frequencies,

$$\frac{R_1}{\sqrt{(R_0 + R_1)^2 + 4R_0^2 Q^2 \dfrac{\Delta f_{3\text{dB}}^2}{f_0^2}}} = \frac{R_1}{\sqrt{2}(R_0 + R_1)} \tag{5.A5}$$

Solving for Q yields

$$Q = \frac{f_0}{2\Delta f_{3\text{dB}}}\left(\frac{R_0 + R_1}{R_0}\right) \tag{5.A6}$$

From V_1/V_0 at resonance, we can express

$$1 - \frac{V_1}{V_0} = 1 - \frac{R_1}{R_0 + R_1} = \frac{R_0}{R_0 + R_1} \tag{5.A7}$$

Therefore,

$$Q = \frac{f_0}{2\Delta f_{3\text{dB}}\left(1 - \dfrac{V_1}{V_0}\right)} \tag{5.A8}$$

where V_1 is measured at resonance, and $2\Delta f_{3dB} = f_U - f_L$, with f_U and f_L being the upper and lower 3 dB frequencies for V_1.

Regarding the envelope time constant

$$\tau = \frac{2Q}{\omega} \tag{5.A9}$$

we mention that for $Q = 300$ and $f = 100$ kHz, $\tau = 0.955$ ms. This can be rounded to 1 ms if high precision is not needed. We do not measure Q through this time constant. It is only mentioned as a limiter of modulation bandwidth.

The quality factor is difficult to predict accurately. Estimated Q from ANSYS HFSS-circuit co-simulation typically comes out about 10–30% higher than in the real circuit. How much higher depends on the level of refinement of the FEM model. The important parameter is the actual achieved Q. ANSYS can still be used for rough coil Q optimization, as long as the level of refinement is kept about the same.

References

1. The Americian Academy of Pain Medicine. (2011). *Relieving pain in America: A blueprint for transforming prevention, care, education, and research.* Institute of Medicine. Committee on Advancing Pain Research, Care, and Education. Washington (DC): National Academies Press (US). ISBN 978–0–309-21484-1.
2. The American Academy of Pain Medicine. *AAPM Facts and Figures on Pain.* Online: http://www.painmed.org/patientcenter/facts_on_pain.aspx#america
3. Woolf, C. J. (2011). Central sensitization: implications for the diagnosis and treatment of pain. *Pain, 152*, S2–S15. https://doi.org/10.1016/j.pain.2010.09.030.
4. Staud, R., Craggs, J. G., Robinson, M. E., Perlstein, W. M., & Price, D. D. (2007). Brain activity related to temporal summation of C- fiber evoked pain. *Pain, 129*, 130–142. https://doi.org/10.1016/j.pain.2006.10.010.
5. Nijs, J., Kosek, E., Van Oosterwijck, J., & Meeus, M. (2012). Dysfunctional endogenous analgesia during exercise in patients with chronic pain: to exercise or not to exercise? *Pain Physician, 1*, 205–214.
6. Kuppens, K., Hans, G., Roussel, N., Struyf, F., Fransen, E., Cras, P., Van Wilgen, C. P., & Nijs, J. (2018). Sensory processing and central pain modulation in patients with chronic shoulder pain: A case-control study. *Scandinavian Journal of Medicine & Science in Sports, 28*(3), 1183–1192. https://doi.org/10.1111/sms.12982.
7. Nakamura, M., Nishiwaki, Y., Sumitani, M., Ushida, T., Yamashita, T., Konno, S., Taguchi, T., & Toyama, Y. (2014). Investigation of chronic musculoskeletal pain (third report): with special reference to the importance of neuropathic pain and psychogenic pain. *Journal of Orthopaedic Science, 19*(4), 667–675. https://doi.org/10.1007/s00776-014-0567-6.
8. Dunne, F. J., Getachew, H., Cullenbrooke, F., & Dunne, C. (2018). Pain and pain syndromes. *British Journal of Hospital Medicine (London), 79*(8), 449–453. https://doi.org/10.12968/hmed.2018.79.8.449.
9. Lerman, S. F., Rudich, Z., Brill, S., Shalev, H., & Shahar, G. (2015). Longitudinal associations between depression, anx-iety, pain, and pain-related disability in chronic pain patients. *Psychosomatic Medicine, 77*(3), 333–341. https://doi.org/10.1097/PSY.0000000000000158.
10. Hall-Flavin DK. Is there a link between pain and depression? Can depression cause physical pain? March 11, 2016. Mayo Clinic. Online: https://www.mayoclinic.org/diseases-conditions/depression/expert-answers/pain-and-depression/faq-20057823.

11. Snekkevik, H., Eriksen, H. R., Tangen, T., Chalder, T., & Reme, S. E. (2014). Fatigue and depression in sick-listed chronic low back pain patients. *Pain Medicine, 15*(7), 1163–1170. https://doi.org/10.1111/pme.12435.
12. Freburger, J. K., Holmes, G. M., Agans, R. P., Jackman, A. M., Darter, J. D., Wallace, A. S., Castel, L. D., Kalsbeek, W. D., & Carey, T. S. (2009). The rising prevalence of chronic low back pain. *Archives of Internal Medicine, 169*(3), 251–258. https://doi.org/10.1001/archinternmed.2008.543.
13. National Inst. of Neurological Disorders and Stroke. *Low back pain fact sheet.* 2018-08-07. NIH Publication No. 15–5161.
14. Krath, A., Klüter, T., Stukenberg, M., Zielhardt, P., Gollwitzer, H., Harrasser, N., Hausdorf, J., Ringeisen, M., & Gerdesmeyer, L. (2017). Electromagnetic transduction therapy in non-specific low back pain: A prospective randomised controlled trial. *Journal of Orthopaedics, 14*(3), 410–415. https://doi.org/10.1016/j.jor.2017.06.016.
15. Nayback-Beebe, A. M., Yoder, L. H., Goff, B. J., Arzola, S., & Weidlich, C. (2017). The effect of pulsed electromagnetic frequency therapy on health-related quality of life in military service members with chronic low back pain. *Nursing Outlook, 65*(5S), S26–S33. https://doi.org/10.1016/j.outlook.2017.07.012.
16. Outcalt, S. D., Kroenke, K., Krebs, E. E., Chumbler, N. R., Wu, J., Yu, Z., & Bair, M. J. (2015). Chronic pain and comorbid mental health conditions: Independent associations of posttraumatic stress disorder and depression with pain, disability, and quality of life. *Journal of Behavioral Medicine, 38*(3), 535–543. https://doi.org/10.1007/s10865-015-9628-3.
17. Alsaadi, S. M., McAuley, J. H., Hush, J. M., & Maher, C. G. (2011). Prevalence of sleep disturbance in patients with low back pain. *European Spine Journal, 20*(5), 737–743. https://doi.org/10.1007/s00586-010-1661-x.
18. Iizuka, Y., Iizuka, H., Mieda, T., Tsunoda, D., Sasaki, T., Tajika, T., Yamamoto, A., & Takagishi, K. (2017). Prevalence of chronic nonspecific low back pain and its associated factors among middle-aged and elderly people: an analysis based on data from a musculoskeletal examination in Japan. *Asian Spine Journal, 11*(6), 989–997. https://doi.org/10.4184/asj.2017.11.6.989.
19. Dworkin, R. H., O'Connor, A. B., Backonja, M., Farrar, J. T., Finnerup, N. B., Jensen, T. S., Kalso, E. A., Loeser, J. D., Miaskowski, C., Nurmikko, T. J., Portenoy, R. K., Rice, A. S., Stacey, B. R., Treede, R. D., Turk, D. C., & Wallace, M. S. (2007). Pharmacologic management of neuropathic pain: Evidence-based recommendations. *Pain, 132*(3), 237–251. https://doi.org/10.1016/j.pain.2007.08.033.
20. Steele, A. (2014). Opioid use and depression in chronic pelvic pain. *Obstetrics and Gynecology Clinics of North America, 41*(3), 491–501. https://doi.org/10.1016/j.ogc.2014.04.005.
21. Stanos, S. P., & Galluzzi, K. E. (2013). Topical therapies in the management of chronic pain. *Postgraduate Medicine, 125*(4 Suppl 1), 25–33. https://doi.org/10.1080/00325481.2013.1110567111.
22. Johnson MI, Claydon LS, Herbison GP, Jones G, and Paley CA. Transcutaneous electrical nerve stimulation (TENS) for fibromyalgia in adults. *Cochrane Database of Systematic Reviews* 2017, Issue 10. Art. No.:CD012172. doi: https://doi.org/10.1002/14651858.CD012172.pub2.
23. Gibson W, Wand BM, O'Connell NE. Transcutaneous electrical nerve stimulation (TENS) for neuropathic pain in adults. *Cochrane Database of Systematic Reviews* 2017, Issue 9. Art. No.: CD011976. doi: https://doi.org/10.1002/14651858.CD011976.pub2.
24. Koca, I., Boyaci, A., Tutoglu, A., Ucar, M., & Kocaturk, O. (2014). Assessment of the effectiveness of interferential current therapy and TENS in the management of carpal tunnel syndrome: a randomized controlled study. *Rheumatology International, 34*, 1639–1645. https://doi.org/10.1007/s00296-014-3005-3.
25. Brosseau, L., Judd, M. G., Marchand, S., Robinson, V. A., Tugwell, P., Wells, G., & Younge, K. (2003). Transcutaneous electrical nerve stimulation (TENS) for the treatment of rheumatoid arthritis in the hand. *Cochrane Database of Systematic Reviews*, (3). https://doi.org/10.1002/14651858.CD004287.pub2.

26. Khadilkar, A., Odebiyi, D. O., Brosseau, L., & Wells, G. A. (2008). Transcutaneous electrical nerve stimulation (TENS) versus placebo for chronic low-back pain. *Cochrane Database of Systematic Reviews*, (4). https://doi.org/10.1002/14651858.CD003008.pub3.
27. Dowswell, C., Bedwell, C., Lavender, T., & Neilson, J. P. (2009). Transcutaneous electrical nerve stimulation (TENS) for pain management in labour. *Cochrane Database of Systematic Reviews*, (2). https://doi.org/10.1002/14651858.CD007214.pub2.
28. Johnson, M. I., Mulvey, M. R., & Bagnall, A. M. (2015). Transcutaneous electrical nerve stimulation (TENS) for phantom pain and stump pain following amputation in adults. *Cochrane Database of Systematic Reviews*, (8). https://doi.org/10.1002/14651858.CD007264.pub3.
29. Makarov, S. N., Bogdanov, G., Makarov, V. S., Noetscher, G. M., & Deng, Z-D. (2018). A whole body non-contact electrical stimulation device with variable parameters. US Patent Application #15868038 of Jan. 11th 2018. Pub # 20180196113; Pub Date July 12th 2018. Revision Oct 2018.
30. Makarov, S. N., Bogdanov, G., Noetscher, G. M., Allpleyard, W., Ludwig, R., Joutsa, U. T., & Deng, Z.-D. (2018). Design and analysis of a whole body non-contact electromagnetic stimulation device with field modulation at low and medium power levels. *bioRxiv, 416065*. https://doi.org/10.1101/416065.
31. InTENSity™ Twin Stim® III Datasheet. Current Solutions, LLC, Austin, TX. Online: https://www.tenspros.com/assets/images/manuals/InTENSity-Twin-Stim-III-di3717-Manual.pdf
32. TENS (transcutaneous electrical nerve stimulation). *National Health Service (NHS) in England*. 08/10/2018. Online: https://www.nhs.uk/conditions/transcutaneous-electrical-nerve-stimulation-tens/
33. Schmerzen lindern ohne Pillen: Elektrotherapie. NDR.de - Ratgeber – Gesundheit. Online: https://www.ndr.de/ratgeber/gesundheit/index.html
34. Leung, A., Fallah, A., & Shukla, S. (2014). Transcutaneous magnetic stimulation (TMS) in alleviating post-traumatic peripheral neuropathic pain states: a case series. *Pain Medicine, 15*(7), 1196–1199. https://doi.org/10.1111/pme.12426.
35. Leung, A., Shukla, S., Lee, J., Metzger-Smith, V., He, Y., & Chen, J. (2015). Golshan. Effect of low frequency transcutaneous magnetic stimulation on sensory and motor transmission. *Bioelectromagnetics, 36*(6), 410–419. https://doi.org/10.1002/bem.21921.
36. O'Reardon, J. P., Solvason, H. B., Janicak, P. G., Sampson, S., Isenberg, K. E., Nahas, Z., McDonald, W. M., Avery, D., Fitzgerald, P. B., Loo, C., Demitrack, M. A., George, M. S., & Sackeim, H. A. (2007). Efficacy and safety of transcranial magnetic stimulation in the acute treatment of major depression: a multisite randomized controlled trial. *Biological Psychiatry, 62*(11), 1208–1216. https://doi.org/10.1016/j.biopsych.2007.01.018.
37. Bajbouj, M., Merkl, A., Schlaepfer, T. E., Frick, C., Zobel, A., Maier, W., O'Keane, V., Corcoran, C., Adolfsson, R., Trimble, M., Rau, H., Hoff, H. J., Padberg, F., Müller-Siecheneder, F., Audenaert, K., van den Abbeele, D., Matthews, K., Christmas, D., Eljamel, S., & Heuser, I. (2010). Two-year outcome of vagus nerve stimulation in treatment-resistant depression. *Journal of Clinical Psychopharmacology, 30*(3), 273–281. https://doi.org/10.1097/JCP.0b013e3181db8831.
38. Trevino, K., McClintock, S. M., & Husain, M. M. (2010). A review of continuation electroconvulsive therapy: application, safety, and efficacy. *The Journal of ECT, 26*(3), 186–195. https://doi.org/10.1097/YCT.0b013e3181efa1b2.
39. Chakraborty, D., Truong, D. Q., Bikson, M., & Kaphzan, H. (2018). Neuromodulation of axon terminals. *Cerebral Cortex, 28*(8), 2786–2794. https://doi.org/10.1093/cercor/bhx158.
40. Jefferys, J. G. (1981). Influence of electric fields on the excitability of granule cells in guinea-pig hippocampal slices. *The Journal of Physiology, 319*, 143–152.
41. Bikson, M., Inoue, M., Akiyama, H., Deans, J. K., Fox, J. E., Miyakawa, H., & Jefferys, J. G. R. (2004). Effects of uniform extracellular DC electric fields on excitability in rat hippocampal slices in vitro. *The Journal of Physiology, 557*.(Pt 1, 175–190. https://doi.org/10.1113/jphysiol.2003.055772.
42. Rahman, A., Reato, D., Arlotti, M., Gasca, F., Datta, A., Parra, L. C., & Bikson, M. (2013). 2013. Cellular effects of acute direct current stimulation: somatic and synaptic

terminal effects. *The Journal of Physiology, 591*(10), 2563–2578. https://doi.org/10.1113/jphysiol.2012.247171.

43. Kerezoudis, P., Grewal, S. S., Stead, M., Lundstrom, B. N., Britton, J. W., Shin, C., Cascino, G. D., Brinkmann, B. H., Worrell, G. A., & Van Gompel, J. J. (2018). Chronic subthreshold cortical stimulation for adult drug-resistant focal epilepsy: safety, feasibility, and technique. *Journal of Neurosurgery, 129*(2), 533–543. https://doi.org/10.3171/2017.5.JNS163134.
44. Stagg, C. J., Antal, A., & Nitsche, M. A. (2018). Physiology of transcranial direct current stimulation. *The Journal of ECT, 34*(3), 144–152. https://doi.org/10.1097/YCT.0000000000000510.
45. Aspart, F., Remme, M. W. H., & Obermayer, K. (2018). Differential polarization of cortical pyramidal neuron dendrites through weak extracellular fields. *PLoS Computational Biology, 14*(5), e1006124. https://doi.org/10.1371/journal.pcbi.1006124.
46. Toloza, E. H. S., Negahbani, E., & Fröhlich, F. (2018). LH interacts with somato-dendritic structure to determine frequency response to weak alternating electric field stimulation. *Journal of Neurophysiology, 119*(3), 1029–1036. https://doi.org/10.1152/jn.00541.2017.
47. Khatoun, A., Asamoah, B., & Mc Laughlin, M. (2017). Simultaneously excitatory and inhibitory effects of transcranial alternating current stimulation revealed using selective pulse-train stimulation in the rat motor cortex. *The Journal of Neuroscience, 37*(39), 9389–9402. https://doi.org/10.1523/JNEUROSCI.1390-17.2017.
48. Fava, M., Freeman, M. P., Flynn, M., Hoeppner, B. B., Shelton, R., Iosifescu, D. V., Murrough, J. W., Mischoulon, D., Cusin, C., Rapaport, M., Dunlop, B. W., Trivedi, M. H., Jha, M., Sanacora, G., Hermes, G., & Papakostas, G. I. (2018). Double-blind, proof-of-concept (POC) trial of low-field magnetic stimulation (LFMS) augmentation of antidepressant therapy in treatment-resistant depression (TRD). *Brain Stimulation, 11*(1), 75–84. https://doi.org/10.1016/j.brs.2017.09.010.
49. Rohan, M., Parow, A., Stoll, A. L., Demopulos, C., Friedman, S., Dager, S., Hennen, J., Cohen, B. M., & Renshaw, P. F. (2004). Low-field magnetic stimulation in bipolar depression using an MRI-based stimulator. *The American Journal of Psychiatry, 161*(1), 93–98. https://doi.org/10.1176/appi.ajp.161.1.93.
50. Rohan, M. L., Yamamoto, R. T., Ravichandran, C. T., Cayetano, K. R., Morales, O. G., Olson, D. P., Vitaliano, G., Paul, S. M., & Cohen, B. M. (2014). Rapid mood-elevating effects of low field magnetic stimulation in depression. *Biological Psychiatry, 76*(3), 186–193. https://doi.org/10.1016/j.biopsych.2013.10.024.
51. Carlezon, W. A., Rohan, M. L., Mague, S. D., Meloni, E. G., Parsegian, A., Cayetano, K., Tomasiewicz, H. C., Rouse, E. D., Cohen, B. M., & Renshaw, P. F. (2005). Antidepressant-like effects of cranial stimulation within a low-energy magnetic field in rats. *Biological Psychiatry, 57*, 571–576. https://doi.org/10.1016/j.biopsych.2004.12.011.
52. Aksoz, E., Aksoz, T., Bilge, S. S., Ilkaya, F., Celik, S., & Diren, H. B. (2008). Antidepressant-like effects of echo-planar magnetic resonance imaging in mice determined using the forced swimming test. *Brain Research, 1236*, 191–196. https://doi.org/10.1016/j.brainres.2008.08.011.
53. Rokni-Yazdi, H., Sotoudeh, H., Akhondzadeh, S., Sotoudeh, E., Asadi, H., & Shakiba, M. (2007). Antidepressant-like effect of magnetic resonance imaging-based stimulation in mice. *Progress in Neuro-Psychopharmacology & Biological Psychiatry, 31*, 503–509. https://doi.org/10.1016/j.pnpbp.2006.11.021.
54. Volkow, N. D., Tomasi, D., Wang, G. J., Fowler, J. S., Telang, F., Wang, R., Alexoff, D., Logan, J., Wong, C., Pradhan, K., Caparelli, E. C., Ma, Y., & Jayne, M. (2010). Effects of low-field magnetic stimulation on brain metabolism. *NeuroImage, 51*, 623–628. https://doi.org/10.1016/j.neuroimage.2010.02.015.
55. Shamji, M. F., De Vos, C., & Sharan, A. (2017). The advancing role of neuromodulation for the management of chronic treatment-refractory pain. *Neurosurgery, 80*(3S), S108–S113. https://doi.org/10.1093/neuros/nyw047.
56. Pelot, N. A., Behrend, C. E., & Grill, W. M. (2017). Modeling the response of small myelinated axons in a compound nerve to kilohertz frequency signals. *Journal of Neural Engineering, 14*(4), 046022. https://doi.org/10.1088/1741-2552/aa6a5f.

57. Provenzano, D. A., Rebman, J., Kuhel, C., Trenz, H., & Kilgore, J. (2017). The efficacy of high-density spinal cord stimulation among trial, implant, and conversion patients: A retrospective case series. *Neuromodulation, 20*(7), 654–660. https://doi.org/10.1111/ner.12612.
58. De Jaeger, M., van Hooff, R. J., Goudman, L., Valenzuela Espinoza, A., Brouns, R., Puylaert, M., Duyvendak, W., & Moens, M. (2017). High-density in spinal cord stimulation: virtual expert registry (DISCOVER): Study protocol for a prospective observational trial. *Anesthesiology and Pain Medicine, 7*(3), e13640. https://doi.org/10.5812/aapm.13640.
59. Al-Kaisy, A., Palmisani, S., Pang, D., Sanderson, K., Wesley, S., Tan, Y., McCammon, S., & Trescott, A. (2018). Prospective, randomized, sham-control, double blind, crossover trial of subthreshold spinal cord stimulation at various kilohertz frequencies in subjects suffering from failed back surgery syndrome (SCS frequency study). *Neuromodulation, 21*(5), 457–465. https://doi.org/10.1111/ner.12771.
60. Zhang, L., Lu, Y., Sun, J., Zhou, X., & Tang, B. (2016). Subthreshold vagal stimulation suppresses ventricular arrhythmia and inflammatory response in a canine model of acute cardiac ischaemia and reperfusion. *Experimental Physiology, 101*(1), 41–49. https://doi.org/10.1113/EP085518.
61. Slotty, P. J., Bara, G., Kowatz, L., Gendolla, A., Wille, C., Schu, S., & Vesper, J. (2015). Occipital nerve stimulation for chronic migraine: a randomized trial on subthreshold stimulation. *Cephalalgia, 35*(1), 73–78. https://doi.org/10.1177/0333102414534082.
62. Zheng, Y., & Hu, X. (2018). Reduced muscle fatigue using kilohertz-frequency subthreshold stimulation of the proximal nerve. *Journal of Neural Engineering*. https://doi.org/10.1088/1741-2552/aadecc.
63. Hotta, H., Onda, A., Suzuki, H., Milliken, P., & Sridhar, A. (2017). Modulation of calcitonin, parathyroid hormone, and thyroid hormone secretion by electrical stimulation of sympathetic and parasympathetic nerves in anesthetized rats. *Frontiers in Neuroscience, 11*, 375. https://doi.org/10.3389/fnins.2017.00375.
64. Vargas Luna, J. L., Mayr, W., & Cortés-Ramirez, J. A. (2018). Sub-threshold depolarizing prepulses can enhance the efficiency of biphasic stimuli in transcutaneous neuromuscular electrical stimulation. *Medical & Biological Engineering & Computing*. https://doi.org/10.1007/s11517-018-1851-y.
65. Gulyaev, Y. V., Bugaev, A. S., Indursky, P. A., Shakhnarovich, V. M., & Dementienko, V. V. (2017). Improvement of the night sleep quality by electrocutaneous subthreshold stimulation synchronized with the slow wave sleep. *Doklady Biological Sciences, 474*(1), 132–134. https://doi.org/10.1134/S0012496617030139.
66. Makarov, S. N., Noetscher, G. M., & Nazarian, A. (2015). *A low-frequency electromagnetic modeling for electrical and biological systems using MATLAB*. New York: Wiley.
67. Hayes, C. E., Edelstein, W. A., Schenck, J. F., Mueller, O. M., & Eash, M. (1985). An efficient, highly homogeneous radiofrequency coil for whole-body NMR imaging at 1.5 T. *Journal of Magnetic Resonance, 63*, 622–628. https://doi.org/10.1016/0022-2364(85)90257-4.
68. Hayes, C. E. (2009). The development of the birdcage resonator: A historical perspective. *NMR in Biomedicine, 22*, 908–918. https://doi.org/10.1002/nbm.1431.
69. Lempka, S. F., McIntyre, C. C., Kilgore, K. L., & Machado, A. G. (2015 Jun). Computational analysis of kilohertz frequency spinal cord stimulation for chronic pain management. *Anesthesiology, 122*(6), 1362–1376. https://doi.org/10.1097/ALN.0000000000000649.
70. Reddy, C. G., Dalm, B. D., Flouty, O. E., Gillies, G. T., Howard, M. A., & Brennan, T. J. (2016). Comparison of conventional and kilohertz frequency epidural stimulation in patients undergoing trialing for spinal cord stimulation: clinical considerations. *World Neurosurgery, 88*, 586–591. https://doi.org/10.1016/j.wneu.2015.10.088.
71. Shechter, R., Yang, F., Xu, Q., Cheong, Y. K., He, S. Q., Sdrulla, A., Carteret, A. F., Wacnik, P. W., Dong, X., Meyer, R. A., Raja, S. N., & Guan, Y. (2013). Conventional and kilohertz-frequency spinal cord stimulation produces intensity- and frequency-dependent inhibition of mechanical hypersensitivity in a rat model of neuropathic pain. *Anesthesiology, 119*(2), 422–432. https://doi.org/10.1097/ALN.0b013e31829bd9e2.
72. Thomson, S. J., Tavakkolizadeh, M., Love-Jones, S., Patel, N. K., Gu, J. W., Bains, A., Doan, Q., & Moffitt, M. (2018). Effects of rate on analgesia in kilohertz frequency spinal cord stim-

ulation: Results of the proco randomized controlled trial. *Neuromodulation, 21*(1), 67–76. https://doi.org/10.1111/ner.12746.

73. Billet, B., Wynendaele, R., & Vanquathem, N. E. (2017). Wireless neuromodulation for chronic back pain: delivery of high frequency dorsal root ganglion stimulation by a minimally invasive technique. *Case Reports in Medicine, 2017*, 4203271. https://doi.org/10.1155/2017/4203271.
74. Huygen, F., Liem, L., Cusack, W., & Kramer, J. (2018). Stimulation of the L2-L3 dorsal root ganglia induces effective pain relief in the low back. *Pain Practice, 18*(2), 205–213. https://doi.org/10.1111/papr.12591.
75. Simopoulos, T., Yong, R. J., & Gill, J. S. (2018). Treatment of chronic refractory neuropathic pelvic pain with high-frequency 10-kilohertz spinal cord stimulation. *Pain Practice, 18*(6), 805–809. https://doi.org/10.1111/papr.12656.
76. Haider, N., Ligham, D., Quave, B., Harum, K. E., Garcia, E. A., Gilmore, C. A., Miller, N., Moore, G. A., Bains, A., Lechleiter, K., & Jain, R. (2018). Spinal cord stimulation (SCS) trial outcomes after conversion to a multiple waveform SCS system. *Neuromodulation, 21*(5), 504–507. https://doi.org/10.1111/ner.12783.
77. Amirdelfan, K., Yu, C., Doust, M. W., Gliner, B. E., Morgan, D. M., Kapural, L., Vallejo, R., Sitzman, B. T., Yearwood, T. L., Bundschu, R., Yang, T., Benyamin, R., Burgher, A. H., Brooks, E. S., Powell, A. A., & Subbaroyan, J. (2018). Long-term quality of life improvement for chronic intractable back and leg pain patients using spinal cord stimulation: 12-month results from the SENZA-RCT. *Quality of Life Research, 27*(8), 2035–2044. https://doi.org/10.1007/s11136-018-1890-8.
78. Kim, Y., Cho, H. J., & Park, H. S. (2018). Technical development of transcutaneous electrical nerve inhibition using medium-frequency alternating current. *Journal of Neuroengineering and Rehabilitation, 15*(1), 80. https://doi.org/10.1186/s12984-018-0421-8.
79. Reichstein, L., Labrenz, S., Ziegler, D., & Martin, S. (2005). Effective treatment of symptomatic diabetic polyneuropathy by high-frequency external muscle stimulation. *Diabetologia, 48*(5), 824–828. https://doi.org/10.1007/s00125-005-1728-0.
80. Humpert, P. M., Morcos, M., Oikonomou, D., Schaefer, K., Hamann, A., Bierhaus, A., Schilling, T., & Nawroth, P. P. (2009). External electric muscle stimulation improves burning sensations and sleeping disturbances in patients with type 2 diabetes and symptomatic neuropathy. *Pain Medicine, 10*(2), 413–419. https://doi.org/10.1111/j.1526-4637.2008.00557.x.
81. Hidmark, A., Spanidis, I., Fleming, T. H., Volk, N., Eckstein, V., Groener, J. B., Kopf, S., Nawroth, P. P., & Oikonomou, D. (2017). Electrical muscle stimulation induces an increase of VEGFR2 on circulating hematopoietic stem cells in patients with diabetes. *Clinical Therapeutics, 39*(6), 1132–1144.e2. https://doi.org/10.1016/j.clinthera.2017.05.340.
82. The HiToP® Models 184–191: *Polyneuropathy high frequency stimulator*. gbo Medizintechnik AG, Rimbach, Germany, Online: https://www.gbo-med.de/pnp.html
83. Borsboom, H. M., Claasen-Vujcic, T., Gaykema, H. J. G., & Mehlkopf, T. (1997). Low-frequency quadrature mode birdcage resonator. *Magma, 5*(1), 33–37.
84. Claasen-Vujcic, T., Borsboom, H. M., Gaykema, H. J. G., & Mehlkopf, T. (1996). Transverse low-field RF coils in MRI transverse low-field RF coils in MRI. *Magnetic Resonance in Medicine, 36*(1), 111–116.
85. Galante, A., Sinibaldi, R., Conti, A., De Luca, C., Catallo, N., Sebastiani, P., Pizzella, V., Romani, G. L., Sotgiu, A., & Della Penna, S. (2015). Fast room temperature very low field-magnetic resonance imaging system compatible with magnetoencephalography environment. *PLoS One, 10*(12), e0142701. https://doi.org/10.1371/journal.pone.0142701.
86. Savukov, I., Karaulanov, T., Castro, A., Volegov, P., Matlashov, A., Urbatis, A., Gomez, J., & Espy, M. (2011). Non-cryogenic anatomical imaging in ultra-low field regime: Hand MRI demonstration. *Journal of Magnetic Resonance, 211*(2), 101–108. https://doi.org/10.1016/j.jmr.2011.05.011.
87. Tsai, L. L., Mair, R. W., Rosen, M. S., Patz, S., & Walsworth, R. L. (2008). An open-access, very-low-field MRI system for posture-dependent 3He human lung imaging. *Journal of Magnetic Resonance, 193*(2), 274–285. https://doi.org/10.1016/j.jmr.2008.05.016.

88. Kuzmin, V. V., Bidinosti, C. P., Hayden, M. E., & Nacher, P. J. (2015). An improved shielded RF transmit coil for low-frequency NMR and MRI. *Journal of Magnetic Resonance, 256*, 70–76. https://doi.org/10.1016/j.jmr.2015.05.001.
89. Makarov, S., Noetscher, G. M., Raij, T., & Nummenmaa, A. (2018). A quasi-static boundary element approach with fast multipole acceleration for high-resolution bioelectromagnetic models. *IEEE Transactions on Biomedical Engineering*. https://doi.org/10.1109/TBME.2018.2813261.
90. Smythe, W. R. (1950). *Static and dynamic electricity* (2nd ed.). New York: McGraw-Hill.
91. International Commission on Non-Ionizing Radiation Protection. (2010). *Guidelines for limiting exposure to time-varying electric and magnetic fields (1 Hz to 100 kHz)*. Health Physics Society. https://doi.org/10.1097/HP.0b013e3181f06c86.
92. Zhang, B., Yen, Y. F., Chronik, B. A., McKinnon, G. C., Schaefer, D. J., & Rutt, B. K. (2003). Peripheral nerve stimulation properties of head and body gradient coils of various sizes. *Magnetic Resonance in Medicine, 50*(1), 50–58. https://doi.org/10.1002/mrm.10508.
93. Davids, M., Guérin, B., Schad, L. R., & Wald, L. L. (2017). Predicting magneto-stimulation thresholds in the peripheral nervous system using realistic body models. *Scientific Reports, 7*, 5316. https://doi.org/10.1038/s41598-017-05493-9.
94. Davids, M., Guérin, B., Vom Endt, A., Schad, L. R., & Wald, L. L. (2018). Prediction of peripheral nerve stimulation thresholds of MRI gradient coils using coupled electromagnetic and neurodynamic simulations. *Magnetic Resonance in Medicine*. https://doi.org/10.1002/mrm.27382.
95. International Electrotechnical Commission. (2010). International standard, *Medical equipment – IEC 60601-2-33: Particular requirements for the safety of magnetic resonance equipment*, 3rd Ed., Amendment 1 (2013), Amendment 2 (2015). Online: https://webstore.iec.ch/publication/22705
96. Homann, H. (2012). SAR prediction and SAR management for parallel transmit MRI. *Karlsruhe Translations on Biomedical Engineering, 16*, 1–124.
97. Makarov, S. N., Noetscher, G. M., Yanamadala, J., Piazza, M. W., Louie, S., Prokop, A., Nazarian, A., & Nummenmaa, A. (2017). Virtual human models for electromagnetic studies and their applications. *IEEE Reviews in Biomedical Engineering, 10*, 95–121. https://doi.org/10.1109/RBME.2017.2722420.
98. Makarov, S. N., Yanamadala, J., Piazza, M. W., Helderman, A. M., Thang, N. S., Burnham, E. H., & Pascual-Leone, A. (2016). Preliminary upper estimate of peak currents in transcranial magnetic stimulation at distant locations from a TMS coil. *IEEE Transactions on Biomedical Engineering, 63*(9), 1944–1955. https://doi.org/10.1109/TBME.2015.2507572.
99. Kozlov, M., Tankaria, H., Noetscher, G. M., & Makarov, S. N. (2017). Comparative analysis of different versions of a human model located inside a 1.5T MRI whole body RF coil. *Conference Proceedings: Annual International Conference of the IEEE Engineering in Medicine and Biology Society, 2017*, 1477–1480. https://doi.org/10.1109/EMBC.2017.8037114.
100. Tran, A. L., & Makarov, S. N. (2017). Degree of RF MRI coil detuning for an anatomically realistic respiratory cycle modeled with the finite element method. *Conference Proceedings: Annual International Conference of the IEEE Engineering in Medicine and Biology Society, 2017*, 1405–1408. https://doi.org/10.1109/EMBC.2017.8037096.
101. Noetscher, G. M., Yanamadala, J., Tankaria, H., Louie, S., Prokop, A., Nazarian, A., & Makarov, S. N. (2016). Computational human model VHP-FEMALE derived from datasets of the National Library of Medicine. *Conference Proceedings: Annual International Conference of the IEEE Engineering in Medicine and Biology Society, 2016*, 3350–3353. https://doi.org/10.1109/EMBC.2016.7591445.
102. Tankaria, H., Jackson, X. J., Borwankar, R., Srichandhru, G. N., Le Tran, A., Yanamadala, J., Noetscher, G. M., Nazarian, A., Louie, S., & Makarov, S. N. (2016). VHP-Female full-body human CAD model for cross-platform FEM simulations: recent development and validations. *Conference Proceedings: Annual International Conference of the IEEE Engineering in Medicine and Biology Society, 2016*, 2232–2235. https://doi.org/10.1109/EMBC.2016.7591173.

103. Yanamadala, J., Noetscher, G. M., Rathi, V. K., Maliye, S., Win, H. A., Tran, A. L., Jackson, X. J., Htet, A. T., Kozlov, M., Nazarian, A., Louie, S., & Makarov, S. N. (2015). New VHP-Female v. 2.0 full-body computational phantom and its performance metrics using FEM simulator ANSYS HFSS. *Conference Proceedings: Annual International Conference of the IEEE Engineering in Medicine and Biology Society, 2015*, 3237–3241. https://doi.org/10.1109/EMBC.2015.7319082.
104. U.S. National Library of Medicine. *The Visible Human Project®* Online: http://www.nlm.nih.gov/research/visible/visible_human.html
105. Hasgall PA, Neufeld E, Gosselin MC, Klingenböck A, Kuster N, Klingenbock A, et al. 2015 IT'IS Database for thermal and electromagnetic parameters of biological tissues. Online: www.itis.ethz.ch/database.
106. Tsai, M. T., Li, W. J., Tuan, R. S., & Chang, W. H. (2009). Modulation of osteogenesis in human mesenchymal stem cells by specific pulsed electromagnetic field stimulation. *Journal of Orthopaedic Research, 27*, 1169–1174. https://doi.org/10.1002/jor.20862.
107. Assiotis, A., Chalidis, B. E., & Sachinis, N. P. (2012 Jun 8). Pulsed electromagnetic fields for the treatment of tibial delayed unions and nonunions. A prospective clinical study and review of the literature. *Journal of Orthopaedic Surgery and Research, 7*, 24. https://doi.org/10.1186/1749-799X-7-24.
108. Wu, S., Yu, Q., Sun, Y., & Tian, J. (2018). Synergistic effect of a LPEMF and SPIONs on BMMSC proliferation, directional migration, and osteoblastogenesis. *American Journal of Translational Research, 10*(5), 1431–1443.
109. Ehnert, S., Fentz, A. K., Schreiner, A., Birk, J., Wilbrand, B., Ziegler, P., Reumann, M. K., Wang, H., Falldorf, K., & Nussler, A. K. (2017). Extremely low frequency pulsed electromagnetic fields cause antioxidative defense mechanisms in human osteoblasts via induction of O2- and H2O2. *Scientific Reports, 7*(1), 14544. https://doi.org/10.1038/s41598-017-14983-9.
110. Li, S., Jiang, H., Wang, B., Gu, M., Bi, X., Yin, Y., & Wang, Y. (2018). Magnetic resonance spectroscopy for evaluating the effect of pulsed electromagnetic fields on marrow adiposity in postmenopausal women with osteopenia. *Journal of Computer Assisted Tomography*. https://doi.org/10.1097/RCT.0000000000000757.
111. Ashraf, A. M., Abdelaal, P. D., Mona, M. T., Doaa, I. A., & Amira, H. D. (2017). Effect of pulsed electromagnetic therapy versus low-level laser therapy on bone mineral density in the elderly with primary osteoporosis: a randomized, controlled trial. *Bulletin of Faculty of Physical Therapy, 22*(1), 34–39.
112. Yuan, J., Xin, F., & Jiang, W. (2018). Underlying signaling pathways and therapeutic applications of pulsed electromagnetic fields in bone repair. *Cellular Physiology and Biochemistry, 46*(4), 1581–1594. https://doi.org/10.1159/000489206.
113. Woods, B., Manca, A., Weatherly, H., Saramago, P., Sideris, E., Giannopoulou, C., Rice, S., Corbett, M., Vickers, A., Bowes, M., MacPherson, H., & Sculpher, M. (2017). Cost-effectiveness of adjunct non-pharmacological interventions for osteoarthritis of the knee. *PLoS One, 12*(3), e0172749. https://doi.org/10.1371/journal.pone.0172749.
114. Bilgin, H. M., Çelik, F., Gem, M., Akpolat, V., Yıldız, İ., Ekinci, A., Özerdem, M. S., & Tunik, S. (2017). Effects of local vibration and pulsed electromagnetic field on bone fracture: A comparative study. *Bioelectromagnetics, 38*(5), 339–348. https://doi.org/10.1002/bem.22043.
115. Sert, C., Mustafa, D., Düz, M. Z., Akşen, F., & Kaya, A. (2002). The preventive effect on bone loss of 50-Hz, 1-mT electromagnetic field in ovariectomized rats. *Journal of Bone and Mineral Metabolism, 20*, 345–349. https://doi.org/10.1007/s007740200050.
116. Chang, K., & Chang, W. H. S. (2003). Pulsed electromagnetic fields prevent osteoporosis in an ovariectomized female rat model: A prostaglandin E2-associated process. *Bioelectromagnetics, 24*, 189–198. https://doi.org/10.1002/bem.10078.
117. Jing, D., Cai, J., Shen, G., Huang, J., Li, F., Li, J., Lu, L., Luo, E., & Xu, Q. (2011). The preventive effects of pulsed electromagnetic fields on diabetic bone loss in streptozotocin-treated rats. *Osteoporosis International, 22*, 1885–1895. https://doi.org/10.1007/s00198-010-1447-3.

118. Jing, D., Cai, J., Wu, Y., Shen, G., Li, F., Xu, Q., Xie, K., Tang, C., Liu, J., Guo, W., Wu, X., Jiang, M., & Luo, E. (2014). Pulsed electromagnetic fields partially preserve bone mass, microarchitecture, and strength by promoting bone formation in hindlimb-suspended rats. *Journal of Bone and Mineral Research, 29*, 2250–2261. https://doi.org/10.1002/jbmr.2260.
119. Lei, T., Li, F., Liang, Z., Tang, C., Xie, K., Wang, P., Dong, X., Shan, S., Liu, J., Xu, Q., Luo, E., & Shen, G. Effects of four kinds of electromagnetic fields (EMF) with different frequency spectrum bands on ovariectomized osteoporosis in mice. *Scientific Reports, 7*(1), 553. https://doi.org/10.1038/s41598-017-00668-w.
120. Li, J., Zeng, Z., Zhao, Y., Jing, D., Tang, C., Ding, Y., & Feng, X. Effects of low-intensity pulsed electromagnetic fields on bone microarchitecture, mechanical strength and bone turnover in type 2 diabetic db/db mice. *Scientific Reports, 7*, 10834. https://doi.org/10.1038/s41598-017-11090-7.
121. Aksoy, M. Ç., Topal, O., Özkavak, H. V., Doğuç, D. K., Ilhan, A. A., & Çömlekçi, S. (2017). Effects of pulsed electromagnetic field on mineral density, biomechanical properties, and metabolism of bone tissue in heparin-induced osteoporosis in male rats. *Biomedical Research, 28*(6), 2724–2729.
122. Andrew, C., Bassett, C. A., Pawluk, R. J., & Pilla, A. A. (1974). Augmentation of bone repair by inductively coupled electromagnetic fields. *Science, 184*, 575–577.
123. Bassett, C. A., Mitchell, S. N., & Gasto, S. R. (1982). Pulsing electromagnetic field treatment in ununited fractures and failed arthrodeses. *Journal of the American Medical Association (JAMA), 247*, 623–628.
124. Tabrah, F., Hoffmeier, M., Gilbert, F., Batkin, S., & Bassett, C. A. (1990). Bone density changes in osteoporosis-prone women exposed to pulsed electromagnetic fields (PEMFs). *Journal of Bone and Mineral Research, 5*, 437–442. https://doi.org/10.1002/jbmr.5650050504.
125. Ramos, L. A. V., Callegari, B., França, F. J. R., Magalhães, M. O., Burke, T. N., Carvalho E Silva, A. P. M. C., Almeida, G. P. L., Comachio, J., & Marques, A. P. (2018). Comparison between transcutaneous electrical nerve stimulation and stabilization exercises in fatigue and transversus abdominis activation in patients with lumbar disk herniation: A randomized study. *Journal of Manipulative and Physiological Therapeutics, 41*(4), 323–331. https://doi.org/10.1016/j.jmpt.2017.10.010.
126. Namuun, G., Endo, Y., Abe, Y., Nakazawa, R., & Sakamoto, M. (2012). The effect of muscle fatigue using short term transcutaneous electrical nerve stimulation. *Journal of Physical Therapy Science, 24*(5), 373–377. https://doi.org/10.1589/jpts.24.373.
127. Morrison, S. J., & Scadden, D. T. (2014). The bone marrow niche for haematopoietic stem cells. *Nature, 505*(7483), 327–334. https://doi.org/10.1038/nature12984.

Part II
Tumor Treating Fields (TTFs)

Chapter 6
Simulating the Effect of 200 kHz AC Electric Fields on Tumour Cell Structures to Uncover the Mechanism of a Cancer Therapy

Kristen W. Carlson, Jack A. Tuszynski, Socrates Dokos, Nirmal Paudel, and Ze'ev Bomzon

6.1 Introduction

Tumor-treating fields (TTFields) are 100–500 kHz electric fields with intensities of about 1–4 V/cm, which are known to exert an anti-mitotic effect on cancer cells. TTFields have been observed to kill virtually all tumour cells in vitro and in some animal preparations [1, 2]. They have minor side effects, are increasingly prescribed for brain cancer and are in development or clinical trials for a number of other aggressive malignant cell types in humans [3, 4]. It was initially believed that TTFields act on highly polar sub-cellular structures, such as tubulin dimers, septin, actin etc., thereby disrupting spindle formation [1]. However, calculations show that TTFields-interaction energy is several orders of magnitude too low to directly disrupt the functionality of these structures, while disruption of other structures such as microtubules (MTs) and motor proteins is possible [6]. We are analyzing various hypotheses for TTFields' mechanism of action toward optimizing its clinical efficacy using numerical analysis, such as finite element modeling (FEM).

K. W. Carlson (✉)
Department of Neurosurgery, Beth Israel Deaconess Medical Ctr/Harvard Medical School, Boston, MA, USA
e-mail: kwcarlso@bidmc.harvard.edu

J. A. Tuszynski
Department of Physics, University of Alberta, Edmonton, AB, Canada

S. Dokos
Graduate School of Biomedical Engineering, University of New South Wales, Sydney, NSW, Australia

N. Paudel
IEEE Applied Computational Magnetics Society, Greenville, SC, USA

Z. Bomzon
Novocure Ltd, Haifa, Israel

S. Makarov et al. (eds.), *Brain and Human Body Modeling*,
https://doi.org/10.1007/978-3-030-21293-3_6

6.2 Overview of the Models

6.2.1 Why Computer Modelling?

Cell studies, such as Kirson et al. [1, 2], Gela et al. [7] and Giladi et al. [8], measure empirical outcomes. Unlike modeling, cell studies generally do not reveal low-level intra-cellular mechanisms although they may engender mechanism hypotheses that can be tested by modeling and further cell studies. But cell studies are also expensive and generally take many months, whereas computer simulations are relatively cheap and can provide results in much shorter time frames. Models can also be parameterized to analyze many scenarios with batch parameter sweeps (e.g. Wenger et al. [9]).

6.2.2 Axiomatizing the Underlying Systems Level

Computer models can be constructed at any biological systems level. Fundamental results at the underlying systems level are taken as assumptions or 'axioms' at the level being modelled [10]. For example, molecular dynamics (MD) simulations are currently conducted on a smaller microcosmic scale than FEM and for nanosecond simulated durations. Our FEM simulations can incorporate the output of MD. Figure 6.1 shows a map of electric potential on the surface of a microtubule (MT) predicted by an MD simulation, and Fig. 6.2 shows the map as imported into COMSOL FEM (Burlington, MA) using an interpolation function. The MT potential map is taken as axiomatic at the higher systems level of the cell and used in turn by FEM to predict responses of polarized intracellular structures to the MT surface.

6.3 Clues to the Mechanisms Are Constraints on the Models

Over the past 15 years, researchers have produced valuable empirical results about TTFields including those summarized in Table 6.1 [1, 2, 8, 9, 12, 13].

While each cell study result constitutes an important constraint on the model, the question remains: which observed effects are causative versus downstream or epiphenomenal?

Cells follow multiple pathways when exposed to TTFields, including apoptosis during interphase, mitotic arrest and death and normal progression into the next cell cycle, possibly indicating multiple mechanisms of action [8]. It also appears TTFields have an effect on the immune system, which may be involved in TTFields' mechanism [14].

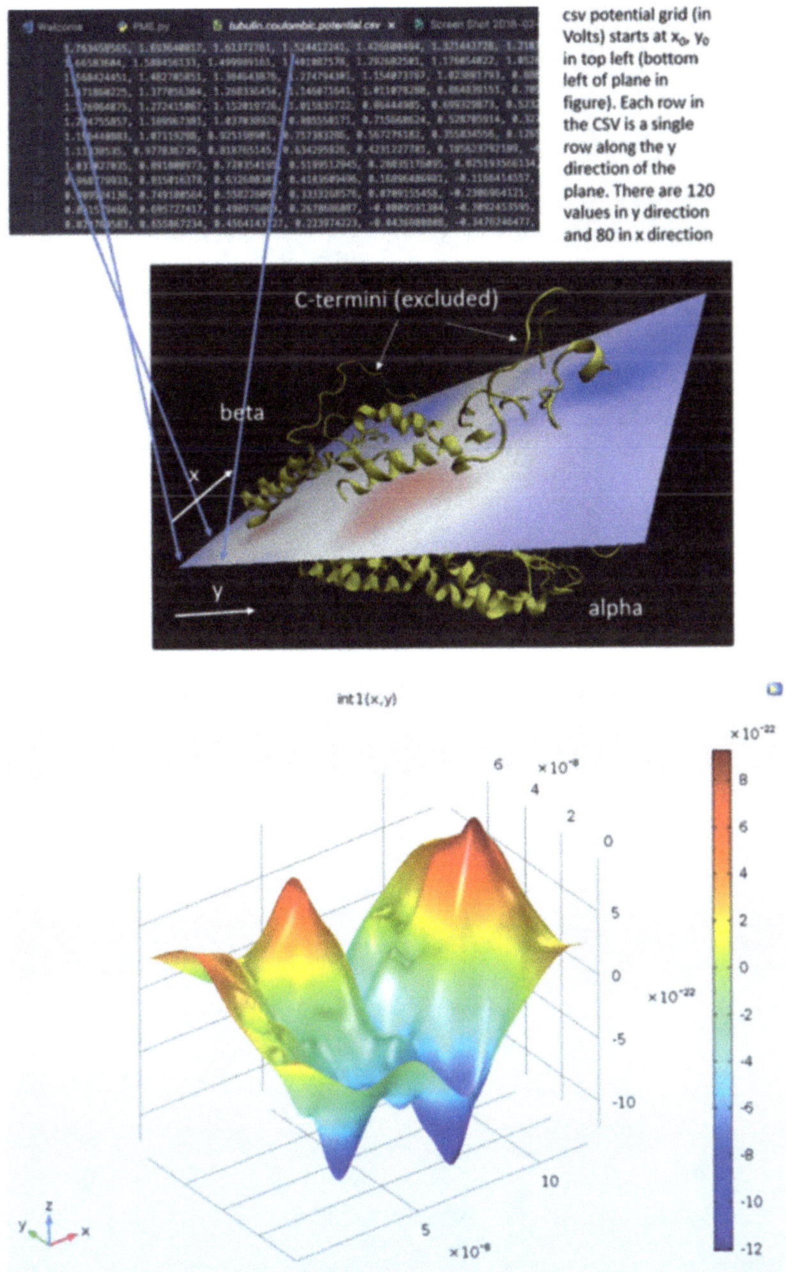

Fig. 6.1 Top: A 2D map of electric potential on a microtubule (MT) surface created in a molecular dynamics simulation of the MT's underlying tubulin dimers and then used as input to a finite element analysis of the interaction of the MT surface potential with charged C-termini tails and the counter-ion layer they attract [11]. (VMD (National Institutes of Health, Bethesda MD USA) courtesy Josh Timmons, Wong lab, Harvard Medical School). Bottom: The same map imported into a COMSOL finite element model, where the z-axis represents potential in volts and x and y are surface coordinates on the MT

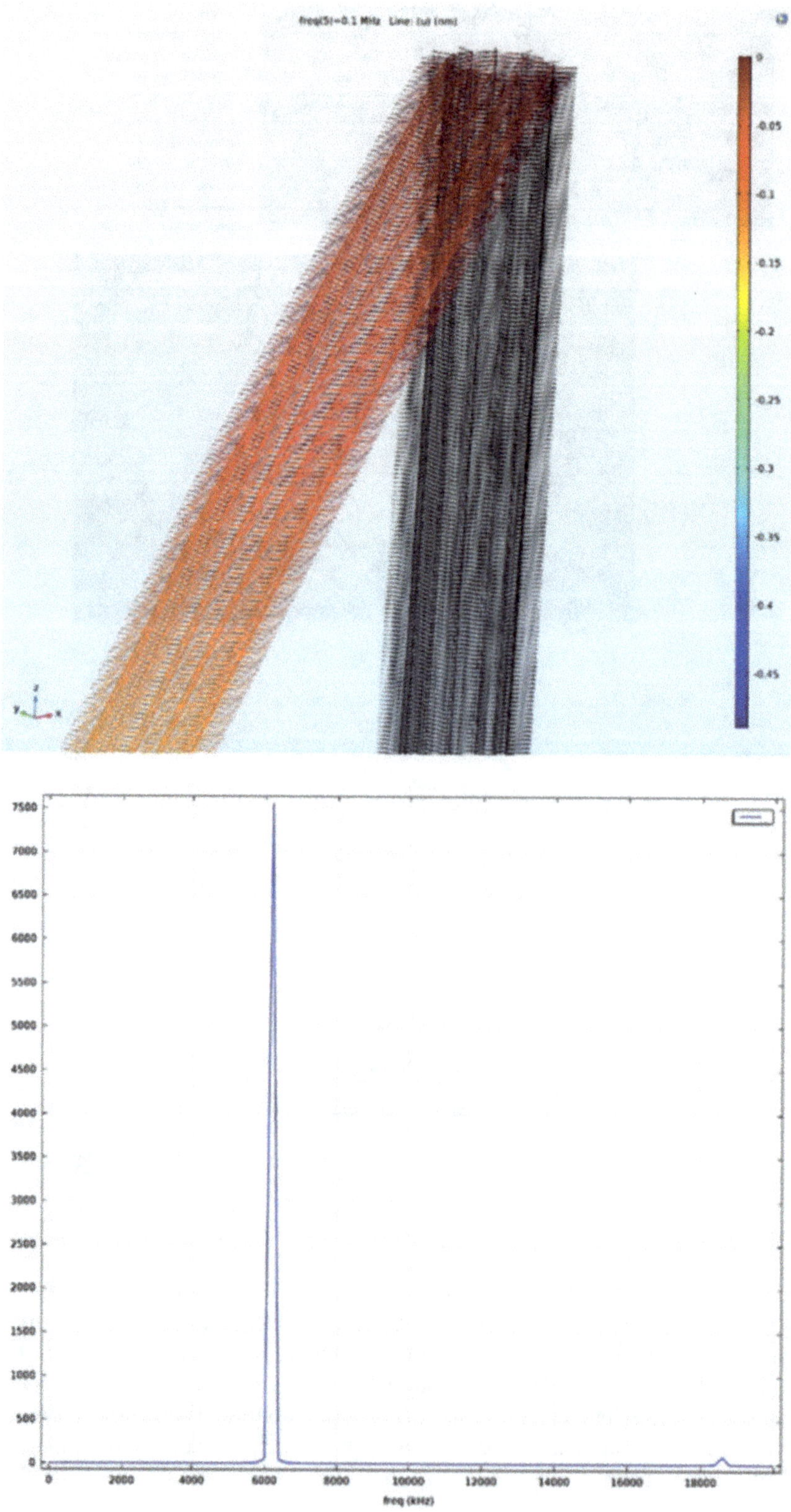

Fig. 6.2 Analysis of resonance peaks of MTs using electro-mechanical FEM. Top: Displacement of the MT along its length. Bottom: Maximum peak kinetic energy plotted against frequency showing a 6 GHz peak

Table 6.1 Key empirical aspects of TTFields

~100% efficacy in vitro and some in vivo preparations
1–4 V/cm electric field strength minimum therapeutic floor
Frequency-sensitive: 100–300 kHz = ~3–10 μs period
Strongest effects with field aligned or orthogonal to cell axis
Does not affect non-tumour cells
Longer exposure = greater efficacy, e.g. after 1st interphase
Strongest effects when applied in early mitosis (prophase)
Increased free vs. polymerized tubulin
Aberrant spindle formation
Cell blebbing
Aneuploidy
Chromosome mis-segregation
Multiple nucleation
Decreased septin concentration at cell midline
Immune system effects

6.4 Candidates for TTFields Mechanisms

We have collected a prioritized list of hypothesized TTFields mechanisms of action to model and test. These include:

4.1. TTFields disrupt the 'walk' of the motor protein kinesin, which carries cargo throughout the cell.
4.2. TTFields disrupt the metastable state transitions of MTs' Carboxyl-termini (C-termini), which signal protein cargo transport 'on' and 'off', i.e. enabling or disabling protein transport along the MT.

Both 4.1 and 4.2 are examples of the general hypothesis that energy imparted by TTFields to sub-cellular structures, in these cases, the counter-ion layer surrounding the C-termini, exceeds a disruption metric, such as the energy contributed by ATP to release the kinesin foot (4.1) and one or two multiples of cellular free energy (4.2) [6]. These possibilities are further explored below.

4.3. The electric field or current density edge effects disrupt MT length dynamics.
4.4. TTFields cause torsion or rotation of septin fibers around their longitudinal axis, which interferes with septin assembly.
4.5. Structural deformation of MTs disrupt signaling.
4.6. DEP forces accelerate organelles to furrow, causing cell blebbing.

6.5 Disruption Metrics Derived from Signal-to-Noise Ratio

Cellular processes constitute signaling that evolved to exceed several background noise levels. Processes that use the least energy are the 'low-hanging fruit' of possible TTFields mechanisms, since the least amount of externally imposed energy may disrupt them. Thus, two important disruption metrics are low multiples of these parameters ('low' since higher multiples pose higher energy barriers):

5.1. Cellular background thermal energy, given by kT = 4.2 x 10^{-21} Joule-nanometer
5.2. Cellular free energy, given by −54–101 × 10^{-19} Joule-nanometer = ~25 kT

where k is Boltzmann's constant, 1.38 × 10^{-23} m^2 kg s^{-2} T^{-1} and T is the absolute temperature in Kelvin [15]. Note that the second metric requires more disruption energy than the first.

5.3. There are conditions under which energy could accumulate or be amplified, exceeding the cellular disruption metrics. Examples include: (1) the amplification of electric field strength at the cellular furrow during late mitosis [9], which we have ruled out for the moment since its timing conflicts with Kirson et al. finding that TTFields' strongest effect occurs during early mitosis (i.e. prophase) [2], (2) hypotheses that incorporate a resonance effect generated by the unique efficacious frequency range of TTFields, 100–300 kHz [16] (Sect. 6.1), (3) a hypothesis that regions of higher conductivity would shunt higher currents through the cell (Section "Electromechanical Model") and (4) that edge effects produced when the field is orthogonal to the MT result in current density being amplified at the ends of the MT, which we have found using FEM.

6.6 Models and Results

6.6.1 MT Resonance

The hypothesis that the ~200 kHz frequency of TTFields may produce a resonance in MTs is in accordance with the empirical finding that TTFields' second-strongest effect occurs when the field is applied orthogonally to the cell axis, to which many MTs are aligned during mitosis. To test this hypothesis, we performed numerical simulations evaluating the magnitude of the electric current along MTs exposed to TTFields at 200 kHz.

Electromechanical Model

An electromechanical model coupling constitutive relations in structural mechanics and electromagnetics was developed in COMSOL. Structural parameters such as several measures of elasticity were calibrated according to Tuszynski et al. [17], and sinusoidal electric fields of 1 V/cm were applied transversely to model MTs 20–960 nm in length and at frequencies ranging from kHz to THz. Both free and one end fixed boundary conditions were examined at both ends of the MT 'beam'. Two disruption metrics were examined: (1) maximum deformation and (2) maximum kinetic energy (KE) along the MT.

The net result of these studies was that, without including a viscosity factor for the ambient medium in which the MT was embedded (e.g. cytosol, admittedly the apparent viscosity on a nanometer scale is unclear), we found significant (in terms of KE) resonance peaks in the GHz but not kHz range (Fig. 6.2), supporting others' studies [18]. Furthermore, we found that incorporating virtually any viscosity factor damped the resonant peaks to a level below the potential for significant disruptive KE.

6.6.2 MT Conductivity

Santelices et al. recently found that MT conductivity considerably exceeds that of the ambient cytosol by as much as two orders of magnitude [19]. This result suggested that MTs could act as electrical cables shunting relatively high currents through the cell. A theoretical basis for assigning conductivities to the various components of a MT is, however, still unknown.

MT as a Multi-Layered Cable

Accordingly, we modeled the MT as a layered cylinder with an inner lumen radius of 15 nm, helix and component protofilament thicknesses of 4.5 nm, C-termini thickness of 3.5 nm, counter-ions of 2 nm thickness and a 3-nm-thick outer Bjerrum (insulative) layer. The counter-ion layer conductivity was significantly higher than that of cytosol [5, 6, 19].

Not surprisingly, the highest current was found to flow through the counter-ion layer surrounding the C-termini. The current density in this layer may exceed the level required to disrupt intra-cellular processes, such as the motor protein kinesin 'walk' along the MT, MT's C-termini state transitions, or MT polymerization. Current density is highest when the field is aligned with the MT, which is in accordance with in vitro experiments showing that the TTFields' effect is strongest when aligned with the cell axis, and that overall MT alignment with the cell axis increases during mitosis [1].

6.6.3 C-Termini State Disruption

If C-termini state transitions were disrupted by TTFields, critical motor protein transport along MTs, and perhaps other critical functions, would be crippled, likely delaying the silencing of a mitotic cell cycle checkpoint that allows cancer cell division to proceed [8].

The energy required for C-termini state transitions was computed in a series of MD simulations by Priel et al. [5], where the 'up' or 'on' state was lowest energy, and at which background thermal energy of ~25 meV would buffet the C-termini within a 40° cone at physiological temperatures (e.g. 300 K) (Fig. 6.3a, b), while

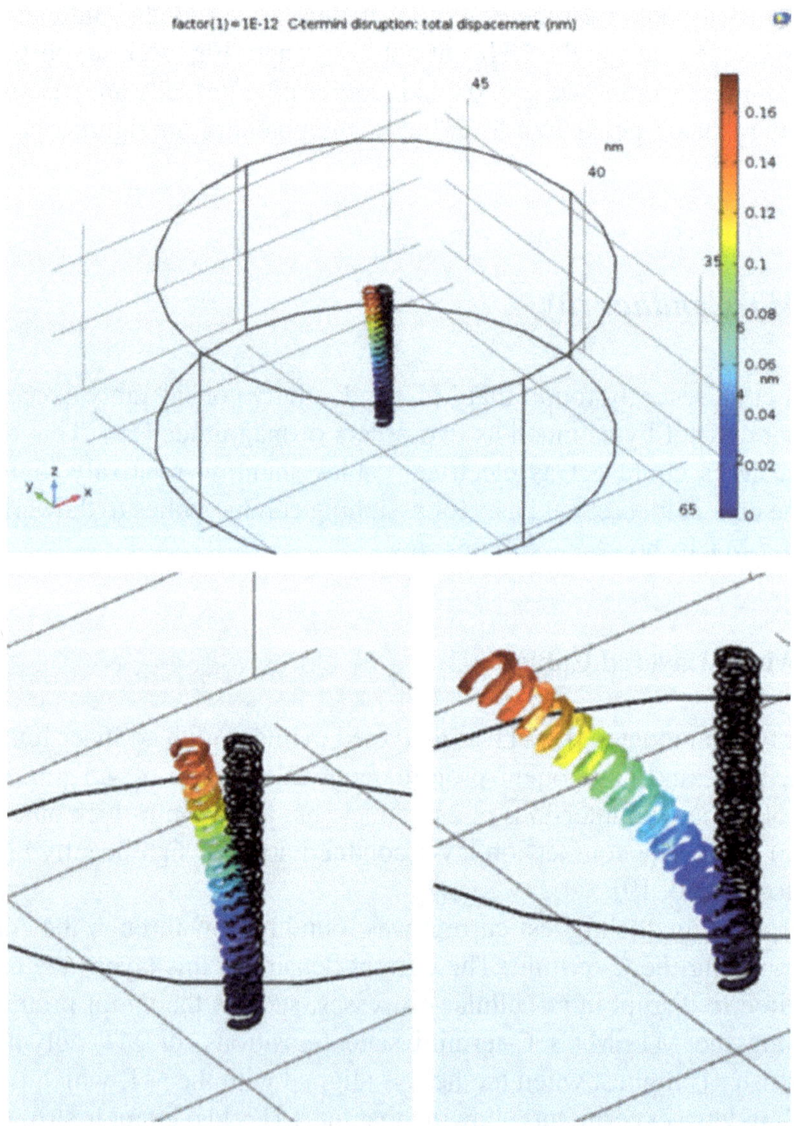

Fig. 6.3 Calibration of C-terminus state transitions according to Priel et al. [5]. Top: Background thermal energy displaces the C-terminus minimally. Bottom: Applied boundary loads of 4 pN and 8 pN displace the C-terminus up to 40°

50 meV would push it beyond 40° (Fig. 6.3c), and 160 meV was required to push a C-termini over a saddle point into its 'off' state along the MT surface with energy of 100 meV. We translated these deformations into forces acting on the C-terminus surface to calibrate the model according to the Priel results (Fig. 6.3).

Model Calibration

The Young's modulus of the MT was initally set to 2 GPa [15] and adjusted until the following constraints were met, assuming a C-terminus length of 3.5 nm [5]:

- kT (25 meV) calibration: a force of 4 pN acting through ~1 nm should induce thermal energy like motions of the C-terminus tip.
- 50 meV calibration: a force of 8 pN should displace the C-terminus tip by ~40° (2.4 nm).
- 120 meV calibration: a force of 16 pN should displace the C-terminus tip by ~80° (4.9 nm).

6.6.4 *Kinesin Walk Diffusion Hypothesis*

Recent studies hypothesize a highly sensitive phase in the motor protein, kinesin, walk along MTs [20]. The back 'foot' of kinesin is released from its MT bond by ATP (10^{-19} Joules). The kinesin molecule's neck then lurches forward over a 10 ms period, skipping over where the forward foot is attached to the MT, and placing the new forward foot two tubulin dimers ahead of its previous position (~16 nm total). The final phase of the walk takes place when thermal buffeting randomly positions the forward foot near enough to the dimer for electrostatic forces to bind it. We plan to use modelling to further examine this diffusion phase, wherein a stall force 10^{-19} N $\leq$ F $\leq$ 10^{-16} N from TTFields would prevent diffusion and disrupt the kinesin walk.

Note that the duration of the diffusion phase is estimated at 4 μs [20] and therefore corresponds to a frequency of 250 kHz, which may indicate a connection to the TTFields' maximum efficacy frequency of 200 kHz.

6.7 Conclusion

Numerical modeling is a necessary complement to cell studies since it can examine underlying mechanisms of action relatively quickly and inexpensively. We are systematically using models to analyze hypothesized mechanisms responsible for TTFields efficacy in killing tumour cells. Such an understanding will facilitate moving TTFields' clinical efficacy toward the 100% ideal achieved in vitro.

Acknowledgements Thanks to Cornelia Wenger, Eric Wong, Ken Swanson, Josh Timmons, Jeffrey E. Arle and Rohit Ketkar, and the editors for valuable insights and feedback. Funding was provided by Novocure Ltd.

References

1. Kirson, E. D., et al. (2004). Disruption of cancer cell replication by alternating electric fields. *Cancer Research, 64*, 3288–3295.
2. Kirson, E. D., et al. (2007). Alternating electric fields arrest cell proliferation in animal tumor models and human brain tumors. *Proceedings of the National Academy of Sciences of the United States of America, 104*, 10152–10157.
3. Mun, E.J., et al. (2018) Tumor-treating fields: A fourth modality in cancer treatment. *Clin Cancer Research, 24*(2), 266–275.
4. Novocure Ltd. (2018). https://www.novocure.com/our-pipeline/. Accessed 23 Oct 2018.
5. Priel, A., et al. (2005). Transitions in microtubule C-termini conformations as a possible dendritic signaling phenomenon. *European Biophysics Journal, 35*, 40–52.
6. Tuszynski, J. A., et al. (2016). An overview of sub-cellular mechanisms involved in the action of TTFields. *International Journal of Environmental Research and Public Health, 13*, 1128.
7. Coque, L., et al. (2011). Specific role of VTA dopamine neuronal firing rates and morphology in the reversal of anxiety-related, but not depression-related Behavior in the *Clock*Δ19 mouse model of mania. *Neuropsychopharmacology, 36*, 1478–1488.
8. Giladi, M., et al. (2015). Mitotic spindle disruption by alternating electric fields leads to improper chromosome segregation and mitotic catastrophe in cancer cells. *Scientific Reports, 5*, 18046.
9. Wenger, C., et al. (2015) Modeling Tumor Treating Fields (TTFields) application in single cells during metaphase and telophase. *Conference Proceedings: IEEE Engineering in Medicine and Biology Society. New York: IEEE, (pp. 6892–6895).*
10. Newman, J. R. (1956) *The world of mathematics; a small library of the literature of mathematics from A'h-mosé the Scribe to Albert Einstein*. New York: Simon & Schuster, (Vol. 4, p. xviii, 2535).
11. Tuszyński, J. A., et al. (2005). Molecular dynamics simulations of tubulin structure and calculations of electrostatic properties of microtubules. *Mathematical and Computer Modelling, 41*, 1055–1070.
12. Gera, N., et al. (2015). Tumor treating fields perturb the localization of septins and cause aberrant mitotic exit. *PLoS One, 10*, e0125269.
13. Giladi, M., et al. (2014). Mitotic disruption and reduced clonogenicity of pancreatic cancer cells in vitro and in vivo by tumor treating fields. *Pancreatology, 14*, 54–63.
14. Wong, E. T., et al. (2015). Dexamethasone exerts profound immunologic interference on treatment efficacy for recurrent glioblastoma. *British Journal of Cancer, 113*, 232–241.

15. Howard, J. (2001). *Mechanics of motor proteins and the cytoskeleton* (p. 367, xvi). Sunderland: Sinauer Associates, Publishers.
16. Porat, Y., et al. (2017) Determining the optimal inhibitory frequency for cancerous cells using Tumor Treating Fields (TTFields). *J Vis Exp,* 123, 1–8.
17. Tuszynski, J. A., et al. (2005). Anisotropic elastic properties of microtubules. *The European Physical Journal. E, Soft Matter, 17*, 29–35.
18. Fels, D., et al. (2015). *Fields of the cell.* Kerala: Research Signpost.
19. Santelices, I. B., et al. (2017). Response to alternating electric fields of tubulin dimers and microtubule ensembles in electrolytic solutions. *Scientific Reports, 7*, 9594.
20. Sozanski, K., et al. (2015). Small crowders slow down kinesin-1 stepping by hindering motor domain diffusion. *Physical Review Letters, 115*, 218102.

Chapter 7
Investigating the Connection Between Tumor-Treating Fields Distribution in the Brain and Glioblastoma Patient Outcomes. A Simulation-Based Study Utilizing a Novel Model Creation Technique

Noa Urman, Shay Levy, Avital Frenkel, Doron Manzur, Hadas Sara Hershkovich, Ariel Naveh, Ofir Yesharim, Cornelia Wenger, Gitit Lavy-Shahaf, Eilon Kirson, and Ze'ev Bomzon

7.1 Introduction

Tumor-treating fields (TTFields) are alternating electric fields in the intermediate frequency range (~100–500 kHz) known to exert an anti-mitotic effect on cancer cells [1–3]. The Optune™ device (Novocure, Ltd, Haifa, Israel) utilizes TTFields to treat glioblastoma multiforme (GBM). A pivotal clinical trial, EF-14, showed a significant benefit in overall survival in newly diagnosed GBM patients who received TTFields in addition to standard chemoradiation compared to patients who only received standard chemoradiation [4] (Fig. 7.1). The results of this trial led to the approval of the Optune™ device for the treatment of newly diagnosed GBM patients in multiple regions, including the USA, Canada, Europe, and Japan [5].

The Optune™ device (see Fig. 7.1) is designed to deliver TTFields at a frequency of 200 kHz to the brain. 200 kHz coincides with the frequency at which the cytotoxic effect of TTFields on glioma cells is maximal [2]. Optune™ is a portable device comprising a battery-operated field generator, which is connected to transducer arrays through which TTFields are delivered (Fig. 7.1). Because TTFields affect cells dividing in parallel to the generated field more than other directions [1],

N. Urman · S. Levy · A. Frenkel · D. Manzur · H. S. Hershkovich · A. Naveh · O. Yesharim
G. Lavy-Shahaf · E. Kirson · Z. Bomzon (✉)
Novocure Ltd, Haifa, Israel
e-mail: zbomzon@novocure.com

C. Wenger
Novocure GmbH, Root D4, Switzerland

S. Makarov et al. (eds.), *Brain and Human Body Modeling*,
https://doi.org/10.1007/978-3-030-21293-3_7

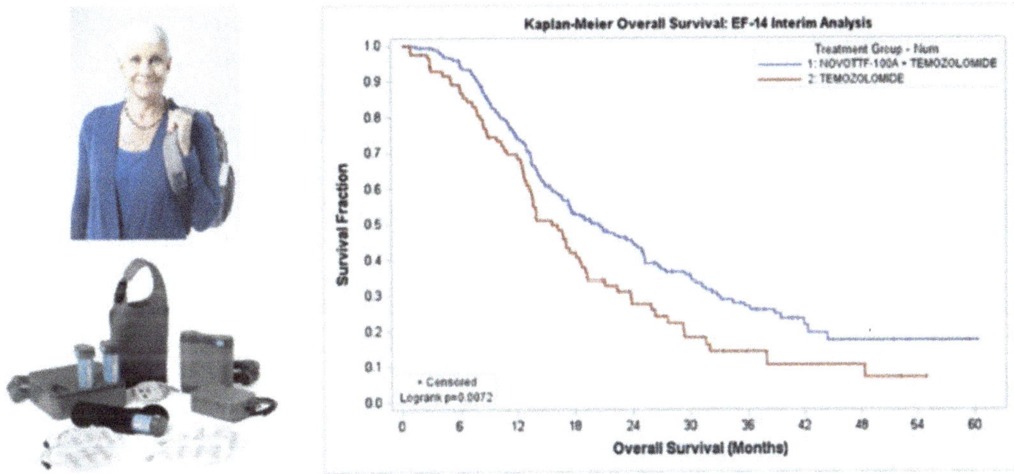

Fig. 7.1 Components of the Optune™ device (bottom left) and model wearing the device (top left), as well as Kaplan-Meier curves showing overall survival for the treatment (blue line) and control (red line) groups of the EF-14 trial. The Optune™ device comprises a battery-operated field generator connected to four transducer arrays that are placed on the patient's head. A backpack and carry-bag for the field generator as well as a battery charger and power supply are also shown in the photographs

Optune™ delivers TTFields in two orthogonal directions via two pairs of transducer arrays. The arrays are placed to deliver two electric fields in roughly orthogonal directions (see Fig. 7.1 (top left) and Fig. 7.4 (top row)). The device switches the field between the two sets of arrays once every second. The transducer arrays comprise nine circular disks each 1 cm in diameter. They are made from a ceramic with a relative dielectric constant >10,000. The disks make contact with the skin through a thin layer of conductive medical gel (~1 mm thickness). The disks are connected to one another using a flexible electric circuit (see bottom left in Fig. 7.1) and are geometrically arranged as shown in Fig. 7.4 (bottom row). The effect of TTFields is time dependent: the more time cells are exposed to the field, the stronger the effect. Therefore, to maximize the effect of treatment, patients are advised to maintain active therapy for at least 18 hr/day on average [6].

Preclinical results [1–3] have shown that the effect of TTFields is intensity dependent, and that the higher the intensity of the field, the stronger the cytotoxic effect of TTFields. The threshold intensity for observing the effect of TTFields is about 1 V/cm amplitude. Several simulation-based studies have shown that it is possible to maximize field intensity in the tumor by carefully selecting the position of the arrays on the scalp [7, 8]. Indeed, the NovoTAL™ system is a software-based system that utilizes morphometric measurements of head size, tumor size, and position (which are determined from a patient MRI) in order to optimize the position of the arrays on the head [9].

The dose-dependent nature of TTFields has been established in preclinical studies [1–3]. However, in order to fully understand the effect of dose and develop effective treatment planning strategies, it is important to establish the connection between

TTFields distribution in the brain and patient outcome in a rigorous manner. Studies addressing this connection require estimating TTFields distribution within a large cohort of patients treated with TTFields over the course of their disease. Since physical measurement of the field distribution is highly invasive, and therefore challenging, numerical simulations utilizing realistic computational models of actual patients are the only practical means for performing such studies.

In order to perform such a study, two challenges need to be addressed:

1. The availability of a dataset encompassing a large number of patients treated with TTFields. The dataset needs to include imaging data from which realistic head models of patients, including the tumor(s), can be obtained, and records of patient outcome including progression and survival.
2. Development of a method for creating realistic computational patients in a robust and rapid manner.

The EF-14 clinical trial dataset includes detailed clinical data on over 400 patients treated with TTFields. Therefore, this dataset is well suited for this study. However, estimating TTFields distribution within these patients requires algorithms that enable construction of realistic head models of patients from MRI scans in a rapid and robust manner. Various pipelines have been adapted for the purpose of simulating the delivery of TTFields to realistic head models. Wenger et al. [8] utilized a pipeline that relied on FSL FLIRT [9], SimNibs [10], and Brainsuite [11] in order to create a realistic head model of a healthy individual into which artificial tumors of various shapes were inserted. Korsheoj et al. [12, 13] presented a different pipeline, utilizing SimNibs to create realistic head models of cancer patients. A different approach was presented by Timmons et al. [14], who utilized SPM8 [15] and ScanIP [16] to create realistic patient models. All of these approaches require a significant amount of human intervention and are therefore time-consuming and not suitable for a study requiring the creation of a large number of head models. An alternative approach could be to estimate the tissue conductivity from imaging data. Indeed, diffusion tensor imaging (DTI) [17] or alternatively water content-based electric property tomography (wEPT) [see Chap. 20 on wEPT in this book] could potentially be used to create realistic head models. However, both these sequences require specifically-adapted image series not available for all patients in the EF-14 study.

To overcome these challenges, we developed a novel method for creating realistic head models of patients utilizing a model of a healthy individual which serves as a deformable template. In this chapter, we present a detailed description of our method and demonstrate its utility by investigating the connection between TTFields field distribution and patient outcome in 119 patients treated with TTFields as part of the EF-14 trial.

7.2 Methods

7.2.1 MRI Data Used for the Study

MRI datasets analyzed in this study were obtained from the patient records of the EF-14 trial participants. This trial was a multi-center, open-label, randomized clinical phase 3 trial, which recruited 695 patients at 83 sites. Patients were randomized at the end of radiotherapy at a ratio of 2:1 to receive standard maintenance temozolomide chemotherapy with or without the addition of TTFields. To create patient models, T1-postcontrast MRIs at baseline (postsurgery and postradiation therapy) of 119 patients from the treatment arm of the trial were selected. Only patients who received TTFields therapy for over 2 months were selected. In general, for all patients, the baseline data contained T1-postcontrast data acquired from at least two of the possible three orientations (axial, sagittal, and coronal).

7.2.2 Image Preprocessing

Patient data was retrieved from the trial records in DICOM format. The DICOM data was imported and converted to NIfTI file format. The header of the NIfTI files was manipulated so that the origin of the file matched the origin of the template tissue probability maps (which are described in Sect. 7.2.6). This step ensures that the MRI images can be registered into the deformable template space. The NIfTI data was padded to add margins to the 3D image and resliced to a uniform grid of $1 \times 1 \times 1$ mm.

7.2.3 MRI Full Head Completion

In order to create a head model from MRI data, it is important that the field of view of the MRI image show the entire head. However, in the clinic, where the focus is to image the tumor, the field of view of the full image set does not always show the entire head. Therefore, in cases where a single T1-contrast image showing the entire head was not available, we combined T1-contrast images acquired at different orientations in order to complete the field of view. In many cases, the image set acquired at one orientation had higher quality than the images acquired at other orientations. Therefore, the image with the highest quality was used as an anchor, and the other images were rigidly registered to it, followed by a histogram matching of the additional images to the anchor image. In order to create a new image, all voxels in which the original image contained MRI data were assigned the same value as the original image in the corresponding voxel. In the area of missing data

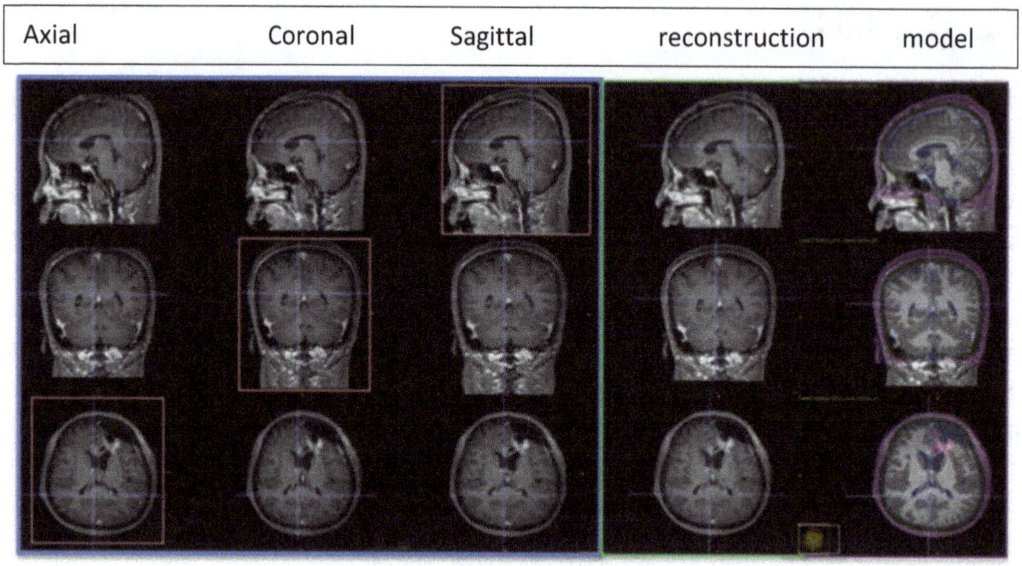

Fig. 7.2 Example of an image set for which the MRI head completion algorithm was applied. For this patient, T1-postcontrast image datasets were captured at axial, coronal, and sagittal orientations existed (first three columns). However, the fields of view in all three image sets did not cover the entire head. An image depicting the full head was created by combining these image datasets using the field of view completion algorithm (column 4). Column 5 shows an overlay of the final patient model on the image of the complete head

of the anchor image (outside the borders of the original MRI series), the value of the voxels were set as the average of all nonzero values of the additional images. Figure 7.2 shows an example of an image set to which this algorithm was applied along with an output image in which the whole head is visible.

7.2.4 *High-Resolution Reconstruction*

The original T1-contrast MRI data is often of lower resolution, which affects the tumor segmentation quality and the accuracy of the head model. A super-resolution algorithm that combines several T1-contrast images of the patient acquired at different orientations into a single high-resolution image was implemented. Based on [18], the best-quality image was set as the anchor image, followed by affine registration of the additional images to it. All images were resliced to a uniform grid using trilinear interpolation and a gray-scale intensity normalization was performed. The value of the voxel in the reconstructed image was a weighted average of the values of the corresponding voxels in the images used for reconstruction. When averaging, a higher weight was given to voxels from slices that were present in an original image versus voxels originating from slices obtained through interpolation.

7.2.5 Background Noise Reduction

A thresholding method was used to remove background noise and aliasing, both of which were found to deteriorate the quality of the head model created using deformable templates. In particular, when background noise was present, the contour of the skull obtained during model creation was often inaccurate and included part of the background. The background noise reduction was performed using a semi-automatic method in which the user selected a single value representing the background noise and the software applied this value as a threshold to automatically detect the contour of the scalp in the MRI image and set the intensity of the background to zero.

7.2.6 Patient Model Creation

Figure 7.3 is a schematic describing the pipeline used to create head models of glioblastoma patients using a deformable template. A prerequisite for this technique is to create a realistic head model of a healthy individual. In order to create a patient model from MRI images, the user first segments the tumor using manual or semiautomatic procedures. The region of the tumor is masked, and a nonrigid registration algorithm is used to register the MRIs to the template space, yielding a transformation from the patient space to the template space. The inverse transformation is then calculated and applied to the deformable template to yield an approximation of the patient head in absence of the tumor. Finally, the tumor is transplanted into the deformed template model to yield the final patient model.

To reduce the schematic to practice, first a deformable template needs to be created:

The deformable template is represented using tissue probability maps (TPMs), which are a set of six 3D matrices that assign to each voxel a probability of belonging to a predefined tissue (white matter, gray matter, CSF, skull, scalp, and air). For the creation of the deformable template, the MRI of a healthy male based in MNI space [19] was segmented as TPMs. The procedure was performed using an algorithm that simultaneously registers and segments the base MRI using an existing set of TPMs (built in a standard space) and applied using the MATLAB toolbox SPM8 [15] and its extension MARS [20]. Manual corrections were made to the deformable template TPMs by manipulating the combination of probabilities in a specific voxel. Manual corrections included mainly adjustments to the regions of skull and scalp, so that a better match between the deformable template and patient MRI data was obtained in these the regions. A final step in creating deformable template TPMs from these probability maps is to apply a smoothing filter to the individual maps. Smoothing is important to allow adjustments to an MRI of any individual. The smoothing was performed using a Gaussian filter with a smoothing kernel of 4 × 4 × 4 mm FWHM (full width half maximum). It is noteworthy that the TPMs generated using this procedure is sharper than the original atlas TPMs of SPM/MARS. This method

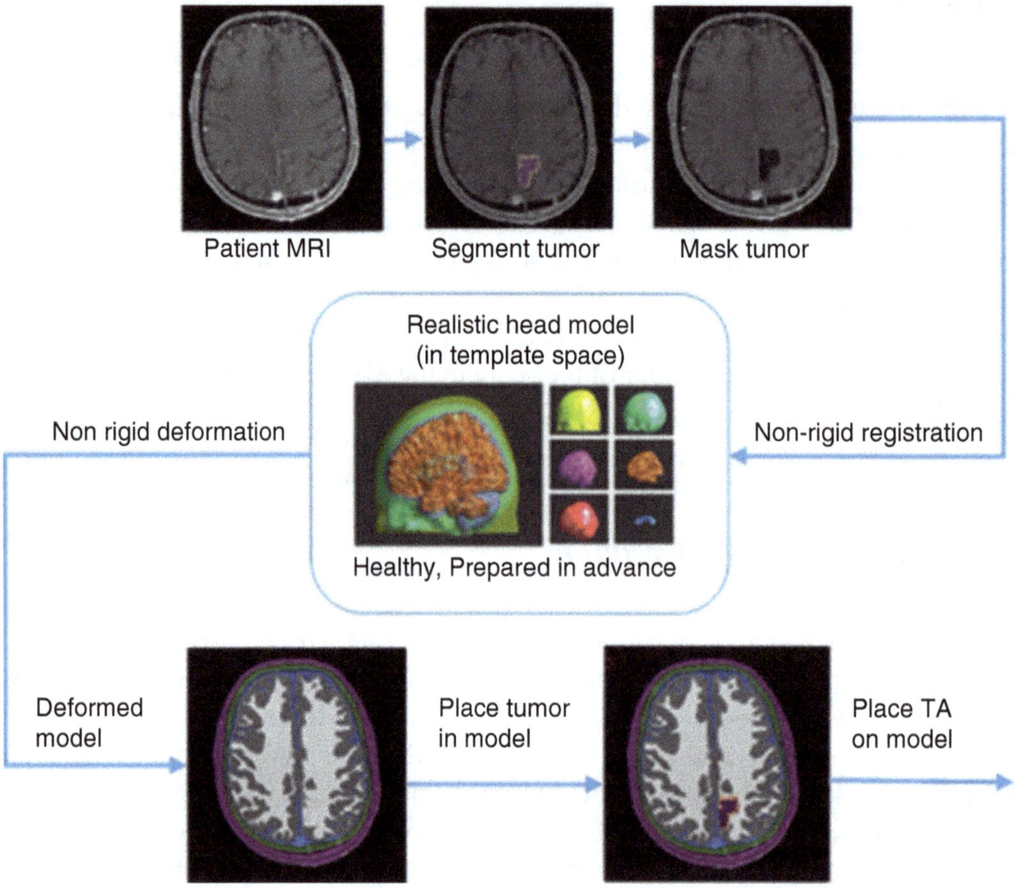

Fig. 7.3 Schematic showing the process used to create patient-specific models from a deformable template. A prerequisite is the creation of a realistic head model of a healthy individual, which serves as the template. To create a patient model, first, the tumor is manually segmented using ITK-SNAP. The segmented region is masked, and the masked MRI registered into the template space using a nonrigid registration algorithm. The registration yields a transformation from patient space to template space as well as an inverse transformation, which is applied to deform the template into the patient space. This yields a model which resembles the patient brain in the absence of a tumor. Finally, the tumor is implanted into the deformed head model to yield the final patient model

ensures that the segmented model has very high resemblance to the patient's MRI data. Deformable template TPMs creation is performed once for the entire database.

T1-postcontrast MRIs were used to create the patient models. Pre-processing operations were performed when image quality or image resolution were insufficient, as described in Sects. 7.2.2, 7.2.3, 7.2.4, and 7.2.5. Following preprocessing, tumors and abnormal tissue were segmented in a semiautomatic manner using ITK-SNAP [21]. The following tumor and abnormal tissue types were segmented: enhancing tumor, necrotic core, enhancing nontumor (coincides with scarring), resection cavity, and skull defects. Furthermore, the following abnormal tissue regions were contoured/segmented: hematoma, ischemia, atrophy, and non-GBM tumor. Abnormal tissue regions were masked in the patient MRI, and the masked images were registered into MNI space using MARS/SPM and the TPMs created from the deformable template. This registration process yielded a nonrigid transformation from the patient space into the template space as well as the inverse transformation from template space to the patient space. The inverse transformation was applied to the probability map for each tissue type independently. A model approximating the patient in the absence of a tumor was then created by assigning to each voxel the tissue type for which the probability value within the voxel was highest. Finally, the manually segmented abnormal tissues were inserted to yield the final patient model.

7.2.7 Placement of Transducer Arrays on the Model

Optune™ transducer arrays comprise a set of nine ceramic disks, which make contact with the skin through a thin layer of medical gel. The disks are arranged in a well-defined geometry as shown in Fig. 7.4. Treatment planning to determine the optimal layout of the transducer arrays on the patient's scalp is performed prior to beginning TTFields therapy. Treatment planning for patients that received TTFields as part of the EF-14 trial was performed. The treatment plan (optimal layout) for each patient was recorded in the patient's clinical record. When simulating delivery of TTFields to a specific patient, arrays were placed on the model in a manner that matched the treatment plan saved in the patient record.

It is important to note that the optimal array layout for a patient was selected from a library of predefined layouts, an example of which can be seen in Fig. 7.4. The position of each array in each layout can be demarcated relative to well-defined anatomical landmarks. In addition, the arrays are constructed from nine discrete disks, set out in a well-defined and rigid geometry. These two observations led to the following procedure for placing the virtual arrays on the patient models.

Automatic Identification of Landmarks and Determination of the Array Positions

In order to determine the location and orientation of the disk array on the head, it is important to identify landmarks, as well as rotate the head model to a well-defined orientation relative to which the orientation of the array is known. An iterative algorithm detecting the orientation of the head and the landmarks was therefore used. The anatomical landmarks that were automatically identified were the centers of the eyes, brainstem, frontal sinuses, and the box bounding the head. Detection of landmarks was performed on the segmented model. The first step was to find the two CSF circles of the eyes in 2D axial slices. For each eye, the slice in which the area of the circle was maximal was found and its center identified. These two points correspond to the centers of the eyes. Using the center of the eyes, an initial correction of the head orientation was performed. The next step was to detect the brainstem (segmented as a round region of white matter in the base of the brain), and to perform full rotation of the model to a predetermined

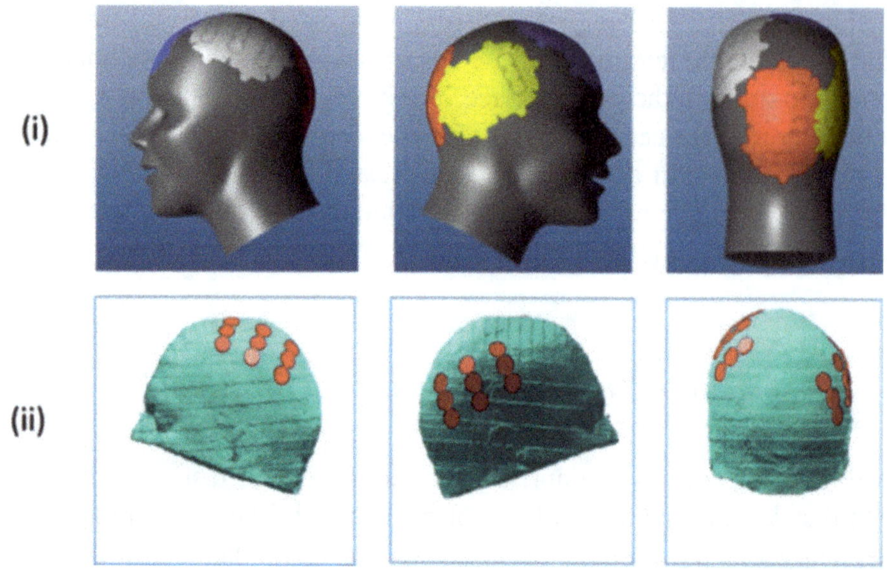

Fig. 7.4 Example showing an array layout as stored in the patient record (top row), along with a patient model onto which arrays matching this layout have been placed (bottom row). Each array layout comprises four transducer arrays, shown as patches of blue, red, yellow, and white in the top row. The arrays are arranged into pairs and electric fields generated between each pair. One pair of arrays delivers an electric field with an anterior-posterior orientation (red and blue arrays in top row), and one pair delivers a field oriented left-right on the patient (white and yellow arrays in top row) Each transducer array comprises nine ceramic disks arranged in a rectangular orientation Each red disk on the model in the bottom row represents a ceramic disk. The bottom row shows a model onto which a pair of transducer arrays has been placed. One array was placed on the left aspect of the model, and the other on the right aspect

orientation. Once the orientation was completed, the frontal sinuses (segmented as air), and bounding box were found. These landmarks were then used to define the positions of the central disks of each array on the scalp as planned for each patient by the NovoTAL™ system.

Positioning of Anchor Points to Assist with Array Placement

After finding the position of each central disk of an array on the scalp, four anchor points at predetermined distances from the central disk were located on the scalp to assist with placement of the remaining disks. The anchor points were positioned to form a cross-like pattern with the central disk of the array at the center. One axis of the cross was vertically oriented, and the second axis of the cross was horizontally oriented. Each anchor point was found using the following iterative algorithm:

(a) Calculate the tangent surface to the scalp at the current point using singular value decomposition (SVD).
(b) Project the vector pointing in the direction of the anchor point onto the tangent surface.
(c) Calculate a point at a predefined (small) distance from the current point along the direction defined by the vector calculated in step (b).

(d) Define a new point on the scalp, as the point on the scalp closest to point calculated in step (c).
(e) If the geodesic distance between the new point and the center point is close enough to the desired distance between center of the central disk and the anchor point, then set the current point as the anchor point, else perform another step towards the anchor point by repeating steps a–e.

Finding the Center of All Disks in an Array

The geometry of the transducer array suggests that the geodesic distances between the centers of the disks are constrained and remain constant when the array is placed on the scalp. Consequently, in order to find the center of a specific disk in the array, geodesic circles with appropriate radii were drawn around the central disk and the anchor points. The approximate intersection of these circles (defined as the point at which the sum of distances from the circles is minimum) corresponds to the center of that specific disk.

Creating Cylinders Representing the Ceramic Disks and the Medical Gel

When placing the arrays on the head, the ceramic disks are tangent to the scalp. The following process was used to ensure that the virtual disks are tangent to the scalp in the patient model. First, the normal direction to the body closest to the disk was calculated. The calculation was performed by finding all points on the phantom skin that were within a distance of one disk radius from the designated point. The coordinates of these points were arranged into the columns of a matrix and SVD performed on the matrix. The normal to the surface is the eigenvector that corresponds to the smallest eigenvalue. To ensure good contact between the disk and the body, the thickness of the gel needs to be in contact with all points under the disk. This was determined by fitting a cylinder to all points on the skin under the disk. Calculation of the positions, orientations, and gel thicknesses associated with the disks was performed in MATLAB 2013b (Mathworks, USA) and the Sim4Life (ZMT-Zurich, Switzerland) Python API was used to generate the transducer arrays in the model.

7.2.8 *Simulations*

Following creation of the patient models, transducer arrays were placed on the models to match the transducer array layouts assigned to the patients as recorded in their medical records. In addition, patient's average compliance (defined as the fraction of time a patient was on active treatment) and the average electrical current delivered to each patient were calculated from log files of the TTFields generators stored in the patient records. The Sim4Life (ZMT Zurich, Switzerland) quasi-electrostatic solver was used to simulate the delivery of TTFields at 200 kHz. Electrical properties were assigned to the various tissue types and materials in the model according to average values reported in the literature [7, 17]. Boundary conditions were set so that the total current delivered to the patient was equal to the average current delivered to the patient during the first 6 months of treatment. It is important to note that a separate simulation was performed for each of the two pairs of arrays placed on the patient.

7.2.9 *Analysis*

For this study, a total of $n = 119$ cases (of 466 patients in the trial treatment arm) were simulated. All patients analyzed in this study were on active treatment for more than 2 months. For each patient, field intensity distributions in a tumor bed comprising the gross tumor volume (GTV) and a proximal boundary zone (PBZ) extending 1 cm from the GTV were derived. To account for compliance, field values for each patient were multiplied by their average compliance over the first 6 months of treatment.

To test the hypothesis that patient outcome correlates with field intensities, two quantities were derived:

1. E95, which is the value such that 95% of the combined volume of the GTV and PBZ, received field intensities (multiplied by compliance) above a specified value.
2. $E_{average}$, which is the average intensity (multiplied by compliance and average current) in the combined volume of the GTV and PBZ.

Patients were divided into two groups based on threshold values of both E95 and $E_{average}$, and the overall survival (OS) and progression-free survival (PFS) of the groups were compared. The threshold values of these parameters were chosen to yield the most statistically significant difference in OS between the groups.

7.3 Results

Demographics of all groups were similar to the demographics of the entire EF-14 trial population and similar to each other (Table 7.1). When dividing the 119 patients into two groups based on E95, the median OS and PFS were superior when E95 > 1.3 *V/cm*: OS (E95 > 1.3 V/cm: 33.0 months vs. E95 < 1.3 V/cm:21.9 months, $p = 0.009$, HR = 0.46) and PFS (E9 > 1.3 V/cm 11.9 months vs E95 < 1.3 V/cm 7.5 months, $p = 0.06$, HR = 0.49). Similar results were seen when dividing the patients into two groups based on a threshold of $E_{average} > 1.0$ *V/cm*: OS ($E_{average} > 1.0$ *V/cm*: 26.1 months vs. $E_{average} < 1.0$ *V/cm:* 21.6 months, $p = 0.025$, HR = 0.50) and PFS

Table 7.1 Demographics of patients split into groups according to a threshold of E95 = 1.3 V/cm

Characteristics	E95 > 1.3 (*N* = 33)	E95 ≤ 1.3 (*N* = 86)	*p*-value	All simulation patients (*N* = 119)	Other TTFields patients (*N* = 347)	*p*-value
Age (Years)						
Mean (SD)	57.9 (10.50)	52.3 (11.27)	0.0156	53.9 (11.30)	55.0 (11.52)	0.354
Sex, No. (%)						
Male	23 (69.7%)	60 (69.8%)	0.994	83 (69.7%)	233 (67.1%)	0.6
Female	10 (30.3%)	26 (30.2%)		36 (30.3%)	114 (32.9%)	
Region, No. (%)						
United States	16 (48.5%)	37 (43.0%)	0.592	53 (44.5%)	168 (48.4%)	0.465
Rest of world	17 (51.5%)	49 (57.0%)		66 (55.5%)	179 (51.6%)	
Extent of Resection, No. (%)						
Biopsy	6 (18.2%)	8 (9.3%)	0.212	14 (11.8%)	46 (13.3%)	0.912
Partial resection	8 (24.2%)	33 (38.4%)		41 (34.5%)	116 (33.4%)	
Gross Total resection	19 (57.6%)	45 (52.3%)		64 (53.8%)	185 (53.3%)	
MGMT Tissue available and tested, No. (%)	28 (84.8%)	75 (87.2%)		103 (86.6%)	283 (81.6%)	
Methylated	10 (35.7%)	30 (40.0%)	0.919	40 (38.8%)	97 (34.3%)	0.709
Unmethylated	15 (53.6%)	38 (50.7%)		53 (51.5%)	156 (55.1%)	
Invalid	3 (10.7%)	7 (9.3%)		10 (9.7%)	30 (10.6%)	
Karnofsky Performance Score						
Mean (SD)	88 (10)	89 (9)	0.6084	89 (9)	87 (11)	0.137
Median (range)	90 (70–100)	90 (70–100)		90 (70–100)	90 (60–100)	

Table also shows demographics of all 119 patients, as well as demographics for all $n = 347$ patients treated with TTFields, but not included in this study. Demographics of all groups are similar. It is worth noting that the median age in the group for which E95 > 1.3 V/cm is significantly higher than in the group for which E95 < 1.3 V/cm

($E_{average} > 1.0$ *V/cm*: 9.9 months vs. $E_{average} < 1.0$ *V/cm:* 6.1 months, $p = 0.026$, HR = 0.54). Progression-free at 6 months and 24-months survival rate were superior when E95 > 1.3 *V/cm* (E95 > 1.3 *V/cm*: 76% vs. E95 < 1.3 *V/cm:* 56%, $p = 0.031$), and (E95 > 1.3 *V/cm*: 66% vs. E95 < 1.3 *V/cm:* 42%, $p = 0.0156$), respectively. Within the E95 > 1.3 *V/cm* group, a higher percentage of patients showed clinical

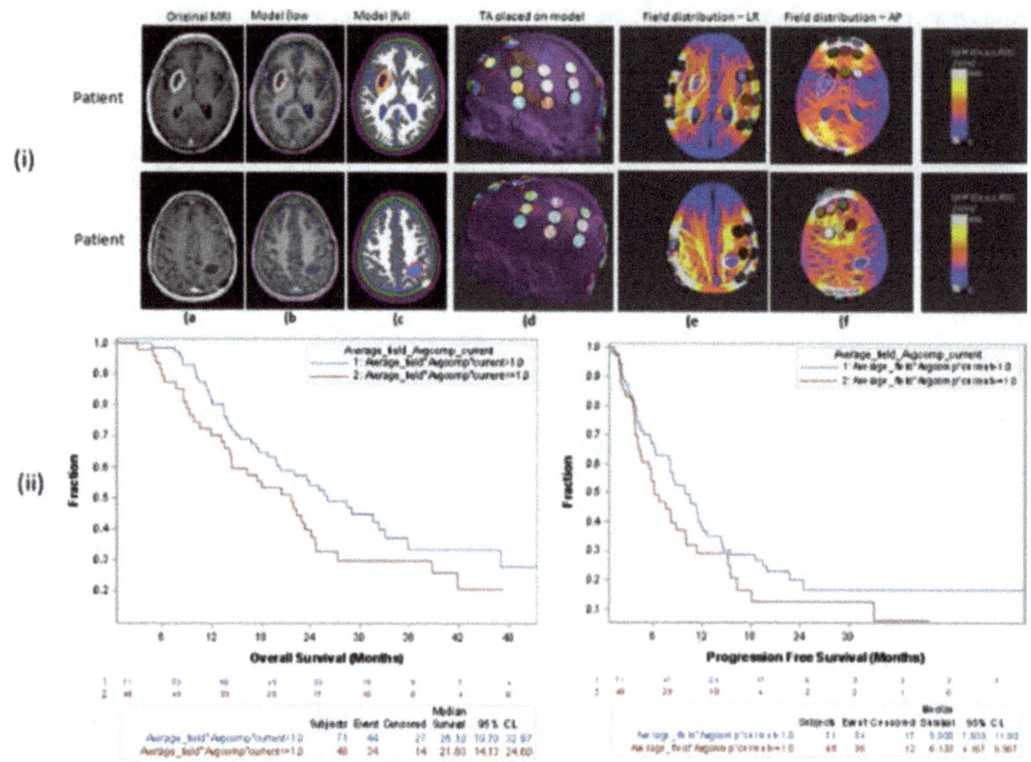

Fig. 7.5 (i) Examples of patient models and calculated field distributions within each patient head. Each row shows a different patient. The figure shows axial slices from patient MRI (1st column), an overlay of the patient model on the slice (2nd column) and the final patient model in the same slice (3rd column), 3D representation of the patient model with virtual arrays (4th column) and field intensity distributions generated by the pair of arrays on the left and right aspects of the head (5th column) and on the anterior and posterior of the head (6th column). (ii) Kaplan-Meier curves for overall survival when splitting patients according to threshold values of $E_{average}$ (left) and E95 (right)

benefit (stable disease/partial response, 97% vs. 83% $p = 0.039$). OS in the *E95* < 1.3 *V/cm* group was superior to OS in the EF-14-control-arm (16.0 months, $p = 0.009$), indicating that patients benefited from treatment even when field intensities were lower. The Kaplan-Meier curves for OS are shown in Fig. 7.5.

7.4 Discussion and Conclusion

In this work, we summarize a simulation-based study suggesting a strong connection between TTFields dose at the tumor bed and patient outcome. A robust and rapid semiautomatic procedure for creating realistic patient models was developed in order to perform this study. This method enabled the efficient creation of patient-specific models and simulation of TTFields delivery to over 100 patients. Simulating delivery of TTFields to such a large cohort is necessary in order to gain a sample size large enough to establish statistically significant connections between dose/

field intensity and patient outcome. To our knowledge, this is the first study in which delivery of low-frequency electromagnetic energy to patients has been performed on such a large cohort. In the future, our modeling process could be adapted to investigate potential connections between electromagnetic field distributions and patient outcomes for other electrotherapeutics. A study utilizing this method to analyze the connection between TTFields dose distribution and patient outcome in 340 patients that participated in the e EF-14 trial was recently published [22].

The head model creation procedure used in this study utilizes a healthy head model, which serves as a deformable template from which the patient models are created. Consequently, the anatomy of the resulting head models bear a large degree of resemblance to the template and may not accurately capture fine anatomical features such as the exact shape of the gyri and sulci of the patient's brain. The frequency of TTFields and the method of delivery (large transducer arrays) dictate that TTFields distribution is broad and coarse in geometry and distributes over the entire brain. Consequently, fine details in the anatomy are unlikely to affect the overall shape of the field distribution in the brain, and are therefore expected to have minimal effect on the analysis and conclusions presented in this study. This is in contrast to applications such as transcranial direct current stimulation (TDCS) [23], or deep brain stimulation [24], where it is important to accurately deliver the electric field to specific and often finely shaped anatomical structures.

The analysis performed in this study suggests a statistically significant connection between TTFields intensity at the tumor bed (multiplied by compliance) and patient outcome. This suggests that treatment planning in which array placement on the scalp is optimized in order to maximize TTFields intensity at the tumor bed could be important for improving patient outcome. Indeed, the NovoTAL system utilizes simple geometrical rules to optimize array placement. In the future, the algorithms presented in this chapter could form the basis for treatment planning software that uses realistic simulations of TTFields distribution in the body for treatment planning. TTFields treatment planning may even be extended beyond a calculation of positioning the arrays on the head. For example, planning may encompass surgical procedures, such as cranial remodeling, which have been shown to increase TTFields distribution at the tumor by over 50% in some cases [12].

In this study, we assumed homogeneous tissue properties for all tissue types based on average values reported in the literature, with the conductivity values assigned to the various compartments of the tumor. The assumption of homogeneity is reasonable when considering healthy tissue [see chapter on wEPT in this book]. However, glioblastoma tumors are structurally heterogeneous. Therefore, their electric properties are likely heterogeneous as well. However, very little literature investigating the electric properties of tumors exists. Such heterogeneity will affect TTFields distribution in the tumor bed. Future studies investigating the electric properties of tumors are planned so that tumor heterogeneity can be accounted for in simulations and treatment planning. Techniques such as wEPT [25] or MR-based electrical impedance tomography (MREIT) [26] provide spatial maps of the electric properties in tissue that could be extremely useful in this regard. However, their applicability to mapping electric properties in the 100 kHz-1 MHz frequency range should first be established. It is important to note that in the context of this study, the effect of heterogeneity in the tumor bed on TTFields dose, and its connection to outcome is effectively accounted for in the analysis through the averaging performed on field intensity in the tumor bed and the large number of patients in the study.

To conclude, this chapter presents novel methods for creating realistic head models suitable for the simulation of TTFields. The approach has enabled a groundbreaking study in which simulation and clinical data have been combined to establish a connection between TTFields dose at the tumor and patient outcome. Future studies based on our approach will lead to a better understanding of TTFields therapy, ultimately leading to improved patient outcomes.

References

1. Kirson, E. D., et al. (2004). Disruption of cancer cell replication by alternating electric fields. *Cancer Research, 64*(9), 3288–3295.
2. Kirson, E. D., et al. (2007). Alternating electric fields arrest cell proliferation in animal tumor models and human brain tumors. *Proceedings of the National Academy of Sciences of the United States of America, 104*(24), 10152–10157.
3. Giladi, M., et al. (2015). Mitotic spindle disruption by alternating electric fields leads to improper chromosome segregation and mitotic catastrophe in cancer cells. *Scientific Reports, 5*, 18046.
4. Stupp, R., et al. (2017). Effect of tumor-treating fields plus maintenance temozolomide vs maintenance temozolomide alone on survival in patients with glioblastoma: A randomized clinical trial. *JAMA, 318*(23), 2306–2316.
5. www.novocure.com
6. Kanner, A. A., et al. (2014). Post Hoc analyses of intention-to-treat population in phase III comparison of NovoTTF-100A™ system versus best physician's choice chemotherapy. *Seminars in Oncology, Suppl 6*, S25–S34.
7. Wenger, C., et al. (2015). The electric field distribution in the brain during TTFields therapy and its dependence on tissue dielectric properties and anatomy: A computational study. *Physics in Medicine and Biology, 60*(18), 7339.
8. Korshoej, A. R., et al. (2018). Importance of electrode position for the distribution of tumor treating fields (TTFields) in a human brain. Identification of effective layouts through systematic analysis of array positions for multiple tumor locations. *PLoS One, 13*(8), e0201957.
9. Chaudhry, A., et al. (2015). NovoTTFTM-100A system (tumor treating fields) transducer array layout planning for glioblastoma: A NovoTALTM system user study. *World Journal of Surgical Oncology, 13*, 316.10. FSL FLIRT.
10. Thielscher, A., et al. (2015). *Field modeling for transcranial magnetic stimulation: A useful tool to understand the physiological effects of TMS?* Milano: IEEE EMBS.
11. www.brainsuite.org
12. Korshoej, A. R., et al. (2016). Enhancing predicted efficacy of tumor treating fields therapy of glioblastoma using targeted surgical craniectomy: A computer modeling study. *PLoS One, 11*(10), e0164051.
13. Korshoej, A. R., Hansen, F. L., Thielscher, A., Von Oettingen, G. B., Christian, J., & Hedemann, S. (2017). Impact of tumor position, conductivity distribution and tissue homogeneity on the distribution of tumor treating fields in a human brain: A computer modeling study. *PLoS One, 12*(6), e0179214.
14. Timmons, J. J., et al. (2017). End-to-end workflow for finite element analysis of tumor treating fields in glioblastomas. *Physics in Medicine and Biology, 62*(21), 8264–8282.

15. Ashburner, J. (2012). SPM: A history. *NeuroImage, 62*, 791–800.
16. https://www.synopsys.com/simpleware/products/software/scanip.html
17. Wenger, C., et al. (2016). Improving tumor treating fields treatment efficacy in patients with glioblastoma using personalized array layouts. *International Journal of Radiation Oncology, Biology, Physics, 94*(5), 1137–1143.
18. Woo, J., et al. (2012). Reconstruction of high-resolution tongue volumes from MRI. *IEEE Transactions on Biomedical Engineering, 59*(12), 3511–3524.
19. Holmes, C. J., et al. (1998). Enhancement of MR images using registration for signal averaging. *Journal of Computer Assisted Tomography, 22*(2), 324–333.
20. Huang, Y., et al. (2015). Fully automated whole-head segmentation with improved smoothness and continuity, with theory reviewed. *PLoS One, 10*(5), e0125477.
21. Yushkevich, P. A., Piven, J., Hazlett, H. C., Smith, R. G., Ho, S., Gee, J. C., & Gerig, G. (2006). User-guided 3D active contour segmentation of anatomical structures: Significantly improved efficiency and reliability. *NeuroImage, 31*(3), 1116–1128.
22. Ballo, M., Urman, N., Lavy-Shahaf, G., Grewal, J., Bomzon, Z., & Toms, S. (2019). Correlation of tumor treating fields dosimetry to survival outcomes in newly diagnosed glioblastoma: A large-scale numerical simulation-based analysis of data from the phase 3 EF-14 randomized trial. *International Journal of Radiation Oncology, Biology Physics*, In press.
23. Miranda, P. C., et al. (2017). Optimizing electric-field delivery for tdcs: Virtual humans help to design efficient, noninvasive brain and spinal cord electrical stimulation. *IEEE Pulse, 8*(4), 42–45.
24. Arle, J. E., et al. (2016). High-frequency stimulation of dorsal column axons: Potential underlying mechanism of paresthesia-free neuropathic pain relief. *Neuromodulation, 19*(4), 385–397.
25. Michel, E., Hernandez, D., & Lee, S. Y. (2017). Electrical conductivity and permittivity maps of brain tissues derived from water content based on T1 -weighted acquisition. *Magnetic Resonance in Medicine, 77*, 0094–1103.
26. Zhang, X., Liu, J., & He, B. (2014). Magnetic resonance based electrical properties tomography: A review. *IEEE Reviews in Biomedical Engineering, 7*, 87–96.

Chapter 8
Insights from Computer Modeling: Analysis of Physical Characteristics of Glioblastoma in Patients Treated with Tumor-Treating Fields

Edwin Lok, Pyay San, and Eric T. Wong

8.1 Introduction

Tumor-treating fields (TTFields) are intermediate frequency electric fields at 200 kHz (kHz or 10^3 Hz) that have anti-tumor activity. In a pivotal phase III clinical trial for newly diagnosed glioblastoma patients, the addition of TTFields to maintenance temozolomide, which was administered after initial radiotherapy and concurrent daily temozolomide, was found to improve both progression-free survival and overall survival when compared to those who only received maintenance temozolomide [1, 2]. This positive trial result led the US FDA to approve the use of TTFields for these patients and the incorporation of this therapy into the National Comprehensive Cancer Network (NCCN) guidelines for malignant gliomas [3].

8.2 TTFields Is Another Treatment Modality from the Electromagnetic Spectrum

Energies from different parts of the electromagnetic spectrum are utilized to treat various types of malignancies. For glioblastoma, ionizing radiation from external beam radiotherapy is the mainstay of treatment at initial diagnosis. The frequency is in the ectahertz (EHz or 10^{18} Hz) range and the high energy results in direct DNA damage, such as double-strand DNA breaks; indirect effects are also a result, including the generation of oxygen radicals causing secondary tissue damage [4]. Major advances of the past decades consist of improving the conformality of the beam, as

E. Lok · P. San · E. T. Wong (✉)
Brain Tumor Center & Neuro-Oncology Unit, Department of Neurology, Beth Israel Deaconess Medical Center, Harvard Medical School, Boston, MA, USA
e-mail: ewong@bidmc.harvard.edu

S. Makarov et al. (eds.), *Brain and Human Body Modeling*,
https://doi.org/10.1007/978-3-030-21293-3_8

in intensity-modulated radiation therapy and stereotactic radiosurgery, resulting in less radiation scatter to the surrounding tissue adjacent to the tumor [5, 6].

Laser interstitial thermal therapy (LITT) is another treatment for brain tumors that utilizes a specific part of the electromagnetic spectrum at the microwave frequency or 100 MHz range. A probe is stereotactically inserted into the tumor target under MRI guidance, and the laser at the tip of the probe emits microwave energy to heat the tumor tissue while the temperature is being monitored [7]. The primary application is for the treatment of brain metastasis and radiation necrosis, but surgically inaccessible glioblastomas can also be treated.

TTFields therapy operates at a frequency of 200 kHz for glioblastoma, which is based on preclinical data on mitotic interference of glioma cells in tissue culture [8]. This frequency is below radio frequency for the transmission of AM signals. There are a number of biological effects that can be exploited to treat malignancies. These include disruption of tubulin and septin, both of which are intracellular macromolecules possessing large dipole moments and are necessary for the orderly progression of mitosis in dividing cells [8, 9]. Tubulin monomers coalesce to form microtubules and the mitotic spindle during metaphase and anaphase. These higher order structures are needed to align the 23 pairs of chromosomes in the mitotic plate and to guide the subsequent migration of the corresponding sister chromatids to the respective centrioles. Septin heterotrimers are needed for cytokinesis, or the contraction of the plasma membrane along the equatorial plane of the dividing cell, that eventually produces two daughter cells. Notable phenomena of violent membrane blebbing and asymmetric chromosome segregation have been observed under the influence of TTFields, resulting in aneupoloidy, mitotic arrest, and/or cellular stress that may trigger an immunogenic response [8–10].

8.3 Quantifying Electric Field Delivery in the Brain

The amount of electric field delivered to the tumor target can be quantified. However, the brain has a complex geometry consisting of multiple layers of folded tissues (such as dura, gray matter, white matter, and subcortical nuclei) and asymmetric spaces (ventricles and the subarachnoid space on the cerebral convexity). In addition, the glioblastoma is situated within the white matter with extension to the adjacent gray matter and displacement of the subarachnoid or ventricular space. This tumor is usually seen as an enhancing lesion on MRI, and its gross tumor volume (GTV) can be contoured and delineated by appropriate software. When solving for the electric field distribution at the GTV, the solutions to the differential equations are not straightforward and only representative numerical approximations can be computed using techniques such as finite element analysis. The accuracy of the solution depends on several factors: the smoothness of interface between structures and the resolution of the finite elements that form the approximate solid geometry of the tissue. There is always a trade-off between accuracy of the model and the computational requirements needed to generate a solution. In general, a larger number of elements that are used to represent a more accurate geometry will require more time and increased computational capability to generate a solution.

The workflow to generate a prediction for the numerical value of the electric field at the GTV requires a number of steps [11]. It begins with the conversion of the Digital Imaging and Communications in Medicine (DICOM) dataset from MRI into Neuroimaging Informatics Technology Initiative (NIfTI) image format using the conversion function from Statistical Parametric Mapping 8 (SPM8) and then co-registering to a template in Montreal Neurological Institute space. The default workflow generates ten binarized masks, including white matter, gray matter, cerebellum/brainstem, cerebrospinal fluid, orbits, skull, scalp, gel, electrodes, and air; it has been customized for the finite element analysis of TTFields. The binary masks are then imported into a 3D image processing software such as ScanIP (Synopsis, Mountain View, CA) for post-processing of generated errors. The boundary conditions are then specified and the presence of island cavities and artifacts are manually corrected to ensure no overlapping or missing boundaries are present. The Dice coefficient can then be used to measure the degree of overlap before and after manual processing. The optimized masks are then imported into a simulation software such as COMSOL Multiphysics (Stockholm, Sweden). The output of this process consists of visual maps of the distributed electric fields and the specific absorption rate (SAR) within the brain and the GTV.

The electric field-volume histogram (EVH) and specific absorption rate-volume histogram (SARVH) are indispensable for the comparison of the electric fields and the SAR at the GTV between different models [12] (Fig. 8.1). The histograms are also applicable to other intracranial structures for monitoring side effects at regions adjacent to the GTV. Both EVH and SARVH also facilitate the comparison of TTFields in individual patients over time or between patients who are receiving the same treatment. A number of standardized parameters can be used, including E_{AUC}, VE_{150}, $E_{95\%}$, $E_{50\%}$, and $E_{20\%}$ for electric field quantification, as well as SAR_{AUC}, $VSAR_{7.5}$, $SAR_{95\%}$, $SAR_{50\%}$, and $SAR_{20\%}$ for specific absorption rate representation. In our prior modeling work, the scalp is the site having the highest overall electric field intensity and SAR because the transducer arrays are adjacent to the scalp. However, within the brain, the distribution is highly variable, with the frontal horns of the lateral ventricles and the genu of the corpus callosum having a higher electric field intensity and SAR due to their geometries and juxtaposition to cerebrospinal fluid [12, 13].

A number of factors can alter the intracranial distribution of TTFields. First, the thickness of the cerebrospinal fluid layer located in the subarachnoid space on the surface of the brain can change the penetration of TTFields into the GTV, which is located deeper within the white matter of the brain. In one of our models, a contraction of the cerebrospinal fluid layer by 1.0 mm increased the E_{AUC} by 15–20%, while an expansion by 0.5 mm can reduce the E_{AUC} by about 10% [12]. Second, changes in the conductivity of GTV affect the distribution of TTFields inside the brain, but variations in permittivity do not. In particular, the presence of a tumor-associated necrotic core, which contains fluid and probably has a higher conductivity, may attenuate the penetration of TTFields into the tumor, likely by diverting the electric

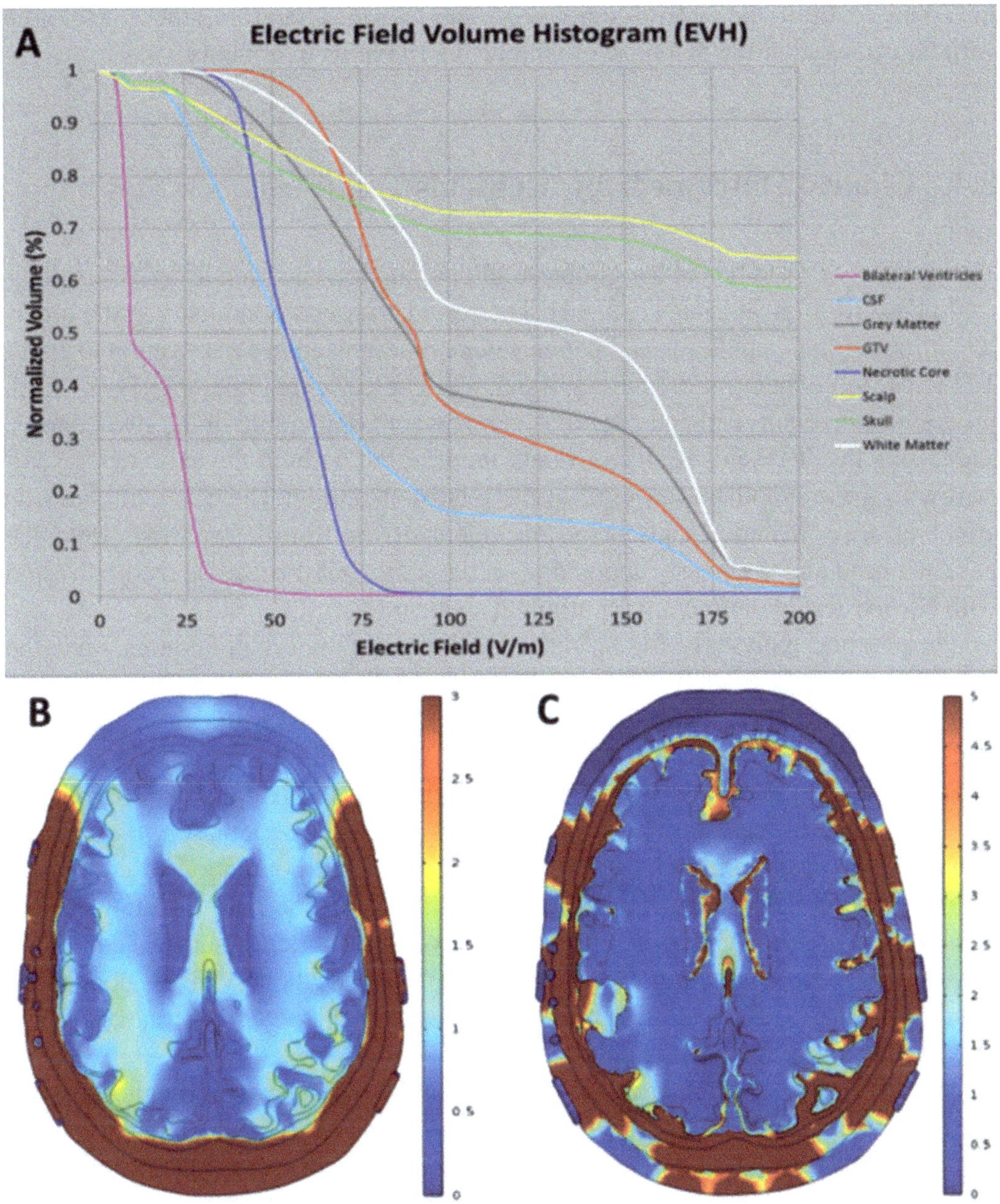

Fig. 8.1 (**a**) EVH for various tissue structures and GTV. Simulation results for (**b**) electric field and (**c**) SAR distributions on an axial plane within the brain, showing higher intensity near the GTV and the corpus callosum

fields away from the active regions of the tumor. Therefore, these anatomic variations may alter the response of patients to TTFields treatments.

8.4 Clinical Outcome from TTFields Treatment

The efficacy of TTFields in the glioblastoma population has been tested in randomized phase III clinical trials [1, 2, 14]. The EF-11 trial was conducted in the recurrent glioblastoma population and patients were randomized in a 1:1 fashion to either TTFields monotherapy or best physician's choice chemotherapy [14]. Both progression-free survival and overall survival were comparable in the two cohorts, suggesting that TTFields is an equivalent therapy, but without the toxicities associated with the chemotherapies used in the recurrent setting. There is a unique side effect of scalp irritation caused by placement of the transducer arrays onto the patient's head and the energy deposition on the scalp, but this can be readily treated with corticosteroid ointment with minimal sequela.

The post hoc analyses of EF-11 yielded a number of important findings. First, the Patient Registry Dataset (PRiDe) contains treatment and health information data from patients who underwent TTFields therapy in real-world, clinical practice settings in the USA from 2011 to 2013 after the initial United States FDA approval. PRiDe revealed that patients lived longer when their TTFields usage compliance averages 75% or greater per day [15]. Furthermore, dexamethasone is an anti-inflammatory drug commonly used in the glioblastoma population to counteract cerebral edema caused by the tumor. A separate post hoc analysis showed that patients who used a higher dose of ≥4.1 mg/day had a shortened survival compared to those who used <4.1 mg/day [16]. Therefore, TTFields therapy in this population requires maximizing treatment compliance while minimizing dexamethasone usage in order to facilitate its anti-glioblastoma effect.

The EF-14 trial is another randomized phase III trial of newly diagnosed glioblastoma patients [2]. All eligible patients received initial radiation and daily temozolomide and were then randomized prior to entry into the post-radiotherapy adjuvant phase of temozolomide treatment. Patients were randomized to receive TTFields plus temozolomide or temozolomide alone, and the TTFields cohort had longer progression-free survival and overall survival compared to the control. Mild to moderate scalp toxicity underneath the transducer arrays was seen in 52% of patients who received TTFields and temozolomide compared to none in the control cohort. Systemic toxicities and health-related quality-of-life measures were comparable between the two groups [2, 17].

8.5 Conclusion

TTFields therapy has established anti-cancer efficacy and the utilization of this treatment for glioblastoma has resulted in prolongation of patient survival. However, the distribution of TTFields within the brain is still shrouded in mystery. Finite element analysis provides a means for the numerical approximation of the distribution of TTFields within the brain and particularly within the GTV. Furthermore, EVH and SARVH permit the quantification of changes in electric fields and specific absorption rate over time, as well as the comparison of these parameters between individuals. Ultimately, the goal is to develop personalized TTFields treatment for each glioblastoma patient.

Acknowledgments This research was supported in part by *A Reason To Ride* research fund.

References

1. Stupp, R., Tallibert, S., Kanner, A. A., et al. (2015). Maintenance therapy with tumor-treating fields plus temozolomide vs temozolomide alone for glioblastoma: A randomized clinical trial. *JAMA, 314*, 2535–2543.
2. Stupp, R., Tallibert, S., Kanner, A. A., et al. (2017). Effect of tumor-treating fields plus maintenance temozolomide vs maintenance temzolomide alsone on survival in patients with glioblastoma: A randomized clinical trial. *JAMA, 318*, 2306–2316.
3. https://www.nccn.org/professionals/physician_gls/pdf/cns.pdf
4. Baskar, R., Dai, J., Wenlong, N., et al. (2014). Biological response of cancer cells to radiation treatment. *Frontiers in Molecular Biosciences, 1*, 24.
5. Bortfeld, T. (2006). IMRT: A review and preview. *Physics in Medicine and Biology, 51*, R363–R379.
6. Taylor, M. L., Kron, T., & Franich, R. D. (2011). A contemporary review of stereotactic radiotherapy: Inherent dosimetric complexities and the potential for detriment. *Acta Oncologica, 50*, 483–508.
7. Sharma, M., Balasubramanian, S., Silva, D., et al. (2016). Laser interstitial thermal therapy in the management of brain metastasis and radiation necrosis after radiosurgery: An overview. *Expert Review of Neurotherapeutics, 16*(2), 223–232.
8. Kirson, E.D., Gurvich, Z., Schneiderman, R., et al. (2004). Disruption of cancer cell replication by alternating electric fields. *Cancer Research, 64*(9), 3288–3295.
9. Gera, N., Yang, A., Holtzman, T. S., et al. (2015). Tumor treating fields perturb the localization of septins and cause aberrant mitotic exit. *PLoS One, 10*, e0125269.
10. Kirson, E. D., Dbalý, V., Tovaryš, F., et al. (2007). Alternating electric fields arrest cell proliferation in animal tumor models and human brain tumors. *Proceedings of the National Academy of Sciences, 104*(24), 10152–10157.
11. Timmons, J. J., Lok, E., San, P., et al. (2017). End-to-end workflow for finite element analysis of tumor treating fields in glioblastomas. *Physics in Medicine and Biology, 62*, 8264–8282.
12. Lok, E., San, P., Hua, V., et al. (2017). Analysis of physical characteristics of tumor treating fields for human glioblastoma. *Cancer Medicine, 6*, 1286–1300.
13. Lok, E., Hua, V., & Wong, E. T. (2015). Computed modeling of alternating electric fields therapy for recurrent glioblastoma. *Cancer Medicine, 4*, 1697–1699.

14. Stupp, R., Wong, E. T., Kanner, A. A., et al. (2012). NovoTTF-100A versus physician's choice chemotherapy in recurrent glioblastoma: A randomized phase III trial of a novel treatment modality. *European Journal of Cancer, 48*, 2192–2202.
15. Mrugala, M. M., Engelhard, H. H., Tran, D. D., et al. (2014). Clinical practice experience with NovoTTF-100A™ system for glioblastoma: The patient registry dataset (PRiDe). *Seminars in Oncology, 42*, S4–S13.
16. Wong, E. T., Lok, E., Gautam, S., & Swanson, K. D. (2015). Dexamethasone exerts profound immunologic interference on treatment efficacy for recurrent glioblastoma. *British Journal of Cancer, 113*, 232–241.
17. Taphoorn, M. J. B., Dirven, L., Kanner, A. A., et al. (2018). Influence of treatment with tumor-treating fields on health-related quality of life of patients with newly diagnosed glioblastoma: A secondary analysis of a randomized clinical trial. *JAMA Oncology, 4*, 495–504.

Chapter 9
Advanced Multiparametric Imaging for Response Assessment to Tumor-Treating Fields in Patients with Glioblastoma

Suyash Mohan, Sumei Wang, and Sanjeev Chawla

9.1 Introduction

Glioblastoma (GBM) is the most malignant tumor in the brain, representing 30% of all central nervous system tumors (CNST) and 70% of primary malignant CNST. GBM is the cause of 225,000 deaths per year throughout the world. It has an incidence of 5 per 100,000 persons, affects 1.5 times more men than women, and is diagnosed at an average age of 64 [1]. The current standard of care treatment for patients with GBM includes maximal safe resection followed by radiotherapy and chemotherapy using temozolomide (TMZ). However, the prognosis of GBM is miserable with a median overall survival (OS) of only 12–18 months following diagnosis [2]. Because of continuing progress in the quest for an effective treatment, the therapeutic armamentarium for patients with GBM has grown significantly over the past decade. However, newly tested adjuvant strategies such as the addition of bevacizumab [3], modified TMZ dosing [4], or use of other targeted therapies [5, 6] have failed to significantly improve OS. These limitations necessitate the investigation of novel therapies to treat patients with GBMs.

9.2 Tumor-Treating Fields: *Scientific Basis*

Recently, the US Food and Drug Administration (FDA) approved the use of alternating electric fields, also known as tumor-treating fields (TTFields), as a novel modality for the treatment of patients with newly diagnosed and recurrent GBM. It consists of a portable, noninvasive, and in-home use battery-operated medical

S. Mohan (✉) · S. Wang · S. Chawla
Department of Radiology, Division of Neuroradiology, Perelman School of Medicine at the University of Pennsylvania, Philadelphia, PA, USA
e-mail: suyash.mohan@uphs.upenn.edu

S. Makarov et al. (eds.), *Brain and Human Body Modeling*,
https://doi.org/10.1007/978-3-030-21293-3_9

device involving four insulated transducer arrays composed of biocompatible ceramic discs (9 discs per array) that are applied to the shaved scalp of a patient. The position and size of the transducer arrays can be adjusted depending upon patient head size, tumor dimensions, and location. TTFields deliver oscillating electric energy at low intensity (1–3 V/cm) and at an intermediate frequency (200–300 kHz) as a loco-regional intervention. TTFields produce antimitotic effects by physically interacting with highly charged macromolecules and organelles in rapidly dividing cancer cells to disrupt their proper alignment during the metaphase and/or anaphase stages of mitotic cell division, mainly sparing the effect of oscillating electric fields on normal quiescent cells [7, 8]. In one study [9], treating U-118 glioma cells with TTFields in combination with standard chemotherapeutic drugs (Paclitaxel, Doxorubicin, Cyclophosphamide) resulted in the destruction of most living cells after 70 hours of treatment, while the drugs or TTFields alone only slowed down cancer cell proliferation, suggesting that TTFields should be combined with another treatment modalities to reach optimal effectiveness. In another study [10], rats bearing intracranial GBM were treated with TTFields for 6 days, leading to smaller tumors compared with untreated rats. Interestingly, this study underlined the necessity of applying TTFields in several directions to yield antitumor efficacy. There also appears to be a time-dependent treatment effect, with optimal efficacy being observed when wearing the treatment mask at least 18 hours per day (75%) [11].

9.3 Tumor-Treating Fields: *Clinical Application in GBM Patients*

TTFields have been widely used in the treatment of a variety of cancers, for example: glioma, melanoma, and adenocarcinoma, with favorable safety profiles and without significant adverse effects in patients [8, 12]. Previously, promising findings of large-scale multinational clinical trials have also been reported in patients with GBM. In particular, in a phase III clinical study [11] involving 466 patients, the addition of TTFields to standard therapy was shown to increase median OS from 15.6 to 20.5 months (hazard ratio = 0.64, $p = 0.0042$). The 2-year survival rate was approximately 50% greater with TTFields plus TMZ versus TMZ alone: 43% versus 29%. Additionally, improved quality of life with better cognitive and emotional functions was observed in TTFields treated cohorts of patients [11]. Moreover, the treatment had limited adverse events, mainly restricted to mild or moderate skin irritations beneath the transducer arrays from wearing the device. Therefore, these results were exciting for both physicians and patients alike.

9.4 Tumor-Treating Fields: *Advanced Neuroimaging Techniques*

We at the University of Pennsylvania are investigating the utility of advanced neuroimaging techniques in monitoring treatment-related temporal characteristics and assessing response to this unique treatment modality. Anatomic magnetic resonance imaging (MRI) of the brain provides excellent soft tissue contrast and is routinely used for the noninvasive characterization of brain tumors. However, conventional imaging utilizing "Response Assessment in Neuro-Oncology" (RANO) criteria is usually not reliable for assessment of the treatment response in patients with GBM due to the lack of specificity [13]. Consequently, there is an urgent need to develop increasingly accurate quantitative imaging biomarkers for early evaluation of treatment response. These biomarkers are the premise of personalized treatment, enabling change or discontinuation of therapy to prevent ineffective treatment or unfavorable events. Moreover, identification of treatment failure may help reduce adverse economic consequences. This is highly relevant because the cost of TTFields therapy is considerably high at $21,000 per month [14]. Advanced MR imaging techniques such as diffusion-tensor imaging (DTI) [15], dynamic susceptibility contrast (DSC)-perfusion weighted imaging (PWI) [16, 17], and proton MR spectroscopy (^{1}H MRS) [18, 19] have shown great potential in evaluating treatment response to different therapeutic regimens in GBM patients. DTI is an MR imaging technique used to noninvasively investigate the cyto-architectural integrity of brain structures by measuring the anisotropy of microscopic water diffusivity. Along with more commonly used DTI parameters such as mean diffusivity (MD) and fractional anisotropy (FA), geometrical DTI indices such as the coefficients of linear anisotropy (CL) and planar anisotropy (CP) can be helpful in characterizing tissue organization and orientation of white matter tracts in the brain. Relative cerebral blood volume (rCBV) derived from PWI reflects tumor angiogenesis and vascularity. ^{1}H MRS is a method that measures metabolic markers of neoplastic activity [20]. Spectra from brain tumors have increased choline (Cho), which correlates with membrane biosynthesis by proliferating cells, and reduced N-acetylasparate (NAA), which indicates loss of neuronal integrity due to tumor cell infiltration [21]. 3D-Echo planar spectroscopic imaging (EPSI) allows acquisition of volumetric metabolite maps with high spatial resolution, minimizing partial-volume averaging effects [22, 23]. Thus, 3D-EPSI may be helpful in providing metabolite information from the entire volume of a neoplasm. The potential of 3D-EPSI has been reported in characterizing glioma grades [24], mapping glycine distribution in gliomas [25], planning radiation therapy for GBM patients [26], identifying residual tumors following radiation therapy [27], evaluating response to epigenetic modifying agents in recurrent GBM [28], in assessing the effect of whole brain radiation therapy on normal brain parenchyma in patients with metastases [29] and in distinguishing true progression (TP) from pseudoprogression (PsP) in GBM patients [30].

9.5 Tumor-Treating Fields: *Initial Experience*

We have recently reported our initial experience of assessing short-term (up to 2 months) response to TTFields in a newly diagnosed patient with left thalamic GBM using physiological and metabolic MR imaging techniques [31]. In addition to conventional imaging, the patient also underwent DTI, PWI, and 3D-EPSI on a 3 T MRI scanner prior to initiation of TTFields (baseline) and at 1- and 2-month follow-ups. The values of various advanced imaging parameters such as MD, FA, rCBV, and choline/creatine (Cho/Cr) were measured from the contrast-enhancing region of the neoplasm at each time point using a previously described method [32]. Tumor size decreased from 32.5 × 27.7 mm (baseline) to 25.8 × 24.9 mm (2nd follow up), as seen on the postcontrast T1-weighted images (Fig. 9.1). Tumor volume also steadily declined at the 1st (~12%) and 2nd (~34%) follow-up periods relative to the baseline. Representative images and parametric maps from baseline and follow-up time-points are presented in Fig. 9.2. Percent changes in volume, MD, FA, $rCBV_{max}$, and Cho/Cr from baseline to post-TTFields at 1- and 2-month follow-up periods are shown in Fig. 9.3. We also found a moderate increase in MD (~11%) along with decreases in FA (~23%) from enhancing regions of neoplasms. Previous studies [33, 34] have reported increased MD and reduced FA from the tumor in patients with gliomas treated with chemoradiation therapy. However, the interpretation of changes in MD following radiation therapy and adjuvant chemotherapy is complex because of co-localization of treatment-induced gliosis, necrosis, and edema [35]. In an earlier study from our group, Wang et al. [36] reported higher MD and significantly lower FA in post-treatment GBM patients with PsP compared with those with TP, suggesting that elevated MD and reduced FA are associated with favorable treatment response. Our DTI results are in agreement with these studies and imply that DTI can assess therapeutic response to TTFields. We believe that cellular growth inhibition and associated cell death at 2 months might have accounted for the large increase in MD observed in our patient. It has been widely reported that organized microstructures secondary to closely packed proliferating tumor cells in gliomas results in high FA [34, 37]. A 23% reduction in FA in the current case may be due to reduced cell density and incoherent orientation of neoplastic cells.

We also observed a moderate decline in rCBVmax (6.21%) at 2 months relative to baseline. Rich capillary networks secondary to angiogenesis are a common feature of GBMs, responsible for high rCBV [38]. Several studies [16, 17, 39] have reported reduced rCBV in gliomas following radiotherapy and anti-angiogenetic therapy. Fibrinoid necrosis, endothelial injury, and occlusion of blood vessels have been proposed as potential reasons for decreased rCBV levels in treated GBMs [40]. In agreement with these studies, reduced rCBV were also noted in the present case, suggesting reduced vascularity and tissue perfusion within the tumor bed. A previous study reported substantial decrease in the levels of CD34 (an immunohistochemical marker of micro-vessel density) and downregulation of vascular endothelial growth factors (VEGF) in murine melanomas exposed to intermediate frequency alternating electric field compared to the control group [41]. While it is not clear

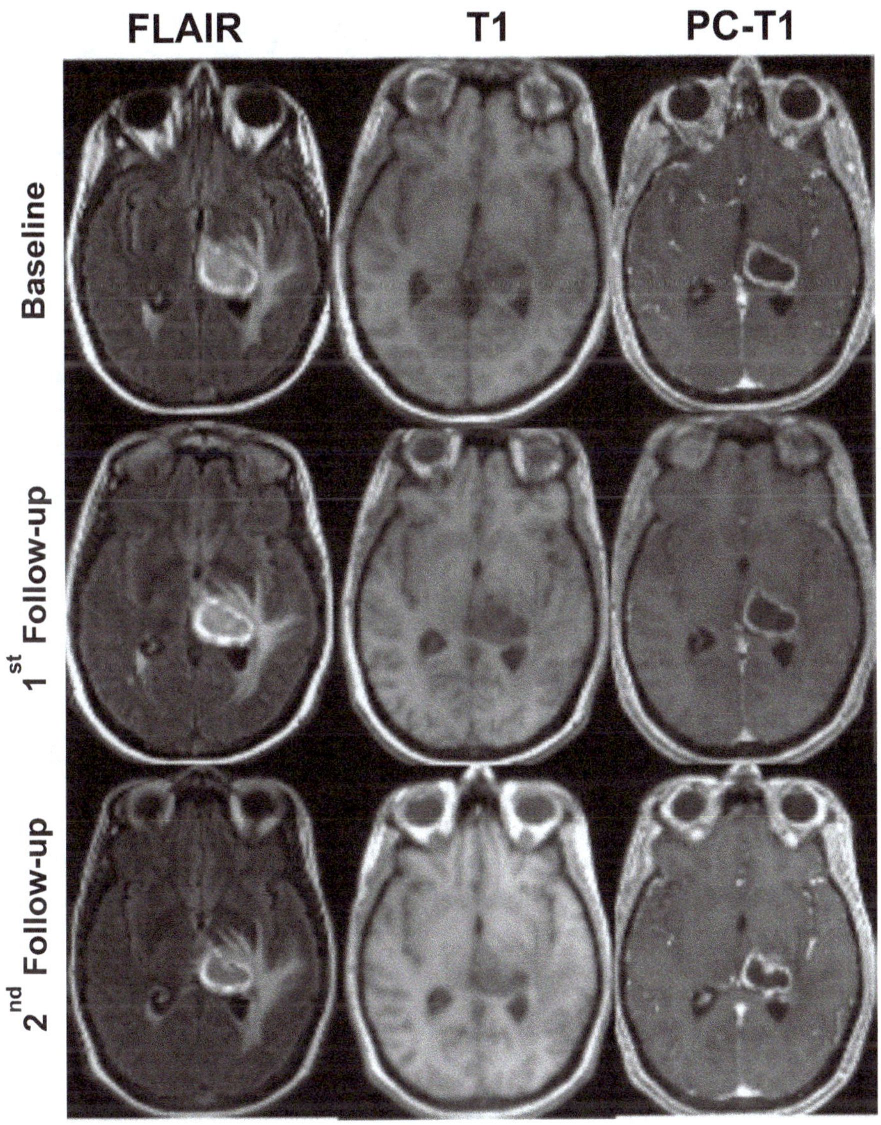

Fig. 9.1 A 51-year-old patient with newly diagnosed GBM treated with TTFields plus TMZ. Axial FLAIR images at three time points demonstrate a heterogeneous mass centered in the left thalamus with surrounding signal abnormality. This mass appears hypointense on the corresponding T1-weighted images and demonstrates heterogeneous peripheral enhancement with central necrotic core on the corresponding postcontrast T1-weighted images. (Reprinted with permission from Ref. [31])

how a combination of TTFields and TMZ chemotherapy modulates tumor vasculature of gliomas, it may be speculated that inhibited angiogenesis might have caused decreased perfusion in our case.

Several prior ^{1}H MRS studies [18, 19, 39] have reported decreased levels of Cho as a surrogate marker of positive treatment response in patients with brain tumors. In accordance with these previous studies, we also observed decreased levels of Cho/Cr at the 2-month period following treatment (Fig. 9.4). It is well documented that Cho content correlates with cell density and with indices of cellular proliferation [42]. We believe that reduction in Cho in our case was most likely a direct consequence of the combined antiproliferative effect of TTFields and TMZ on cellular metabolism of gliomas.

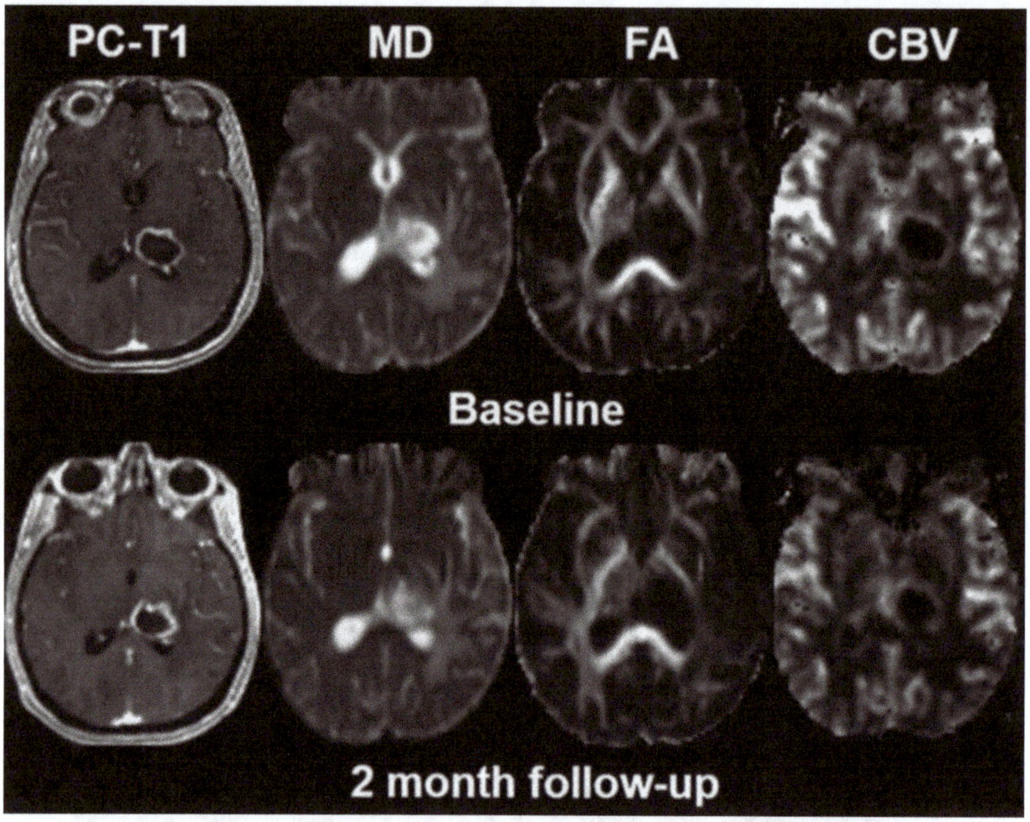

Fig. 9.2 Axial co-registered contrast-enhanced T1-weighted image and corresponding MD, FA, and CBV maps are shown at baseline and at a 2-month follow-up period. (Reprinted with permission from Ref. [31])

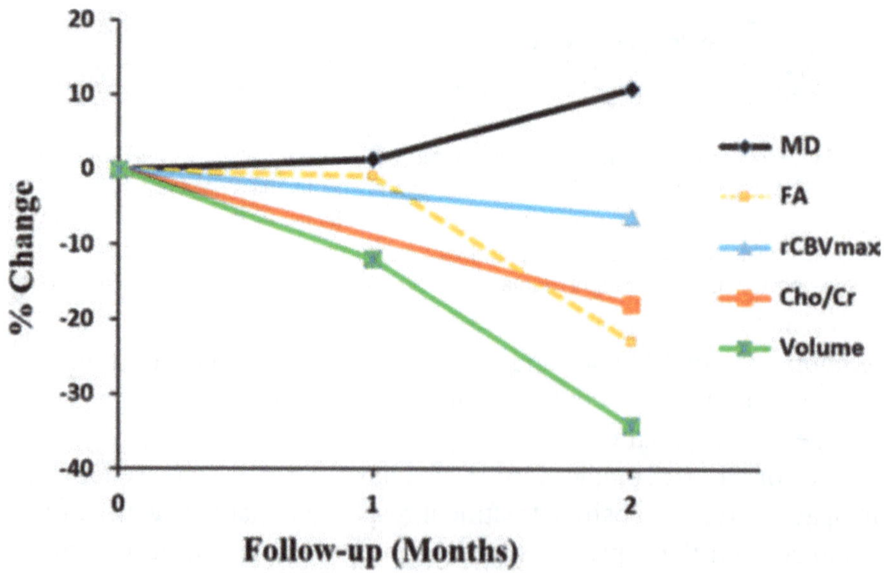

Fig. 9.3 Percentage change in parameters from baseline to 1- and 2-month follow-up periods. Trends towards decreased tumor volume, $rCBV_{max}$, Cho/Cr, and FA along with an increased MD were observed at follow-up relative to baseline indicating tumor growth arrest. (Reprinted with permission from Ref. [31])

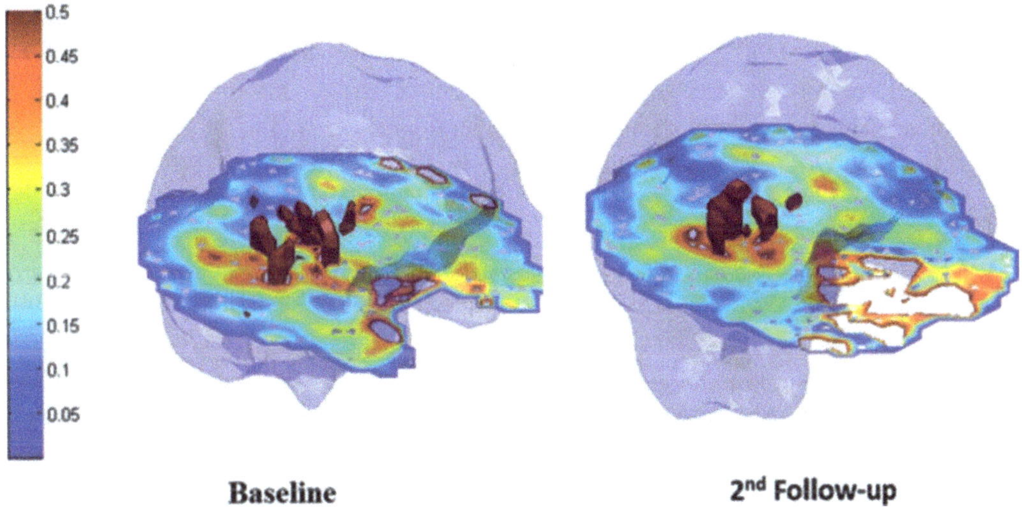

Fig. 9.4 Red volumes in Cho/Cr maps correspond to voxels that exceed a threshold value of 0.55 at baseline and at 2-month follow-up period. The total number of voxels that exceed the threshold value of 0.55 were 50 at baseline and 34 at 2nd follow-up, suggesting reduced levels of Cho/Cr relative to baseline. (Reprinted with permission from Ref. [31])

Taken together, our initial observations [31] indicate that a multiparametric approach utilizing the unique strengths of advanced imaging techniques as performed in the present case may provide a comprehensive assessment of treatment response. Our work is in progress and we are currently recruiting and evaluating patients with newly diagnosed, as well as recurrent GBM treated with TTFields in an ongoing clinical trial.

9.6 Conclusion

The identification of novel image-based biomarkers may be helpful in determining early and true therapeutic response to TTFields in patients with GBM. However, it is difficult to compare the results of individual studies because of methodological differences and varying clinical endpoints. Analytical methods of advanced MR imaging techniques can vary, including subjective/qualitative evaluation of parametric maps, user-defined region of interest values (using mean, median, maximum, or minimum), histogram analysis, and voxel-wise analysis (i.e., PRMs and fDMs). It should be noted that there is a need to establish universal quantitative imaging biomarker thresholds to evaluate treatment response to TTFields in GBMs. We believe that adequately powered, randomized, placebo-controlled, multicenter studies using optimal acquisition parameters of advanced MR imaging techniques, along with standardized postprocessing methods, are warranted to comprehensively determine the potential efficacy of TTFields in patients with GBM. This approach will enhance the decision-making process in the use of this novel treatment modality.

Acknowledgments The authors would also like to thank the University of Pennsylvania radiology research team, Lisa Desiderio, Lauren Karpf, and MRI technicians, for their valuable contributions to this project.

Disclosures This study was funded in part by a grant from Novocure Ltd., Haifa, Israel.

References

1. Bush, N. A., Chang, S. M., & Berger, M. S. (2017). Current and future strategies for treatment of glioma. *Neurosurgical Review, 40*, 1–14.
2. Stupp, R., Mason, W. P., van den Bent, M. J., et al. (2005). Radiotherapy plus concomitant and adjuvant temozolomide for glioblastoma. *The New England Journal of Medicine, 352*, 987–996.
3. Chinot, O. L., & Reardon, D. A. (2014). The future of antiangiogenic treatment in glioblastoma. *Current Opinion in Neurology, 27*, 675–682.
4. Walker, G. V., Gilbert, M. R., Prabhu, S. S., et al. (2013). Temozolomide use in adult patients with gliosarcoma: An evolving clinical practice. *Journal of Neuro-Oncology, 112*, 83–89.
5. Batchelor, T. T., Gerstner, E. R., Emblem, K. E., et al. (2013). Improved tumor oxygenation and survival in glioblastoma patients who show increased blood perfusion after cediranib and chemoradiation. *Proceedings of the National Academy of Sciences of the United States of America, 110*, 19059–19064.
6. Westphal, M., Heese, O., Steinbach, J. P., et al. (2015). A randomised, open label phase III trial with nimotuzumab, an anti-epidermal growth factor receptor monoclonal antibody in the treatment of newly diagnosed adult glioblastoma. *European Journal of Cancer, 51*, 522–532.
7. Stupp, R., Wong, E. T., Kanner, A. A., et al. (2012). NovoTTF-100A versus physician's choice chemotherapy in recurrent glioblastoma: A randomised phase III trial of a novel treatment modality. *European Journal of Cancer, 48*(14), 2192–2202.
8. Pless, M., & Weinberg, U. (2011). Tumor treating fields: Concept, evidence and future. *Expert Opinion on Investigational Drugs, 20*(8), 1099–1106.
9. Kirson, E. D., Schneiderman, R. S., Dbalý, V., et al. (2009). Chemotherapeutic treatment efficacy and sensitivity are increased by adjuvant alternating electric fields (TTFields). *BMC Medical Physics, 8*(9), 1.
10. Kirson, E. D., Gurvich, Z., Schneiderman, R., et al. (2004). Disruption of cancer cell replication by alternating electric fields. *Cancer Research, 64*(9), 3288–3295.
11. Stupp, R., Taillibert, S., Kanner, A. A., et al. (2015). Maintenance therapy with tumor-treating fields plus temozolomide vs temozolomide alone for glioblastoma: A randomized clinical trial. *Journal of the American Medical Association, 314*(23), 2535–2543.
12. Wong, E. T., Lok, E., & Swanson, K. D. (2015). An evidence-based review of alternating electric fields therapy for malignant gliomas. *Current Treatment Options in Oncology, 16*(8), 40. https://doi.org/10.1007/s11864-015-0353-5.
13. Jackson, E. F., Barboriak, D. P., Bidaut, L. M., & Meyer, C. R. (2009). Magnetic resonance assessment of response to therapy: tumor change measurement, truth data and error sources. *Translational Oncology, 2*(4), 211–215.
14. Bernard-Arnoux, F., Lamure, M., Ducray, F., et al. (2016). The cost-effectiveness of tumor-treating fields therapy in patients with newly diagnosed glioblastoma. *Neuro-Oncology, 18*, 1129–1136.
15. Saraswathy, S., Crawford, F. W., Lamborn, K. R., et al. (2009). Evaluation of MR markers that predict survival in patients with newly diagnosed GBM prior to adjuvant therapy. *Journal of Neuro-Oncology, 91*(1), 69–81.
16. Schmainda, K. M., Prah, M., Connelly, J., et al. (2014). Dynamic-susceptibility contrast agent MRI measures of relative cerebral blood volume predict response to bevacizumab in recurrent high-grade glioma. *Neuro-Oncology, 16*(6), 880–888.

17. Aquino, D., Di Stefano, A. L., Scotti, A., et al. (2014). Parametric response maps of perfusion MRI may identify recurrent glioblastomas responsive to bevacizumab and irinotecan. *PLoS One, 9*(3), e90535. https://doi.org/10.1371/journal.pone.0090535.
18. Jeon, J. Y., Kovanlikaya, I., Boockvar, J. A., et al. (2012). Metabolic response of glioblastoma to superselective intraarterial cerebral infusion of bevacizumab: A proton MR spectroscopic imaging study. *American Journal of Neuroradiology, 33*(11), 2095–2102.
19. Muruganandham, M., Clerkin, P. P., Smith, B. J., et al. (2014). 3-Dimensional magnetic resonance spectroscopic imaging at 3 Tesla for early response assessment of glioblastoma patients during external beam radiation therapy. *International Journal of Radiation Oncology, Biology, Physics, 90*(1), 181–189.
20. Chawla, S., Wang, S., Wolf, R. L., et al. (2007). Arterial spin labelling and magnetic resonance spectroscopy in differentiation of gliomas. *American Journal of Neuroradiology, 28*, 1683–1689.
21. Chawla, S., Oleaga, L., Wang, S., et al. (2010). Role of proton magnetic resonance spectroscopy in differentiating oligodendrogliomas from astrocytomas. *Journal of Neuroimaging, 20*, 3–8.
22. Ebel, A., Soher, B. J., & Maudsley, A. A. (2001). Assessment of 3D proton MR echo-planar spectroscopic imaging using automated spectral analysis. *Magnetic Resonance in Medicine, 46*, 1072–1078.
23. Maudsley, A. A., Darkazanli, A., Alger, J. R., et al. (2006). Comprehensive processing, display and analysis for in vivo MR spectroscopic imaging. *NMR in Biomedicine, 19*(4), 492–503.
24. Roy, B., Gupta, R. K., Maudsley, A. A., et al. (2013). Utility of multiparametric 3-T MRI for glioma characterization. *Neuroradiology, 55*, 603–613.
25. Maudsley, A. A., Gupta, R. K., Stoyanova, R., et al. (2014). Mapping of glycine distributions in gliomas. *American Journal of Neuroradiology, 35*, S31–S36.
26. Parra, N. A., Maudsley, A. A., Gupta, R. K., et al. (2014). Volumetric spectroscopic imaging of glioblastoma multiforme radiation treatment volumes. *International Journal of Radiation Oncology, Biology, Physics, 90*, 376–384.
27. Lin, D., Lin, Y., Link, K., et al. (2016). Echoplanar magnetic resonance spectroscopic imaging before and following radiation therapy in patients with high-grade glioma. *International Journal of Radiation Oncology, Biology, Physics, 96*(2S), E133–E134.
28. Shim, H., Holder, C. A., & Olson, J. J. (2013). Magnetic resonance spectroscopic imaging in the era of pseudoprogression and pseudoresponse in glioblastoma patient management. *CNS Oncology, 2*, 393–396.
29. Chawla, S., Wang, S., Kim, S., et al. (2015). Radiation injury to the normal brain measured by 3D-echo-planar spectroscopic imaging and diffusion tensor imaging: Initial experience. *Journal of Neuroimaging, 25*, 97–104.
30. Verma, G., Chawla, S., Mohan, S., et al. (2019). Differentiation of true progression from pseudoprogression in patients with glioblastoma using whole brain echo-planar spectroscopic imaging. *NMR in Biomedicine, 32*(2), e4042. https://doi.org/10.1002/nbm.4042. Epub 2018 Dec 17.
31. Mohan S, Chawla S, Wang S, Verma G, Skolnik A, Brem S, Peters KB, Poptani H. (2016). Assessment of early response to tumor-treating fields in newly diagnosed glioblastoma using physiologic and metabolic MRI: initial experience. *CNS Oncology, 5*(3):137–144. https://doi.org/10.2217/cns-2016-0003. Epub 2016 Apr 14. PubMed PMID: 27076281; PubMed Central PMCID: PMC6042635.
32. Wang, S., Kim, S., Chawla, S., et al. (2009). Differentiation between glioblastomas and solitary brain metastases using diffusion tensor imaging. *NeuroImage, 44*(3), 653–660.
33. Zhang, J., van Zijl, P. C. M., Laterra, J., et al. (2007). Unique patterns of diffusion directionality in rat brain tumors revealed by high-resolution diffusion tensor MRI. *Magnetic Resonance in Medicine, 58*, 454–462.
34. Beppu, T., Inoue, T., Shibata, Y., et al. (2005). Fractional anisotropy value by diffusion tensor magnetic resonance imaging as a predictor of cell density and proliferation activity of glioblastomas. *Surgical Neurology, 63*, 56–61.

35. Tomura, N., Narita, K., Izumi, J.-I., et al. (2006). Diffusion changes in a tumor and peritumoral tissue after stereotactic irradiation for brain tumors: Possible prediction of treatment response. *Journal of Computer Assisted Tomography, 30*(3), 496–500.
36. Wang, S., Martinez-Lage, M., Sakai, Y., et al. (2016). Differentiating tumor progression from pseudoprogression in patients with glioblastomas using diffusion tensor imaging and dynamic susceptibility contrast MRI. *American Journal of Neuroradiology, 37*(1), 28–36.
37. Kinoshita, M., Hashimoto, N., Goto, T., et al. (2008). Fractional anisotropy and tumor cell density of the tumor core show positive correlation in diffusion tensor magnetic resonance imaging of malignant brain tumors. *NeuroImage, 43*(1), 29–35.
38. Lee, S. J., Kim, J. H., Kim, Y. M., et al. (2001). Perfusion MR imaging in gliomas: Comparison with histologic tumor grade. *Korean Journal of Radiology, 2*, 1–7.
39. Khan, M. N., Sharma, A. M., Pitz, M., et al. (2016). High-grade glioma management and response assessment-recent advances and current challenges. *Current Oncology, 23*(4), e383–e391. https://doi.org/10.3747/co.23.3082.
40. Liu, X. J., Duan, C. F., Fu, W. W., et al. (2015). Correlation between magnetic resonance perfusion weighted imaging of radiation brain injury and pathology. *Genetics and Molecular Research, 14*(4), 16317–16324.
41. Chen, H., Liu, R., Liu, J., & Tang, J. (2012). Growth inhibition of malignant melanoma by intermediate frequency alternating electric fields, and the underlying mechanisms. *The Journal of International Medical Research, 40*, 85–94.
42. Miller, B. L., Chang, L., Booth, R., et al. (1996). In vivo 1H MRS choline: Correlation with in vitro chemistry/histology. *Life Sciences, 58*, 1929–1935.

Chapter 10
Estimation of TTFields Intensity and Anisotropy with Singular Value Decomposition: A New and Comprehensive Method for Dosimetry of TTFields

Anders Rosendal Korshoej

10.1 Introduction

Tumor-treating fields (TTFields) are a new and effective treatment against glioblastoma (GBM) [1–3]. The treatment uses alternating fields (200 kHz for GBM) to inhibit cancer cell division and tumor growth. TTFields are induced by two electrical sources, each connected to its own pair of 3×3 transducer arrays, which are placed on the patient's body surface in the vicinity of the tumor [4]. Recently, finite element (FE) methods have been used to calculate the distribution of TTFields in realistic human head models in efforts to estimate the treatment dose of TTFields [5–8]. This has provided important information about how the TTFields distribution is affected by human head morphology [9], tumor position [10, 11], tissue dielectric properties [9, 10, 12, 13], and transducer array layout [11, 14]. In addition, FE methods have been used to provide preclinical proof of concept for a new implementation of TTFields in which individual computational modeling is used to plan a surgical skull remodeling procedure that enhances the efficacy of TTFields by creating small holes in the skull at selected positions, facilitating current flow into the tumor [15, 16]. However, state-of-the-art approaches only use the intensity of TTFields as a surrogate "dose" estimate. This is motivated by in vitro studies showing that increasing TTFields intensity decreases tumor growth rate [17]. However, it is also known that the antitumor effects of TTFields depend on the treatment exposure time as well as the direction of the induced fields relative to the direction of cell division. Specifically, longer exposure time kills more cancer cells [18], and cells dividing along the direction of the active field are damaged to a greater extent than

A. R. Korshoej (✉)
Department of Neurosurgery, Aarhus University Hospital, Aarhus, Denmark

Department of Clinical Medicine, Aarhus University, Aarhus, Denmark

Department of Neurosurgery, Odense University Hospital, Odense, Denmark
e-mail: andekors@rm.dk

S. Makarov et al. (eds.), *Brain and Human Body Modeling*,
https://doi.org/10.1007/978-3-030-21293-3_10

cells dividing perpendicularly to the field [17, 19]. This observation is supported by the fact that two sequential fields induced by layouts placed in orthogonal directions on the scalp enhance the efficacy of the treatment in vivo by approximately 20% compared to a single field [20]. This illustrates the notion that multidirectional TTFields are able inhibit a larger fraction of cells in a volume because the effect is distributed more uniformly across cells dividing in random directions [20]. By a similar notion, TTFields are currently applied using two array pairs, which are activated in an even 50% duty cycle of 2-second duration (Optune®, Novocure, Ltd.). The arrays are positioned so that the field intensity in the tumor is maximized, while the arrays are maintained in approximately orthogonal orientations [4]. However, because of the complex conductivity distribution of the head and individual differences in anatomy and tumor morphology, the induced fields are not necessarily orthogonal throughout the exposed volume. This problem has not been addressed in TTFields modeling until now, which may give a biased or incomplete foundation for determining the actual efficacy of TTFields. This chapter presents a new method, which potentially resolves this limitation by quantifying both the average field intensity and the amount of unwanted spatial correlation between the induced fields. The chapter is based on results published by Korshoej et al. [21], and further elaborates on the underlying modeling methods. The new dosimetry approach is based on FE computations and principal component analysis (PCA). I will describe how significant field anisotropy can occur in GBM patients and how this potentially affects layout planning and clinical implementation. Finally, I will briefly discuss how unwanted field anisotropy can potentially be reduced using activation cycle optimization.

10.2 Preparation of Computational Models and Calculation of the Electrical Field

In the following sections, I will describe the methods used to perform FE calculations of the TTFields distribution in a realistic, patient-based head model. I will focus mainly on the basic physics of TTFields, as well as the general concept of FE computation.

10.2.1 Laplace's Equation: The Electro-quasistatic Approximation of Maxwell's Equations

The physical effects of TTFields are governed by Maxwell's equations of electrodynamics [22]. To describe the interaction of TTFields with a volume conductor, e.g., the human head or another body region, the goal is to approximate a solution to Maxwell's equations under a particular set of boundary conditions. These typically represent constraints applied to the functional values at particular regions of the model. For TTFields, we may assume the electrodynamic behavior to be quasistationary, which simplifies the problem [23, 24]. Quasistationary systems satisfy particular conditions regarding the frequency of the current/field, the dielectric properties of the system materials, and the size of the system. Specifically, quasistationarity requires that the magnetic permeabilities and inductive effects are negligible. Furthermore, we require the ratio $\varepsilon\omega / \sigma$ to be low, i.e., $\varepsilon\omega / \sigma \ll 1$, where ε is the real-valued permittivity of the system, ω the angular frequency of the field, and σ the real-valued conductivity, which implies that capacitive effects are also negligible. Therefore, the induced currents are mainly Ohmic. Magnetization currents and displacement currents do not contribute notably and so local changes in the field are propagated throughout the physical system without time delay and produce synchronous field variations in the system. TTFields satisfies these electro-quasistatic assumptions when applied to the head. This is because of the dielectric properties of biological tissues, the low/intermediate frequency (200 kHz) of TTFields, and the small dimension of the physical system, i.e., head (0.2 m). Furthermore, this implies that the electric potential φ can be approximated with Laplace's equation $\nabla \bullet (\sigma \nabla \varphi) = 0$, where σ is the real Ohmic conductivity [5, 22–25]. The requirement of $\varepsilon\omega / \sigma \ll 1$ is supported by Wenger et al. [12], who reported a low sensitivity of the TTFields towards permittivity variations. Similar results were obtained by Lok et al. [26], which further supports that the electro-quasistatic assumption is valid within the range of parameters relevant for TTFields. In the following sections, I will describe the basics of FE approximation to Laplace's equation.

10.2.2 The Finite Element Framework for TTFields

In the data presented here, we used finite element methods to solve Laplace's equation of the electrostatic potential, as defined in the previous section. We used Dirichlet boundary conditions given by the geometrical boundaries of the head surface and the desired choice of electrostatic potential at the electrode interface. The finite elements had tetrahedral geometry adapted in shape and size to approximate the individual volumes and surfaces of the patient's head. In addition to providing a close anatomical approximation, the mesh was dense enough to allow for detailed variations in dielectric properties. Using first-order finite elements, we formulated the electric potential at any point in the model as a linear function of the electric potentials at the nodes of the tetrahedron containing the point. These linear "basis" functions were then used to build a system of linear equations, which was solved using a conjugate gradient solver (GetDP, http://getdp.info/) with the residual tolerance set to <1E-9. Potentials were fixed at the top of the individual electrodes in an array. All electrodes belonging to one array were thereby set to a potential of 1 V and the electrodes of the corresponding array to −1 V. We then calculated the electric field as the numerical gradient of the electric potential. The current density vectors were calculated with Ohm's law and we then rescaled the potentials, fields, and current densities to obtain a total current of 1.8 A through each array pair. This was computed as the numerical integral of current density components normal to the arrays over the entire transducer array area. This approach was chosen over Neumann boundary conditions because it enabled us to model the actual situation in which all of the nine transducers in an array were connected to the same current source and therefore had the same electric potential.

10.2.3 Creation of Personalized Head Models

A number of different approaches have been used to create computational head models for TTFields [7]. We created a patient-specific head model based on T1- and T2-weighted MRI sequences from a male patient with GBM in the left parietal region. The images were processed using SimNIBS (simnibs.org) to produce a 3D volume head mesh consisting of five tissue types, namely skin, skull, cerebrospinal fluid (CSF), gray matter (GM), and white matter (WM). A detailed description of the SimNIBS workflow is given in Windhoff et al. [27]. In summary, segmentation is based on the initial extraction of tissue boundary surfaces, which are then processed and tessellated to produce a tetrahedral volume mesh with variable resolution. Surface meshes from WM, GM, cerebellum, and the brain stem are extracted from the T1 MRI data using Freesurfer algorithms (http://surfer.nmr.mgh.harvard.edu/) [28], while skin, skull, and CSF boundaries are extracted from both T1 and T2 MRI data using the FSL toolbox (https://fsl.fmrib.ox.ac.uk/fsl) [29]. The surface meshes are then postprocessed in MeshFix [30] to repair self-intersections, remove duplicate triangles, and enhance triangle uniformity and regularity. Surfaces are then decoupled to ensure that all tissues are nonoverlapping. The tumor region was outlined manually from a gadolinium-enhanced T1 MRI sequence, following the initial automated segmentation, shown in Fig. 10.1. The resulting binary volume mask was subsequently smoothed and transformed into a triangulated surface mesh using custom scripts based on MeshFix, FSL, and Freesurfer. The tumor surface mesh was then merged with the GM and WM surfaces so that the entire tumor was included in the GM volume. This procedure was conducted using Meshfix by joining the inner part of the tumor surface and the outer part of the white matter surface and equivalently joining the outer part of the tumor surface with the outer part of the GM surface. The surfaces were then re-optimized. Following surface optimization, individual tissue volumes were tessellated using Gmsh [31] (http://gmsh.info) to produce a tetrahedral finite element mesh with five tissue volumes (Fig. 10.1). The quality of the resulting mesh was optimized [32] and the anatomical accuracy of the final segmentation was evaluated by visually inspecting the overlay of the mesh on the structural MRI images. The cerebellum was included as WM volume and the ventricles in the CSF volume. The tumor volume was finally defined by selecting the tetrahedra in the GM volume, which also lay within the binary mask created by manual outlining of the tumor. A peritumoral border zone was defined by automatic selection of all GM and WM tetrahedra within 1 cm of the outermost tumor border.

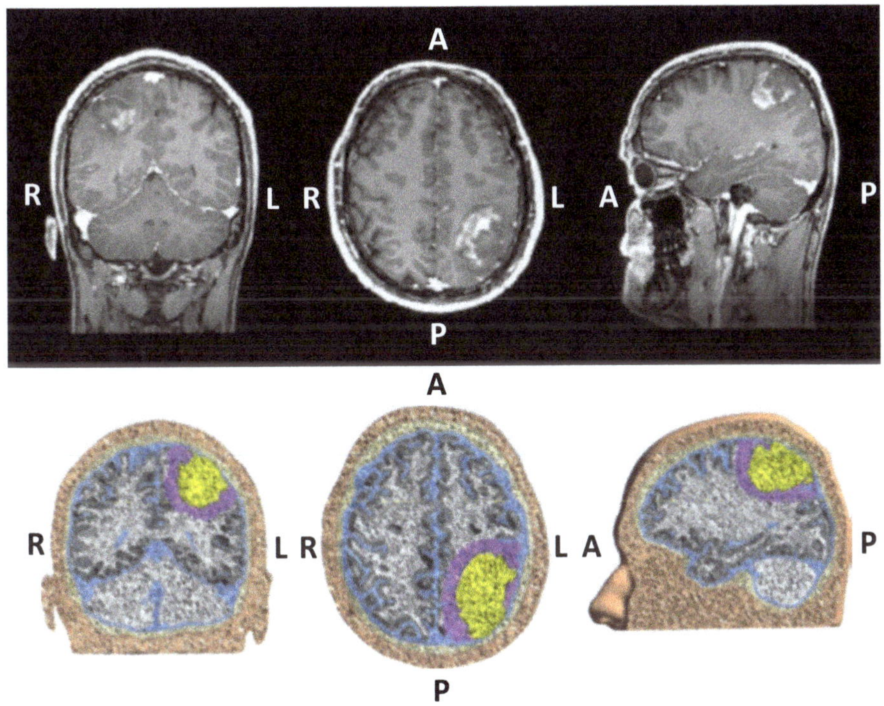

Fig. 10.1 *Top row*, gadolinium-enhanced T1 MRI from a GBM patient. *Bottom row*, the resulting volume segmentation based on the MRI data shown above and the SimNIBS software. The model is composed of skin, skull, CSF, GM, WM, tumor (yellow), and peritumoral volumes (magenta)

The final mesh consisted of seven tissue volumes and provided an accurate morphological representation based on 4,014,379 tetrahedra with a mean volume of 0.80 mm^3.

10.2.4 Placement of Transducer Arrays

To place the arrays in the model, we defined appropriate centers and orientations of each array using the SimNIBS graphical user interface and the specific head model. The orientation was given as a unit vector along the short axis of the array. The central transducer of the array was placed at the defined array center. We then constructed a longitudinal orthonormal vector as the cross product of the directional vector and the unit vector normal to the skin surface at the defined center. Two points were then defined at 45 mm and −45 mm from the center of the transducer along the line of the longitudinal vector, respectively. Each point was projected onto the closest triangle of the skin surface, which was then defined as the center of the middle transducer of the corresponding column in the array. In this way, all three transducers of the longitudinal center row were placed. A similar approach was adopted to place the transducers of the first and third row in the array. However, in this case, the originally defined directional vector of the short axis was used to

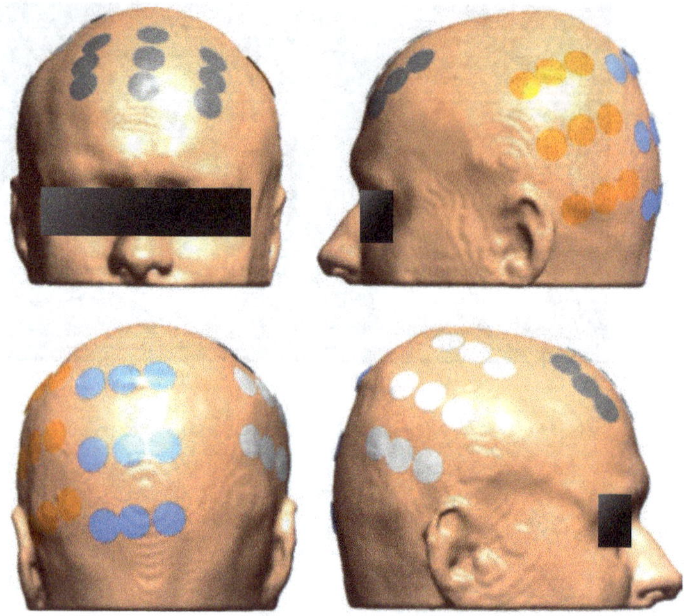

Fig. 10.2 Surface representation of the head model after placement of transducer arrays. The layout shown here corresponds to the layout proposed by the NovoTAL® algorithm used for clinical treatment. (Courtesy of Novocure, Ltd.)

define points at 22 mm and −22 mm distances from the centers of each transducer in the middle row. Using this approach, it was possible to place all transducers automatically and without significant undesirable overlap. Four different, clinically relevant layouts were tested. The procedure was implemented in a custom MATLAB script (Mathworks, Inc.). Figure 10.2 shows an example array layout.

10.2.5 Assignment of Tissue Conductivity

For the skin, skull, and CSF volumes, uniform and isotropic scalar conductivity values were assigned to all nodes belonging to the corresponding tissue volume in the mesh. Values were taken from the literature and based on in vitro and in vivo measurements at comparable frequencies (skin 0.25 S/m; bone 0.010 S/m; and CSF 1.654 S/m [33–39]). Electrodes were modeled with a 0.5 mm layer of conductive gel (1.0 S/m conductivity) between the electrode and the scalp. For GM, WM, and tumor tissues, we used an individualized anisotropic conductivity estimation technique, *direct conductivity mapping*, based on diffusion MRI (dMRI) data [40, 41]. The technique is based on the cross-property relation between general classes of transport tensors, e.g., diffusion and conductivity tensors, and the underlying microstructure of the transport medium. The general principle is that different "transport processes" will share the same eigenvectors of the corresponding transport tensors when taking place in the same medium. This allows for a simplified representation of the transport process through calculation of the eigenvalues specific for the given process. In the case of conductivity and diffusion, these eigenvalues are approximately linearly related so that the anisotropic conductivity tensor, required for accurate approximation of a solution to Laplace's equation, can be directly inferred from

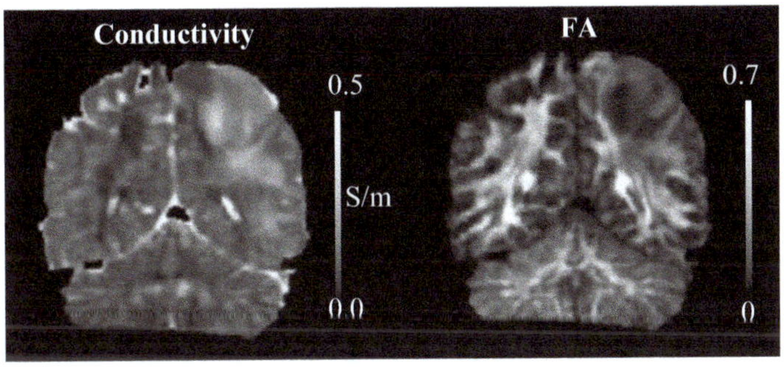

Fig. 10.3 Coronal views of the mean conductivity and FA of the conductivity tensor obtained using direct conductivity mapping. Significant conductivity variations occur in the tumor and peritumoral regions. WM tissue is highly anisotropic and FA values also vary within the region of the tumor. (Adapted from Wenger et al. [26])

diffusion MRI data using the same diffusion tensor eigenvectors and linearly scaled eigenvalues. Specifically, conductivity eigenvalues were calculated by fitting a linear relation with no intercept to the mean of diffusion eigenvalues, thereby ensuring that the distribution of the geometrical mean of conductivities (scaled eigenvalues) was centered at the in vivo mean estimates for the corresponding tissue in a least squares sense [41, 42]. The linear relationship was given by $\sigma_v = s \bullet d_v$, where σ_v is the conductivity along a given eigenvector v, d_v are the diffusion eigenvalues in the same direction, and s is a linear scale factor given by $s = \left(\overline{d_{\mathrm{WM}}}\sigma_{\mathrm{WM}}^{\mathrm{iso}} + \overline{d_{\mathrm{GM}}}\sigma_{\mathrm{GM}}^{\mathrm{iso}}\right) / \left(\overline{d_{\mathrm{WM}}}^2 + \overline{d_{\mathrm{GM}}}^2\right)$. In the latter expression, $\sigma_{\mathrm{WM}}^{\mathrm{iso}}$ and $\sigma_{\mathrm{GM}}^{\mathrm{iso}}$ are the uniform isotropic conductivity values (in vivo mean) of WM and GM, respectively, and $\overline{d_{\mathrm{WM/GM}}} = \sqrt[3]{\sum_N d_1 \cdot d_2 \cdot d_3 / N}$ represent the "average" value over N voxels in the corresponding tissues (GM and WM separately) of the geometric mean of the diffusion eigenvalues [42].

The scale factor was fitted using diffusion data within the GM and WM tissues of the healthy right hemisphere, and the scale factor was then applied to the entire diffusion tensor to extrapolate the conductivity estimates for the GM, WM, and tumor region in both the left and right hemispheres. The calculated voxel conductivities were assigned to mesh nodes of the corresponding tissue using nearest-neighbor interpolation [42]. The direct mapping procedure was implemented using the *dwi2cond* algorithm in SimNIBS. The resulting conductivity tensor is shown in Fig. 10.3 along with a topographical map of the fractional anisotropy.

10.3 Dosimetry of TTFields

10.3.1 The Problem

As described previously, Optune® therapy (TTFields) is performed by sequential activation of two electrode array pairs. This means that the field distribution throughout an activation cycle is composed of two consecutive fields, which are active for

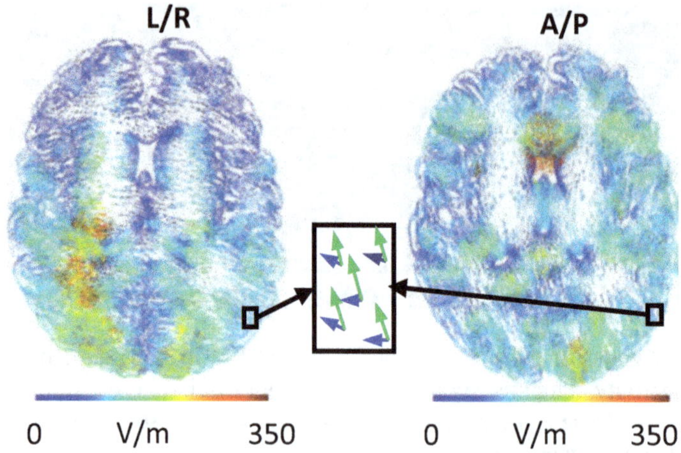

Fig. 10.4 Topographical field distribution for the left-right (L/R, left panel) and anterior-posterior (A/P, right panel) array pairs. The middle panel shows an enlarged schematic view of the two sequential field vectors in a small subvolume of the tumor. The illustration shows that the two fields are not entirely orthogonal and that they do not have equal magnitude

an equal amount of time (1 s). In the present context, the objective is to quantify the average effect of TTFields based on all sequential fields in the duty cycle. Previously, studies have presented the field distributions induced by each active pair, as originally described by Miranda et al. [5]. This approach is demonstrated in Fig. 10.4.

Although illustrative for many purposes, this approach does not account for spatial field correlation. This unwanted correlation causes the average antitumor effect to vary between cells depending on their direction of cell division, as previously explained. Figure 10.4 illustrates schematically how the induced sequential field vectors of each electrode pair are not orthogonal but rather have a variable extent of spatial correlation. In addition, fields may be orthogonal, but have different magnitude and so the average efficacy will still be biased and have variable antitumor activity towards cells dividing in different directions. In the following sections, I will describe a method for estimating the strength of the *uncorrelated* field components in any small tissue region and over one entire activation cycle of TTFields. The method provides individual measures of [1] the mean field intensity experienced by a cell dividing in any random direction in a local volume and [2] of the directional preference of efficacy caused by spatial field correlation. The calculations are based on the field distributions obtained using the FE methods described above.

10.3.2 The Basic Framework

First, we will consider TTFields at a specified point in the head model. We can assume that the fields are constant within a small volume surrounding the point. Although Optune® therapy is currently applied using two array pairs, TTFields can potentially be applied using any number $n \in N$ of pairs. We will denote the field vector generated by the ith array pair as $\boldsymbol{E}_i$ $(i = 1, 2, \ldots, n)$ and define $\boldsymbol{\varepsilon}$ to be the *field matrix* with transposed sequential field vectors $\boldsymbol{E}_i$ in each row, i.e.,

$$\varepsilon = \begin{bmatrix} \boldsymbol{E}_1^T \\ \vdots \\ \boldsymbol{E}_n^T \end{bmatrix} \in \mathbb{R}^{n \times 3} \tag{10.1}$$

We will now define the relative activation time α_i of $\boldsymbol{E}_i$ as

$$0 < \alpha_i = \frac{t_i}{\sum_{j=1}^{n} t_j} < 1, \tag{10.2}$$

where $t_i \geq 0$ is the absolute "on-time" of $\boldsymbol{E}_i$ during an activation cycle. We see that $\alpha_1 = \alpha_2 = \frac{1}{2}$ for Optune®, corresponding to a 50% duty cycle. If we denote $\boldsymbol{A} \in R^{n \times n}$ as the diagonal *activation time matrix* with entries $a_{ii} = \sqrt{\alpha_i}$ for all i, then the matrix

$$\boldsymbol{P} = \boldsymbol{A}\varepsilon \in \mathbb{R}^{n \times 3} \tag{10.3}$$

defines an "activation-time-weighted" field matrix. We now want to quantify the activation-time-weighted field intensities and evaluate whether they are distributed isotropically over the three directions in space. This can be estimated using principal component analysis of $\boldsymbol{P}$ to represent this matrix by up to three orthonormal basis vectors (principal components) collected in a matrix $\boldsymbol{W} \in R^{3 \times 3}$, and these will be uncorrelated over the dataset:

$$\boldsymbol{T} = \boldsymbol{P}\boldsymbol{W} \in \mathbb{R}^{n \times 3}. \tag{10.4}$$

The matrix $\boldsymbol{T}$ is equivalent to $\boldsymbol{P}$ after a change of basis has been performed. Estimating $\boldsymbol{W}$ is equivalent to fitting an ellipsoid to the data, and the axes of the ellipsoid are defined by the principal components (see Fig. 10.5).

There are a number of ways to estimate the principal components. Here we use the singular value decomposition (SVD)

$$\boldsymbol{P} = \boldsymbol{U}\Sigma\boldsymbol{W}^T \in \mathbb{R}^{n \times 3}. \tag{10.5}$$

The matrix $\boldsymbol{W}$ contains the orthonormal right singular basis vectors and the matrix $\boldsymbol{\Sigma} \in R^{n \times 3}$ contains the singular values σ_k, $k \leq 3$, which correspond to the lengths of the semiaxes of the fitted ellipsoid. This gives us the opportunity to estimate the average intensity of TTFields and the local field correlation for arbitrary configurations with n array pairs.

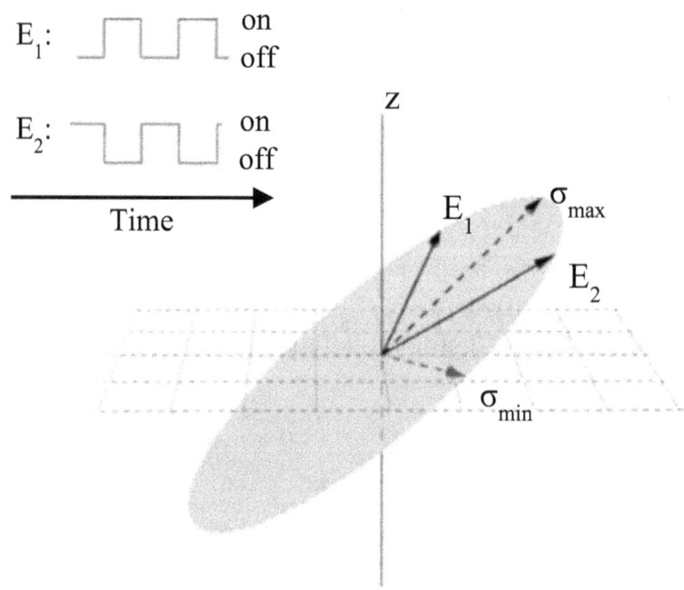

Fig. 10.5 Geometrical interpretation of the singular value decomposition. $\mathbf{E}_1$ and $\mathbf{E}_2$ are the two sequential field vectors, and the corresponding duty cycle is shown on the left. The parameters $\sigma_{\min}$ and $\sigma_{\max}$ are the minimum and maximum nonzero principal components corresponding to the semiaxes of best ellipsoid fit to the field data

10.3.3 Estimation of the TTFields Intensity

In light of the above derivations, we can express the average field intensity using singular value notation. First, we will define the average field intensity E_{avr} as the Frobenius norm of $\boldsymbol{P}$:

$$E_{\text{avr}} = \boldsymbol{P}_F = \sqrt{\sum_{i=1}^{n} \alpha_i \boldsymbol{E}_i^2} = \sqrt{\sum_{k=1}^{3} \sigma_k^2} \tag{10.6}$$

We see that E_{avr} is the square root of the activation-time-weighted contributions of energy from each sequential field in the activation cycle. This definition is convenient because it is linked with both the decorrelated principal components and the actual applied fields, i.e., it can be calculated from both parameters and it has a direct physical interpretation. We note that σ_k is the magnitude of the average field vector in the direction of the corresponding principal component. Since $rank(\boldsymbol{P}) = 2$ for the current Optune™ device, this configuration can induce a maximum of two nonzero principal components because it uses only two field directions. So, at least one singular value of $\boldsymbol{P}$ will be zero. This implies that the cells dividing in the direction of any zero eigenvector will likely experience little or no inhibiting effect of TTFields. Also, cells dividing at a positive angle to the plane of the induced fields will experience reduced average effects proportional to the projection of the field onto the subspace of nonzero principal components. Based on this notion, it would be necessary to alternate between at least three linearly independent field vectors throughout the activation cycle and thereby use at least three sets of electrode pairs in order to avoid this problem and induce inhibitory effects on all cells in the volume. It is also notable that the maximum average field intensity will be equal to the largest singular value and the minimum average field intensity equal to the minimum singular value. The directions in which these extreme values occur are given by the corresponding right-singular vectors. This further prompts the consideration that optimization of the activation cycle and the current settings may be employed to induce a more effective field distribution created as a linear combination of the sequential fields. This topic is briefly described below in this chapter.

Also, it is important to note that experimental data shows that there is a nonlinear dependence between the antitumor effects of TTFields and the Euclidean magnitude of the field. Specifically, low effects occur when the field is below a threshold [17]. Compared to a linear weighting of individual field strengths, the above definition of average field intensity accounts for circumstance to some extent by assigning higher weight to stronger fields. However, alternative definitions can also be used and additional studies are needed to determine which norm best represents the efficacy of TTFields. Recently, the local minimum field intensity (LMiFI, V/cm) and the local minimum power density (LMiPD) were proposed as appropriate dose estimates. The LMiFI is the lower of the two sequential field intensities delivered at a given point in the model. The LMiPD is the lower of the two power densities delivered at a given point in the model (mW/cm^3), where the power density is calculated as $P = 1/2\left(\sigma \boldsymbol{E}^2\right)$. The authors found that both estimates correlated with clinical outcome [43, 44], such that patients who had a high LMiFI and LMiPD during the course of treatment (when accounting for the average device on-time, compliance) lived longer. Although promising, this approach is also based on field intensity estimates alone, and does not account for activation cycle variations or spatial field correlations, so it should be explored in similar studies as to whether these parameters represent independent predictors of treatment efficacy.

10.3.4 Estimating the Spatial Correlation of TTFields Using the Fractional Anisotropy (FA) Measure

Having characterized the average field intensity, the next objective is to define a measure of field correlation, which generalizes to multiple field directions. To do this, we will adopt the FA estimate, which has been used extensively in diffusion tensor imaging [45]:

$$\mathrm{FA} = \sqrt{1/2}\,\frac{\sqrt{\left(\sigma_1-\sigma_2\right)^2+\left(\sigma_2-\sigma_3\right)^2+\left(\sigma_3-\sigma_1\right)^2}}{\sqrt{\sigma_1^2+\sigma_2^2+\sigma_3^2}}. \tag{10.7}$$

FA is calculated from the singular values, and it estimates the fractional deviation of $\boldsymbol{P}$ from the condition in which all principal components have equal magnitude, i.e., the time-averaged field over an activation cycle is the same in all directions and therefore the same for all cells in the small volume of interest. For implementations of TTFields with less than three singular values, such as the current Optune®

technology, the missing values are assigned a value of zero. FA can generally take values between zero and one, where higher values represent a higher degree of unwanted spatial field correlation and lower values the opposite. Since two pairs of arrays can induce a maximum of two linearly independent fields, and therefore a maximum of two nonzero principal components, the lowest value of FA which can be achieved with Optune® is $\frac{1}{\sqrt{2}} \oplus 0.71$. Lower values of FA would require that at least three field components (i.e., three array pairs) be used, as described above.

It must be noted that other generalized measures on spatial correlation may also be used, e.g., relative anisotropy, volume ratio, etc. [45]. Furthermore, simpler measures such as the scalar product, cross product, or angle between field vectors can be used, but these measures are only well defined for configurations with exactly two field directions.

10.3.5 Step-by-Step Framework for Calculation of FA and E_{avr}

In this section, I will briefly recapitulate the framework for estimation of FA and E_{avr}.

Step 1: Calculate the field distribution for each array layout. This is done using the methods described in Sect. 10.2, although alternative approaches may also be used [7]. Here, we calculated the field distributions for two different scenarios, namely before and after resection of the tumor in a realistic GBM head model. Resection was modeled by assigning isotropic CSF conductivity to the tumor volume, which is equivalent to a realistic resection cavity.

Step 2: Build the field matrix for each element in the model. Each matrix is composed of the calculated field vectors for the element, and the collection of matrices defines a tensor field equivalent for the computational model.

Step 3: Define the diagonal activation time matrix describing the relative activation times given by the TTFields activation cycle. In the case of Optune®, this will be a 2×2 matrix with the diagonal entries given by 0.5.

Step 4: Calculate the activation-time-weighted field matrix $\boldsymbol{P}$ as the product of the field matrix and activation-time matrix for each element in the model.

Step 5: Calculate the singular value decomposition of $\boldsymbol{P}$ for all elements in the model. This yields tensor fields for the left-singular matrices, the singular value matrices, and the right-singular matrices.

Step 6: For all elements, calculate FA and E_{avr}, as given by Eqs. (10.6) and (10.7), to obtain scalar fields of these estimates in all compartments of interest in the model.

Step 7: Postprocess the estimates of FA and E_{avr}, e.g., to visualize the data, obtain average estimates, distribution functions, or other outputs of interest.

10.4 Results from Example Calculations

In this section, I will present the results obtained using the proposed approach and the patient-based GBM head model prepared as described in Sect. 10.2.

10.4.1 *Topographical Distributions of FA and* $\mathbf{E}_{avr}$

Figure 10.6 (middle panels) shows the distribution of the maximum and minimum singular values given in a particular sequence of field distributions (shown in Fig. 10.6, left panels).

The field distribution in each finite element is anisotropic with a notable difference between the two principle components throughout the brain. This notion is further illustrated in the right-most lower panel of Fig. 10.6, which shows the topographical map of FA. Although FA was reasonably low in the tumor region, it was considerable in the peritumoral border zone. We see that the use of two field directions was not entirely able to distribute the inhibiting fields equally among the different directions is the plane of the two fields, as was in fact intended.

Following resection, the observed field anisotropy was significantly more pronounced (Fig. 10.7). The two array pairs induced high field intensities in different areas of the brain and resection border (Fig. 10.7, left). This caused significant differences between the principal components (i.e., $\sigma_{max} \gg \sigma_{min}$, Fig. 10.7 middle

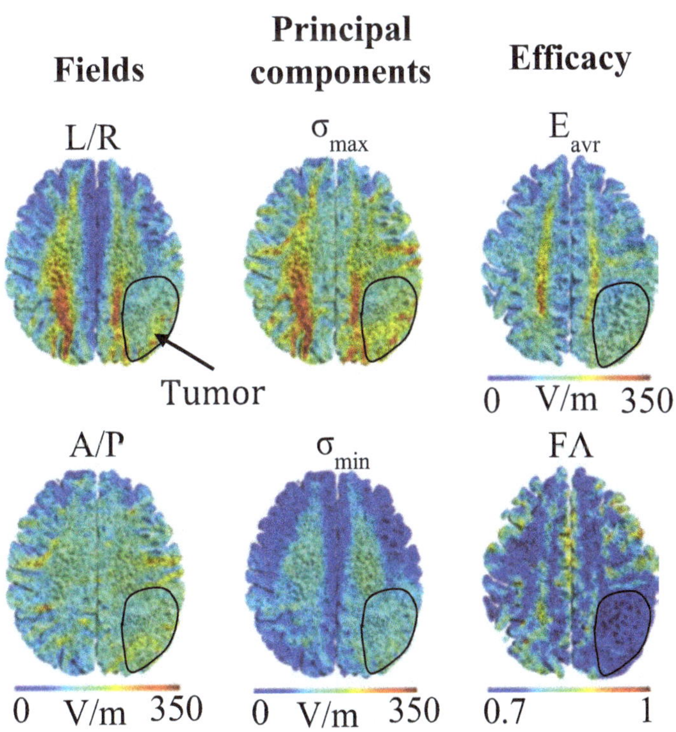

Fig. 10.6 The left panels show the field intensity distributions induced by the L/R and A/P array pairs. The tumor is outlined by the solid line. The middle panels show the corresponding distribution of minimum (bottom) and maximum (top) singular values. The right panels show the efficacy parameters E_{avr} (top) and FA (bottom). (Adapted from Korshoej et al. [21])

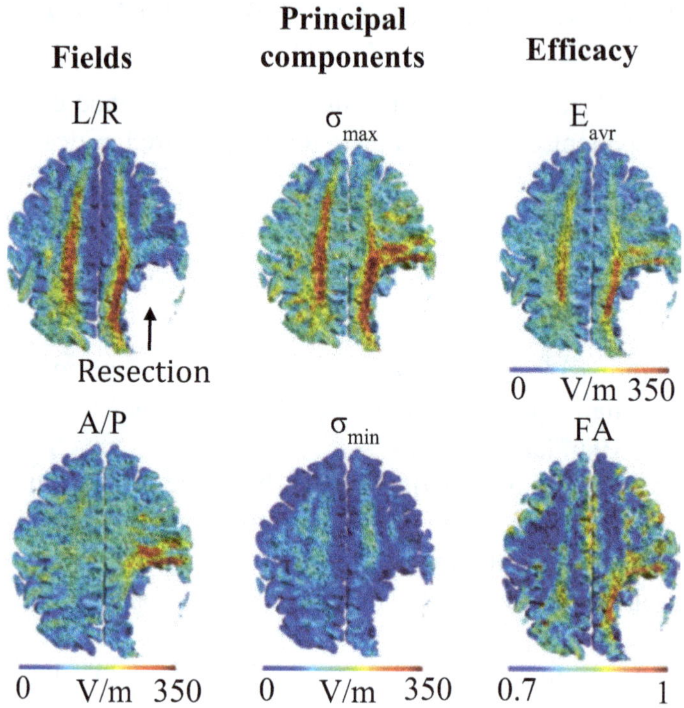

Fig. 10.7 The left panels show the field intensity distributions induced by the L/R and A/P array pairs after tumor resection. The resection cavity is indicated. The middle panels show the corresponding distribution of minimum (bottom) and maximum (top) singular values. The right panels show the efficacy parameters E_{avr} (top) and FA (bottom). (Adapted from Korshoej et al. [21])

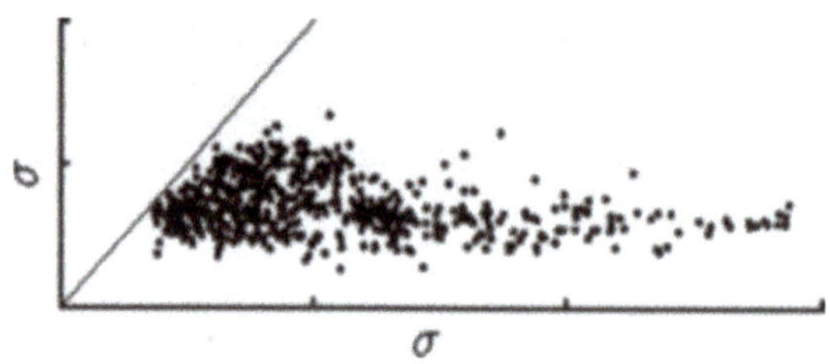

Fig. 10.8 Scatterplot of the corresponding minimum (*ordinate*) and maximum (*abscissa*) nonzero singular values (V/m) in the peritumoral region. The identity line (solid black) represents field isotropy. (Figure adapted from Korshoej et al. [21])

panels) and pronounced FA (Fig. 10.7, right bottom panel) in these regions. Specifically, FA was high in the region surrounding the resection cavity, though E_{avr} was also high in this region (Fig. 10.7, right top panel).

Figure 10.8 illustrates the extent of anisotropy as a scatterplot of paired singular values from elements in the peritumoral region around the resection cavity. The maximum achievable extent of isotropy using two array pairs would imply that both nonzero singular values were the same. However, as evident from Fig. 10.8, a large number of elements show considerable deviation from this condition.

The field redistribution observed following resection was caused by increased shunting of current through the CSF-filled resection cavity, which further caused high field strengths in the regions where the resection border was perpendicular to current and the applied field [6, 10]. In this case, the use of two field directions arguably served the purpose of distributing the field across the whole region, rather than inducing inhibiting fields in multiple directions. With this in mind, it is worth considering if highly anisotropic cases would be treated more efficiently with TTFields configurations designed to induce a high average field intensity and satisfactory pathology coverage, albeit at high anisotropy. Such implementations may be exemplified by "single-field" configurations. Alternatively, configurations may potentially be optimized to maintain an acceptable coverage and field intensity, albeit at a lower extent of anisotropy. The latter topic is highly interesting from a duty cycle optimization perspective, as discussed below.

10.4.2 Variations in FA and E_{avr} for Different Array Layouts

To examine a clinically relevant aspect of TTFields dosimetry, we investigated the field decomposition of four different layouts (Fig. 10.9).

Without tumor resection, E_{avr} indicated the following order of layout performance: Layout 4 > 3 > 2 > 1 (Fig. 10.10a, c). The FA estimate, however, indicated a reverse order (Fig. 10.10b, d), which raises questions about the true efficacy of the layouts. The median FA in the tumor and the peritumoral regions were 0.715 and

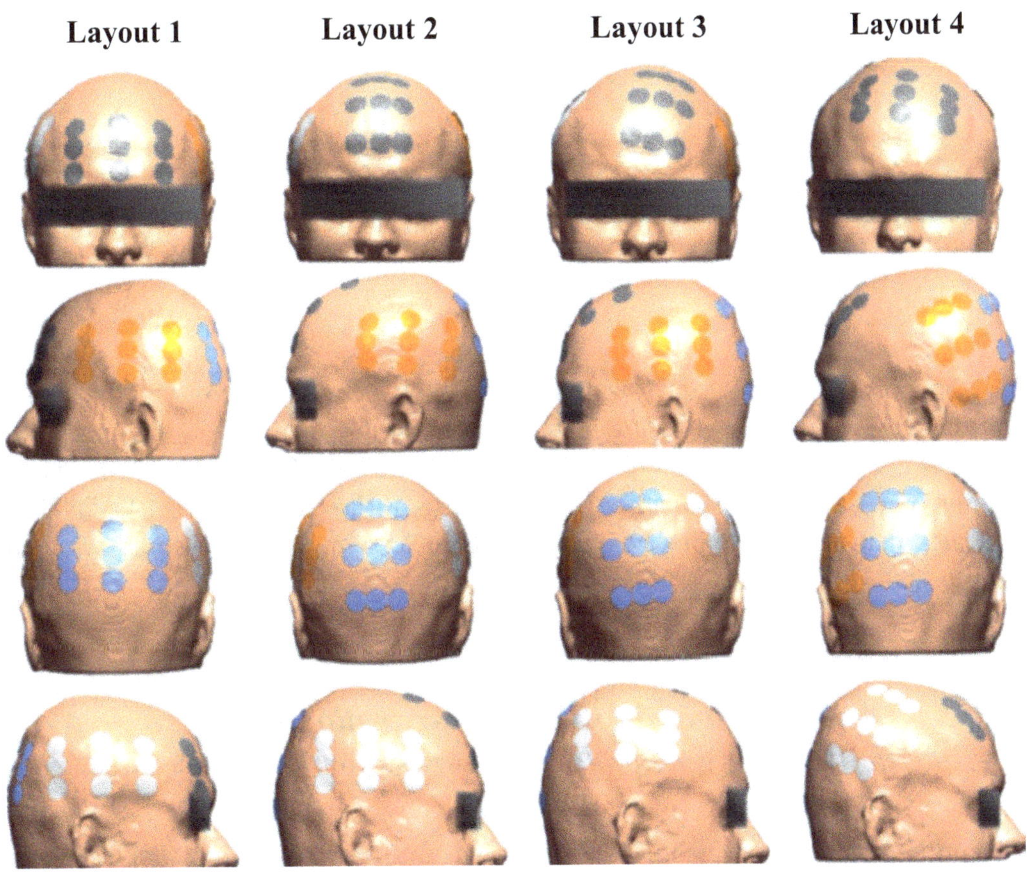

Fig. 10.9 Surface plot of four different array layouts tested. Gray and blue represent one pair and orange and white another. (Adapted from Korshoej et al. [21])

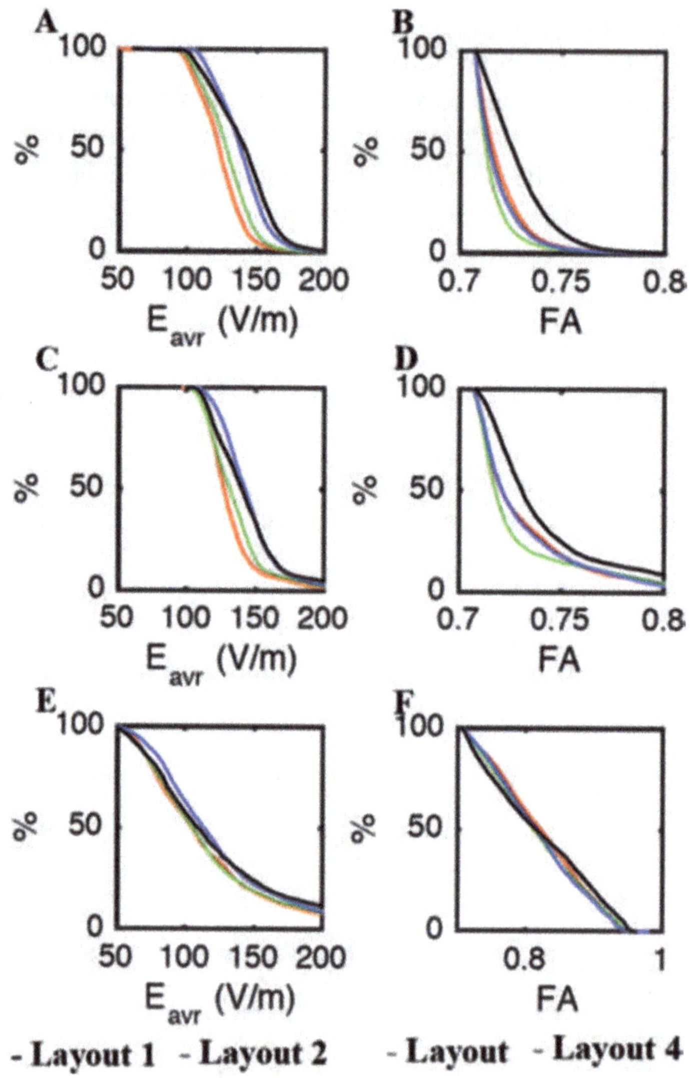

Fig. 10.10 Cumulative probability density functions of the mean intensity E_{avr} (left column) and FA (right column). The graphs show the percentage of elements in the selected volume that are equal to or higher than the corresponding value on the *abscissa*. Panels **a** and **b** represent values from the tumor volume before resection, panels **c** and **d** represent values from the peritumoral border zone before resection, and panels **e** and **f** represent the peritumoral region after resection

0.73, respectively. Resection caused a significant increase in FA for all layouts (median FA >0.82, Fig. 10.10f). Contrarily, resection reduced the median field intensity E_{avr} from ~120–150 V/m to ~100–110 V/m (Fig. 10.10c, e), and the distributions of FA and E_{avr} were close to identical for all layouts (Fig. 10.10e, f). This indicates that the positioning of arrays may be less important in some resected cases.

10.4.3 *Optimization of the TTFields Activation Cycle to Reduce Unwanted Field Anisotropy*

Given the possibility of quantifying FA, it is natural to consider whether TTFields therapy can be optimized to reduce this unwanted parameter. For instance, it might be desirable to plan the treatment array layout to maximize field intensity in the tumor, while simultaneously reducing FA. However, this may not always be possible, as some high-field configurations also produce high FA and vice versa. Recently, we proposed an approach in which FA can be reduced for an arbitrary layout without compromising the field intensity. The principle is based on individualizing the activation cycle of TTFields for each patient and each given array layout, rather than using a standard even 50% cycle. Specifically, the activation cycle is altered such that both pairs are activated simultaneously with a balanced intensity, so that the two fields are combined linearly to produce two resulting sequential fields, which are minimally anisotropic on average (in the volume of interest). Throughout the volume of interest, the field will thus be relatively orthogonal and have approximately equal magnitude. The factors determining the balance can be considered as linear gain factors to be applied to the standard current settings of each source in the system. In an alternative embodiment, the derived scale factors are used to modify the on-time of the given source. Although promising, the proposed activation cycle procedure is not currently supported by the Optune® device. Furthermore, it is important to note that reduced FA comes at the expense of increased total current density if the average field intensity is maintained. If FA is reduced at unchanged total current settings, then the mean field intensity will be reduced. So in all cases, there will be a trade-off, which highlights the importance of clarifying which efficacy parameters are more significant. For further details, the reader is kindly referred to Korshoej et al. [46].

10.5 Summary

In this chapter, I have described an extended framework to estimate the antitumor "dose" of TTFields. The approach is based on principal component decomposition of average field vectors induced over an activation cycle, and it quantifies both the mean intensity (E_{avr}) and unwanted spatial correlation (FA) of TTFields. These measures have a physical interpretation and generalize to an arbitrary number of array pairs. Furthermore, they account for all factors known to affect TTFields efficacy and provide a more comprehensive method for dose estimation than the current art. Computations show that significant unwanted FA occurs in the entire brain and tumor, which potentially affects the treatment effect as well as the approach to treatment planning. Without resection of the tumor, we found that E_{avr} and FA varied significantly for different layouts. Layouts that induced a high mean intensity also caused considerable unwanted anisotropy of the average field components. This effect may influence the overall efficacy, and therefore it should be incorporated in future dose estimation methods to improve accuracy. As a general observation, FA could only be reduced at the expense of reduced E_{avr}. Future experiments are necessary to determine the optimal balance between E_{avr} and FA, and the two measures may potentially be combined into a single measure of clinical efficacy. When characterizing the effect of tumor resection on the TTFields dose, we found that resection changes the topographical distribution of E_{avr} and FA. Furthermore, it almost nullified the differences in field distribution that we observed before resection for different array layouts. This suggests that accurate positioning may be less important after tumor resection. Resection also increased FA significantly, particularly at the resection border. This implies that multiple fields may not always be able to distribute the effect of TTFields sufficiently to target cells dividing in different directions, which is otherwise the intended purpose of using sequential and orthogonal fields. Instead, multiple fields may serve the main purpose of ensuring that all tumor-infiltrated regions are exposed to high mean field values. This suggests that it may be better to plan the array layout in such a way that good pathology coverage is obtained, even if the macroscopic orientation of the layout is not orthogonal. It is clear that the use of only one electrode pair that induces the highest average field intensity in the tumor will maximize E_{avr} across the activation cycle. Such configurations could be used if good field coverage of the tumor can be obtained, as it would be expected for smaller lesions or resections. Finally, the singular value decomposition approach allows for a direct linear optimization of the activation cycle for each patient and each layout, with the objective of reducing FA while maintaining high field intensities in the tumor.

Acknowledgments I kindly thank Novocure, Ltd. for providing technical information about the Optune® technology. Dr. Axel Thielscher and MSc Guilherme B. Saturnino from the Danish Research Centre for Magnetic Resonance and Dr. Frederik Lundgaard Hansen from Aarhus University Hospital, Dept. of Neurosurgery, have contributed greatly to the modeling experiments. Professor Jens Chr. Sørensen and Dr. Gorm von Oettingen from Aarhus University Hospital, Dept. of Neurosurgery, have contributed greatly to providing clinical perspective to the research, along with Professor Frantz Rom Poulsen from Odense University Hospital, Dept. of Neurosurgery.

References

1. Stupp, R., Taillibert, S., Kanner, A., Read, W., Steinberg, D. M., Lhermitte, B., et al. (2017). Effect of tumor-treating fields plus maintenance Temozolomide vs maintenance Temozolomide alone on survival in patients with glioblastoma: A randomized clinical trial. *JAMA, 318*(23), 2306–2316.
2. Mun, E. J., Babiker, H. M., Weinberg, U., Kirson, E. D., & Von Hoff, D. D. (2017). Tumor treating fields: A fourth modality in cancer treatment. *Clinical Cancer Research, 24*(2), 266–275.
3. Stupp, R., Wong, E. T., Kanner, A. A., Steinberg, D., Engelhard, H., Heidecke, V., et al. (2012). NovoTTF-100A versus physician's choice chemotherapy in recurrent glioblastoma: A randomised phase III trial of a novel treatment modality. *European Journal of Cancer, 48*(14), 2192–2202.
4. Trusheim, J., Dunbar, E., Battiste, J., Iwamoto, F., Mohile, N., Damek, D., et al. (2017). A state-of-the-art review and guidelines for tumor treating fields treatment planning and patient follow-up in glioblastoma. *CNS Oncology, 6*(1), 29–43.
5. Miranda, P. C., Mekonnen, A., Salvador, R., & Basser, P. J. (2014). Predicting the electric field distribution in the brain for the treatment of glioblastoma. *Physics in Medicine & Biology, 59*(15), 4137.

6. Using computational phantoms to improve delivery of Tumor Treating Fields (TTFields) to patients. Engineering in medicine and biology society (EMBC), 2016 IEEE 38th annual international conference of the: IEEE; 2016, DOI: https://doi.org/10.1109/EMBC.2016.7592208.
7. Wenger, C., Miranda, P., Salvador, R., Thielscher, A., Bomzon, Z., Giladi, M., et al. (2018). A review on tumor treating fields (TTFields): Clinical implications inferred from computational modeling. *IEEE Reviews in Biomedical Engineering, 11*, 195–207. https://doi.org/10.1109/RBME.2017.2765282.
8. Timmons, J. J., Lok, E., San, P., Bui, K., & Wong, E. T. (2017). End-to-end workflow for finite element analysis of tumor treating fields in glioblastomas. *Physics in Medicine & Biology, 62*(21), 8264.
9. Lok, E., San, P., Hua, V., Phung, M., & Wong, E. T. (2017). Analysis of physical characteristics of tumor treating fields for human glioblastoma. *Cancer Medicine, 6*(6), 1286–1300.
10. Korshoej, A. R., Hansen, F. L., Thielscher, A., von Oettingen, G. B., & Sørensen, J. C. H. (2017). Impact of tumor position, conductivity distribution and tissue homogeneity on the distribution of tumor treating fields in a human brain: A computer modeling study. *PLoS One, 12*(6), e0179214.
11. Wenger, C., Salvador, R., Basser, P. J., & Miranda, P. C. (2016). Improving TTFields treatment efficacy in patients with glioblastoma using personalized array layouts. *International Journal of Radiation Oncology Biology Physics, 94*(5), 1137–1143.
12. Wenger, C., Salvador, R., Basser, P. J., & Miranda, P. C. (2015). The electric field distribution in the brain during TTFields therapy and its dependence on tissue dielectric properties and anatomy: A computational study. *Physics in Medicine & Biology, 60*, 7339–7357.
13. Korshoej, A. R., Hansen, F. L., Mikic, N., Thielscher, A., von Oettingen, G. B., & Sørensen, J. C. H. (2017). Exth-04. Guiding principles for predicting the distribution of tumor treating fields in a human brain: A computer modeling study investigating the impact of tumor position, conductivity distribution and tissue homogeneity. *Neuro-Oncology, 19*(suppl_6), vi73.
14. Korshoej, A. R., Hansen, F. L., Mikic, N., von Oettingen, G., Sørensen, J. C. H., & Thielscher, A. (2018). Importance of electrode position for the distribution of tumor treating fields (TTFields) in a human brain. Identification of effective layouts through systematic analysis of array positions for multiple tumor locations. *PLoS One, 13*(8), e0201957.
15. Korshoej, A. R., Saturnino, G. B., Rasmussen, L. K., von Oettingen, G., Sørensen, J. C. H., & Thielscher, A. (2016). Enhancing predicted efficacy of tumor treating fields therapy of glioblastoma using targeted surgical craniectomy: A computer modeling study. *PLoS One, 11*(10), e0164051.
16. Korshoej, A., Lukacova, S., Sørensen, J. C., Hansen, F. L., Mikic, N., Thielscher, A., et al. (2018). ACTR-43. Open-label phase 1 clinical trial testing personalized and targeted skull remodeling surgery to maximize TTFields intensity for recurrent glioblastoma–interim analysis and safety assessment (OptimalTTF-1). *Neuro-Oncology, 20*(suppl_6), vi21–vi21.
17. Kirson, E. D., Gurvich, Z., Schneiderman, R., Dekel, E., Itzhaki, A., Wasserman, Y., et al. (2004). Disruption of cancer cell replication by alternating electric fields. *Cancer Research, 64*(9), 3288–3295.
18. Giladi, M., Schneiderman, R. S., Voloshin, T., Porat, Y., Munster, M., Blat, R., et al. (2015). Mitotic spindle disruption by alternating electric fields leads to improper chromosome segregation and mitotic catastrophe in cancer cells. *Scientific Reports, 5*, 18046.
19. Wenger, C., Giladi, M., Bomzon, Z., Salvador, R., Basser, P. J., & Miranda, P. C. (2015). *Modeling tumor treating fields (TTFields) application in single cells during metaphase and telophase*. Milano: IEEE EMBC.
20. Kirson, E. D., Dbaly, V., Tovarys, F., Vymazal, J., Soustiel, J. F., Itzhaki, A., et al. (2007). Alternating electric fields arrest cell proliferation in animal tumor models and human brain tumors. *Proceedings of the National Academy of Sciences of the United States of America, 104*(24), 10152–10157.
21. Korshoej, A. R., & Thielscher, A. (2018). *Estimating the intensity and anisotropy of tumor treating fields using singular value decomposition. Towards a more comprehensive estimation*

of anti-tumor efficacy. In: 2018 40th annual international conference of the IEEE Engineering in Medicine and Biology Society (EMBC) (pp. 4897–4900). IEEE. https://doi.org/10.1109/EMBC.2018.8513440.

22. Haus, H. A., & Melcher, J. R. (1989). *Electromagnetic fields and energy*. Englewood Cliffs: Prentice Hall.
23. Humphries, S. (1997). *Finite element methods for electromagnetics*. Albuquerque: Field Precision LLC.
24. Plonsey, R., & Heppner, D. B. (1967). Considerations of quasi-stationarity in electrophysiological systems. *Bulletin of Mathematical Biology, 29*(4), 657–664.
25. Wong, E. T. (2016). *Alternating electric fields therapy in oncology: A practical guide to clinical applications of tumor treating fields*. Springer.
26. Lok, E., Swanson, K. D., & Wong, E. T. (2015). Tumor treating fields therapy device for glioblastoma: Physics and clinical practice considerations. *Expert Review of Medical Devices, 12*(6), 717–726.
27. Windhoff, M., Opitz, A., & Thielscher, A. (2013). Electric field calculations in brain stimulation based on finite elements: An optimized processing pipeline for the generation and usage of accurate individual head models. *Human Brain Mapping, 34*(4), 923–935.
28. Dale, A. M., Fischl, B., & Sereno, M. I. (1999). Cortical surface-based analysis: I. segmentation and surface reconstruction. *NeuroImage, 9*(2), 179–194.
29. Smith, S. M., Jenkinson, M., Woolrich, M. W., Beckmann, C. F., Behrens, T. E., Johansen-Berg, H., et al. (2004). Advances in functional and structural MR image analysis and implementation as FSL. *NeuroImage, 23*, S208–S219.
30. Attene, M. (2010). A lightweight approach to repairing digitized polygon meshes. *The Visual Computer, 26*(11), 1393–1406.
31. Geuzaine, C., & Remacle, J. (2009). Gmsh: A 3-D finite element mesh generator with built-in pre-and post-processing facilities. *International Journal for Numerical Methods in Engineering, 79*(11), 1309–1331.
32. Schöberl, J. (1997). NETGEN an advancing front 2D/3D-mesh generator based on abstract rules. *Computing and Visualization in Science, 1*(1), 41–52.
33. Latikka, J., Kuurne, T., & Eskola, H. (2001). Conductivity of living intracranial tissues. *Physics in Medicine & Biology, 46*(6), 1611.
34. Gabriel, C., Peyman, A., & Grant, E. (2009). Electrical conductivity of tissue at frequencies below 1 MHz. *Physics in Medicine & Biology, 54*(16), 4863.
35. Geddes, L., & Baker, L. (1967). The specific resistance of biological material—A compendium of data for the biomedical engineer and physiologist. *Medical and Biological Engineering and Computing, 5*(3), 271–293.
36. Geddes, L. A. (1987). Optimal stimulus duration for extracranial cortical stimulation. *Neurosurgery, 20*(1), 94–99.
37. Gabriel, C. (2006). Dielectric properties of biological materials. *Bioengineering and Biophysical Aspects of Electromagnetic Fields, 1*, 87–136.
38. Miranda, P. C., Lomarev, M., & Hallett, M. (2006). Modeling the current distribution during transcranial direct current stimulation. *Clinical Neurophysiology, 117*(7), 1623–1629.
39. Holton, K. S., & Walker, C. F. (1990). Correlation of a magnetic resonance brain image and tissue impedance. Engineering in Medicine and Biology Society. *Proceedings of the Twelfth Annual International Conference of the IEEE*. Philadelphia: IEEE.
40. Tuch, D. S., Wedeen, V. J., Dale, A. M., George, J. S., & Belliveau, J. W. (2001). Conductivity tensor mapping of the human brain using diffusion tensor MRI. *Proceedings of the National Academy of Sciences of the United States of America, 98*(20), 11697–11701.
41. Rullmann, M., Anwander, A., Dannhauer, M., Warfield, S. K., Duffy, F. H., & Wolters, C. H. (2009). EEG source analysis of epileptiform activity using a 1 mm anisotropic hexahedra finite element head model. *NeuroImage, 44*(2), 399–410.
42. Opitz, A., Windhoff, M., Heidemann, R. M., Turner, R., & Thielscher, A. (2011). How the brain tissue shapes the electric field induced by transcranial magnetic stimulation. *NeuroImage, 58*(3), 849–859.

43. Ballo, M., Bomzon, Z., Urman, N., Lavy-Shahaf, G., & Toms, S. (2018). ACTR-46. Higher doses of TTFields in the tumor are associated with improved patient outcome. *Neuro-Oncology, 20*(suppl_6), vi21–vi22.
44. Ballo, M. T., Urman, N., Lavy-Shahaf, G., Grewal, J., Bomzon, Z. E., & Toms, S. (2019). Correlation of Tumor Treating Fields Dosimetry to Survival Outcomes in Newly Diagnosed Glioblastoma: A Large-Scale Numerical Simulation-Based Analysis of Data from the Phase 3 EF-14 Randomized Trial. International Journal of Radiation Oncology* Biology* Physics.
45. Basser, P. J., & Pierpaoli, C. (2011). Microstructural and physiological features of tissues elucidated by quantitative-diffusion-tensor MRI. *Journal of Magnetic Resonance, 213*(2), 560–570.
46. Korshoej, A. R., Sørensen, J. C. H., Von Oettingen, G., Poulsen, F. R., & Thielscher, A. (2019). Optimization of tumor treating fields using singular value decomposition and minimization of field anisotropy. Physics in Medicine & Biology, 64(4), 04NT03.

Chapter 11
The Bioelectric Circuitry of the Cell

Jack A. Tuszynski

11.1 Introduction

The study of electrical field effects on cells dates back to 1892 when Wilhelm Roux observed pronounced stratification of the cytoplasm of animal eggs when exposed to electric fields. Over the many decades since, a number of electric field effects have been implicated in the functioning of living cells, in particular in the cytoskeletal or cytoplasmic self-organization processes. For example, electrotherapies and wound healing have been hypothesized to involve ionic current flows. At the cell level, cytochrome oxidase enzyme has been linked to electric current action [1] and cell division coherent polarization waves have been proposed as playing a major role in chromosome alignment and subsequent segregation [2]. In addition, endogenous electric currents have been detected in animal cells. In the phase between fertilization and the first cleavage, a steady current enters the animal pole and leaves the vegetal pole. In the silkmoth oocyte-nurse complex, the oocyte cytoplasm is slightly more positive (by 10 mV) than the nurse cell cytoplasm, which allows for the passing of a small electric current on the order of 5×10^{-8} A [3]. A steady current enters the prospective cleavage furrow in both frog and sea urchin eggs during the initial period prior to cleavage formation, but after initiation, this current reverses its direction and leaves the furrow region [3].

Various plant and animal cells have been observed [3] to undergo significant changes when subjected to steady-state weak electric fields, including changes in their regeneration growth rates. A substantial reduction in the mitotic index was found in pea roots exposed to 60-Hz electric fields at a 430 V/m intensity and after

J. A. Tuszynski (✉)
Department of Oncology, University of Alberta, Edmonton, AB, Canada

Department of Physics, University of Alberta, Edmonton, AB, Canada

DIMEAS, Politecnico di Torino, Turin, Italy
e-mail: jackt@ualberta.ca

S. Makarov et al. (eds.), *Brain and Human Body Modeling*,
https://doi.org/10.1007/978-3-030-21293-3_11

4 hours of exposure [4]. The effect of 50-Hz electric fields of a 50 kV/m intensity on the mitotic index of cultured human embryo fibroblastoid cells was also found [5]. More recently, AC electric fields in the frequency range between 100 and 300 kHz and an intensity of only 1–2.5 V/cm have been shown to arrest cancer cells in mitosis [6], which is an astonishing effect in view of the weak intensity of the field. This discovery has led to an FDA approved treatment for the deadly brain cancer form, glioblastoma multiforme (GBM) [7]. It has been speculated that these field effects act on microtubules (MTs) as a primary mechanism of action [8]. However, what aspects of MT behavior in the presence of electric fields are involved is still not clear (depolymerization, rotation, electric conduction, etc.). This latter development provides strong motivation to elucidate the response of MTs in cytoplasm or buffer solution to externally applied AC electric fields. Beyond this, the overriding question still remains: if living cells are sensitive to electric fields and even exhibit electric current effects, then which structures within the cell perform the functions of bioelectric circuit elements?

The idea that the building blocks of living cells, especially proteins, may exhibit electric conduction properties should be credited to Albert Szent-Györgyi who viewed them as semiconducting devices [9, 10]. However, they were considered in their monomeric form, which results in a large energy gap between valence and conduction bands, making electronic conductivity of single proteins very challenging. Moreover, protein conductivity is also largely dependent on their hydration state [11]. What was missing in these early studies of biological conductivity was the role of ionic species, which are abundant in living cells, and an examination of polymeric forms of proteins and DNA, which makes a major difference to both electronic and ionic conduction. Significant experimental challenges of measuring electric fields and currents at a subcellular level persist today and studies of cellular components in isolation provide a proxy for intracellular measurements. Specific interest in the electrical properties of microtubules, actin filaments, DNA, and, of course, ion channels, has produced a number of interesting results that merit close examination, especially in terms of frequency dependence for AC conductivity analysis. Since most living cells are composed of 70% water molecules by weight, the role of water in the transmission of electrical pulses [12] is undoubtedly crucial in these processes. In general, electric charge carriers involved in protein and DNA conduction can be electrons and protons, as well as ions of various types surrounding proteins in the cytoplasm. Actin filaments (AFs) and microtubules have been implicated in numerous forms of electrical processes involving mainly positive counterions due to their net negative charge localized largely on their surfaces [13, 14].

The presence of several types of ionic species (especially K^+ at 140 mM, Na^+ at 10 mM, Cl^- at 10 mM, Mg^{2+} at 0.5 mM, and Ca^{2+} at 0.1 μM typical concentrations) as well as positively charged protons at a typical pH of 7 provides the cell with intrinsic ionic conductivity properties, which can be affected by the transmembrane potential and the action of ion channels. These ions can either diffuse freely in the cytoplasm or be directed to move along the electric field lines that can follow well-defined polymeric pathways in the cell. While cell membranes support strong electric fields on the order of 10^7 V/m, due to Debye screening, these fields decay exponentially away from the membrane. Dielectric studies of biological cells and their constituent macromolecules in solution have been conducted for almost a century [15, 16] and have revealed a wealth of information about transmembrane potentials, macromolecular charges, their dipole moments, and polarizabilities [17].

For example, Lima et al. [18] recently measured the electric impedance Z'(f) of NaCl and KCl solutions. They observed a large plateau between 10 Hz and 400 kHz, increasing in the low frequency range and decreasing at high frequencies. The value of the plateau decreased with increasing salt concentration, yielding the maximum value of resistance R ~ 10^5 Ω at very low frequencies (~10 mHz) that was independent of salt concentration. The imaginary part of impedance, Z''(f), showed anionic relaxation with a precipitous drop in the 100 kHz range. This is important in the context of ionic solutions present in the living cells and their concentration dependence of conductivity as a function of frequency.

The cytoplasm has a high concentration of proteins with actin (2–8 mg/mL) and tubulin (4 mg/mL) being the most abundant cytoplasmic proteins. Both actin and tubulin exist in either polymerized (actin filaments and microtubules, respectively) and unpolymerized states. It is the polymerized state of these proteins that exhibits interesting conducting properties. These properties are due to the fact that AFs and MTs have a very high density of uncompensated electric charges (on the order of 100,000 per micron of polymer length). In an ionic solution, most of these charges are compensated by counterions, but this leads to a large dielectric moment and nonlinear electro-osmotic response [19–22]. As discussed below, AFs and MTs are nonlinear electric conduction transmission lines. These cytoskeletal protein networks propagate signals in the form of ionic solitons [23–25] and traveling conformation transformations [26–28]. Experiments with polarized bundles of AFs and MTs demonstrated propagation of solitary waves with a constant velocity and without attenuation or distortion in the absence of synaptic transmission [25].

While DNA has been shown to also act as a nanowire [29–31], no transformation of signals was observed in experiments with DNA as opposed to MTs, which showed signal amplification [13]. In terms of using these structures as bioelectric wires, there are not only conductive but also mechanical differences, which can lead to different electromechanical arrangements into micro-scale circuits. In contrast to MTs and AFs, which are the most rigid structures in a cell, DNA is mechanically flexible and undergoes coiling transformations including its packaging into chromosomes [32, 33]. Therefore, DNA circuits can be packed and unpacked depending on the ionic environment while MT circuits can be polymerized and depolymerized using magnesium and calcium signals, for example. MTs can be stabilized by microtubule-associated protein (MAP) interconnections, while AFs have the ability to branch out using ARP2/3 constructs. Consequently, each of these bioelectric elements has different abilities to form complex and dynamic circuits.

11.2 Ion Channel Conduction Effects

Each cell has numerous ion channels embedded in its membrane, with specialized roles in terms of their selectivity and the rate of ion flows. Since ions are charged, these ion flows can be viewed as electric conduction events. A single ion takes approximately 5 ps to traverse an ion channel, whose length is on the order of 5 nm, resulting in an average speed of 1000 m/s. In specific ion channels, such as the bacterial KcsA channel, one K^+ ion crosses the channel per 10–20 ns under physiological conductance conditions of roughly 80–100 pS [34]. This allows for a maximum conduction rate of about 10^8 ions/s. Estimating the distances between the center of the channel pore and the membrane surface to scale as 5×10^{-9} m (5 nm), and assuming the most simple watery-hole and continuum electro-diffusion model of channels, this would provide an average speed of 5×10^{-1} m/s per ion (0.5 m/s). All these numbers for KcsA channels are consistent with our generic estimates except for the speed, which is lowered by the inclusion of the refractory period. In fact, while it is known that the ion flow rate per channel is on the order of 10^5 ions/ms, giving a clock time of approximately 10 ns per ion, one must conclude that a 5 ps active event of traversing a channel is separated by a 2000 times longer refractory interval of 10 ns during which there is no electrical signal propagation taking place. Since the value of a typical transmembrane potential is on the order of 100 mV and a flow of singly charged ions like sodium or potassium leads to an electrical current on the order of 10 pA, the Ohmic resistance of an ion channel can be approximated as 10 GΩ. Note that for a given cell, its ion channels can be viewed as resistors in parallel with each other. Liu et al. [35] reported activation of a Na^+ ion channel's pumping mode with an oscillating electric field of 200 V/m, at a frequency of approximately 1 MHz. Channel types and number per cell (densities) strongly vary among different cellular phenotypes. For example, in mammalian medial entorhinal cortex cells (MECs), an average of 5×10^5 fast-conductance Na^+ and delayed-rectifier K^+ channels per neuron have been estimated to exist [36]. In unmyelinated squid axons, counts can reach up to 10^8 channels per cell. Therefore, these numbers would proportionately reduce the overall electrical resistance of a cell compared to a single ion channel value. In more detailed studies, it has been demonstrated that ion transitions occur through a sequence of stable multi-ion configurations through the filter region of the channels, which allows rapid and ion-selective conduction [37]. The corresponding kinetic energy together with the electrostatic potential energy equals 2×10^{-20} J, which is very similar to an estimate of the ATP energy, hence justifying an active transport requirement as opposed to a thermally activated process.

Finally, in connection with biological relevance of ion channels and ionic currents flowing through them, Levin [38] has extensively investigated ionic signals in regard to such phenomena as morphogenesis and cancer. Ionic currents in cells associated with injury have been shown to be both necessary and sufficient for regeneration [39]. Patterning structural information during embryogenesis and regenerative repair has been shown to be influenced by bioelectric ionic signals

[40]. Moreover, ionic electrical signals, and endogenous voltage gradients affected by ionic flows, have been associated with key cellular processes such as proliferation, cell cycle progression, apoptosis, migration and orientation, and differentiation and de-differentiation [41]. Therefore, it can safely be stated that ion channels and ionic currents are at the center of cellular activities. The question remains whether there is additional electrical activity downstream from ion channels, namely in the cytoplasm. As discussed in the following sections, the complex and well-organized structure of the cytoskeleton lends itself to such interactions, especially since the filaments of the cytoskeleton are now known to be electrically conductive. We next discuss the particular case of actin filaments followed by microtubules.

11.3 Actin Filament Conductivity

Actin filaments, also referred to as F-actin or microfilaments, are approximately 7 nm in diameter and form a helical structure with a pitch of approximately 37 nm. They are highly electrostatically charged [20, 42]. Within an AF, actin monomers arrange themselves head-to-head to form actin dimers, resulting in an alternating distribution of electric dipole moments along the filament [43]. We assume, therefore, that there is a helical distribution of ions winding around the filament at approximately one Bjerrum length. Experimental studies demonstrated that they conduct ionic currents via the surrounding counterion cloud-like layer [20]. The ionic charge distribution along an AF has been modeled as an electrical circuit with the following elementary components representing the functional role of each actin monomer: (a) a nonlinear (saturable) capacitor associated with the spatial charge distribution between the ions located in the outer and inner regions of the polymer, (b) an inductance due to helical nature of the ionic current flow, and (c) a resistor due to the viscosity of the medium opposing the ionic flows. This representation provided the basis for a physical model of F-actin as a conducting polyelectrolyte, where ion flows are expected to occur at a radial distance from the surface of the filament approximately equal to the Bjerrum length and follow a solenoidal geometry due to the actin's double stranded helical structure. Using Kirchhoff's equations and taking the continuum limit for a long transmission line results in nonlinear inhomogeneous partial differential equations for the propagating nonlinear waves of ions along and around the AF. These ionic waves, in the form of elliptic Jacobi functions and solitary waves of the kink-type, have been described as the solutions of the above nonlinear partial differential equations [23].

The objective of this model was to explain the experimental results of Lader et al. [44], who applied an input voltage pulse with amplitude of approximately 200 mV and duration of 800 ms to an AF, and measured electrical signals at the opposite end of the AF. The obtained results showed that AFs support ionic waves in the form of axial nonlinear currents that maintain their amplitude and hence are not dissipative. These data supported an earlier experiment [20] in which the observed wave patterns in electrically stimulated single AFs were remarkably similar to those found in the recorded solitary waveforms for electrically stimulated nonlinear transmission lines [45]. In view of the fact that the AFs are highly nonlinear complex biophysical structures acting under the influence of thermal fluctuations and supporting the counterionic cloud hypothesis [46], the observation of soliton-like ionic waves is consistent with the idea of AFs functioning as biological transmission lines. Based on the continuum transmission line model, ionic currents along AFs have been estimated to have a velocity of propagation between 1 and 100 m/s [23]. This model has been later updated to include more realistic estimates of model parameters [47, 48]. Interestingly, but not surprisingly, actin filaments can be manipulated by external electric fields [49], which opens the door to electric field manipulations of actin cytoskeleton geometry, resulting in a dynamically flexible electric circuitry within the cell. For a filament with n monomers, the following numbers have been obtained for the electric circuit parameters of each monomer, labeled i, as a fundamental unit of the circuit: an effective resistance (longitudinal and radial, respectively), capacitance, and inductance, where $R_{1,i} = 6.11 \times 10^6\ \Omega$, and $R_{2,i} = 0.9 \times 10^6\ \Omega$, $C_i = 10^{-4}$ pF, and $L_i = 2$ pH. Hence, for a 1 μm length of an actin filament, we find the following corresponding values characterizing it as a conducting bioelectric wire: $R_{eff} = 1.2 \times 10^9\ \Omega$, $L_{eff} = 340 \times 10^{-12}$ H, $C_{eff} = 0.02 \times 10^{-12}$ F.

We can also easily find for a single actin monomer and an AF what characteristic time scales apply to their electrical circuit properties. For a single monomer, the time scale for LC oscillations is very fast, namely $\tau_0 = (LC)^{1/2} = 6 \times 10^{-14}$ s. The decay time for longitudinal ionic waves is also very fast, $\tau_1 = R_1C = 6 \times 10^{-10}$ s, while the corresponding time for radial waves is $\tau_2 = R_2C = 0.9 \times 10^{-10}$ s. As an example of a typical AF, we consider a 1 μm polymer and find the following characteristic time scales in a similar manner to the calculations above: $\tau_0 = 10^{-11}$ s, which is still very short but $\tau_1 = R_1C = 2.4 \times 10^{-5}$ s for longitudinal electric signal propagation is in the range for interactions with AC electric fields in the 100 kHz range.

If actin filaments support ionic conduction, even lossless transmission of electric signals in the cell, it is also natural to expect unusual behavior of microtubules under electric stimulation. This can be inferred from the known structural and electrostatic properties of MTs, which are highly electrostatically charged, even more so than AFs, larger than AFs and they exhibit a cylindrical geometry with a helical pattern of protofilaments wrapping around the cylinder surface.

11.4 Microtubule Conductivity

MTs are a major part of the cell's cytoskeleton. The building block of a MT is a tubulin dimer that contains approximately 900 amino acid residues comprising some 14,000 atoms with an overall mass of 110 kDa (1 Da = 1.7×10^{-27} kg). Each tubulin dimer in an MT has an approximate length of 8 nm, along the MT cylinder axis, a width of 6.5 nm and a thickness along the radial direction of an MT of 4.6 nm. The outer diameter of an MT is 25 nm, while the inner core of the cylinder, i.e., its lumen, is approximately 15 nm in diameter. A microtubule is a highly asymmetric electrolyte since each tubulin monomer has a charge of −47 elementary charges ($e = 1.6 \times 10^{-19}$ C) and is surrounded by a cloud of neutralizing cations. Based on the physical properties of tubulin, MTs have been theorized to possess intrinsic electronic conductivity as well as ionic conductivity along their length. Their electronic conductivity is envisaged to occur through the macromolecule itself, with mobile (conduction band) electrons hopping through the periodic structure of acceptor sites along the MT [50]. Due to the large electric charges on tubulin, MTs have a highly electronegatively charged outer surface as well as highly flexible C-terminal tails (TTs) whose net charge amounts to 40% of the tubulin's overall charge. This exposed negative charge distribution is predicted to attract a cloud of counterions from the surrounding cytoplasmic environment of the cell. It has been experimentally demonstrated, and later theoretically elucidated, how ionic waves are amplified along MTs [50, 51]. Many diverse experiments were performed to date in order to measure the various conductivities of MTs, with a range of results largely dependent on the experimental method applied. Curiously, Sahu et al. report that intrinsic conductivities along MTs are not length dependent [52], which would indicate at least some of the resistance of this complex system is non-Ohmic, but this conclusion still requires independent confirmation.

MTs have also been implicated in intracellular signaling, communication and even information processing, which would likely be facilitated by the fact that tubulin has a large dipole moment and a large negative charge. Consequently, MTs could be viewed as complex bioelectronic devices with a potential for carrying signal transmission via several independent channels (C-termini states, ionic waves, electronic transitions, conformational changes, etc.). It has also been hypothesized that MTs are involved in information processing, via ionic conductivity effects in neurons, as well as an organism-wide matrix of connected biological wires [28].

Ionic conductivity experiments largely show that MTs are able to increase their ionic conductivity compared to a buffer solution free of tubulin. Minoura and Muto found the conductivity to be increased 15-fold relative to that of the surrounding solution, although the ionic concentration used, at ~1 mM, is much lower than physiological ionic concentrations of just over 0.1 M [53]. Priel et al. demonstrated microtubules' ability to amplify ionic charge conductivity, with current transmission increasing by 69% along MTs [13]. The buffer was close to that of the intracellular ionic concentration, using 135 mM KCl. Ionic current amplification along MTs is explained by the highly negative surface charge density along the outside of the microtubules that creates a counterionic cloud, which allows for amplification of axially transferred signals [13]. From Priel et al.'s conductance data, we approximate the conductivity of their result to be 367 S/m. Next, we quantitatively assess the effect of AC electric fields on MTs in these ionic conductivity experiments, which are expected to be sensitive to the electric field frequencies in the 100 kHz to 1 MHz range.

Measuring intrinsic conductivity of individual MTs has been a major challenge since this requires conducting measurements in solution, which only records the increased ionic conductivity. Fritzsche et al. [54] made electrical contacts to single microtubules following dry-etching of a substrate containing gold microelectrodes. Their results indicate intrinsic resistance of a 12 μm-long microtubule to be in the range of 500 MΩ, giving a value of resistivity of approximately 40 MΩ/μm in their dry state. The same group [55] later attempted to measure dry protein conductivity, but their setup is far removed from MTs native environment and so any results from these experiments may not be indicative of the intrinsic conductivity of MTs in their biological environment. The major concern is that most of the conductivity contribution measured may come from the microelectrodes and not the protein polymer. Nonetheless, MTs adsorbed onto a glass substrate yielded an intrinsic conductivity of less than 3 S/m, which is very high. The same group performed measurements on microtubules [55] covered with a 30 nm layer of gold. The resistance of these metallized MTs was estimated to be below 50 Ω, i.e., it unfortunately originated entirely due to the metallic coating.

Another attempt to measure MT conductivity involved putting MTs in an ultrapure water solution and bridging gold electrodes that were making contacts with the MTs present [56]. As the setup used only two probes, and the conductivity was estimated from the difference in conductance of buffer solution, MT + buffer, and pure water, using an estimated 50 MT contacts between electrodes, the calculation in effect theorizes ionic conductivity indirectly. More recently, Sahu et al. [52] performed four-probe measurements of DC and AC conductivities (instead of intrinsic conductivity) in an attempt to resolve the problem of measuring ionic conductivity along the periphery of MTs. The DC intrinsic conductivities of MTs, from a 200 nm gap, were found to range between 10^{-1} and 10^{2} S/m. Surprisingly, they found that MTs at specific AC frequencies (in several frequency ranges) become approximately 1000 times more conductive, exhibiting MT conductivities in the range of 10^{3}–10^{5} S/m [52]. These effects were referred to as causing ballistic conductivity along MTs. They further claimed that it is in fact the water channel inside the MT lumen that is responsible for the high conductivity of the MT at specific AC frequencies [52].

Minoura and Muto [53] estimated the conductivity and dielectric constant of MTs using an electro-orientation method applying AC electric fields with frequencies below 10 kHz. The normally resultant convection effect was avoided by applying electric fields with a frequency between 10 kHz and 5 MHz and a sufficient field strength (above 500 V/cm) to successfully orient MTs in solution. For example, MTs aligned within several seconds in a 90 kV/m field at 1 MHz [53]. Based on these experiments, MT ionic conductivity was estimated to be 150 mS/m, which is approximately 15 times greater than that of the buffer solution.

Another attempt to measure the conductivity of MTs used radio frequency reflectance spectroscopy [57]. These investigators concluded that the conductivity of MTs was similar to that of lead or stainless steel, which would be on the order of 10^{6} S/m. This number is unrealistically high and cannot be verified by other independent studies. Furthermore, the authors [57] reported measurements of RF reflectance spectroscopy of samples containing the buffer solution, free tubulin in buffer, microtubules in buffer, and finally, microtubules with MAPs in buffer. The concentration of tubulin was 5 mg/mL and the concentration of MAP 2 and tau proteins was 0.3 mg/mL. The average DC resistance reported by these authors was: (a) 0.999 kΩ (buffer), (b) 0.424 kΩ (tubulin), (c) 0.883 kΩ (microtubules), and (d) 0.836 kΩ (MTs + MAPs). It is virtually impossible to translate these results into an estimate of the resistivity of microtubules without making assumptions about their geometrical arrangement and connectivity as resistor networks. However, assuming that all tubulin has been polymerized in case (c) and formed a uniform distribution of MTs with a combination of parallel and series networks, one can find the resistance of a 10 μm long MT, forming a basic electrical element in such a circuit, to have approximately an 8 MΩ value. This compares reasonably well to an early theoretical estimate of MT conductivity, which used the Hubbard model with electron hopping between tubulin monomers [58]. This model predicted the resistance of a 1 μm microtubule to be in the range of 200 kΩ, hence a 10 μm microtubule would be expected to have an intrinsic resistance of 2 MΩ, which is the same order of magnitude as the result reported by Goddard and Whittier [57].

Very recently, Santelices et al. [59] reported the results of precise measurements of the small-signal AC conductance of electrolytic solutions containing MTs and tubulin dimers, with a number of different concentrations, using a microelectrode system. They found that MTs at a 212 nM tubulin concentration in a 20-fold diluted BRB80 electrolyte increased the overall solution conductance by 23% at 100 kHz. This effect was shown to be directly proportional to the concentration of MTs in solution. The frequency response of the measured electrolytes containing MTs was found to exhibit a concentration-independent peak in the conductance spectrum with a maximum at around 110 kHz that decreased linearly with MT concentration. Conversely, tubulin dimers at a concentration of 42 nM were seen to decrease the overall solution conductance by 5% at 100 kHz under similar conditions. When interpreted in terms of the numbers of MTs polymerized in the sample, and assuming their action as a parallel resistor network with a lower resistance than the surrounding solution, we can estimate the conductance of individual MTs as 20 S/m compared to 10 mS/m measured for the buffer itself. This indicates that indeed MTs have electric conductivities which are three orders of magnitude higher than those of the solution. Additional measurements were made of the system's capacitance and it translated into a value of $C = 600$ pF per average 10 μm MT, which is very similar to the earlier theoretical estimates presented in this chapter.

Finally, it is interesting to address the issue of the power dissipated due to a current flowing along a microtubule. Taking a 10 μm long MT as an example, we estimate the average power drain as

$$\langle \mathrm{P} \rangle = (1/2)\,\mathrm{V}_0^{\,2}\left[\mathrm{R}/\left(\mathrm{R}^2 + \mathrm{X}_\mathrm{c}^{\,2}\right)\right]$$

where $\mathrm{X_c} = 1/\omega C$ is the capacitive resistance. Substituting the relevant numbers as per the discussion above, we obtain the dissipated power to be in the 10^{-11} W range, which is comparable to the power generated by a cell in metabolic processes. To elaborate on this conclusion, consider that an average metabolic energy production in the human body is 100 W and there are approximately 3×10^{13} cells in the body. Therefore, the power generation per cell is found to be $P_{cell} = 3 \times 10^{-12}$ W. Neurons are the most energy demanding cells, since the brain consumes 25 W of power, we can estimate the power generation per glial or neuronal cell to be $P_{glia} = 10^{-10}$ W. Consequently, additional heat generated by the processes related to MT conduction caused by externally applied electric field in the range of the peak frequency of 100 kHz may be disruptive to living cells, which could provide a mechanistic explanation of the action of TTFields.

The multiple mechanisms of MT conductance provide ample possibility to explain the varied published reports on MT conductivity. Ionic conductivity along the outer rim of the MT, intrinsic conductivity through the MT itself, and possible proton jump conduction and conductivity through the inner MT lumen has been theorized. The experimental challenge is to simulate in vivo conditions, and the possible significance of structured water, ionic, pH, and temperature conditions, over different time scales and at different frequencies. It is possible that the ionic currents generated by externally applied AC fields in the TTField mechanism may overwhelm the intrinsically generated ionic currents in cells undergoing mitosis where electric current densities, j, were measured to be in the range $0.002 < j < 0.6$ A/m^2 [60]. Since $j = \sigma E$, where $E = 100$ V/m for TTFields, and σ had a large range of values reported between 0.1 and 100 A/m^2, even taking the lower limit of 0.1 would result in ionic currents along MTs that could overwhelm the intrinsic ion flows in a dividing cell. It is entirely possible that these externally stimulated currents cause a major disruption of the process of mitosis.

11.5 Conclusions

An important aspect of the impact of external electric fields on a cell is that their penetration into the cell significantly depends on the cell's shape. Theoretical calculations on the electric field strength in a spherical cell indicate that, assuming the conductivities of the extracellular and intracellular fluids of the cell are the same, due to the small conductivity of the membrane versus these fluids, the electric field strength inside a typical cell is approximately five orders of magnitude lower than that outside the cell [61]. Recently, a COMSOL-based computational model has been developed [62] to better understand the application of TTFields to isolated cells during mitosis. The distribution of the scalar electric potential V for frequencies ranging between 60 Hz and 10 GHz was computed, taking into account the variation in cell shape during mitosis, from perfectly spherical through three stages of cytokinesis. The model demonstrated that the intracellular electric field intensity distribution is nonuniform, peaking at the cleavage plane. It also clearly showed that this effect strongly depends on the applied frequency, with the highest rate of field penetration into the cell occurring for frequencies between 100 and 500 kHz depending on the stage of cytokinesis.

In the presence of either endogenous or externally applied fields (e.g., TTFields), the cytoskeleton and, especially, both actin filaments and microtubules become bioelectric wires conducting ionic currents throughout the cell. It is also possible that proton gradients due to uneven pH distributions within cells, e.g., cancer cells, may also contribute to electrical conduction processes in living cells. This chapter discussed how these processes are critically related to the presence of large net electrostatic charges on tubulin and actin, which are largely but not completely screened by counterions. Both of these proteins are abundant in all eukaryotic cells and form long, rigid polymeric filaments. Actin filaments have been shown to provide conduits for lossless ionic transport, while microtubules have been shown to amplify ionic current flows and be orders of magnitude more conductive than the cytoplasmic medium in which they are bathed. The longer the microtubule, the more pronounced the ionic conduction effect under AC electric field influence. Additionally, it is possible that ionic currents can flow not only along the MT axis but also in the direction perpendicular (i.e., radial with respect to the MT axis) to the MT surfaces (this is also true for actin filaments). With proper initial conditions in place, solenoidal flows of ions and protons can also be induced, leading to the generation of the system's inductance. The resultant complex functional dependence of impedance on frequency is also strongly dependent upon the length of each filament and solution pH.

Moreover, in MTs some of the charges are localized on the highly flexible C-termini, leading to the propensity for oscillating charge configurations. In addition, the presence of large dipole moments on tubulin and MTs can lead to a variety of frequency- and amplitude-dependent responses of these structures to both endogenous and external electric (and electromagnetic) fields. Finally, there can be induced dipole moment contributions to the response of these structures to electric and electromagnetic stimulation, making the problem very complex and simultaneously offering a rich spectrum of possibilities for the cell to utilize in terms of communication within its confines and with other cells. Disentangling the relative importance of the various effects under different conditions is nontrivial and requires careful computational and experimental investigations under controlled conditions.

To summarize, depending on the orientation of the electric fields to the microtubule (or AF) axis, there could in general be three types of ionic waves generated: (a) Longitudinal waves propagating along the protein polymer's surface, the polymer acts like a conduction electrical cable with its inherent resistance R but also capacitance C. (b) Helical waves propagating around and along each protein polymer, for MTs there could be three or five such waves propagating simultaneously corresponding to the 3-start or 5-start geometry of a microtubule. (The effective resistance of such cables would be the individual resistance divided by the number of cables in parallel. Each cable has its own capacitance and inductance.) (c) Radial waves propagating perpendicularly to the protein polymer surface. If an electric field is oriented at an angle to the polymer axis, it is expected that all these wave types may be generated simultaneously.

It is also important to note that elongation of dividing cells facilitates penetration of these fields into cells while spherical cells would largely shield the fields and prevent them from entering into their interiors. Once AC fields generate oscillating ionic flows, these can in turn not only cause electrical currents for the purpose of signaling or communication but also lead to detrimental effects such as: (a) interference with ion flows in the cleavage area of dividing cells, (b) interference with motor protein motion and MAP-MT interactions, (c) perturbations of ion channel dynamics, and (d) changes in the net charge of the cytoplasm. In addition to the above possible subcellular effects of TTFields, there may also arise measurable heating effects in the cytoplasm of the exposed cells due to Ohmic resistance arising from ionic and protonic flows.

Identification of the strength, cause, and function of intracellular electric fields has only recently been experimentally accessible, although speculations in this area have existed for a long time. These insights may assist in devising and optimizing ways and means of affecting cells, especially cancer cells, by the application of external electric or electromagnetic fields. With the advent of nanoprobe technology, which has shown promise in measuring these fields, it is very timely to explore the various physical properties of the cytoplasmic environment including the cytoskeleton and the ionic contents of the cytoplasm.

The research outlined in this chapter promises to contribute to our general understanding of the electroconductive properties of the cytoplasm in living cells and especially the role of microtubules and actin filaments in creating dynamic and structural order in healthy functioning cells. This dynamic order may also involve electrical signal communication within and between cells. Once we are able to properly map the bioelectric circuitry of cell interiors, it should also be possible to identify biophysical differences between normal and cancer cells, which could also lead to the identification of what causes increased metastatic behavior of some cancer cells. Such an understanding may lead to better therapies and to the discovery of specific targets in order to halt metastatic transformation.

Acknowledgments This research was supported by a grant from NSERC (Canada). Additional funding from Novocure, Haifa, Israel, is gratefully acknowledged.

References

1. Cope, F. W. (1975). *Journal of Biological Physics, 3*, 1.
2. Cooper, M. (1981). *Collective Phenomena, 3*, 273.
3. Jaffe, L. F., & Nuccitelli, R. (1977). *Annual Reviews in Biophysics and Bioengineering, 6*, 446.
4. Robertson, D., Miller, M. W., Cox, C., & Davis, H. T. (1981). *Bioelectromagnetics, 2*, 329.
5. Dyshlovoi, V. D., Panchuk, A. S., & Kachura, V. S. (1981). *Cytology and Genetics, 15*, 9.
6. Kirson, E. D., Gurvich, Z., Schneiderman, R., Dekel, E., Itzhaki, A., Wasserman, Y., Schatzberger, R., & Palti, Y. (2004). *Cancer Research, 64*, 3288.
7. Kirson, E. D., Dbalý, V., Tovarys, F., et al. (2007). *Proceeding of the National Academy of Sciences of USA, 104*, 10152.
8. Davies, A. M., & Weinberg U Palti, Y. (2013). Tumor treating fields: a new frontier in cancer therapy. *Annals of the New York Academy of Sciences, 1291*, 86.
9. Szent-Györgyi, A. (1941). *Nature, 148*, 157.

10. Szent-Györgyi, A. (1957). *Bioenergetics*. New York: Academic Press.
11. Gascoyne, P. R. C., Pethig, R., & Szent-Györgyi, A. (1981). *Proceeding of the National Academy of Sciences of USA, 78*, 261.
12. Zheng, J. M., Chin, W. C., Khijniak, E., Khijniak, E. J., & Pollack, G. H. (2006). *Advances in Colloid Interface Science, 127*, 19.
13. Priel, A., Ramos, A. J., Tuszynski, J. A., & Cantiello, H. F. (2006). *Biophysical Journal, 90*, 4639.
14. Sekulić, D. L., Satarić, B. M., Tuszyński, J. A., & Satarić, M. V. (2011). *European Physical Journal E Soft Matter, 34*, 49.
15. Schwan, H. P. (1958). *Advances in biological and medical physics*. New York: Academic Press.
16. Grant, E. H., Sheppard, R. J., & South, G. P. (1978). *Dielectric be havior of biological molecules in solution*. New York: Oxford University Press.
17. Sanabria, H., & Miller, H., Jr. (2006). *Physical Review E, 74*, 051505.
18. Lima, L. F., Veira, A. L., Mukai, H., Andrade, C. M., & Fernandes, P. R. (2017). *Journal of Molecular Liquids, 241*, 530.
19. Chandra Singh, U., & Kollman, P. A. (1984). *Journal of Computational Chemistry, 5*, 129.
20. Lin, E. C., & Cantiello, H. F. (1993). *Biophysical Journal, 65*, 1371.
21. Angelini, T. E., Golestanian, R., Coridan, R. H., Butler, J. C., Beraud, A., Krisch, M., Sinn, H., Schweizer, K. S., & Wong, G. C. L. (2006). *Proceedings of the National Academy of Sciences of USA, 103*, 7962.
22. Raviv, U., Nguyen, T., Ghafouri, R., Needleman, D. J., Li, Y., Miller, H. P., Wilson, L., Bruinsma, R. F., & Safinya, C. R. (2007). *Biophysical Journal, 92*, 278.
23. Tuszynski, J. A., Portet, S., Dixon, J. M., Luxford, C., & Cantiello, H. F. (2004). *Biophysical Journal, 86*, 1890.
24. Sataric, M. V., Sekulic, D., & Zivanov, M. (2010). *Journal of Computational and Theoretical Nanoscience, 7*, 2281.
25. Poznanski, R. R., Cacha, L. A., Ali, J., Rizvi, Z. H., Yupapin, P., Salleh, S. H., & Bandyopadhyay, A. (2017). *PLoS One, 12*, e0183677.
26. Pokorny, J., Jelinek, F., Trkal, V., Lamprecht, I., & Hölzel, R. (1997). *Journal of Biological Physics, 23*, 171.
27. Pokorny, J. (2004). *Bioelectrochemistry, 63*, 321.
28. Friesen, D. E., Craddock, T. J. A., Kalra, A. P., & Tuszynski, J. A. (2015). *Biosystems, 127*, 14.
29. Warman, J. M., de Haas, M. P., & Rupprecht, A. (1996). DNA: A molecular wire? *Chemical Physics Letters, 249*, 319.
30. Beratan, D. N., Priyadarshy, S., & Risser, S. M. (1997). *Chemistry & Biology, 4*, 3.
31. Berlin, Y. A., Burin, A. L., & Ratner, M. A. (2000). *Super-lattices and Microstructures, 28*, 241.
32. Gittes, F., Mickey, B., Nettleton, J., & Howard, J. (1993). *The Journal of Cell Biology, 120*, 923.
33. Sato, M., Schwarz, W. H., & Pollard, T. D. (1987). *Nature, 325*, 828.
34. Roux, B., & Schulten, K. (2004). *Structure, 12*, 1342.
35. Liu, D. S., Astumian, R. D., & Tsong, T. Y. (1990). *Journal of Biological Chemistry, 265*, 7260.
36. White, J. A., Rubinstein, J. T., & Kay, A. R. (2000). *TINS, 23*, 131.
37. Berneche, S., & Roux, B. (2001). *Nature, 414*, 73.
38. Levin, M., & Stevenson, C. G. (2012). *Annual Reviews in Biomedical Engineering, 14*, 295.
39. Levin, M. (2009). *Semininars in Cell & Developmental Biollogy*. Amsterdam: Elsevier.
40. Levin, M. (2012). *Biosystems, 109*, 243.
41. Adams, D. S., & Levin, M. (2012). *Cell Tissue Research, 352*, 95.
42. Cantiello, H. F., Patenaude, C., & Zaner, K. (1991). *Biophysical Journal, 59*, 1284.
43. Kobayasi, S. H., Asai, H., & Oosawa, F. (1964). *Biochimica et Biophysica Acta, 88*, 528.
44. Lader, A. S., Woodward, H. N., Lin, E. C., & Cantiello, H. F. (2000). In V. Faramaz (Ed.), *MEMTMBS* (p. 77). Las Vegas, NV: CRA.

45. Lonngren, K. E. (1978). In K. E. Lonngren & A. Scott (Eds.), *Solitons in action* (p. 127). New York: Academic Press.
46. Oosawa, F. (1971). *Polyelectrolytes*. New York: Marcel Dekker, Inc..
47. Satarić, M. V., Bednar, N., Satarić, B. M., & Stojanovic, G. (2009). *International Journal of Modern Physics B, 23*, 4697.
48. Sataric, M. V., Ilic, D. I., Ralevic, N., & Tuszynski, J. A. (2009). *European Journal of Biophysics, 38*, 637.
49. Arsenault, M. E., Zhao, H., Purohit, P. K., Goldman, Y. E., & Bau, H. H. (2007). *Biophysical Journal, 93*, L42.
50. Tuszynski, J. A., Priel, A., Brown, J. A., Cantiello, H. F., & Dixon, J. M. (2007). In S. E. Lyshevski (Ed.), *CRC nano and molecular electronics handbook* (Vol. 1). London: Taylor and Francis.
51. Priel, A., & Tuszynski, J. A. (2008). *Europhysics Letters, 83*, 68004.
52. Sahu, S., Ghosh, S., Ghosh, B., Aswani, K., Hirata, K., Fujita, D., & Bandyopadhyay, A. (2013). *Biosensensors & Bioelectronics, 47*, 141.
53. Minoura, I., & Muto, E. (2006). *Biophysical Journal, 90*, 3739.
54. Fritzsche, W., Boehm, K., Unger, E., & Koehler, J. M. (1998). *Nanotechnology, 9*, 177.
55. Fritzsche, W., Böhm, K. J., Unger, E., & Köhler, J. M. (1999). *Applied Physics Letters, 75*, 2854.
56. Umnov, M., Palusinski, O. A., Deymier, P. A., Guzman, R., Hoying, J., Barnaby, H., Yang, Y., & Raghavan, S. (2006). *Journal of Materials Science, 42*, 373.
57. Goddard, G., & Whittier, J. E. (2006). *Proceedings of SPIE–Smart Structures and Materials: Smart electronics, MEMS, BioMEMS, and nanotechnology* (Vol. 6172, p. 43).
58. Tuszynski, J. A., Brown, J. A., & Hawrylak, P. (1998). *Transactions of the Royal Society A, 356*, 1897.
59. Santelices, I. B., Friesen, D. E., Bell Lewis, J., Xiao, J., Hough, C. M., Freedman, H., Rezania, V., Shankar, K., & Tuszynski, J. A. (2017). *Scientific Reports, 7*, 9594.
60. Ussing, H. H., & Thorn, N. A. (1973). *Alfred Benzon symposium* (p. 1972). Copenhagen, Denmark: Munksgaard.
61. Hobbie, R. K., & Roth, B. J. (2007). *Intermediate physics for medicine and biology*. Berlin: Springer.
62. Wenger C, Giladi M, Bomzon ZE, Salvador R, Basser PJ, & Miranda PC 2015 In *Engineering in Medicine and Biology Society (EMBC), 2015 37th Annual International Conference of the IEEE* 6892.

Part III
Electromagnetic Safety

Chapter 12
Brain Haemorrhage Detection Through SVM Classification of Electrical Impedance Tomography Measurements

Barry McDermott, Eoghan Dunne, Martin O'Halloran, Emily Porter, and Adam Santorelli

12.1 Introduction

An important medical problem is the accurate and timely detection and diagnosis of the presence of a brain haemorrhage in a patient. Brain haemorrhages can be present in pathologies such as stroke and traumatic brain injury. Stroke (also known as a cerebral vascular accident (CVA)) features a disruption in the flow of blood to an area of the brain and a subsequent sudden loss of neurological function [1]. Stroke is the main cause of adult disability in the United States, the fourth largest killer, and costs the country in the region of $70 billion annually in direct and indirect costs [2]. The aetiology of an incidence of CVA will either be related to a blockage of a blood vessel (ischaemic stroke) or the rupture of a blood vessel and subsequent bleed (haemorrhagic stroke). Crucially, as the treatment is radically different depending on the stroke type, it is vital to differentiate the cause as ischaemic or haemorrhagic [3]. For example, the use of the drug tissue plasminogen activator (tPA) is indicated for ischaemic patients but may be lethal to haemorrhagic patients [3]. Further, the patient outcomes following a CVA are directly linked to the length of interval between stroke onset and the start of treatment, with a worse prognosis associated with a delay. This underlines the need for both accurate and rapid detection of the presence, and equally the absence, of brain haemorrhage in stroke patients. Currently, definitive diagnosis is dependent on imaging modalities such as computed tomography (CT) and magnetic resonance imaging (MRI), which often suffer from accessibility issues for patients [4]. A device based on electrical impedance tomography (EIT), and augmented with machine learning (ML), may result in expedited initial diagnosis for CVA patients.

Traumatic brain injury (TBI) is any of a range of injuries that results from an external force impacting the head with a consequent disruption in brain function.

B. McDermott (✉) · E. Dunne · M. O'Halloran · E. Porter · A. Santorelli
Translational Medical Device Lab, National University of Ireland Galway, Galway, Ireland
e-mail: b.mcdermott3@nuigalway.ie

S. Makarov et al. (eds.), *Brain and Human Body Modeling*,
https://doi.org/10.1007/978-3-030-21293-3_12

TBI results in an annual cost of $61 billion in the United States [5]. Initial triage of TBI usually involves subjective assessment of severity with use of metrics such as the Glasgow Coma Scale [6]. Imaging (usually CT) is indicated for more severe TBI cases including incidents featuring haemorrhage [6, 7]. Better initial triage, including improved early detection of brain haemorrhage, potentially with the use of a modality like EIT coupled with ML, would improve the efficiency of the patient pathway through more objective selection of patients for gold standard imaging like CT. This need is illustrated by the estimation that a 10% reduction in the use of CT for minor TBI patients could save $10 million annually in the United States [8].

It is emphasised that in both of these motivating clinical examples, the imaging of a bleed is unnecessary; it is the definitive ruling in or out of the presence of the bleed that is essential to the progress of the patient in the work-up.

The use of machine learning applied to medical diagnostics and other medical areas has been the scene of significant and important growth recently [9]. The fact that computers can process large amounts of data at high speed, combined with the rapidly increasing ability of machines to learn and improve performance over time, makes the technique amply suited to analysis and interpretation of biological data. A popular biomedical application for ML and the closely related and complimentary area of data mining (DM) has been interpretation of diagnostic imaging which includes data from such modalities as CT, MRI, and ultrasound [10–12]. However, ML and DM are now being used in a range of other areas such as genetic analysis, monitoring of physiology, and the evaluation of disability [13]. In this work, we examine the potential for EIT to be used to assess anatomy and physiology of the body, coupled with the ML technique of support vector machine (SVM) classification to be used in medical diagnostics, denoted as EIT-SVM.

Fundamental to this research is the use of computational (numerical) models. Computational models allow controlled development of a technology or algorithm with the ability to experiment and test parameters resulting in progression and a better final product before translation to patients.

In the next section, the basis behind EIT, including the nature of EIT measurement frames, which are the input to the classifiers, is described. Description of the SVM classifier and the computational modelling techniques used are also presented in Sect. 12.2. Section 12.3 then summarises the application of a linear SVM classifier to raw and minimally pre-processed EIT measurement frames, investigating the performance of the classifier in detecting bleeds in different scenarios, including variations in simulated noise, bleed size, bleed location, electrode positioning, and anatomy of the model. Section 12.4 presents methods to improve the performance and efficiency of the classifier, including changing the kernel function, selective pre-processing of the frames (including the use of sub-frames), dimensionality reduction and selection of specific features (using Laplacian scores and principal component analysis (PCA)), and finally using an ensemble classifier. Section 12.5 ends the chapter with a discussion and conclusion.

The content of this chapter builds on the research presented in [14], which was expanded upon in [15]. This previously published material from [14, 15] forms the core of Sect. 12.3, before new content in Sect. 12.4 is presented, which aims to improve the classifier performance.

12.2 Technologies

This section introduces the core technologies used in this study; EIT and SVM classifiers. In the final part of the section, the computational modelling techniques and tools centred on a two-layer computational model of the head with variants, designed to emulate various test scenarios, is described.

12.2.1 *Electrical Impedance Tomography*

Electrical impedance tomography is an imaging modality and the basis of an ever-increasing and vibrant area of active research with a number of applications in the biomedical sphere [16]. EIT is based on the feature of biological tissue of electrical conductivity, as a result of the ion containing extracellular fluid (ECF) and intracellular fluid (ICF). The ECF bathes and surrounds cells, while the ICF refers to the fluid within cells. The cell membrane that surrounds the individual cells represents the border between the two compartments [17]. The conductivity is characteristic to each particular tissue. For example, blood is a good conductor, owing to the high ion content of the tissue, whereas bone is a poor conductor [16]. The conductivity is quantified in Sm^{-1} and is the inverse of the resistivity. Closely related is the concept of electrical impedance, which is the extension of the idea of resistance to alternating current (AC) circuits with a real (resistance) and complex (reactance) part. A biological tissue can be modelled as a three-part electrical circuit as shown in Fig. 12.1, where R_e is the resistance of the ECF, R_i the resistance of the ICF, and the cell membrane is modelled as a capacitor with capacitance C_m [18]. At low AC frequencies, the capacitive reactance of the cell membrane is high with the result of the overall impedance of the system being effectively R_e. At higher AC frequencies, current can pass through the cell as the capacitive reactance drops and, consequently, the overall impedance of the system drops. This concept is illustrated in Fig. 12.2 [18]. As conductivity is inversely related to impedance, it follows that the electrical conductivity of a tissue will increase with increase in AC frequency. The exact nature of the conductivity profile is a characteristic of the tissue in question.

EIT makes use of the difference in conductivity profiles of tissues. This difference in conductivity profiles is often used to generate an image of the region of

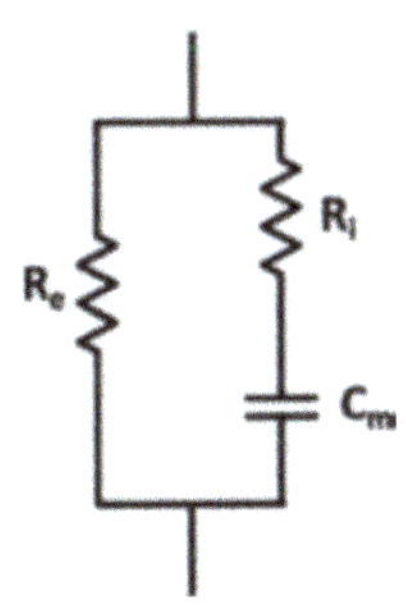

Fig. 12.1 An electrical model of biological tissue with current having two paths of flow. One path is the ECF with resistance R_e, while the other is through the cell which has a capacitor like membrane with capacitance C_m and the ICF with resistance R_i. (Adapted from [18])

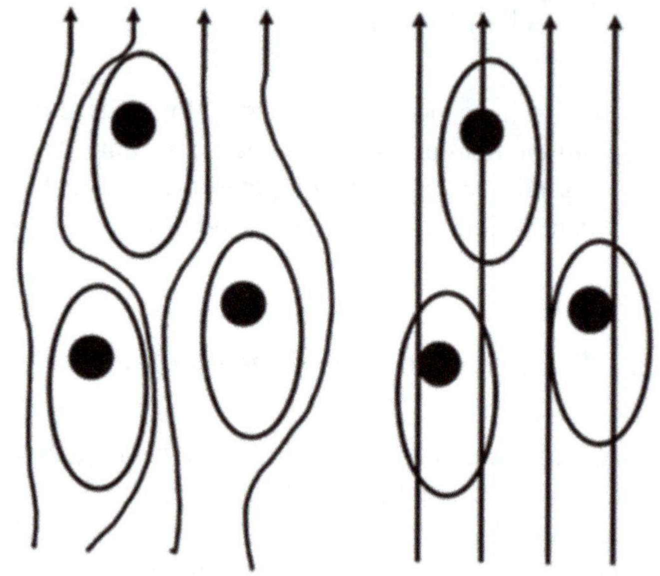

Fig. 12.2 Current movement through tissues at low and high frequencies. At low frequencies, the capacitor effect of the cell membrane impedes current flow through the interior of the cell. At high frequencies, the capacitor effect becomes negligible and current can flow through the ICF. (Adapted from [18])

interest (ROI). EIT characteristically involves an array of electrodes positioned on the boundary of the ROI. A popular electrode configuration is that of a ring of electrodes, typically with 8–64 electrodes surrounding the region [18]. Electrical current is then injected through a pair of electrodes ("stimulation") and the resultant voltages measured at all other electrode pairs. The injection pair is then changed, and voltage measurements are taken between the new measuring pairs. The overall pattern of stimulation and measurement constitutes an EIT "protocol", with each individual measurement referred to as a "channel". The complete set of channels comprises the EIT measurement "frame". EIT systems typically operate in the 1 kHz–2 MHz frequency range, with injected currents of the order of μA to low mA [16]. Importantly, international safety standards limit the current to 100 μA rms for currents up to 1 kHz with the limit rising to an absolute limit of 10 mA when operating above 100 kHz [16, 19]. The electrode configuration and number, protocol, current amplitude, and frequency are application dependent. In Fig. 12.3, a sample EIT measurement channel, with a "skip 2" protocol and the electrodes arranged in a 16-electrode ring surrounding a circular body, is illustrated. In this protocol, each electrode is paired to the electrode three positions away from it (i.e., with 2 in-between electrodes skipped over). The ROI illustrated in Fig. 12.3 is of homogenous tissue with one region of differing conductivity present (illustrated as a red circle). The presence of this tissue affects the voltage at the different measurement electrodes. For example, at 50 kHz, a bleed is more electrically conductive than the surrounding brain parenchyma [15]. Hence, for a given channel with a constant injection current, the measured voltage will be smaller in magnitude if a bleed is present than if there is only healthy brain tissue present. This trend follows from Ohm's law, described in Eq. (12.1) where V is the voltage, I is the current; and σ is the electrical conductivity,

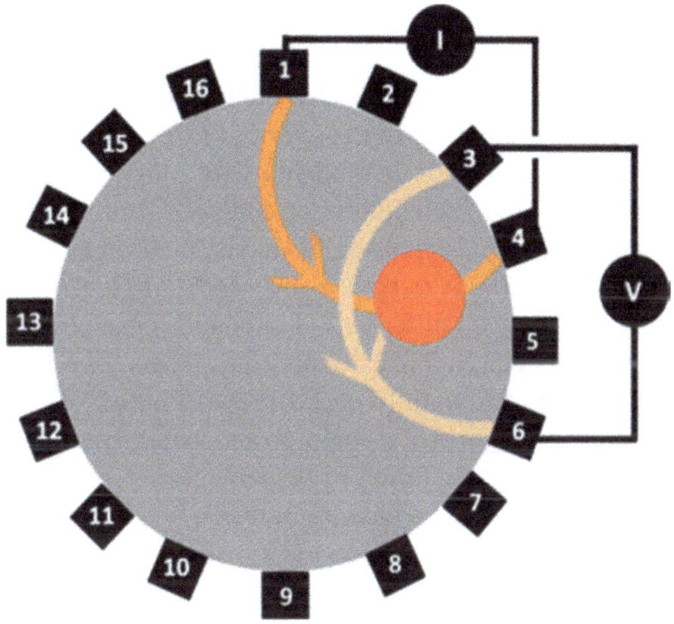

Fig. 12.3 An EIT measurement channel from a "skip 2" protocol involving a 16-electrode ring around a circular ROI. Current is injected between electrodes #1 and #4 (orange arc) with voltage measured between electrodes #3 and #6 (beige arc). A tissue with different conductivity to the bulk tissue is illustrated by the red circle

$$V\sigma = I \tag{12.1}$$

If a bleed is larger, the measured voltages will be smaller. Further, channels nearer to the bleed are affected more by the presence of the bleed than those further away, as EIT is more sensitive to changes where current density is higher [19]. Hence, information regarding the presence, nature, and location of the various tissues in the ROI are theoretically encoded in the final measurement frame.

For a 16-electrode ring, a given injecting pair results in 16 measurement pairs. However, it is common practice not to take measurements from either of the injecting electrodes hence 13 measurements are taken [19]. Over the course of a complete protocol, there will be 16 injecting pairs and so a complete frame will be made up of a total of 208 channels. The number of channels in the frame is summarised in Eq. (12.2):

$$N_{\mathrm{M}} = N_{\mathrm{E}}\left(N_{\mathrm{E}} - 3\right) \tag{12.2}$$

where N_{M} is the number of measurements when using N_{E} electrodes.

The relationship between the conductivity profile of the ROI and the values in EIT measurement frames is given by the EIT forward and inverse problems. The EIT "forward problem" refers to the prediction of the measured values given the complete conductivity profile of the body [19]. In the computational model used in this study (described in Sect. 12.2.3), the finite element method (FEM) was used to solve the forward problem for the geometry of interest, which is that of the human head. An important calculated parameter is the sensitivity matrix (the Jacobian, J). The Jacobian gives the sensitivity of each measurement to a conductivity change within the ROI [19]. The "inverse problem" of EIT involves calculating the conductivity profile of the interior of the body of interest given a set of measurements. This is an ill-posed inverse problem (the number of "voxels" to be assigned conductivity values is typically larger than the number of measurements) with the need for regularisation techniques in order to obtain the most reasonable solution [19]. The result is a conductivity map of the interior of the ROI.

EIT is a non-invasive modality with a high temporal resolution [19]. However, it has drawbacks, including poor spatial resolution, low sensitivity to conductivity changes at a depth from the boundary, and high sensitivity to electrode modelling errors [19, 20]. Attempts to overcome these challenges and reconstruct useful images have seen different EIT modalities established, many of which rely on difference imaging in order to minimise errors. The most successful EIT modality to date is that of time difference EIT (tdEIT), which reconstructs an image based on differencing frames of a "before" and "after" measurement. This modality has been applied to the monitoring of physiological functions in regions such as the thorax where there is a large contrast between inspiration (air in the lungs) and expiration (air emptied from the lungs) [19]. Static scenes are more challenging, without a satisfactory modality for imaging established to date. In a complex region such as the head, where the high impedance of the skull severely dampens the stimulating current, the imaging of static pathologies such as an established bleed has been proven to be difficult [18, 21].

In this work we examine the viability of using EIT measurement frames in a more direct manner, without the mathematically difficult and challenging image-reconstruction step. In scenarios that do not require an immediate image, such as stroke classification or TBI triage, it may be sufficient to definitively rule in or out a bleed. The information relating to the presence or absence of such a perturbation in the body of interest is encoded in the EIT measurement frame. The basis for this is the a priori knowledge that there is a notable difference in conductivity between blood and normal brain parenchyma [22]. ML offers techniques that can potentially learn from raw or processed EIT frames and classify the frame as positive or negative for a bleed. In the next section we examine such a ML technique: SVM classifiers.

12.2.2 Support Vector Machine (SVM) Classifiers

A definition of ML proposed by Mitchell is that of a "computer program that improves its performance at some task through experience" [23]. Different types of "tasks" exist when referring to ML. One of the major task types is classification. In a classification task, each observation is assigned to one of a number of designated classes or labels. Each observation consists of several features (traits) that define it. These features are used as the inputs to the ML algorithm. The algorithm will then use this information to create a trained model that can be used to predict the class that future observations belong to. In the context of the work presented here, the input features are the EIT measurement frames (processed or un-processed) obtained from numerical simulations of the head in which a bleed is or is not present. The two classes defined in this scenario are "bleed" or "normal", denoted as +1 and −1, respectively. The task of the classifier is to use the measurement frames, with the channel measurement values (or equivalent if processed) as features, to correctly predict whether future observations belong to the "bleed" or "normal" class.

SVMs are a group of popular ML algorithms commonly employed for binary classification. They have been used in previous biomedical applications, including the use of microwave signals to classify whether a breast scan is considered healthy or tumourous [24–26], and electrical impedance spectroscopy signals for classification of breast [27–29] and prostate [30] as diseased or normal. The use of EIT measurements in ML algorithms is a relatively new area of research. Some work has been done in the area of bladder volume estimation [31, 32] and the focus of this chapter, brain haemorrhage detection, has been explored by our group [14, 15, 33].

As is typical in the use of SVMs and related classifiers, the basis of the algorithm is the creation of a model using a training set. This training set consists of observations with the true class known (supervised learning) or unknown (unsupervised learning). The performance of the trained classifier can be assessed by analysis of the results of classifying a test set of previously unseen observations. The trained and tested classifier can then be used to classify new observations; assuming the training and testing process was properly implemented, the classifier will perform in-line with expectations even on new observations.

The core of the SVM model is the creation of a hyperplane that best separates observations from the two classes. A representation of a two-dimensional (2-D) hyperplane (a line) separating the observations classified as +1 or −1 is shown in Fig. 12.4. In the training phase, a mathematical model of the hyperplane and margin is developed with the training observations having n-dimensions (n number of features). The hyperplane is used to decide whether future observations belong to either the +1 or −1 class. When the data is not perfectly separable (there exists no margin that guarantees no observations between it and the hyperplane), "soft" margins can be used to ignore those outliers [34]. An important parameter when using SVM classifiers is the kernel, which defines the function used to generate the hyperplane. A linear kernel is the simplest type of kernel, which offers potential advantages including speed, low computational overhead, and an ease of implementation [35]. Other kernel functions, including the non-linear Gaussian Radial Basis Function (RBF), can be used to define the hyperplane [24, 27]. Additional information about the mathematical formulations governing the various SVM algorithms can be found in [34, 36].

The performance of a classifier can be reported by a number of different metrics. A key result is the confusion matrix, which compares the expected and predicted classes. An example of a confusion matrix, for a binary classifier, is shown in Fig. 12.5. As shown, a true positive (TP) refers to observations where the expected and predicted classes are +1, and a true negative (TN) where the expected and predicted classes are −1. A false positive (FP) is where the expected class is −1 but is predicted as +1, with a false negative (FN) the opposite.

Two key metrics of performance derived from the confusion matrix are the sensitivity and specificity. Sensitivity (TP Rate) is the proportion of observations clas-

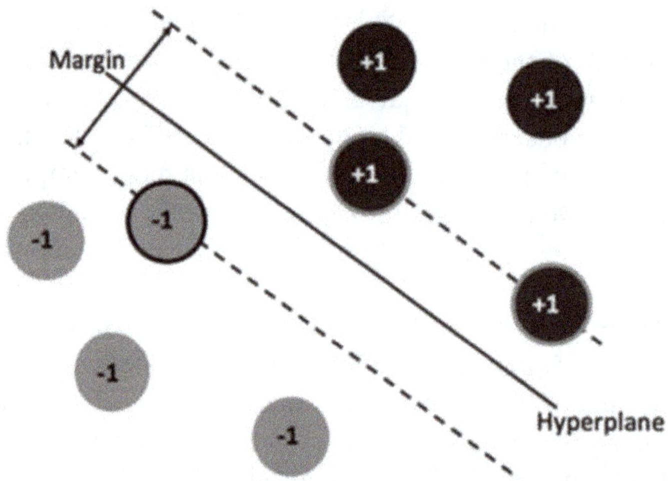

Fig. 12.4 Visualisation of a SVM classifier. The trained SVM classifier model calculates the optimal hyperplane that separates the two classes (shown here as black circles for the +1 class and grey circles for the −1). The margin of the hyperplane is as wide as possible (for a "hard" margin), with the borders of the margins defined by the cases called "support vectors" represented here as circles with a visible outer shell. The hyperplane in this case is 2D (a line)

Fig. 12.5 The confusion matrix for a binary classifier with classes ±1. The expected (true) class and predicted class assigned to cases are compared

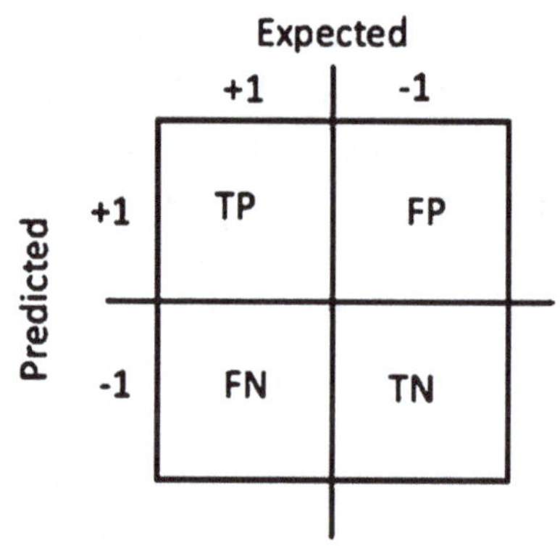

sified as +1 out of the total that are truly +1. Specificity (TN Rate) is the proportion of observations classified as −1 out of the total that are truly −1. Accuracy is the proportion of correctly classified cases out of the total number of cases. These metrics are defined in Eqs. (12.3)–(12.5),

$$\text{Sensitivity} = \frac{\text{TP}}{\text{TP} + \text{FN}} \tag{12.3}$$

$$\text{Specificity} = \frac{\text{TN}}{\text{TN} + \text{FP}} \tag{12.4}$$

$$\text{Accuracy} = \frac{\text{TP} + \text{TN}}{\text{TP} + \text{TN} + \text{FP} + \text{FN}} \tag{12.5}$$

The above Eqs. (12.3)–(12.5) imply the values of sensitivity, specificity, and accuracy range between 0 and 1, with 1 indicating perfect performance for that metric (this range is equivalent to 0–100%).

The receiver operating characteristic (ROC) curve is a plot of sensitivity versus (1 – specificity) [35]. It is a useful tool to illustrate the trade-off between sensitivity and specificity. If a classifier is 100% sensitive and 100% specific, as is ideal, then the ROC curve is said to have an Area Under the Curve (AUC) of 1. In the proposed application of brain haemorrhage detection applied to stroke and TBI, this is the ideal performance of a trained classifier. However, in cases where the performance is imperfect, this is reflected in a ROC curve where the AUC is <1. In such cases, it is possible to adjust the operating point of the classifier with a trade-off between sensitivity and specificity. For brain haemorrhage detection, it could be proposed that sensitivity is more important than specificity. A reduced specificity indicates an increased level of FPs which is not ideal but the alternative of reduced sensitivity with a consequent increased level of FNs would result in patients with bleeds being classified as normal and potentially receiving a dose of lethal tPA in the case of stroke or not receiving timely CT scan in the case of TBI. Hence, for brain haemorrhage detection, the optimal point of operation of the classifier is the point on the ROC curve where sensitivity is 1 while minimising (1 – Specificity). An example of three ROC curves is shown in Fig. 12.6.

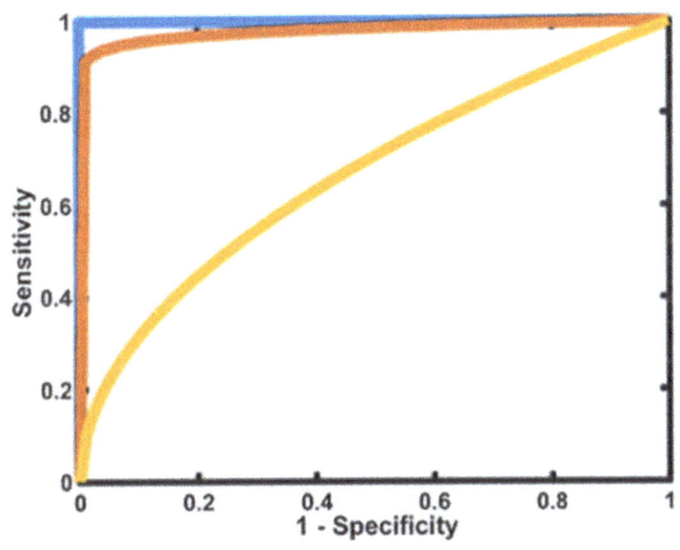

Fig. 12.6 Receiver operating characteristic (ROC) curves are a plot of sensitivity versus (1 – specificity) and show the trade-off possible between sensitivity and specificity. An ideal ROC curve has an area under the curve (AUC) of 1 with an example of this shown as the blue trace. Here, an operating point where both sensitivity and specificity are both 100% is at (0,1). The red and yellow traces show imperfect ROC curves where AUC <1. In this case it is possible to maximise sensitivity by moving to the operating point shown with the penalty of reduced specificity. At any given point, the red curve gives a better sensitivity/specificity trade-off compared to the yellow curve. The yellow curve only offers a sensitivity of 1 where specificity is 0, which would result all observations being classified as +1

12.2.3 Computational Modelling Techniques

The core computational model used in this work was a FEM model of the human head and brain. The head is an anatomically complex and intricate structure [37], but for the purposes of EIT, simplifications can be made by focusing on those tissues that have a significant effect on the conduction of electrical current. Typically, EIT simulations use a four-layer model, which includes the brain as the innermost layer, the electrically conductive cerebrospinal fluid (CSF) layer immediately external to it, the highly resistive skull, and the moderately resistive scalp [18]. Naturally, more complex models exist and may be relevant depending on the research question. For example, physical phantom models which model the differing resistivity across the skull are reported in the literature [38, 39].

In this work the head was designed as a two-layer structure. The layers were anatomically accurate representations of the brain and an aggregate outer layer comprised of the tissues external to the brain (the scalp, skull, and CSF layers), derived from anatomically realistic stereolithography (STL) files of the head [40] and brain [41]. As described in [15], this simplified model facilitated the development of an equivalent physical phantom, allowing comparison between the computational results and the phantom results. Further, it was computationally "light" and allowed rapid development of variant test models.

The STL files were meshed into a FEM model using the software packages EIDORS [42], which itself uses Netgen [43] and Gmsh [44] for meshing. EIDORS is an open source set of tools designed to aid the development of EIT (and the related area of diffuse optical tomography), and is written for use with MATLAB [45] and Octave [46]. Using EIDORS, a 16-electrode ring was placed on the exterior surface of the FEM model at the approximate level of the inion-nasion line symmetrically across the sagittal plane. The electrode ring defined a transverse plane, and a refinement of the mesh at the contact points [47] was carried out. This constituted the "base numerical model". Modifications were made to expand this model to create a total of 243 models of the "normal" (bleed free) head. These 243 models were created by varying the head and brain anatomy (±5% in size in each Cartesian axis), and modifying the electrode position (±2 mm in the positioning of the ring in terms of height). More complete details on these 243 models can be found in [33].

Bleeds were modelled as spheres within the brain layer using the computer-aided design package Autodesk Fusion 360 [48]. The two primary bleed sizes used were 30 ml and 60 ml, with some experiments using bleeds of smaller volume (down to 5 ml). In stroke patients, a 30 ml bleed is a threshold size associated with worse outcomes, with 60 ml a threshold for significant mortality [49, 50]. These bleeds were placed in each of the 243 normal models at each of the 4 cardinal points of north ('N', front), south ('S', back), east ('E', right), and west ('W', left) in the plane of the ring, at the exterior of the brain. This resulted in 1944 "bleed" head models, each model with one bleed of a given size and location. The electrical conductivity, fundamental to EIT, can be assigned to each FEM model element depending on which tissue is being modelled. The realistic conductivity values of 0.1 Sm^{-1},

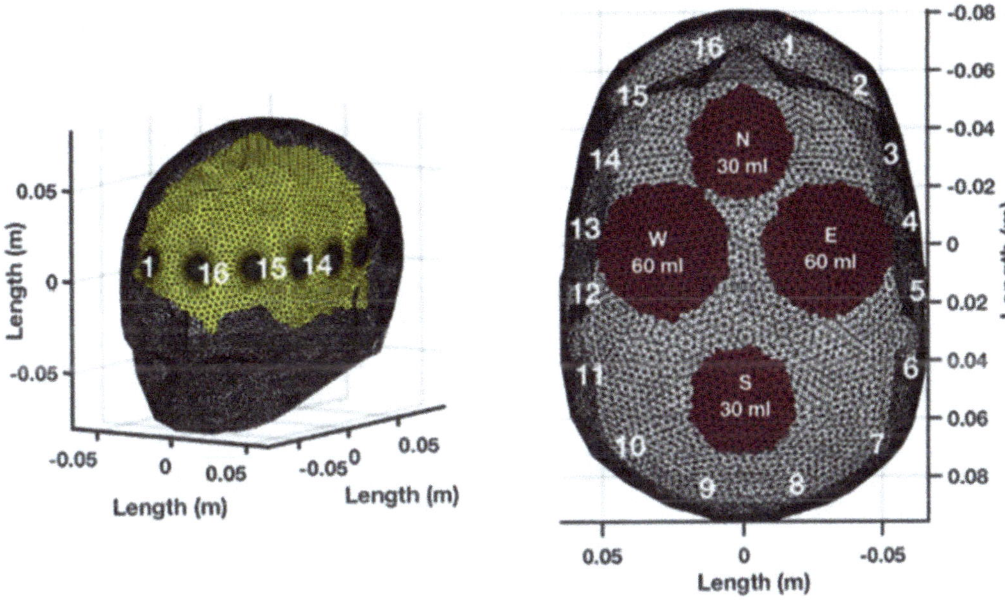

Fig. 12.7 Computational (numerical) model of the head. Left: The base numerical model is an anatomically accurate two-layer model of the brain and aggregated tissues external to the brain. The 16-electrode ring is shown with electrode contact areas in green and white numbering of some electrodes for orientation. Right: Removal of the brain layer to illustrate the size and positioning of the bleeds. The positioning of the electrodes #1–16 are shown as a ring of white numbers. Bleeds of volume 30 ml and 60 ml are positioned in the north, south, east, and west locations as shown. A given model will contain either no bleed or only one bleed. The bleeds are positioned immediately at the exterior of the brain layer in the plane of the ring. The different colouring of the layers represents the different electrical conductivities; 0.1 Sm^{-1} for the aggregate outer layer (white), 0.3 Sm^{-1} for the brain (yellow), and 0.7 Sm^{-1} for bleed (burgundy). (Adapted from [15])

0.3 Sm^{-1}, and 0.7 Sm^{-1} were used for the aggregate outer layer, the brain layer, and the bleeds, respectively [15]. EIDORS allows defining of the EIT protocol (“skip 2” for this work) and the subsequent generation of measurement frames from a FEM model. This suite of 243 normal and 1944 bleed heads allowed the emulation of a wide variety of test situations, with these experiments and results described in later sections. In Fig. 12.7 the base numerical model is shown along with the positioning of the 30 ml and 60 ml bleeds within the model.

12.3 SVM Applied to Raw EIT Measurement Frames with Analysis of the Effect of Individual Variables on SVM Performance

Initial experiments focussed on the effect of individual variables such as measurement noise, bleed size and location, electrode position, and anatomy. These variables constitute important parameters. Understanding the effect they have on EIT measurement frames, and consequent performance of the SVM classifier, can help inform future research experiment decisions. The results and conclusions from these experiments are briefly summarised herein; for more detail, refer to [15].

In each experiment, measurement frames generated from a subset of FEM models were used to train and test a linear SVM with no (raw) or minimal processing of the frames performed prior to use of the classifier. Minimal processing constituted sorting the values in the measurement frames in order of numerical value. This simple pre-processing step was found to aid performance in certain scenarios (see Sect. 12.3.2). In all cases, the training and test sets comprise an equal number of measurement frames from normal models and models with bleeds present. In this section, a linear SVM classifier was implemented for all experiments. The classifier was trained with 80% of the data set and then tested with the remaining, unseen, 20%. The classifier is optimised by generating a ROC curve in training. The generalised accuracy in training is used to choose a point on the ROC curve that maximises sensitivity. The final classifier is re-trained at this operating point and the performance of the trained classifier on the test set data is used to obtain the performance metrics presented in this section.

12.3.1 The Effect of Noise

The amount of noise in a measurement frame can be controlled by adjusting the signal-to-noise ratio (SNR) using tools supplied by EIDORS. The SNR is defined in Eq. (12.6), where the noise is a numerical value in dB,

$$\text{SNR} = 20\text{Log}_{10}(\text{Signal/Noise}) \tag{12.6}$$

In order to add noise to a measurement frame, EIDORS generates a vector (of same size as the measurement frame) of normally distributed random numbers with the values in this vector then scaled by multiplication of the ratio of the Euclidean norms of the measurement frame and noise vector, before further scaling by division by the desired SNR value. This final scaled vector of noise values is added to the measurement frame, resulting in a "noisy" frame.

EIT applications such as thoracic imaging may be successful with a system offering a SNR of 30–40 dB, whereas more demanding neural applications, that may involve smaller changes and issues such as the skull dampening, may require systems capable of 80 dB and higher [51]. In order to study the effect of noise on performance, the base numerical model was used to generate normal frames, with the 30 ml and 60 ml bleeds placed in the north location to generate bleed frames. Noise was added to the measurement frames so that a SNR of 80 dB, 60 dB, 40 dB, and 20 dB was obtained. These measurement frames at the four SNR levels were used as the input features for a linear SVM classifier. Separate experiments were performed with the raw and sorted frames. The results for the sensitivity and specificity are shown in Fig. 12.8. The results show that the classifier performs well at a

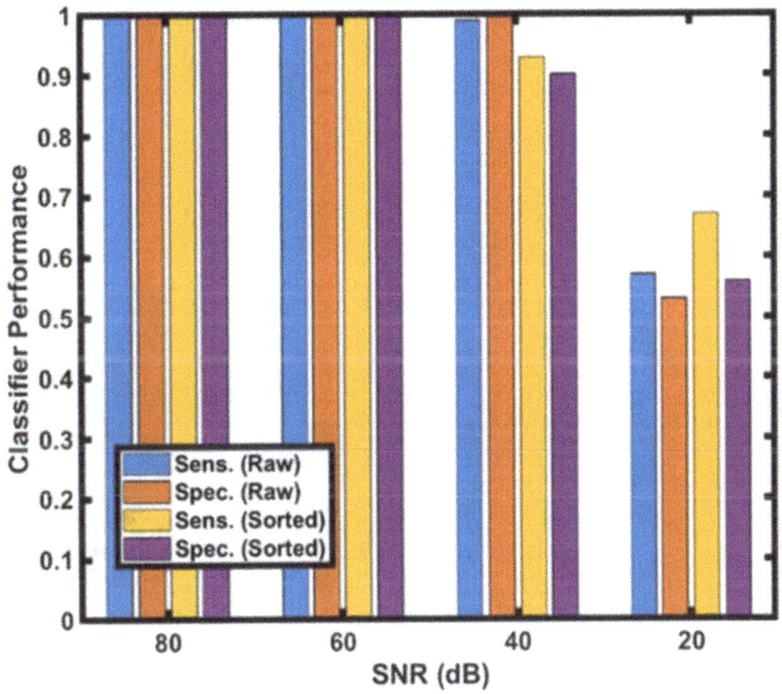

Fig. 12.8 Effect of noise on classifier performance. Measurement frames from normal and bleed cases have SNR levels of 80 dB, 60 dB, 40 dB, and 20 dB. The frames are unaltered (raw) or sorted by numerical values. The results of the classifier in terms of sensitivity (Sens.) and specificity (Spec.) for each scenario are reported above

SNR of 80 dB and 60 dB (sensitivity and specificity at or near 1), with a falloff in performance at 40 dB and poor performance at 20 dB.

12.3.2 *Effect of Bleed Location*

The base numerical model was used to generate normal frames, with 30 ml and 60 ml bleeds placed at the north location in the training set. The test set was created from frames generated by placing 30 ml and 60 ml bleeds at the three other cardinal points. Hence, the test set had novel bleed locations in comparison to the training set. The results for sensitivity and specificity for the raw and sorted frames are reported in Fig. 12.9, with the experiment performed at SNR levels of 80 dB, 60 dB, 40 dB, and 20 dB. The classifier is seen to fail at bleed detection (sensitivity of 0) at 80 dB and 60 dB when using raw measurement frames. This indicates an inability to cope with bleeds in locations different to that of the training set. The specificity is near 1 at 80 dB and 60 dB as expected as it is a measure of the ability to detect normal cases, which are the same in the training and test sets. The sensitivity then paradoxically increases at lower SNR levels, but an explanation may be the introduction of general inability to differentiate normal from bleed at lower SNR levels as evidenced by the drop in specificity. The simple pre-processing step of sorting the frames by channel value helps increase the sensitivity from 0 to 0.33 and 0 to 0.47

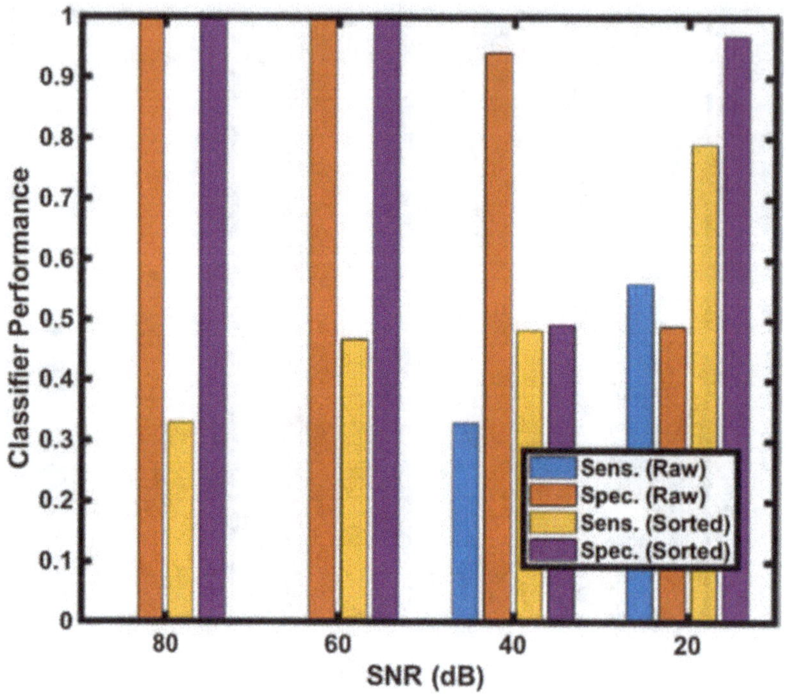

Fig. 12.9 Effect of bleed location on classifier performance. The classifier performs poorly at detecting bleeds, as judged by the sensitivity, in novel locations to that used in the training set at 80 dB and 60 dB with a paradoxical improvement seen at 20 dB SNR. The implementation of a simple pre-processing step, sorting the frames, improves sensitivity from 0 at 80 dB and 60 dB to values of 0.33 and 0.47 respectively. Specificity is not affected as severely, but this is expected as the normal cases are the same in both the training and test sets

at 80 dB and 60 dB, respectively. The sorting results in channels located near the bleed location (with smaller measured voltages as explained in Sect. 12.2.1) to cluster in the same area of the frame regardless of location. In the absence of the bleed, this area of "clustered" channels will have higher values characteristic of the no bleed case.

However, effectively these results suggest that accurate detection of bleeds in unseen locations is challenging. As described in [15], it is possible to improve performance by working at an adjusted point on the ROC curve which improves sensitivity at cost to specificity.

12.3.3 *Effect of Bleed Size*

As described in Sect. 12.2.1, the larger the size of the bleed, the greater the voltage measurements will deviate from normal values. To investigate this effect, measurements from the base numerical model without a bleed and then with bleeds of 60 ml, 30 ml, 20 ml, 10 ml, and 5 ml at each of the four locations were generated at 60 dB SNR. The 60 ml bleed subset was used to train the classifier, which was then tested with each of the smaller volumes in turn at 60 dB SNR. These results are shown in

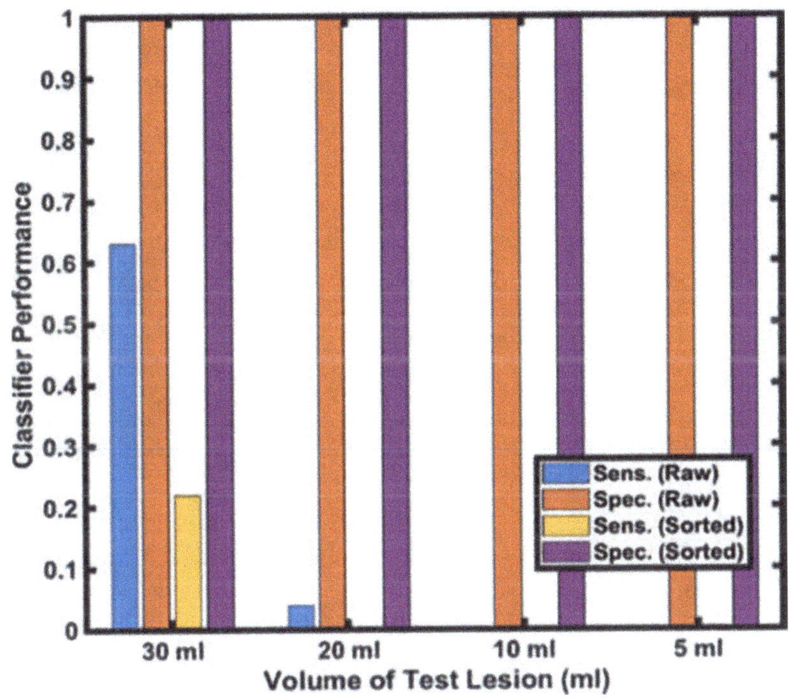

Fig. 12.10 Effect of bleed size on classifier performance. A 60 ml bleed size is used in the training set with the test set comprised of bleeds of smaller volume. All experiments are performed at 60 dB SNR. The SVM classifier is unable to detect smaller bleeds than those trained with; however, the 30 ml bleed is detected with a sensitivity of 0.63 when using the raw measurement frames

Fig. 12.10, which indicates a general inability to detect bleeds smaller than those trained with. The best value for sensitivity observed was 0.63 when using raw frames to detect the 30 ml bleed. Again, the TN rate (specificity) is not affected as the normal cases are the same in both the training and test sets.

Repeating the experiment using the 5 ml bleed in the training set and testing with each of the larger bleeds gives the results shown in Fig. 12.11, which shows generally good performance (sensitivity and specificity near 1) for detection of each of the larger bleed sizes. As discussed in Sect. 12.2.1, the size of voltage measurements is related to bleed size, with larger bleeds affecting measurements more than smaller ones. Hence, training with a small bleed "sensitises" the classifier to the bleed type, with larger bleeds resulting in even more pronounced changes in voltages and hence easier classification as bleeds.

12.3.4 Effect of Electrode Positioning

Recent literature suggests that EIT is sensitive to errors in electrode positioning [52]. In this experiment, the base numerical model is used to generate measurement frames with and without all permutations of the 30 ml and 60 ml bleed at all four positions. The test set then comprises of measurement frames from equivalent

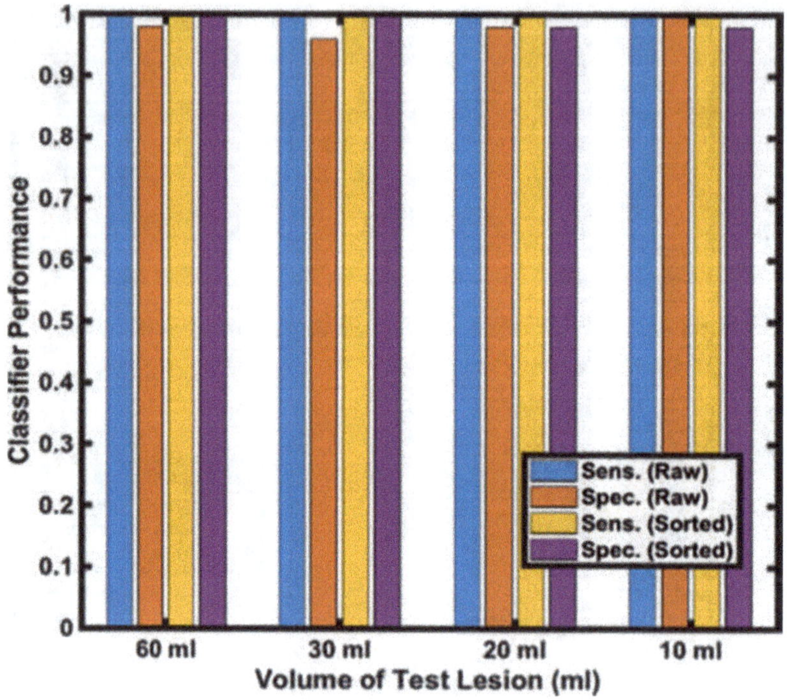

Fig. 12.11 Effect of Bleed Size on Classifier Performance. A 5 ml bleed size is used in the training set with the test set comprised of bleeds of larger volume. All measurement frames have a SNR of 60 dB. The SVM classifier performs well (Sensitivity and Specificity near 1) for all test volumes

models that differ only in the position of the electrode ring, with the ring displaced ±2 mm with respect to the original, parallel to the plane of the original. This was to replicate operator error in placing a ring on a patient's head. This analysis was performed at a SNR of 60 dB. This small error in electrode positioning causes a decrease in the sensitivity by 0.05 and 0.03, for the raw and sorted measurement frames, respectively. There is no impact on the specificity from this small electrode displacement.

12.3.5 Effect of Normal Variation in Between-Patient Anatomy

The ability of the classifier to classify normal from bleed in unseen anatomies is assessed in this experiment. The training set is made up of measurement frames calculated from the base numerical model with and without the 30 ml and 60 ml bleed at all four locations. The test set is comprised of measurement frames from 80 other anatomies that differ in the size of both the aggregate outer layer and brain layer by ±5% in the three Cartesian axes but have the electrode ring in the same position (as described in Sect. 12.2.3). These anatomies are used to generate measurement frames with and without the equivalent bleeds present. Noise is added to all measurement frames, leading to a 60 dB SNR. The results indicate that the classifier struggles with unseen anatomy; the sensitivity and specificity were below 0.60

for both raw and sorted measurement frames, a decrease in over 0.40 from the classifier performance with known anatomies. Further analysis showed that an excess of brain tissue or lack of outer tissue in a test model compared to the training model was often misclassified as a bleed. Conversely, lack of brain tissue or excess outer tissue compared to the training model was often misclassified as normal.

12.4 SVM Applied to EIT Processed Measurement Frames

Section 12.3 examined the use of a linear SVM classifier to classify FEM models of the head and brain as having a bleed or no bleed. The emphasis was on the effect of individual variables such as noise, bleed location and size, electrode positioning, and head anatomy on classifier performance. The section constituted an initial exploratory study with minimal attempt to intelligently select features for input to the classifier or indeed in selection of the best type of SVM classifier. In this section, research into these areas is reported, starting with the effect of a change of kernel on performance. Then, the effect of pre-processing and selecting input features is examined.

In all the experiments in this section, all 243 normal models and 1944 bleed models are used to generate measurement frames. As described in Sect. 12.2.3 (and elaborated on in [15]), the starting STL files of the head and brain are each distorted by ±5% in each Cartesian axis as well as in all three axes simultaneously, giving nine distinct head and nine distinct brain anatomies. FEM models of all combinations of these brain anatomies as well as the electrode ring in one of three heights resulted in 243 normal models. Bleed models were based on every combination of these normal head models combined with one of either the 30 ml or 60 ml bleed in one of the four locations, leading to a total of 1944 bleed models. An equal number of frames from the normal head set and bleed head set were used to generate 155,520 measurement frames.

A consistent method is applied in this section to optimise the performance of the SVM classifiers. First, the data is separated into five separate folds, each with a unique training data set and testing data set that is made up of 80% and 20% of the original data set, respectively. The training data set is used to optimise the SVM classifier hyper-parameters, namely the box constraint and kernel scaling factor. A Bayesian optimisation procedure is implemented to identify the hyper-parameters that lead to the greatest generalised accuracy across fivefold cross-validation. Once identified, a final trained SVM classifier is created with these optimised hyper-parameters. The excluded testing data set is then used to obtain performance metrics for the final classifier. This procedure is then repeated for all five of the unique training-testing data pairs, and final classifier performance is presented as the mean and standard deviation (STD) across these five iterations. This nested testing methodology, which has been used previously in the literature [26, 53], provides a more generalised and robust indication of classifier performance.

12.4.1 Radial Basis Function Kernel Compared to Linear Kernel

The RBF kernel can be used for SVMs when the relationship between the features and labels is non-linear, has less hyperparameters than a polynomial kernel, and has less numerical difficulties [54]. The RBF can be conceptualised as a flexible membrane that fits through sample points while minimising the curvature. Hence, the hyperplane is a "gently varying surface" and is suitable for scenarios where the data points (measurement values) do not change dramatically within a short distance in the n-dimensional hyperspace.

The first investigation of this section involves comparing the use of the linear and RBF kernels with a SVM classifier trained and optimised across all four SNR levels (80 dB, 60 dB, 40 dB, and 20 dB). In Fig. 12.12, the classifier performance, in terms of the sensitivity, specificity, and accuracy, for both the linear-SVM (top) and the RBF-SVM (bottom), is shown. Each dot on the plot denotes the mean classifier performance across the fivefold testing, with error bars representing the standard deviation range. While perfect classifier performance (1.00 ± 0.00 in all metrics) is achieved by both kernel types at 80 dB, it is observed from this figure that use of the RBF kernel can improve the classifier performance, notably at the 60 dB and 40 dB SNR levels; there is an increase in the mean accuracy between approximately 0.03 (3%) and 0.09 (9%), respectively, at these SNR levels when using the RBF kernel. When the SNR decreases to 20 dB, the performance of both classifiers approaches that of guesswork, with the mean accuracy only slightly above 50%, indicating that the changes in impedance due to the presence of the bleed are embedded within the noise. This finding suggests that hardware should guarantee an SNR well above 20 dB. From Fig. 12.12, we can in fact infer that the SNR for a hardware system should be on the order of 60 dB to expect accurate detection of brain bleeds. The improvement with the use of the RBF kernel over the linear kernel provided the motivation for the use of this kernel in all the following sections of Sect. 12.4.

12.4.2 Frame Pre-processing

In the previous sections, the classifier input features were the unprocessed EIT measurement frames, with the injection channels removed. This section will explore the use of various pre-processing techniques, ranging from manually chosen feature-extraction methods, such as taking the mean of sub-frames, to using electrode pair proximity to decide input features, to variance-based methods such as Laplacian scores and PCA. These feature extraction methods are carried out on data at all four SNR levels (80 dB, 60 dB, 40 dB, and 20 dB), with the RBF-SVM classifier optimised as described in Sect. 12.4.1. As before, classifier performance is presented as the results across fivefold testing.

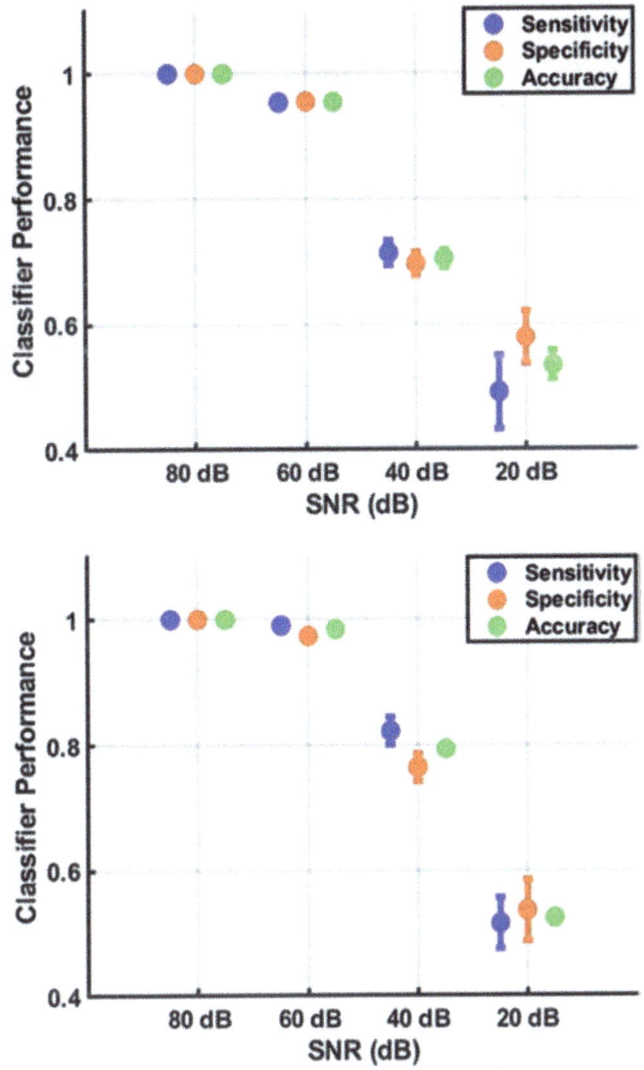

Fig. 12.12 Comparison of classifier performance using the linear-SVM (top) and RBF-SVM (bottom). Each dot on the plot denotes the mean classifier performance across the fivefold testing, with the error bars representing the standard deviation range at the respective SNR level. There is a significant improvement in performance when using the RBF kernel, notably at the 60 dB and 40 dB SNR levels

Sub-frame Means

A sub-frame is defined as the set of measurement channels associated with a given injection pair. A measurement frame from a 16-electrode array using a skip 2 pattern will have 16 such sub-frames, each with 13 channels (three channels are removed as they use either of the injecting electrodes). The 13 voltage measurements in each of the 16 sub-frames are averaged, with the resulting 16 mean-values used as the input features to the classifier. This reduces the dimensionality of the input from 208 features to 16 features. The pre-processing work-flow is shown in Fig. 12.13 below.

The performance of the RBF-SVM classifier using the sub-frame means as inputs is reported in Fig. 12.14 at each SNR level as the mean ± standard deviation of the sensitivity, specificity, and accuracy after fivefold cross validation and Bayesian optimisation. As seen, the performance at 80 dB is excellent, being near 1 ± 0 for all metrics, with a fall off at lower SNRs with, for example, sensitivity at

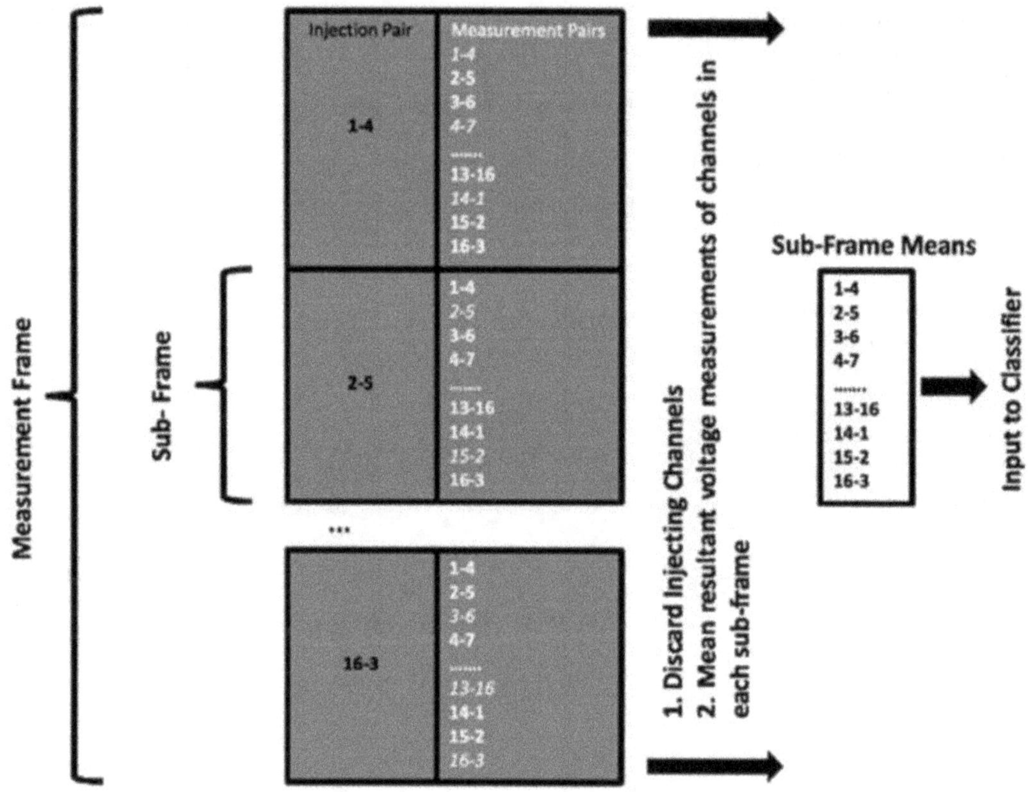

Fig. 12.13 Generating the mean of each sub-frame. Each sub-frame is made up of the 16 channels associated with a given injection pair. Removal of channels involving either of the injection pair electrodes gives 16 sub-frames each with 13 channels. The mean of the voltage measurements from each set of 13 channels in a given sub-frame is used, leaving 16 values, the sub-frame means, which are used as inputs to the classifier

approximately 0.71 ± 0.02 at 60 dB and all metrics at approximately 0.5 at 40 dB and 20 dB. It is noteworthy however that near identical performance is achieved at 80 dB relative to that of using full measurement frames (with a difference of <0.01 (1%) in all metrics), despite the significant drop in the number of features. Such a reduction in dimensionality, with nearly no effect on performance, would result in a less computationally expensive algorithm.

Near and Far Sub-frame Channels

In this section we explore using selected channels of each measurement sub-frame based on the physical locations of the recording electrodes relative to the injection pairs. Specifically, we analyse classifier performance when using "near" sub-frame channels and "far" sub-frame channels. The "near" sub-frame channels are defined as the seven channels nearer in physical location to the injecting pair of a given sub-frame. The "far" sub-frame channels are defined as the six channels further in location from the injecting pair. The complete set of near channels from each sub-frame are amalgamated and used as the input to the classifier with the same process performed to the far channels. This process reduces the input feature size to 112

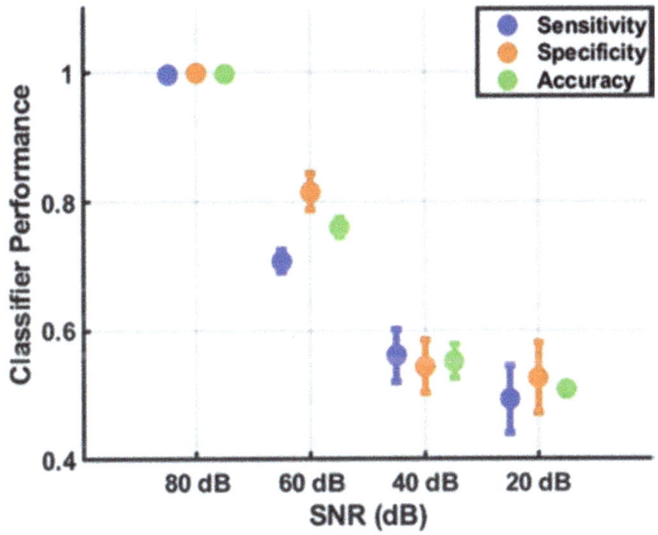

Fig. 12.14 Performance of the RBF-SVM using sub-frame means as input features. Each dot on the plot denotes the mean classifier performance across the fivefold testing, with the error bars representing the standard deviation range at the respective noise level. The performance at 80 dB SNR is near the ideal of 1 ± 0 for all metrics, comparable to the performance achieved when using the complete frames. However, performance falls off quickly at lower SNRs, with all metrics below 0.85 at 60 dB and at approximately 0.5 at 40 dB and 20 dB

features for the near sub-frame channels and to 96 features when using the far sub-frame channels, as compared to 208 for a full measurement frame. It is anticipated that the near sub-frame channels are more informative due to their proximity to the injecting pairs. The near and far sub-frame channels, for one sub-frame (that of the 1–4 injection pair), are shown in Fig. 12.15. The injecting electrode pair is denoted by the red arrow, with the near sub-frame channels shown in orange, and the far sub-frame channels shown in green.

The performance of the RBF-SVM classifier using the near and far sub-frame channels are again reported at each SNR level as the mean ± standard deviation of the sensitivity, specificity, and accuracy. These results are given in Fig. 12.16. Both the near and far sub-frame channels offer perfect performance (sensitivity, specificity, and accuracy of 1.00 ± 0.00) at 80 dB SNR, with a slight drop in performance at 60 dB SNR (but all values are ≥0.99 ± 0.01) before further drops at the 40 dB and 20 dB SNR levels. The near sub-frame channels result in better performance than the far sub-frame channels. Performance at all SNR levels for the near sub-frame channels in particular is equivalent to that of using complete frames despite an almost 50% reduction in dimensionality.

Laplacian Scores

A type of feature selection method is filter-based methods. Filter methods work by analysing the data before classification, giving a ranking to each feature. Then, the number of ranked features that optimises performance can be chosen by the user. In

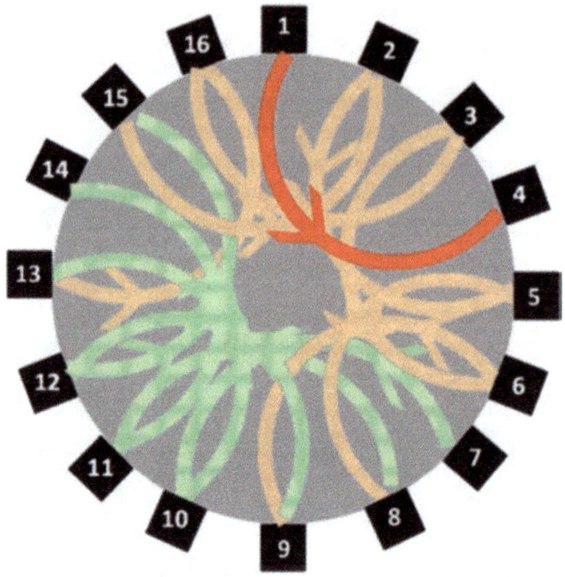

Fig. 12.15 Near and far sub-frame channels. Here the injection pair of 1–4 is shown (red). The 7 nearest channels are shown in orange, with the 6 far channels shown in green. Channels involving the measurement pair are not considered. These near sub-frames and far sub-frames channels are then used as inputs to the classifier

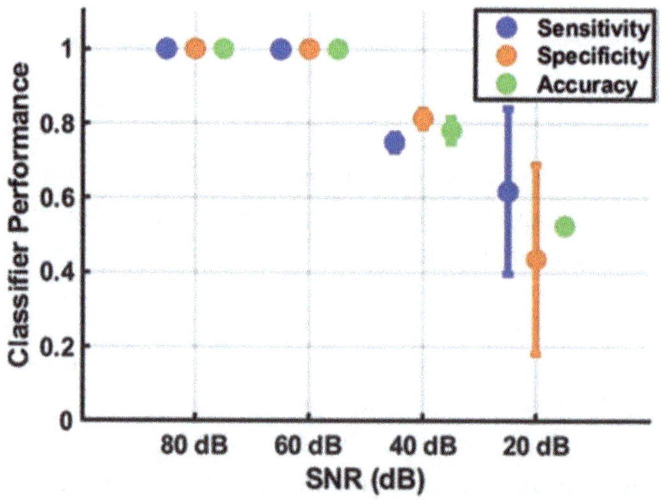

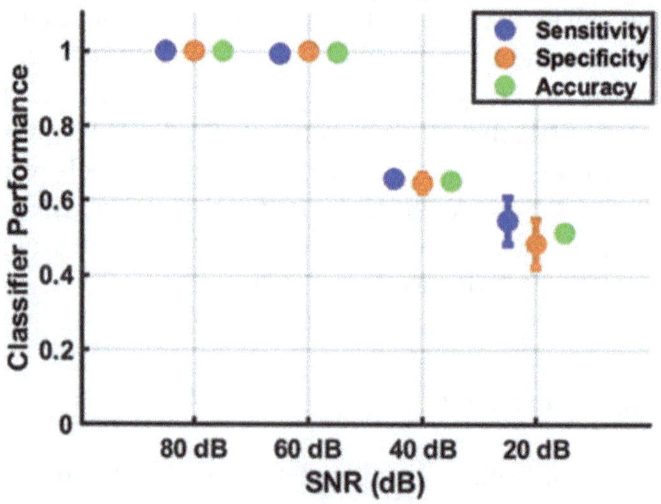

Fig. 12.16 Comparison of RBF-SVM classifier performance with using the near (top) and far (bottom) sub-frame channels as input features. Both the near and far sub-frame channels results in perfect (1.00 ± 0.00) performance at 80 dB SNR and near perfect (≥0.99 ± 0.01) at 60 dB SNR. At lower SNR levels of 40 dB and 20 dB, the near sub-frames outperform the far sub-frames. Of note, the near sub-frames result in equivalent performance to using full measurement frames at all SNR points

the context of this work, features correspond to the measurement channels. Filter methods can be implemented as either supervised or unsupervised methods. Supervised filter methods require both the observations (inputs) and classes (labels) in order to rank the features. In order to avoid any bias or data contamination, it is important to carefully choose a subset of the entire data set for the feature selection process when using supervised filter methods. Alternatively, unsupervised filter methods can use the entire dataset in order to rank the features, without biasing the classification result. An unsupervised feature selection algorithm, the Laplacian Score algorithm [55, 56], was used in this work to rank the features on the measurement sets (datasets). Specifically, the Laplacian Score algorithm works on the assumption that if two data points are close, then the data points most likely share a label [55]. Further detail on the algorithm can be found in [55]. The distance metric used in this work to define the weight matrix of the algorithm was the Euclidian distance. The advantage of using the filter-based feature selection is that after determination of the optimal number of ranked features, the original data can be used as input for the classification, with only the additional computational cost of removal of unnecessary features.

After first standardising the data, the Laplacian score is applied to each data set corresponding to each of the four SNR levels (80 dB, 60 dB, 40 dB, and 20 dB) to obtain a ranking of the 208 features at each SNR level. The optimal number of ranked features is then chosen through finding the number of features that lead to greatest generalised accuracy in the cross-validation training of the SVM classifier. In Fig. 12.17, the generalised accuracy is presented, at each of the four SNR levels, as the number of Laplacian score ranked features is increased. Based on Fig. 12.17, we can determine the optimal number of features, i.e. the best combination between the number of features and the best generalised accuracy; these optimal points are tabulated in Table 12.1.

The performance of the classifier at each SNR level is assessed with the predetermined number of ranked features as given in Table 12.1. The results are shown

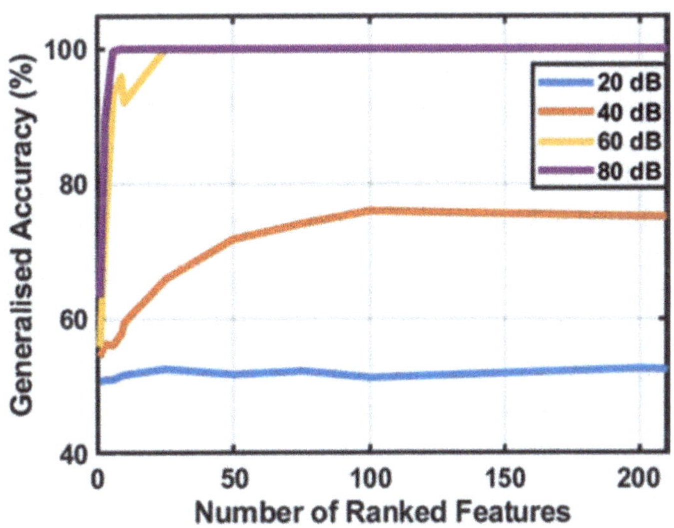

Fig. 12.17 The performance of the RBF-SVM classifier using a different number of ranked features, measured by the generalised accuracy. The ranked features were determined using the Laplacian score. The optimal point at a given SNR the one offering the highest accuracy with the lowest number of features

Table 12.1 The optimum number of ranked features at each SNR level (Maximal accuracy with fewest number of features)

SNR point	Number of ranked features	Generalised accuracy (%)
80 dB	25	100
60 dB	75	100
40 dB	100	75.96
20 dB	208	52.55

in Fig. 12.18. The accuracy, sensitivity, and specificity are perfect (1.00 ± 0.00) at 80 dB SNR, and all are better than 0.97 ± 0.01 at 60 dB SNR. Thus, classification performance is preserved while significantly reducing the input feature size from 208 to 25 and 75 features for the 80 dB and 60 dB SNR levels, respectively. Even at 40 dB SNR, classifier performance was essentially unchanged (compared to using full measurement frames) while reducing the input data set to only 100 features. As with all previous analyses, as the SNR level decreased to 20 dB, classifier performance approaches that of a random guess (metric scores of 0.5).

While unsupervised filter-based feature selection does allow preservation of the captured data to be used as inputs to the classifier in a reduced form, transforming the data with variance techniques such as PCA may enhance the results. The PCA approach is considered next.

Principal Component Analysis

A commonly implemented feature extraction method is PCA [24, 25]. PCA is used to reduce the dimensionality of data by generating new variables that represent the original data. These new variables, referred to as the principal components, are created from a linear combination of the original variables, with each successive component defining an orthogonal axis to the previous components. Thus, the entire set of principal components form an orthogonal basis for the space defined by the original data set. The data set can then be projected onto this new orthogonal basis in such a way that the variance in each axis is maximised, allowing data to be, potentially, better discriminated [57], and only a select few principal components can be used to accurately represent the data. Thus, PCA is used to both extract specific features and reduce the dimensionality of the data.

The projection of the original data on specific principal components can be referred to as the "scores". For every observation, it is these scores that will be used as input features to the RBF-SVM classifier. As PCA is a variance based feature extraction algorithm, it is important to prevent any data contamination; when performing PCA, it is necessary that there is no knowledge of the test data set. In this work, PCA is performed on only the training data, with the transformative coefficients stored and then applied to the test-set data to obtain the projection onto the principal components. Thus, we can ensure that there is no knowledge of the test-set data when performing PCA.

Similar to the previous section, a search for the optimal number of principal components is completed prior to assessing the classifier performance. The opti-

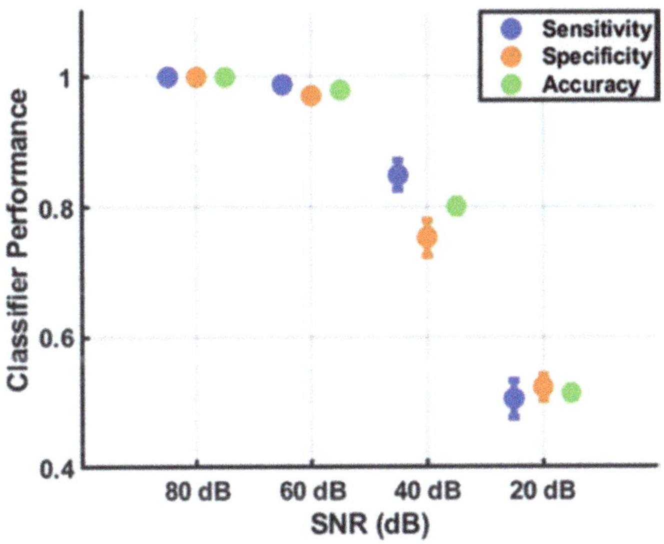

Fig. 12.18 Performance of the RBF-SVM Classifier at each SNR level using features based on Laplacian scores. The number of ranked features offering maximal accuracy is pre-determined with this feature set (selection of channels) used to train and set the classifier. Perfect performance is achieved at 80 dB SNR with 1.00 ± 0.00 in all metrics, with the use of only 25 features, with accuracy >0.97 at 60 dB SNR using 75 features

mal number of principal components is found by finding the best generalised accuracy, for each of the four SNR levels, across the cross-validation training. In Fig. 12.19, a comparison of the generalised accuracy and the number of principal components, for each of the four SNR levels, is shown. From this graph it becomes clear that for each SNR level, there is a range of principal components when performance is maximised prior to a decrease of performance as more principal components are added. This is explained by the fact that each successive principal component explains less and less variance of the original data. Therefore, those final components are simply expressing the noise in the data set, with no meaningful information contained. The optimal number of components chosen for the 80 dB, 60 dB, 40 dB, and 20 dB SNR levels is 10, 10, 11, and 31 principal components, respectively.

The classifier performance is then assessed by projecting the test data set onto the principal components using the stored projection coefficients found in training. In Fig. 12.20, the performance of the classifier is compared at all four of the SNR levels. The use of PCA leads to a marked improvement in comparison to using the entire raw data set (complete measurement frames), while also significantly reducing the input data set to at most 31 features. Most notably, at 40 dB SNR, there is an increase of almost 10% in the mean accuracy compared to using the complete measurement frames, while decreasing the input feature size from 208 features to only 11 features. Also, significantly at 60 dB SNR perfect performance is achieved using only 10 components. However, as in all previous analyses, the classifier is no better than random guesswork at 20 dB SNR.

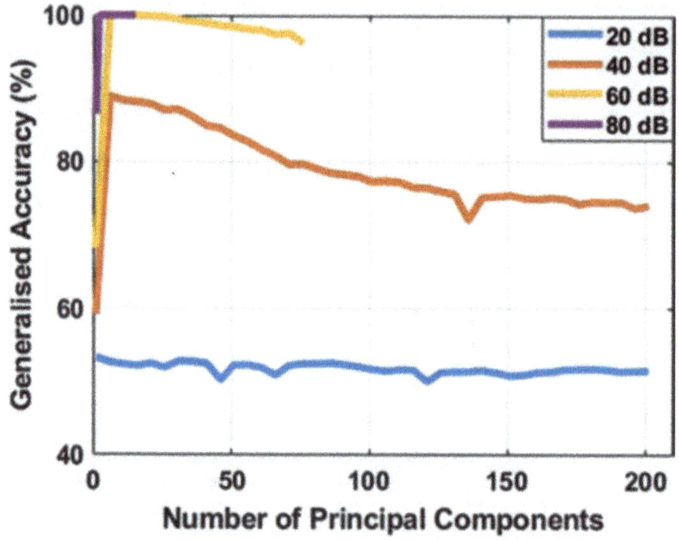

Fig. 12.19 A comparison of the generalised accuracy at the four SNR levels as the number of principal components is increased. The optimal number of principal components at each SNR level is that number giving the highest generalised accuracy which is 10, 10, 11, and 31 principal components for 80 dB, 60 dB, 40 dB, and 20 dB SNR levels, respectively

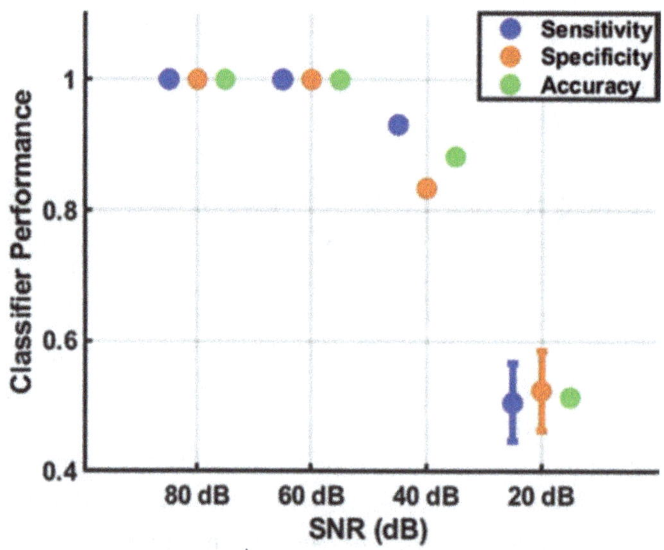

Fig. 12.20 Comparison of the performance at each of the four SNR levels for the classifier after performing PCA. Perfect performance (1.00 ± 0.00 in all metrics) is given at the 80 dB and 60 dB SNR levels despite using only 10 components at each point. A near 0.1 (10%) improvement in accuracy is seen at the 40 dB SNR level compared to the full measurement frames, but performance is approximately 0.5 in all metrics at 20 dB SNR, essentially representing a random classifier

12.4.3 Ensemble Classifier

An ensemble classifier aims to make use of multiple classifiers to make an informed decision. Additionally, these classifiers allow for better control of the sensitivity and specificity of the classifier performance [58]. In this work, an ensemble classifier was created by assigning a classifier to each of the 16 sub-frames for a given complete measurement frame. A voting scheme from each of the 16 classifiers was then used for the final classification decision. The design and implementation of this ensemble classifier is shown in Fig. 12.21.

For each observation, each of the 16 classifiers separately classified the case as ±1 (bleed or normal). Next, the sensitivity, specificity, and accuracy of the ensemble classifier at different threshold points were calculated. A threshold was the minimum number of separate classifiers needed to classify a case as a bleed for it to be classified as such; if the number was below this threshold, then the case was classified as not bleed. The threshold was adjusted from 1 to 16 in steps of 1. This control on the sensitivity and specificity allowed for the generation of a ROC curve. In Fig. 12.22, a comparison of the ROC curve, at each of the four SNR levels, for the ensemble classifier is shown.

For a low threshold (for example 1), the general trend is that the FP (1 – Specificity) rate will be high as the ensemble classifier is very sensitive to bleeds. This translates as a high sensitivity at a cost to specificity if the system is not robust. At a high threshold (for example 16), sensitivity is lost but specificity is maximised as the FN is high, with more classifiers needing to agree on labelling a case as a bleed before it is classified as a bleed. The accuracy will lie in between these two values of specificity and sensitivity at all threshold points. The trade-off in sensitivity and specificity is best illustrated at the lower SNR levels of 40 dB and 20 dB. For the higher SNR values of 80 dB and 60 dB, there is a threshold (or set of thresholds)

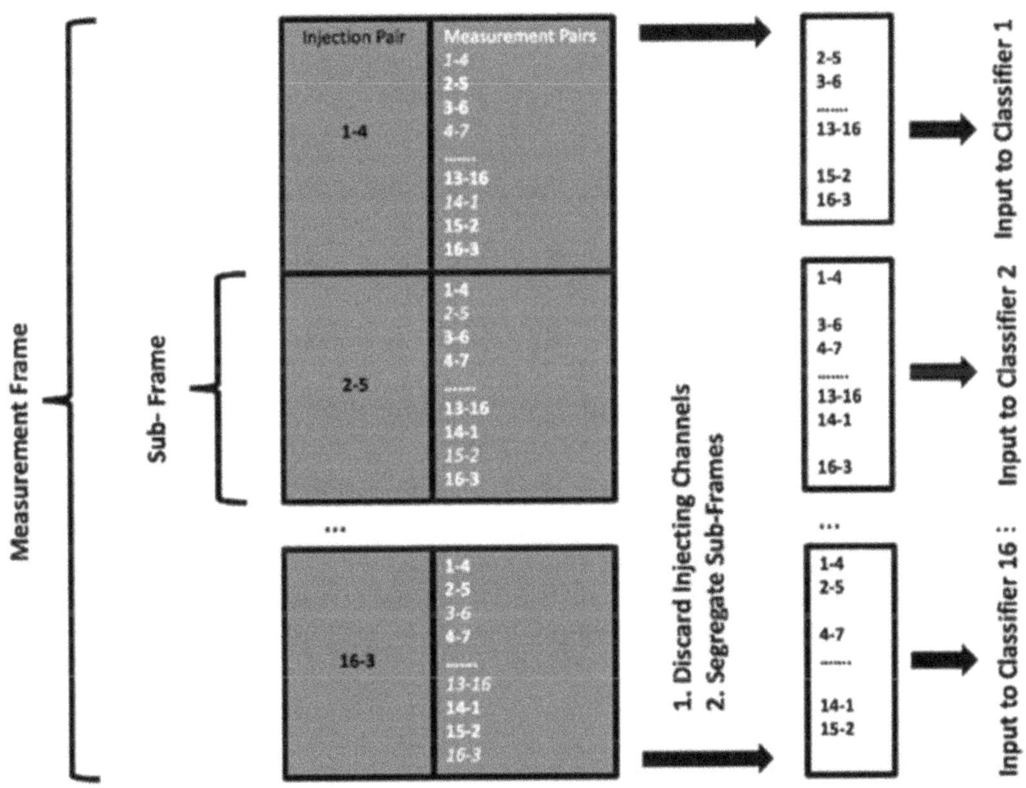

Fig. 12.21 Example of the design and implementation of the ensemble classifier. The measurement frame for a given case can be divided into sub-frames with the channels from each sub-frame used as the input for a separate classifier. The complete set of frames are segregated in this way with 16 classifiers trained and tested. Each classifier separately labels a case as ±1 with the aggregate result calculated according to a threshold which can be adjusted

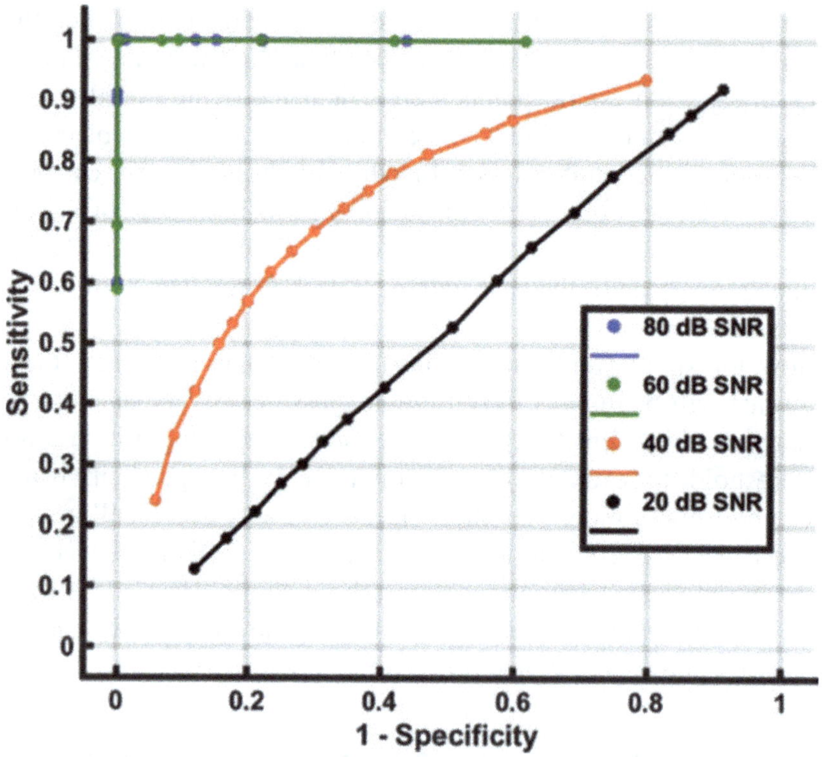

Fig. 12.22 ROC curves for the ensemble classifier at each SNR level. The points on each curve correspond to each discrete threshold value, between 1 and 16 (from right to left), with the corresponding line interpolated between the points. The curves illustrate the trade-offs between sensitivity and specificity possible at each SNR level by changing the operating points. Both the 80 dB and 60 dB plots offer an operating point of perfect performance (0,1). Performance is reduced at 40 dB and is worst at 20 dB SNR as expected. The 20 dB line is approximately that of a random classifier, being a diagonal line passing through the points (0,0) and [1]

in the intermediate area where sensitivity, specificity, and accuracy all are 1 ± 0. For both the 80 dB and 60 dB SNR levels, this area is centred at a threshold of 10. The ROC curve allows the user to select the operating point offering optimal performance, which for the proposed application of bleed detection is maximal sensitivity as justified in Sect. 12.2.2. As shown in Fig. 12.22, the 80 dB and 60 dB SNR levels result in an operating point offering the perfect combination of sensitivity and specificity both equal to 1. At 40 dB SNR, for example, a maximal sensitivity of just over 0.9 is achieved with a reduction in specificity to 0.2, with a worse performance given at 20 dB SNR, which has the performance of a random classifier.

12.5 Discussion and Conclusions

This chapter illustrates the important role that computational modelling tools have in exploring both the feasibility and the challenges in developing technologies that tackle important medical problems such as brain bleed detection. Brain haemorrhages are a medical emergency that require a prompt and accurate diagnosis prior to any appropriate treatment being administered. An ideal technological solution would be portable, non-invasive, cost effective, and crucially feature a sensitivity to the presence of a bleed (with ideally simultaneous high specificity) in the brain. Such a technology may be found in EIT coupled with modern machine learning algorithms. This work examined the feasibility of EIT coupled with ML to develop a bleed/ normal classifier based on EIT measurement frames. The approach removes the image reconstruction steps that are challenging to EIT. Further, it is EIT applied to a static scene where the most successful EIT modality, time difference EIT, cannot be applied. The chapter builds on the material presented in earlier works, including [14] and particularly [15] where, to our knowledge, such an approach with a static scene was investigated for the first time.

The effect of individual variables on performance such as the effect of noise in measurement frames, bleed location, bleed size, electrode positioning, and variations in anatomy was initially summarised in Sect. 12.3. The conclusions drawn from this section are: good performance (sensitivity, specificity, and accuracy at or near 1) is achievable particularly at 80 dB SNR; the technique is sensitive to new bleed locations not seen in the training data (although the simple pre-processing step of sorting the measurement values can improve this); the technique robustly detects bleeds larger than those trained on, but struggles with those smaller; the technique is robust to small changes in electrode positioning; and the technique struggles with unseen anatomies, in this case modelled as deviations in the morphology of the head and brain FEM models.

The simple replacement of the linear kernel with a Gaussian RBF kernel resulted in improved performance. Although both resulted in perfect sensitivity, specificity, and accuracy of 1 ± 0 at 80 dB SNR, the benefit of the RBF kernel is seen at 60 dB and 40 dB SNR levels with an increase in the mean accuracy between approximately 3% and 9%, respectively. This significant improvement in classifier performance highlights the need to explore options related to classifier choice and also the input feature selection process.

The final part of this work examined methods that moved the nature of the classifier input away from raw or minimally processed measurement frames with a view to increasing computational efficiency through intelligent feature selection that reduced dimensionality. Approaches used included processing of the measurement frames to create sub-frame means, near and far sub-frame channels, using Laplacian scores and PCA to extract specific features, and examining an ensemble classifier with thresholding to control the sensitivity to bleeds. A summary of the performance of these different classifiers, at the 60 dB and 40 dB SNR levels, where performance was mostly impacted, is shown in Tables 12.2 and 12.3, respectively. For all classifiers, the 80 dB SNR level yielded perfect classification results, whereas at 20 dB SNR all classifiers performed at essentially a guess level.

Each of the methods described in Sect. 12.4 significantly reduced the dimensionality of the input data to the classifier. The sub-frame means approach reduced the input data size to only 16 features, however suffered from poor performance when the SNR levels dropped below 80 dB, with a decrease in the mean accuracy of almost 25% in comparison to using all 208 features even at 60 dB SNR.

Table 12.2 Summary of different classifier performance at 60 dB SNR (all metrics reported as the mean ± standard deviation of the sensitivity (Sens.), specificity (Spec.), and accuracy (Acc.) with a perfect score being 1.00 ± 0.00)

		Sens.	Spec.	Acc.
Classifier type	Lin.	0.95 ± 0.01	0.96 ± 0.01	0.95 ± 0.01
	RBF	0.99 ± 0.00	0.97 ± 0.00	0.98 + 0.00
	Mean	0.71 ± 0.02	0.82 ± 0.03	0.76 ± 0.01
	Near	1.00 ± 0.00	1.00 ± 0.00	1.00 ± 0.00
	Far	0.99 ± 0.01	1.00 ± 0.00	1.00 ± 0.00
	Laplac.	0.99 ± 0.01	0.97 ± 0.01	0.98 ± 0.00
	PCA	1.00 ± 0.00	1.00 ± 0.00	1.00 ± 0.00
	Ensemb.	0.99 ± 0.00	1.00 ± 0.00	1.00 ± 0.00

All classifiers used RBF kernel except when labelled 'Linear'. Linear (Lin.): Linear kernel with full measurement frames as the classifier input; RBF: RBF kernel with full measurement frames as the classifier input; Mean: Sub-frame means as classifier input; Near: Near sub-frame channels as input; Far: Far sub-frame channels as classifier input; Laplacian (Laplac.): Optimal number of ranked features as determined by Laplacian filtering used as classifier input; PCA: Optimal number of principal components used as classifier input; Ensemble (Ensemb.): Results correspond to the threshold offering maximal sensitivity

Table 12.3 Summary of different classifier performance at 40 dB SNR (all metrics reported as the mean ± standard deviation of the sensitivity (Sens.), specificity (Spec.), and accuracy (Acc with a perfect score being 1.00 ± 0.00)

		Sens.	Spec.	Acc.
Classifier type	Lin.	0.71 ± 0.02	0.70 ± 0.02	0.70 ± 0.01
	RBF	0.82 ± 0.02	0.76 ± 0.02	0.79 ± 0.00
	Mean	0.56 ± 0.04	0.54 ± 0.04	0.55 ± 0.03
	Near	0.75 ± 0.02	0.81 ± 0.02	0.78 ± 0.03
	Far	0.66 ± 0.01	0.65 ± 0.02	0.65 ± 0.01
	Laplac.	0.85 ± 0.02	0.75 ± 0.03	0.80 ± 0.01
	PCA	0.93 ± 0.00	0.83 ± 0.01	0.88 ± 0.00
	Ensemb.	0.61 ± 0.04	0.77 ± 0.05	0.69 + 0.01

All classifiers used RBF kernel except Linear. Abbreviations of the classifier type are consistent with Table 12.3

The near and far sub-frame channels gave an approximate 50% reduction in dimensionality. Using the near sub-frame channels preserved the classifier performance when in comparison to the full data set, whereas the far channels led to a reduction in the mean accuracy of almost 15% at 40 dB SNR. These results imply, as was hypothesised, that the near sub-frame channels are more important for classifier performance.

Using the Laplacian scores to rank and choose features led to similar classifier performance using all 208 features at all SNR levels. However, at 80, 60, and 40 dB, the input features were reduced to only 25, 75, and 100 features respectively.

The use of PCA to extract and select features, in combination with the RBF-SVM classifier, lead to the best overall results, with mean accuracy values of 100%

and 88.26% at the 60 dB and 40 dB SNR levels. This marks a 1.25% and 8.91% improvement over using all 208 features, while only needing the first 10 and 11 components, at 60 dB and 40 dB SNR, respectively.

The ensemble classifier approach offered a trade-off between sensitivity and specificity depending on the threshold used. At 80 dB and 60 dB, a wide region centred around a threshold of 10 offered perfect sensitivity, specificity, and accuracy. However, this method fails to match the performance of using all the input features at 40 dB.

This work has demonstrated promise in the approach of using EIT measurement frames coupled with ML for bleed detection. Careful consideration and experimentation in regard to measurement frame processing, choice of ML algorithm, and parameters can significantly improve performance. These areas alone merit further study as well as the testing with a more realistic multi-layered computational model and physical phantom. Encouragingly, EIT hardware with SNR levels at or near 80 dB exist, which adds to the hope that computational results can be translated into real world models [59]. EIT is already a valuable imaging tool in time changing scenes but has the potential to be a valuable modality in cases with static pathologies such as brain bleeds with innovative methods such as those presented in this set of studies. We encourage researchers to further build on and develop these ideas and paradigms in order to make a measurable impact in tackling important medical problems and improving patient outcomes.

Acknowledgements The research leading to these results has received funding from the European Research Council under the European Union's Horizon 2020 Programme/ERC Grant Agreement BioElecPro n.637780, Science Foundation Ireland (SFI) grant number 15/ERCS/3276, the Hardiman Research Scholarship from NUIG, the charity RESPECT, the Irish Research Council GOIPD/2017/854 fund, and the People Programme (Marie Curie Action) of the European Union's Seventh Framework Programme (FP7/2007-2013) under REA Grant Agreement no. PCOFUND-GA-2013-608728.

References

1. Velayudhan, V. Stroke imaging: Overview, computed tomography, magnetic resonance imaging [Internet]. Medscape. [cited 2016 Oct 19]. Available from: http://emedicine.medscape.com/article/338385-overview
2. Ovbiagele, B., & Nguyen-Huynh, M. N. (2011). Stroke epidemiology: Advancing our understanding of disease mechanism and therapy. *Neurotherapeutics, 8*(3), 319–329.
3. Donnan, G. A., Fisher, M., Macleod, M., & Davis, S. M. (2008). Stroke. *The Lancet, 371*(9624), 1612–1623.
4. Birenbaum, D., Bancroft, L. W., & Felsberg, G. J. (2011). Imaging in acute stroke. *The Western Journal of Emergency Medicine, 12*(1), 67–76.
5. Faul, M., & Coronado, V. (2015). Epidemiology of traumatic brain injury. In *Handbook of clinical neurology [Internet]* (pp. 3–13). Elsevier. [cited 2018 Sep 19]. Available from: http://linkinghub.elsevier.com/retrieve/pii/B9780444528926000015.
6. NICE (National Institute for Health and Care Excellence). Head injury overview [Internet]. nice.org.uk. [cited 2016 Oct 19]. Available from: https://pathways.nice.org.uk/pathways/head-injury

7. Kim, J. J., & Gean, A. D. (2011). Imaging for the diagnosis and management of traumatic brain injury. *Neurotherapeutics, 8*(1), 39–53.
8. Lee, B., & Newberg, A. (2005). Neuroimaging in traumatic brain imaging. *NeuroRx, 2*(2), 372–383.
9. Shen, D., Zhang, D., Young, A., & Parvin, B. (2015). Editorial: Machine learning and data mining in medical imaging. *IEEE Journal of Biomedical and Health Informatics, 19*(5), 1587–1588.
10. Giger, M. L. (2018). Machine learning in medical imaging. *Journal of the American College of Radiology, 15*(3), 512–520.
11. Brattain, L. J., Telfer, B. A., Dhyani, M., Grajo, J. R., & Samir, A. E. (2018). Machine learning for medical ultrasound: Status, methods, and future opportunities. *Abdominal Radiology (NY), 43*(4), 786–799.
12. Shen, D., Wu, G., & Suk, H.-I. (2017). Deep learning in medical image analysis. *Annual Review of Biomedical Engineering, 19*(1), 221–248.
13. Jiang, F., Jiang, Y., Zhi, H., Dong, Y., Li, H., Ma, S., et al. (2017). Artificial intelligence in healthcare: Past, present and future. *Stroke Vascular Neurology, 2*(4), 230–243.
14. McDermott, B., O Halloran, M., Porter, E., & Santorelli, A. (2018). Brain haemorrhage detection through SVM classification of impedance measurements. In *2018 40th annual international conference of the IEEE Engineering in Medicine and Biology Society (EMBC)*. Honolulu, Hawaii, United States: IEEE.
15. McDermott, B., O'Halloran, M., Porter, E., & Santorelli, A. (2018). Brain haemorrhage detection using a SVM classifier with electrical impedance tomography measurement frames. Stoean R, editor. *PLoS One, 13*(7), e0200469.
16. Brown, B. (2003). Electrical impedance tomography (EIT): A review. *Journal of Medical Engineering & Technology, 27*(3), 97–108.
17. Alberts, B. (Ed.). (2002). *Molecular biology of the cell* (4th ed.). New York: Garland Science. 1548 p.
18. Holder, D., & Institute of Physics (Great Britain) (Eds.). (2005). *Electrical impedance tomography: methods, history, and applications*. Bristol/Philadelphia: Institute of Physics Pub. 456 p. (Series in medical physics and biomedical engineering).
19. Adler, A., & Boyle, A. (2017). Electrical impedance tomography: Tissue properties to image measures. *IEEE Transactions on Biomedical Engineering, 64*(11), 2494–2504.
20. Adler, A., Grychtol, B., & Bayford, R. (2015). Why is EIT so hard, and what are we doing about it? *Physiological Measurement, 36*(6), 1067–1073.
21. Horesh, L., Gilad, O., Romsauerova, A., Arridge, S., & Holder, D. (2005). Stroke type differentiation by multi-frequency electrical impedance tomography – a feasibility study. In *Proc IFMBE* (pp. 1252–1256).
22. Dowrick, T., Blochet, C., & Holder, D. (2015). *In vivo* bioimpedance measurement of healthy and ischaemic rat brain: Implications for stroke imaging using electrical impedance tomography. *Physiological Measurement, 36*(6), 1273–1282.
23. Mitchell, T. M. (1997). *Machine learning*. New York: McGraw-Hill. 414 p. (McGraw-Hill series in computer science).
24. Santorelli, A., Porter, E., Kirshin, E., Liu, Y. J., & Popovic, M. (2014). Investigation of classifiers for tumour detection with an experimental time-domain breast screening system. *Progress In Electromagnetics Research, 144*, 45–57.
25. Conceicao, R. C., O'Halloran, M., Glavin, M., & Jones, E. (2010). Support vector machines for the classificaion of early-stage breast cancer based on radar target signatures. *Progress In Electromagnetics Research B, 23*, 311–327.
26. Oliveira, B., Godinho, D., O'Halloran, M., Glavin, M., Jones, E., & Conceição, R. (2018). Diagnosing Breast Cancer with Microwave Technology: Remaining challenges and potential solutions with machine learning. *Diagnostics (Basel), 8*(2), 36.
27. Golnaraghi, F., & Grewal, P. K. (2014). Pilot study: Electrical impedance based tissue classification using support vector machine classifier. *IET Science, Measurement and Technology, 8*(6), 579–587.

28. Gur, D., Zheng, B., Lederman, D., Dhurjaty, S., Sumkin, J., Zuley, M. (2010). A support vector machine designed to identify breasts at high risk using multi-probe generated REIS signals: A preliminary assessment. In: Manning DJ, Abbey CK, editors. [cited 2018 Jan 18]. p. 76271B. Available from: http://proceedings.spiedigitallibrary.org/proceeding.aspx?doi=10.1117/12.844452.
29. Laufer, S., & Rubinsky, B. (2009). Tissue characterization with an electrical spectroscopy SVM classifier. *IEEE Transactions on Biomedical Engineering, 56*(2), 525–528.
30. Shini, M. A., Laufer, S., & Rubinsky, B. (2011). SVM for prostate cancer using electrical impedance measurements. *Physiological Measurement, 32*(9), 1373–1387.
31. Schlebusch, T., Nienke, S., Leonhardt, S., & Walter, M. (2014). Bladder volume estimation from electrical impedance tomography. *Physiological Measurement, 35*(9), 1813–1823.
32. Dunne, E., Santorelli, A., McGinley, B., Leader, G., O'Halloran, M., & Porter, E. (2018). Supervised learning classifiers for electrical impedance-based bladder state detection. *Scientific Reports, 8*(1), 5363.
33. McDermott, B., O'Halloran, M., Santorelli, A., McGinley, B., & Porter, E. (2018). Classification applied to brain haemorrhage detection: Initial phantom studies using electrical impedance measurements. In *Proceeding of the 19th international conference on biomedical applications of electrical impedance tomography*. Edinburgh.
34. Cortes, C., & Vapnik, V. (1995). Support-vector networks. *Machine Learning, 20*(3), 273–297.
35. Cristianini, N., & Shawe-Taylor, J. (2000). *An introduction to support vector machines: And other kernel-based learning methods*. Cambridge; New York: Cambridge University Press. 189 p.
36. Boser, B. E., Guyon, I. M., & Vapnik, V. N. (1992). A training algorithm for optimal margin classifiers. In *Proceedings of the fifth annual workshop on Computational learning theory – COLT '92 [Internet]* (pp. 144–152). Pittsburgh: ACM Press. [cited 2018 Oct 4]. Available from: http://portal.acm.org/citation.cfm?doid=130385.130401.
37. Standring, S., Ananad, N., & Gray, H. (Eds.). (2016). *Gray's anatomy: The anatomical basis of clinical practice* (41st ed.). Philadelphia: Elsevier. 1562 p.
38. Zhang, J., Yang, B., Li, H., Fu, F., Shi, X., Dong, X., et al. (2017). A novel 3D-printed head phantom with anatomically realistic geometry and continuously varying skull resistivity distribution for electrical impedance tomography. *Scientific Reports [Internet], 7*(1). Available from: http://www.nature.com/articles/s41598-017-05006-8.
39. Avery, J., Aristovich, K., Low, B., & Holder, D. (2017). Reproducible 3D printed head tanks for electrical impedance tomography with realistic shape and conductivity distribution. *Physiological Measurement, 38*(6), 1116–1131.
40. Grozny. Thingiverse – Human Head [Internet]. [cited 2017 Feb 15]. Available from: http://www.thingiverse.com/thing:172348
41. Dilmen, N. NIH 3D print exchange- brain MRI [Internet]. [cited 2017 Feb 15]. Available from: https://3dprint.nih.gov/discover/3DPX-002739
42. Adler, A., & Lionheart, W. R. B. (2006). Uses and abuses of EIDORS: An extensible software base for EIT. *Physiological Measurement, 27*(5), S25–S42.
43. Schoeberl, J. Netgen [Internet]. Vienna: Vienna University of Technology; Available from: https://ngsolve.org/
44. Geuzaine, C., & Remacle, J.-F. (2009). Gmsh: A 3-D finite element mesh generator with built-in pre- and post-processing facilities. *International Journal for Numerical Methods in Engineering, 79*(11), 1309–1331.
45. MATLAB 2017A [Internet]. Natick: The MathWorks Inc. Available from: https://uk.mathworks.com/
46. Eaton, J.W., Bateman, D., Hauberg, S., Wehbring, R. GNU Octave version 4.2.2 manual: A high-level interactive language for numerical computations [Internet]. 2018. Available from: https://www.gnu.org/software/octave/doc/v4.2.2/
47. Grychtol, B., Adler, A. FEM electrode refinement for electrical impedance tomography. In: 2013 35th annual international conference of the IEEE Engineering in Medicine and Biology

Society (EMBC) [Internet]. IEEE; 2013. p. 6429–6432. Available from: http://ieeexplore.ieee.org/document/6611026/

48. Autodesk. Fusion 360 [Internet]. Mill Valley: Autodesk. Available from: https://www.autodesk.com/products/fusion-360
49. Broderick, J. P., Brott, T. G., Duldner, J. E., Tomsick, T., & Huster, G. (1993). Volume of intracerebral hemorrhage. A powerful and easy-to-use predictor of 30-day mortality. *Stroke, 24*(7), 987–993.
50. Hemphill, J. C., Bonovich, D. C., Besmertis, L., Manley, G. T., Johnston, S. C., & Tuhrim, S. (2001). The ICH score: A simple, reliable grading scale for intracerebral hemorrhage editorial comment: A simple, reliable grading scale for intracerebral hemorrhage. *Stroke, 32*(4), 891–897.
51. Hun Wi, Sohal, H., McEwan, A. L., Eung Je Woo, & Tong In Oh. (2014). Multi-frequency electrical impedance tomography system with automatic self-calibration for long-term monitoring. *IEEE Transactions on Biomedical Circuits and Systems, 8*(1), 119–128.
52. Jehl, M., Avery, J., Malone, E., Holder, D., & Betcke, T. (2015). Correcting electrode modelling errors in EIT on realistic 3D head models. *Physiological Measurement, 36*(12), 2423–2442.
53. Li, Y., Santorelli, A., Laforest, O., & Coates, M. (2015). Cost-sensitive ensemble classifiers for microwave breast cancer detection. In *2015 IEEE International Conference on Acoustics, Speech and Signal Processing (ICASSP) [Internet]* (pp. 952–956). South Brisbane: IEEE. [cited 2018 Oct 9]. Available from: http://ieeexplore.ieee.org/document/7178110/.
54. Hsu, C.-W., Chang, C.-C., Lin, C.-J. (2010). A practical guide to support vector classification. p. 16.
55. He, X., Cai, D., & Niyogi, P. (2005). Laplacian Score for Feature Selection. In *NIPS'05 Proceedings of the 18th International Conference Neural Information Process System* (pp. 507–514). Vancouver.
56. Dunne, E., Santorelli, A., McGinley, B., Leader, G., O'Halloran, M., & Porter, E. (2018). Image-based classification of bladder state using electrical impedance tomography. *Physiological Measurement, 39*(12), 124001
57. Conceição, R. C., O'Halloran, M., Glavin, M., & Jones, E. (2011). Evaluation of features and classifiers for classification of early-stage breast cancer. *Journal of Electromagnetic Waves and Applications, 25*(1), 1–14.
58. Li, Y., Porter, E., Santorelli, A., Popović, M., & Coates, M. (2017). Microwave breast cancer detection via cost-sensitive ensemble classifiers: Phantom and patient investigation. *Biomedical Signal Processing and Control, 31*, 366–376.
59. Avery, J., Dowrick, T., Faulkner, M., Goren, N., & Holder, D. (2017). A versatile and reproducible multi-frequency electrical impedance tomography system. *Sensors, 17*(2), 280–280.

Chapter 13
Patient-Specific RF Safety Assessment in MRI: Progress in Creating Surface-Based Human Head and Shoulder Models

Mikhail Kozlov, Benjamin Kalloch, Marc Horner, Pierre-Louis Bazin, Nikolaus Weiskopf, and Harald E. Möller

13.1 Introduction

The interaction of radio frequency (RF) electromagnetic (EM) fields with the human body during magnetic resonance imaging (MRI) is complex and subject specific. The specific absorption rate (SAR) used as the safety limit in MRI is also subject specific, especially at RF above 100 MHz [1]. Safety limits based on the SAR in MRI are typically derived from three-dimensional (3D) numerical EM simulations of MRI RF transmit coils loaded with human body models [2–7].

An increasing number of MRI investigations that study the human brain employ multimodal setups, where additional devices are used to record complementary information or manipulate brain states [8–10], examples include electroencephalography (EEG), transcranial magnetic stimulation (TMS), and transcranial direct current stimulation (tDCS). This requires a dedicated setup of wires and electrodes that are in contact with human skin. For example, a tDCS setup includes two external wires and electrodes. The wires enter the MRI RF transmitter coil's effective exposure volume and operate as an antenna, the performance of which depends on the relative positioning of the wires and the human body, patient

M. Kozlov (✉) · N. Weiskopf · H. E. Möller
Max Planck Institute for Human Cognitive and Brain Sciences, Leipzig, Germany
e-mail: kozlov@cbs.mpg.de

B. Kalloch
Max Planck Institute for Human Cognitive and Brain Sciences, Leipzig, Germany

Leipzig University of Applied Sciences, Leipzig, Germany

M. Horner
ANSYS, Inc., Evanston, IL, USA

P.-L. Bazin
Max Planck Institute for Human Cognitive and Brain Sciences, Leipzig, Germany

University of Amsterdam, Amsterdam, The Netherlands

S. Makarov et al. (eds.), *Brain and Human Body Modeling*,
https://doi.org/10.1007/978-3-030-21293-3_13

landmark position, and the quality of the electrical contact between the electrode and skin. Assessing the RF safety of a device that is in electrical contact with the skin during an MRI examination requires the evaluation of RF-induced heating of human tissue located near the contact area.

An increasing number of MRI examinations are being performed on patients with an active implantable medical device (AIMD), for example, a cardiac pacemaker or deep brain stimulator, or a passive implant, such as an orthopedic hip implant [11–13]. One of the major safety concerns for ensuring safe examinations of such patients is the evaluation of in vivo RF-induced heating of tissue near the lead electrode, which can result in tissue damage.

Due to the complexity of assessing MRI RF-induced heating in vivo, 3D EM and transient thermal co-simulation is used to assess RF-induced heating of implanted devices and devices that bring electrodes into contact with human skin [14, 15].

When modeling RF-induced heating during MRI, a computational EM solver is used to compute the absorption of EM energy in different types of human tissue. The volume and surface losses from 3D EM simulations act as thermal sources in tissue heating calculations. Volume losses in human tissue substantially depend on tissue geometries and electrical properties. For example, *(i)* in a patient undergoing an MRI at a head landmark position, the cerebrospinal fluid (CSF) space must be a continuous medium in the numerical domain to excite a significant current; *(ii)* electrical properties of the skin and underlying tissues, especially fat, differ significantly and volume losses depend on tissue geometries, *(iii)* the correct skin thickness is very important when assessing MRI RF safety for devices where electrodes are in contact with human skin.

Different numerical approaches can be applied to perform 3D EM and transient thermal co-simulation for simple geometrical objects, for example, a phantom as defined in ASTM F2182a-11 [16]. However, reliable simulations of realistic human models require a correct match between solver capabilities and geometrical properties of the human models under investigation.

To accurately represent individual tissue structures in a patient-specific human model, they must first be segmented from imaging data. Most imaging data are voxel-based data obtained, for example, from MRI scans, high-resolution cryosection image datasets, or histological sections. Therefore, most available numerical human models are voxel-based geometries [17].

Voxel-based human models are commonly simulated using time-domain solvers, in most cases these are finite-difference time-domain (FDTD) or finite integration technique (FIT) solvers, and use hexahedral meshes. The hexahedral mesh results in a staircased discretization of the surfaces of curved structures.

The size of the hexahedral mesh elements must be substantially smaller than the thickness of the coil's conductive elements, the thickness of relevant thin human tissue (e.g., CSF and skin), the wire diameters, and electrode thickness of EEG or tDCS setups to maintain the precision of the geometric model. Structures that are thinner than the employed resolution are undersampled in the mesh and, therefore, appear as being separated, noncontinuous segments within a space. In this case, correct electrical and thermal contact between anatomically connected tissue parts or between an electrode and human tissue are not ensured.

In most common implementations of time-domain solvers, the size of the hexahedral mesh elements must be uniform for a given mesh line. Thus, using small-sized hexahedral mesh elements for some objects results in meshing practically the entire numerical domain with small mesh elements. Because the simulation time of a time-domain solver is proportional to the number of mesh elements and is inversely proportional to the smallest-sized mesh elements, correct meshing of realistic MRI RF coils with a high-resolution human model and electrodes results in a significant increase in computation time.

Different subgridding approaches are used to overcome this limitation of time-domain solvers. However, these are not very effective for MRI-related simulations due to the geometrical complexity of RF coils for MRI and different types of human tissue, the bent shape of electrodes, and the bent trajectories of the wires.

The aforementioned simulation drawbacks of voxel-based human models can be avoided with anatomically correct surface-based models and solvers based on unstructured meshes. A flexible discretization of human structures can be achieved with tetrahedra, pyramids, and extruded triangles (prisms) as mesh elements. In this way, the complex shape of curved human tissue structures, electrodes, and wires can be maintained. Electrical contact within and between tissues, as well as between an electrode and human tissue, can also be preserved.

For solvers based on unstructured grids, it is beneficial to have only one boundary between adjacent structures. When these boundaries are triangulated, the resulting surfaces are free from intersections or intermediate gaps, and the number of triangles in the entire model is reduced.

Unfortunately, 3D EM frequency-domain solver development has advanced beyond geometry import, pre-processing, and mesh generation capabilities. For most up-to-date solvers based on unstructured meshes, a surface-based model must only include objects that are geometrically error free (no self-intersections, over-connections, etc.), and the number of faces in the model must be limited to approximately 500,000 to be meshed by commercially available packages in an acceptable time interval.

Although the surface-based Virtual Family v2.x models [18] were developed primarily for 3D EM simulations, these models and high-resolution voxel-based human models (less than ~2 mm voxel size) do not meet the aforementioned error-free geometry requirements. Thus, their use with most up-to-date solvers based on unstructured meshes is practically impossible.

Two workhorses for 3D EM investigations are the Virtual Family v1.x models [19] and the Virtual Population 3.0 models [18]. Developed as surface-based anatomical models, they are used in 3D EM simulations as discretized voxel-based geometries. The Virtual Population 3.0 models are integrated within the multiphysics simulation platform Sim4Life, which includes only a time-domain EM solver, or the SEMCAD time-domain EM solver. The Virtual Population 3.0 models cannot be exported to any third-party software. The Virtual Family v1.x models are also compatible with Sim4Life or SEMCAD time-domain EM solvers or can be exported to other solvers only in voxel format, which is not suitable for import into solvers based on unstructured meshes.

Some of the surface-based human body models presented in 3D EM simulation reports, for example, the Chinese reference man [20], have been used only with FDTD solvers. The reasons for this are unknown.

Recent literature includes reports of the development of surface-based models for a variety of applications, for example, emission imaging (namely the 4-D XCAT Phantoms) [21], biomechanics, and injury biomechanics [22]. These application-specific models require an efficient conversion into a format that is compatible with a geometric modeling kernel of a 3D EM solver based on unstructured meshes or its geometrical pre-processor, as well as handling the geometrical complexity of the models at the appropriate level if a geometrical pre-processor cannot be used. The complexity of the direct conversion of an application-specific surface-based geometry to 3D EM suitable surface-based geometry could be a reason why a model should be voxelized as the first step, and new surface meshes should be generated as the second step, as was the case for 4-D XCAT Phantoms [23].

Converting voxel-based data to high-quality surface-based objects and correctly matching contact regions presents a significant challenge. It is even more difficult to meet all the requirements for importing a human model composed of numerous tissue structures into an EM solver in the form of surface-based geometries.

Only a few surface-based full-body human models, for example the NEVA Electromagnetics (Yarmouth Port, Cape Cod, MA, USA) female VHP model [24], developed based on the Visual Human Project® data set [25], and the Aarkid (East Lothian, Scotland) male model [26], have been used successfully with 3D EM solvers based on unstructured meshes. Available models provide different levels of detail of different human tissue types. For example, CSF is rarely included, and there are sometimes multiple levels of model fidelity.

The electrical properties of some types of human tissue are quite similar. Thus, a human model that only includes a subset of human tissue could be sufficient for application-specific MRI EM simulations. For MRI birdcage coil simulations, fat, muscle, bone, and air spaces are especially important to consider [26]. For high-field MRI head coil simulations, a human model should additionally include CSF, white matter (WM), and grey matter (GM) [27].

Generating a correct full-body surface-based model requires great effort throughout each stage. This is why head and torso models such as that developed by the team from NeuroSpin-CEA [28] have become effective solutions for investigating head RF exposure.

We previously introduced a semi-automatic processing pipeline to generate individualized surface-based models of the human head and upper torso from the MR images of individual subjects [29]. A key feature of this workflow is that the resulting models have a single surface between adjacent structures. The comprehensive workflow covers image acquisition, atlas-based segmentation of relevant structures, generation of segmentation masks, and surface mesh generation of the single, external boundary of each structure of interest. Two head and torso models were generated and used for 3D EM simulations using this pipeline [30].

The voxel models derived from the Visual Human Project Visible Man and Visible Woman data sets have formed the basis for a large number of MRI RF safety assessments [27]. See [31] for an example of the HUGO anatomical model. For interlab studies in general, it is beneficial to use voxel- and surface-based models derived from the same dataset. Therefore, we have selected the Visible Man data set as source data for this investigation.

In our case, the generated human models were intended for simulations of head coils in high-field MRI and 3T MRI whole body coils with patients at the head landmark position. For these purposes, a human model can be truncated at the torso without introducing substantial uncertainty.

In this investigation, the pre-segmented AustinMan dataset [32] was used to facilitate fast generation of the surface-based head and torso model of the Visual Human Project, Visible Man.

MRI coil development and the MRI RF safety assessment of a given RF coil require multi-port simulations, and results for only a single frequency in which the MRI scanner is running. The latter eliminates one of the major drawbacks of most frequency-domain solvers—the requirement to simulate a set of frequencies over the bands of interest. The size of the smallest mesh elements do not substantially influence the simulation time of most frequency-domain solvers. Thus, a frequency-domain solver is a good candidate for reliable RF safety assessment in MRI.

ANSYS HFSS (ANSYS, Inc., Canonsburg, PA, USA) was chosen for our 3D EM simulations because of its robustness in handling complex MRI coil geometries and fast multi-port simulations. Therefore, a substantial part of our work was to investigate optimization approaches that ensure successful 3D EM simulations when using surface-based geometry. It is important to note that the geometry kernel and associated functionality vary from solver to solver. Thus, some additional geometrical pre-processing may be required if our head and torso models are used with other 3D EM solvers.

The ANSYS Non-Linear Thermal (NLT) platform will be used for our future investigations into temperature rise for multimodal setups. Thus, the requirements of the ANSYS NLT platform were taken into account during development of the 3D EM model.

13.2 Methods

13.2.1 Surface Mesh Generation

Here, we present a dedicated subset of our previously established workflow, namely the post-processing of segmentation images to so-called segmentation masks followed by surface mesh generation. We applied this sub-part of the pipeline to the segmented AustinMan dataset [30].

We utilized the *Medical Image Processing, Analysis and Visualization* (MIPAV) toolset [33] (v.7.3) in conjunction with the *Java Image Science Toolkit (JIST)* [34] (v.2.0-2013) to automate the segmentation mask generation of the subsequently described procedures. The final surface meshing was realized in ParaView (v.5.0.1, Kitware Inc., New York, USA).

The AustinMan dataset was provided as a set of individual, segmented slices in the MATLAB MAT-file format. We did not select all available slices for further processing. Slices below the bottom of the lungs were discarded to generate a model of the head and upper torso. The in-slice resolution was three times higher than the resolution between slices, yielding a voxel size of $0.33 \times 0.33 \times 1.0$ mm^3. Using MATLAB, we converted the slices to a single volume image in the NIfTI file format, which can be imported into JIST. The structures as represented in the segmented image are unsuitable for surface meshing for several reasons. First, the anisotropic voxel size of the volume leads to an unbalanced level of detail in the three spatial directions. For this reason, the volume must be resampled to an isotropic voxel size. Second, due to their nested arrangement, most structures of the human body exhibit an outer and an inner boundary. However, the inner boundary may resemble the shape of the outer boundary of an adjacent internal structure. If triangulated, these adjacent boundaries are prone to mutual intersections and small gaps, which must be avoided. Third, the number and type of segmented structures exceed the typical level of detail required for EM simulations and can therefore be reduced. The segmented image is post-processed to segmentation masks to account for these requirements.

Using MIPAV, we resampled the image to an isotropic voxel size of 1 mm by reducing the in-slice resolution. We then split the image at a slice located at the chin of the enclosing exterior structure of the head to account for different requirements concerning the type of represented structures and different topological constraints of nested structures in the head and torso. The labels of these two images were integrated into a reduced set of labels comprising only the structures we aimed to consider for our EM simulations, namely enclosing exterior structure, bone, cerebrospinal fluid, the ventricles, cerebral GM and WM, the eyes, fat tissue only of the torso, muscle, and air. With the exception of the vascular system, all remaining structures, for example, the intestines, that could not be clearly assigned to one of these target structures were combined with the class of the muscle. The voxels representing the blood vessels had to be handled differently since the vascular system runs through many structures of the body. Assigning them to a class of musculature would introduce considerable and unreasonable segmentation errors, for example, muscle tissue inside the skull, the bones, or cerebrospinal fluid. Therefore, we cleared the labels of voxels representing the vascular system, that is, we assigned them the background value of 0. This procedure created holes in several structures at locations which were formerly attributed to the large draining veins of the brain. These holes were subsequently closed while the segmentation masks were being created. Finally, each structure was transferred to a separate image file and binarized.

Following these preparations, the segmentation masks for the head and body structures were generated separately. The images comprising the binarized, segmented structures were processed sequentially in a fixed order distinct for the body and the head. Separating the workflow for the head and torso was necessary since different topological constraints apply in each region. For example, air is entirely surrounded by bone in the sinuses of the skull, whereas the air in the lungs is outside any boney structure in the body.

The procedure started with the image representing the innermost structure, for example, the ventricles in the head, to the image representing the outermost structure, namely the enclosing exterior structure. Each image was processed identically: A morphological closing operation followed by a filling operation ensured a continuous outer boundary and eliminated the inner boundary as well as the holes created by the removal of the blood vessels. As a consequence, voxels that had originally been identified as blood vessels now represented the structure these vessels perfused. Small groups of detached voxels (namely less than 100 connected voxels) were identified as connected components and removed to obtain one large object per structure. A morphologically dilated version of the adjacent internal structure was added to the current structure. This way, we ensured a minimum thickness of two voxels for each structure surrounding another structure and avoided intersecting structures. An adapted approach was necessary in two cases: 1) Certain adjacent structures which were not nested still shared a common boundary (e.g., the thorax and the air in the lungs). In these cases the dilated mask image of one structure was subtracted from the mask image of the other, which created a spacing of at least two voxels between both structures. 2) The GM segmentation of AustinMan features very narrow sulci, down to the size of only one voxel, creating small detached islands of sulcal CSF in the GM. To avoid creating a discontinuous surface representation of the CSF structure, we applied a 2D filling operation to the GM mask for every slice independently, thereby eliminating such narrow sulci. The entire process resulted in individual segmentation masks for each tissue class of the head and torso separately, each with only a single external boundary.

The segmentation masks were then imported into ParaView. ParaView provides the so-called Contour Filter to compute a triangulated, polygonal representation of isosurfaces (namely surfaces of identical values in a 3D volume). The Contour Filter implements the "synchronized templates" algorithm [35], an improved version of the Marching Cubes algorithm [36]. As a result, we obtained high-resolution surfaces with a high number of triangles, which closely resembled the outer boundary of the segmentation masks, and the typical voxel grid-like structure of naïve surface reconstructions of structures from voxel-based 3D images was mitigated. These surfaces were then exported to individual files using the stereolithography (STL) format for subsequent processing.

The triangle size of the surface meshes had a side length of approximately 1 mm. This was defined by the resolution of the segmentation masks that ensured: *(i)* correct geometrical representation of inter-cranial tissues, and *(ii)* generated surface meshes to be geometrically error-free (no self-intersections, over-connection, etc.).

The total number of faces was approximately 10 million, thus the human model that was generated with ParaView was unsuitable for simulation with ANSYS HFSS.

Depending on the resolution and the number of structures that are to be included in the model, the automated processing of the workflow took about 45 minutes to 1 hour. An additional 30 minutes of manual preparation was required by a trained and experienced person, for example, to combine some classes of tissue and for splitting the segmentation at the head.

13.2.2 Head and Torso Model Mesh Optimization for 3D EM Simulation in ANSYS HFSS

A human model in ANSYS HFSS must be represented as a set of solid bodies. ANSYS SpaceClaim (ANSYS, Inc., Canonsburg, PA, USA) was used as the geometry preprocessor for: *(a)* importing STL files, *(b)* verifying (and, if necessary, correction) that all surface meshes were error free (watertight, no disconnected regions, no self-intersections, no over-connections, etc.), *(c)* combining the head and torso sections of the exterior structure, *(d)* optimizing the number of mesh elements for each individual object, *(e)* converting surface meshes into solid bodies, and *(f)* exporting each object of the model in ACIS binary format, which is the native file format of the ANSYS HFSS geometric modeling kernel.

Steps *"a"* to *"c"* were implemented using the built-in functionality of the ANSYS SpaceClaim Faceted Data Toolkit. The Faceted Data Toolkit's mesh repair functionality was sufficient for correcting a small number of mesh errors, but it was not suitable for handling the large number of mesh errors that appeared for most objects of our model if surface meshes contained triangles with a side length of more than 1 mm in ParaView. The main reasons for this were: *(i)* some compartments of the human model object were too thin (less than 3 mm) and it reduced the degree of freedom for correct mesh modification because of the low number of triangles and *(ii)* ParaView mesh errors were cascaded (disconnected regions, self-intersections, over-connections, etc.) in these areas.

Two approaches to reduce the number of mesh elements were applied in SpaceClaim: *(i)* All neighboring triangular faces located on the same geometrical plane were combined into a single face, and *(ii)* the number of faces were reduced by generating a triangular faceted wrapper around each model object.

An approach based on generating nonuniform rational B-spline (NURBS) surfaces was not implemented because the ANSYS HFSS geometry kernel operates internally with geometric primitives, for example, different types of facets and tetrahedra. Our tests based on ANSYS HFSS provided strong evidence that the meshing time for surface-based models based on NURBS is significantly longer than the time required for models generated using the previously mentioned approaches.

We also did not apply the ANSYS SpaceClaim Reduce tool to reduce the number of facets in a faceted body, because *(i)* for objects with a small number of facets, it provided a small reduction of the facet if spatial deviation was set to zero, *(ii)* mesh errors often resulted from complex geometric objects with a large number of facets.

Combining All Neighboring Triangular Faces

The "synchronized templates" algorithm generated high-resolution surface meshes that preserved all details of the underlying segmentation masks. Equally sized triangles were used for surface triangulations in the algorithm. This resulted in a redundant number of triangles, especially for large flat areas, because the 3D ACIS geometric modeling kernel (Dassault Systèmes, Vélizy-Villacoublay, France) does not require explicit triangulation to represent a flat area.

ANSYS SpaceClaim can combine all neighboring triangular faces located in the same geometrical plane into one face. The face reduction factor of this approach depends on the size of a given planar surface. For large planar surfaces, for example, those that are repeatedly represented on the enclosing-exterior-structure object, the reduction factor was very high (more than 100) (Fig. 13.1). For relatively small areas commonly observed in bent objects, for example, WM (Fig. 13.2), it was small (on the order of 10). This approach resulted in zero deviation of derived surface meshes from the original geometry.

Although the same 3D ACIS geometry modeling kernel is employed in both ANSYS SpaceClaim and ANSYS HFSS, different behavior was observed for the geometry validation check in ANSYS HFSS and ANSYS SpaceClaim if the face-combining approach was applied. The ANSYS HFSS geometry check reported errors for the objects that were error-free surface meshes in ANSYS SpaceClaim. One reason for this HFSS error is that the ANSYS SpaceClaim face combining procedure produces coincident edges that do not mark the boundaries of new faces. All ends that defined the combined faces are coincident edges. Therefore, the number of coincident edges is quite large, more than 10,000, for most model objects.

Our comprehensive ANSYS SpaceClaim tests of different human model geometries provided strong evidence that the ANSYS SpaceClaim Split Edges tool can detect and successfully merge coincident edges only if the number of coincident edges is relatively small, that is, less than approximately 1000.

Exporting human model objects prepared using ANSYS SpaceClaim to ACIS binary files was fast and problem free.

Faceted Wrapper

Using a second approach, generating a faceted wrapper around each model object decreased the number of faces. The reduction factors that specified the ratio of number of faces in the original object to the faceted wrapper object varied from

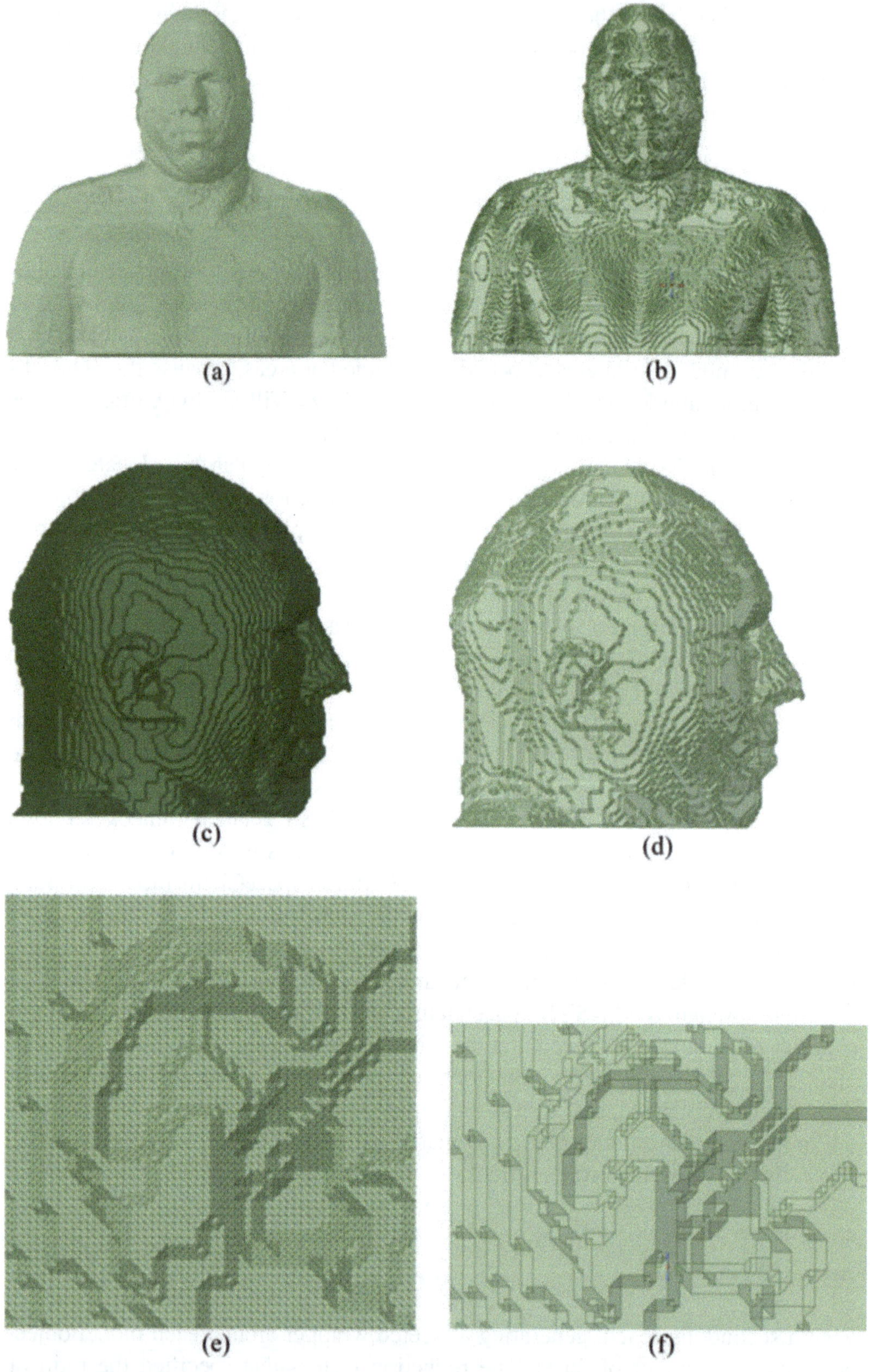

Fig. 13.1 The enclosing exterior structure object of AustinMan (**a**) after mesh generation with ParaView and (**b**) after combining faces. Head section of the skin object of AustinMan (**c**) after mesh generation and (**d**) after combining faces. Close-up view of the enclosing exterior structure object of AustinMan (**e**) after mesh generation and (**f**) after combining faces

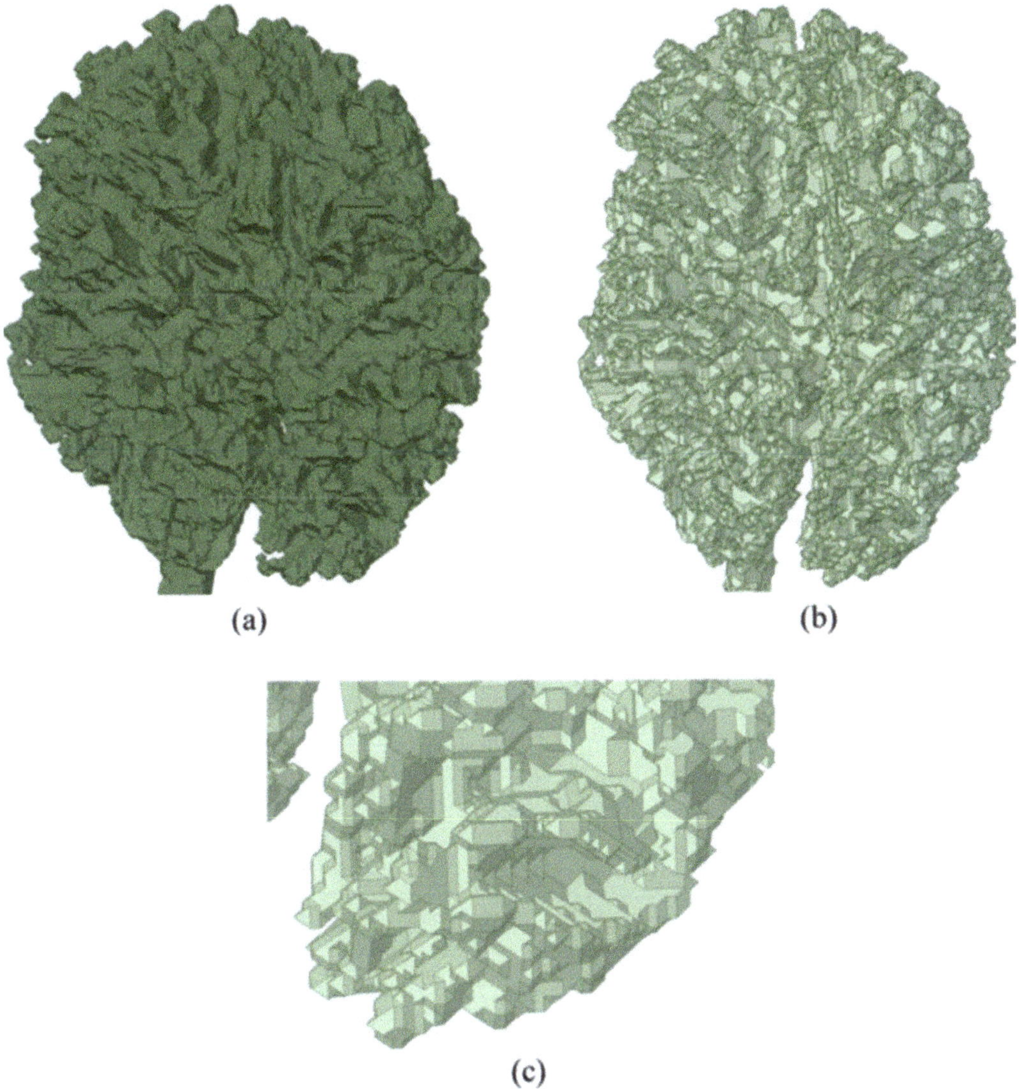

Fig. 13.2 WM object of AustinMan (**a**) after mesh generation, (**b**) after combining faces, and (**c**) shows a close-up view of WM object after combining faces

approximately 5 to 40, depending on tissue importance for 3D EM simulations and geometrical complexity (Table 13.1).

The largest ratio was applied for the enclosing exterior structure (Fig. 13.3). This resulted in a deviation of up to 3 mm in the ear area. The smallest ratio was applied for WM, which resulted in a deviation of less than 0.2 mm between the original mesh and the wrapper (Fig. 13.4).

Generating a faceted wrapper around each model object did not result in geometry errors when the model was imported into ANSYS HFSS. Thus, the ANSYS HFSS healing procedure was not required for these model objects.

Table 13.1 Summary of geometrical properties of AustinMan object

Model object	ParaView, faces	Combined faces approach, faces	Space-SpaceClaim wrap size, mm	Faceted wrapper approach, faces
Enclosing exterior structure	1,656,526	67,871	5	57,004
Head muscle	450,380	20,372	5	13,033
Skull	463,320	29,322	3	40,369
CSF	257,716	14,650	4	12,503
GM	236,868	15,202	4	11,527
WM	498,984	43,382	2	96,586
Air head top part	99,774	8128	2.5	10,728
Air head mouth part	17,744	1079	2.5	2198
Vent	17,384	1285	3	1772
Thorax fat	1,349,632	46,823	5	34,172
Thorax muscle	1,415,676	72,889	5	39,063
Lung	709,100	41,697	4	31,996
Thorax bone	1,039,264	77,366	3	90,764

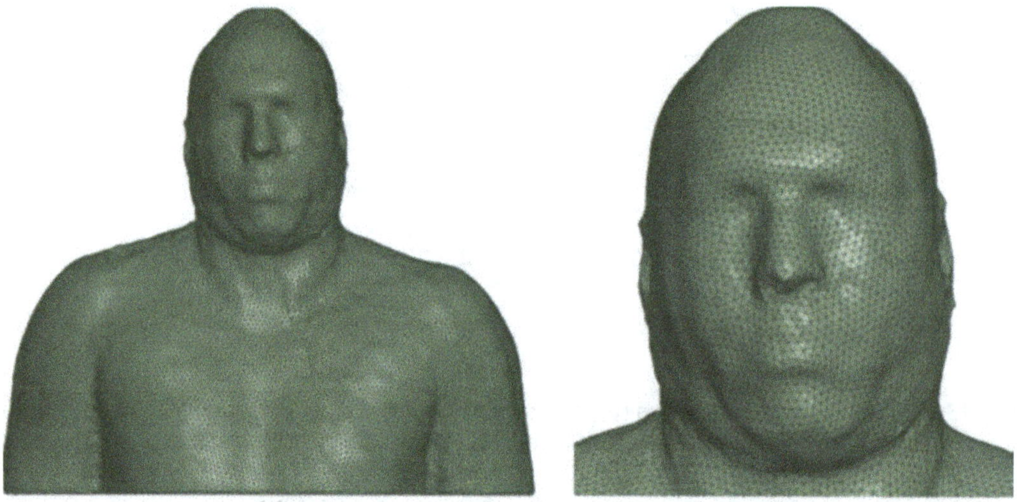

(a) (b)

Fig. 13.3 The enclosing exterior structure object of AustinMan after generating a faceted wrapper. (**a**) Entire object. (**b**) Head section

Comparison of Different Tissues

The shapes of the CSF, skull, and ribcage after geometric preprocessing for both approaches are shown in Figs. 13.5, 13.6, and 13.7.

Similar human model mesh optimization for 3D EM simulation in ANSYS HFSS was applied to prepare human Models 1 and 2 (Fig. 13.8) from surface meshes developed in our previous study [30]. One problem with most MRI scanners is that the maximum field of view is only 50 cm wide, and the patient on the patient table can only be moved in the axial direction, thus the subjects' arms and shoulders

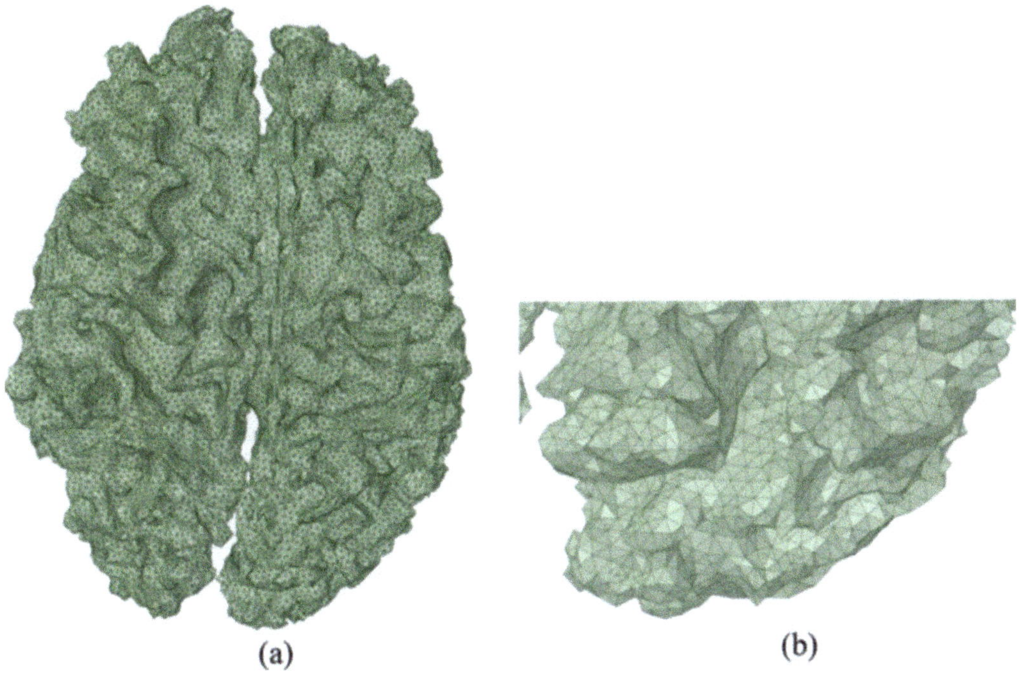

Fig. 13.4 WM object of AustinMan after generating a faceted wrapper. (**a**) Entire object. (**b**) Close-up view of WM object

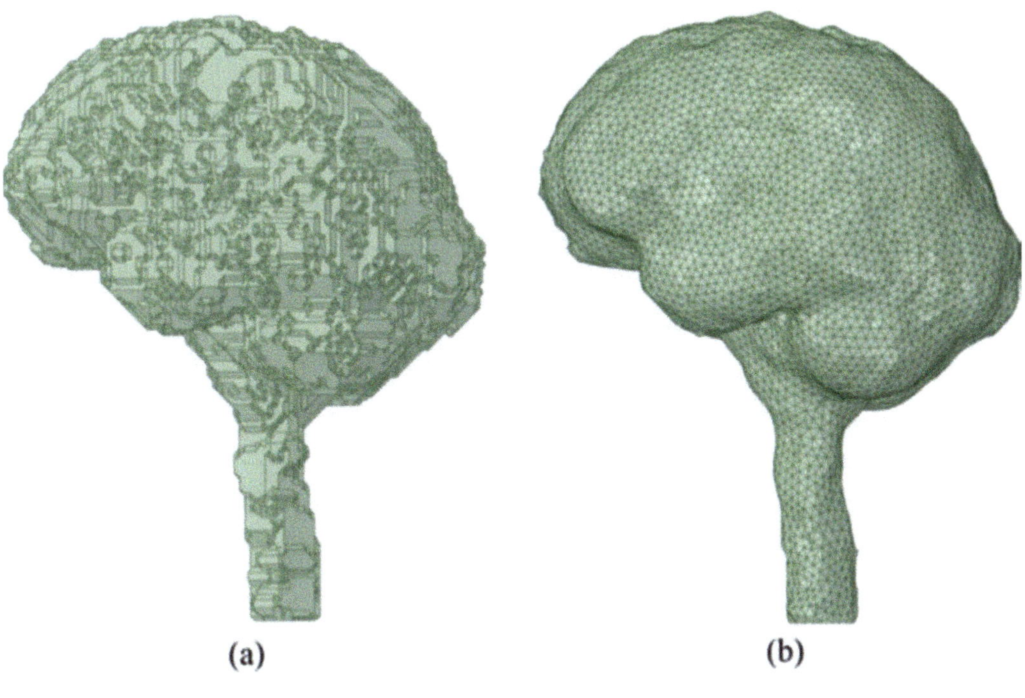

Fig. 13.5 CSF object of AustinMan (**a**) after combining faces and (**b**) after generating a faceted wrapper

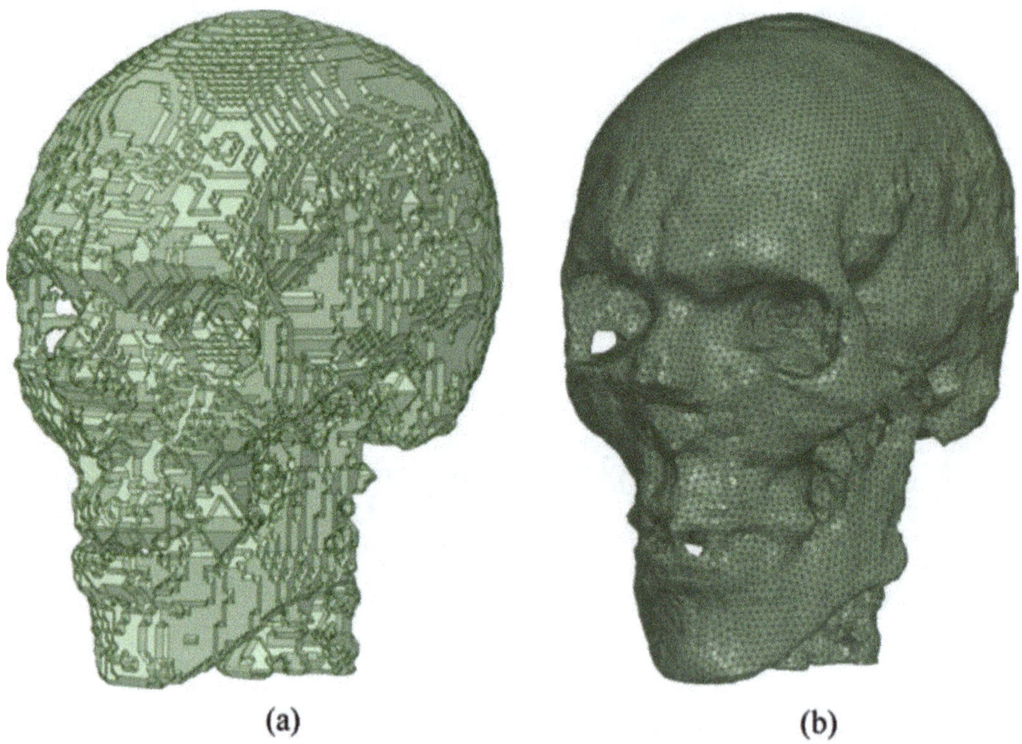

Fig. 13.6 Skull object of AustinMan (**a**) after combining faces and (**b**) after generating a faceted wrapper

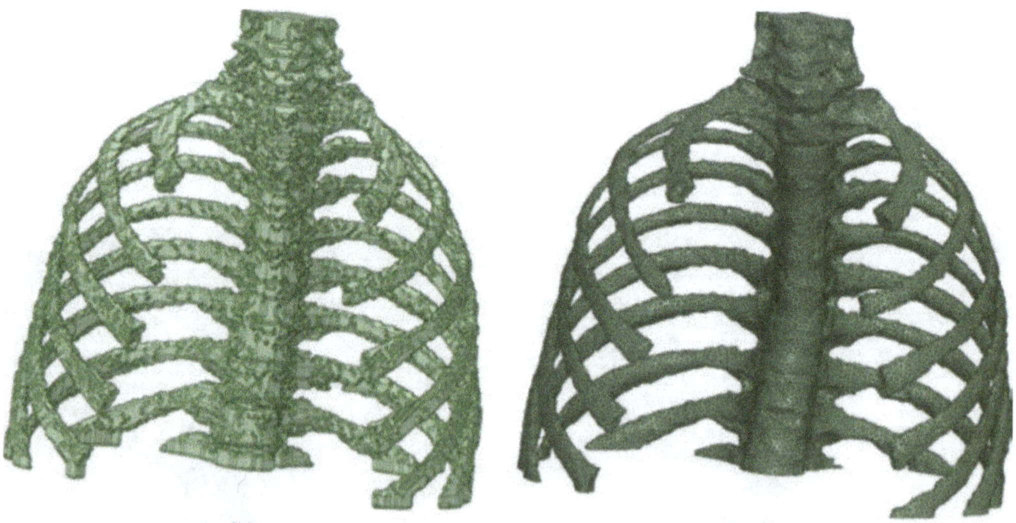

Fig. 13.7 Rib cage object of AustinMan (**a**) after combining faces and (**b**) after generating a faceted wrapper

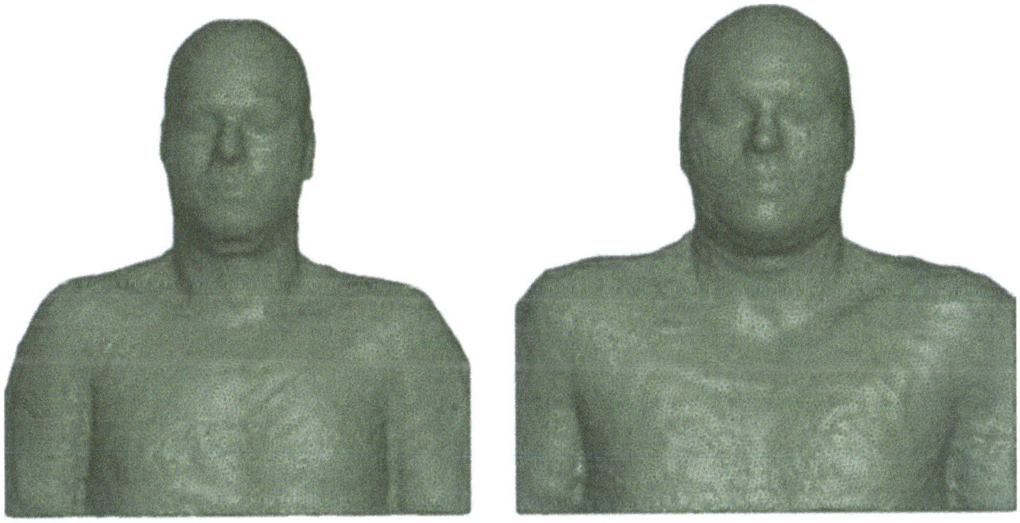

(a) (b)

Fig. 13.8 The enclosing exterior structure object of human models after generating a faceted wrapper. (**a**) Model 1. (**b**) Model 2

can be truncated in the image data if a default imaging protocol is applied. This problem is most noticeable in Model 2.

13.2.3 A Test of Entire Body Mesh Optimization for 3D EM Simulation in ANSYS HFSS

To investigate the performance and limitations of our workflow for entire body model generation, the enclosing exterior structure object of the AustinMan model was generated using a segmentation mask resolution of 2 mm (Fig. 13.9a), while the extents and resolution of all other structures remained the same as in the head and torso model. The enclosing exterior structure object of the entire body resulted in 1,530,456 facets. The number of faces was reduced to 618,737 facets using a face-combining operation (Fig. 13.9b). The reduction ratio for a surface mesh where the side of a triangle was 2 mm was substantially smaller than for a surface mesh in which the side of a triangle was 1 mm. Use of a faceted wrapper for the enclosing exterior structure object of the entire AustinMan model with the same settings as the faceted wrapper for the enclosing exterior structure object of the AustinMan model's head and torso resulted in significant spatial modification of areas between the model's body and arms (Fig. 13.9c).

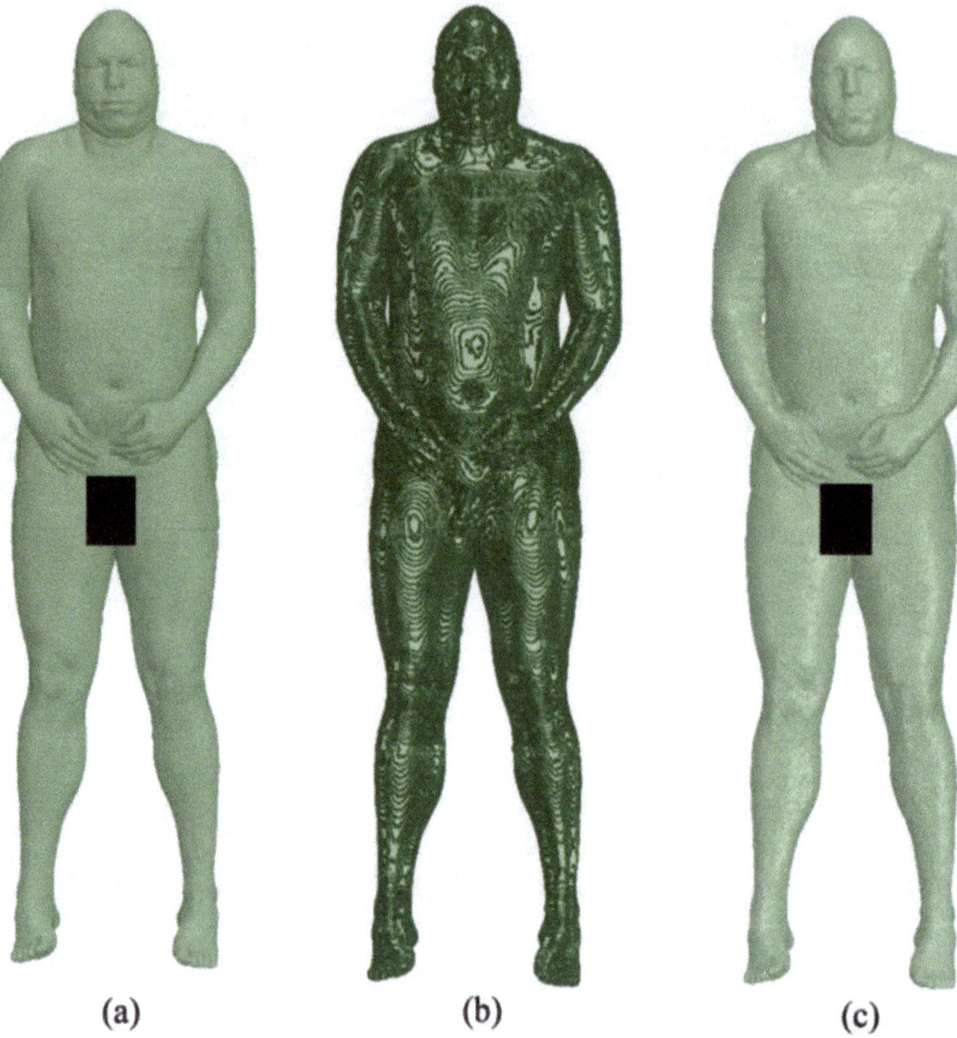

Fig. 13.9 (**a**) The enclosing exterior structure of the entire AustinMan model after mesh generation with ParaView. (**b**) The enclosing exterior structure of the entire AustinMan model after combining faces. (**c**) The enclosing exterior structure of entire AustinMan after generating a faceted wrapper

13.2.4 Finalizing the HFSS Model

Importing ACIS binary files exported from ANSYS SpaceClaim was fast and problem free in ANSYS HFSS. After import, each object underwent the ANSYS HFSS geometry validation check. A healing procedure was automatically applied if the ANSYS HFSS geometry check reported errors. This was a time-consuming process and took up to 10 hours to eliminate geometric errors per model object for complicated geometries if the face-combining approach was applied. ANSYS HFSS was not able to generate an error-free object after several days of healing the enclosing exterior structure of AustinMan.

An imported and healed (if necessary) model in ANSYS HFSS consisted of *(i)* the enclosing exterior structure, *(ii)* objects located above the chin slice created in the surface mesh generation step, primary head objects, and *(iii)* objects located below the chin slice, as well as primary torso objects.

The faceted wrapper slightly modified the geometries of model objects. However, an ANSYS HFSS simulation of a human model can consist of objects with both geometrical preprocessing approaches, because *(i)* deviation of any faceted wrapper from the original geometry is less than a quarter of the thickness of the given object, *(ii)* only one boundary between adjacent structures exists in areas outside the chin slice, and *(iii)* the intersection of adjacent structures in the chin slice area can be eliminated according to requirements.

To prevent intersections of objects in the area of the chin slice, all objects except the enclosing exterior structure underwent a boolean "split" operation in ANSYS HFSS. If the split operation resulted in two objects, only the primary object located above (for head objects) or below (for torso objects) the chin slice split plane was kept in the numerical domain.

13.2.5 Human Model Electrical Properties

Electrical properties of tissues were adopted from the IT'IS database [37]. Electrical property maps for electrical conductivity and relative electrical constant at 297.2 MHz provide a reasonable representation of human structures (Figs. 13.10, 13.11, and 13.12).

13.2.6 7T MRI Application-Specific Case Study

We performed 3D EM simulations of dual-row 7T head transmit array coil loaded with either the AustinMan model or Model 1 in ANSYS HFSS to evaluate the impact of human models on the spatially averaged 10-gram specific absorption rate (SAR_{10g}), which is used as the RF power deposition safety limit in 7T head MRI transmission and safety efficiencies. The coil consisted of 16 identical rectangular loops (100×102.25 mm^2) arranged in two rows of eight elements each (Fig. 13.13) [38]. A gap of 10 mm was applied between elements that were in the same row as well as between the two rows. The lower row elements were rotated by 22.5° with respect to the upper row. All adjacent elements were inductively decoupled.

The 3D EM model of the array included: *(a)* all array construction details for the resonant elements, *(b)* the load, namely, the surface-based human model, *(c)* the array environment, including the MRI scanner's gradient shield and magnet bore, all simulated with precise dimensions and material electrical properties, and *(d)* inductive decoupling of all adjacent elements. However, neither RF cable traps nor coax cable interconnection wiring were included in the model.

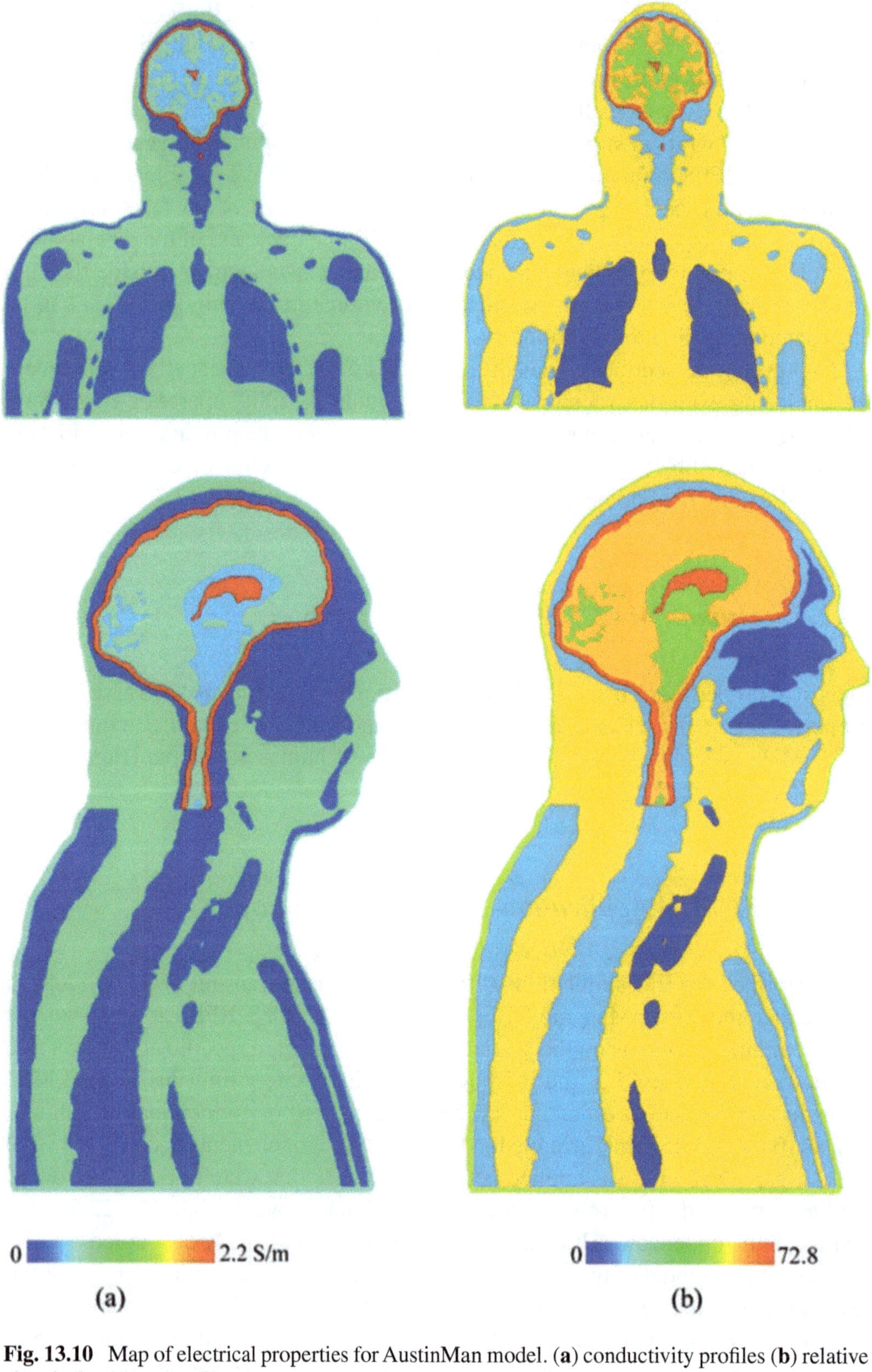

Fig. 13.10 Map of electrical properties for AustinMan model. (**a**) conductivity profiles (**b**) relative electrical constant profiles

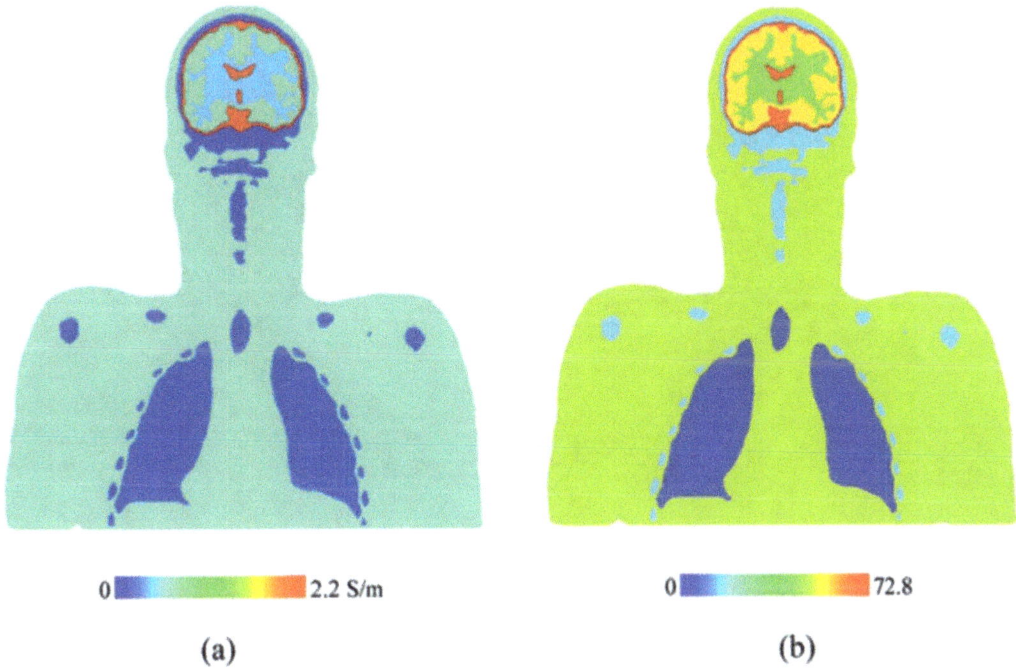

Fig. 13.11 Map of electrical properties for Model 1. (**a**) conductivity profiles (**b**) relative electrical constant profiles

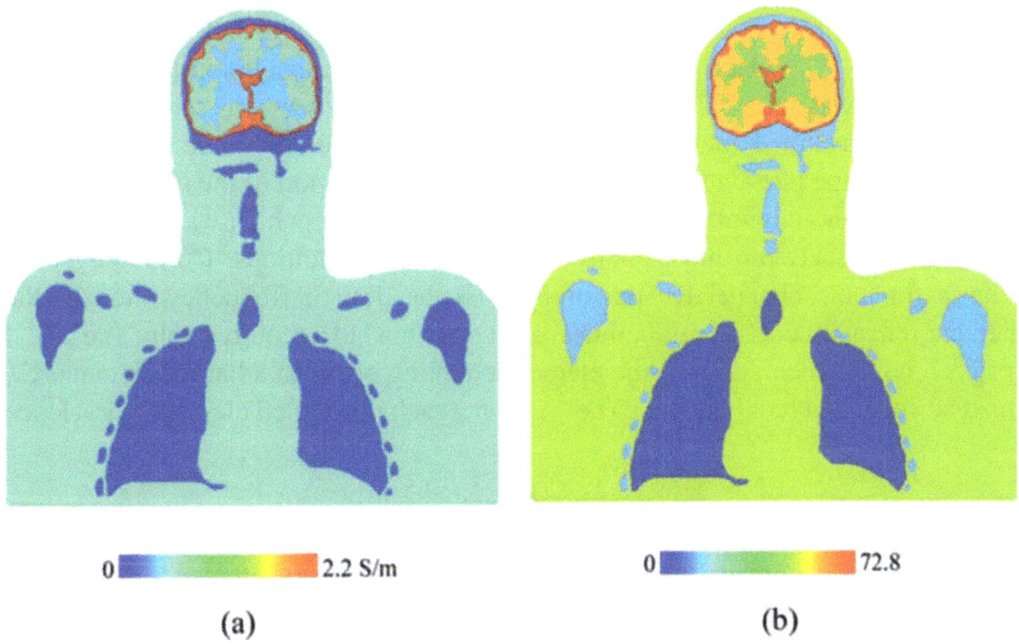

Fig. 13.12 Map of electrical properties for Model 2. (**a**) conductivity profiles (**b**) relative electrical constant profiles

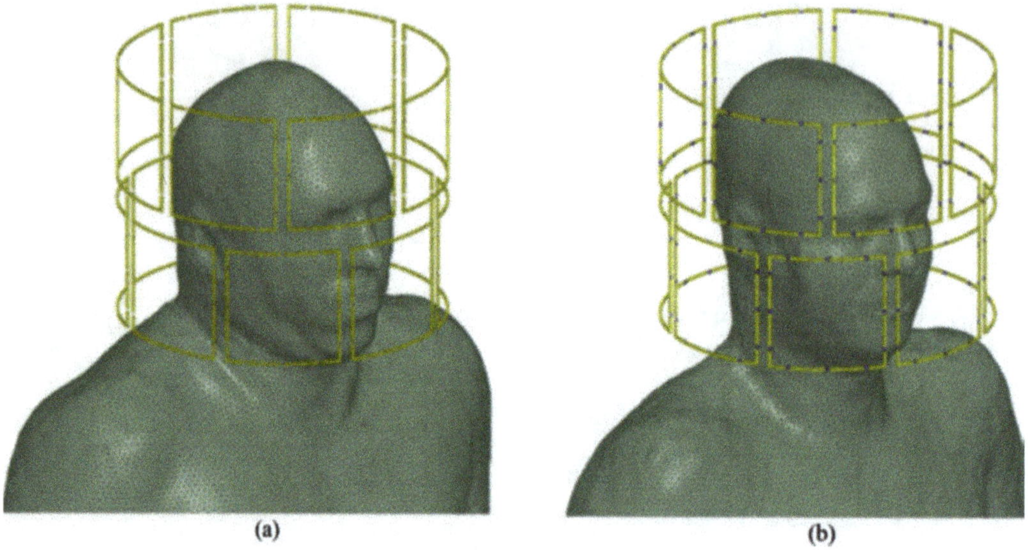

Fig. 13.13 7T dual row coil geometry and loads: (**a**) AustinMan model, (**b**) human Model 1

Twelve distributed capacitors were inserted in each radiative element to provide feed, tune, shunt, and distributed capacitor functionality. One PIN diode with a resistance of 0.18 Ω was placed in series with one of the distributed capacitors. This diode was used for decoupling transmit-only radiative elements during MRI signal reception.

The decoupling networks were defined by inductors, with inductance L_{ind} and coupling factor K_{ind}, placed in series with the distributed capacitors. The Q factor of all capacitors was set to 324, and the Q factor of all inductors was set equal to 400.

The coil was tuned, matched, and decoupled for the single tissue phantom with an external shape like a human model [38]. The optimization of the transmitter coil was based on the minimization of an error function (EF), which was a measure of the difference between the actual and desired coil conditions. Commonly used criteria for multi-channel RF transmitters, at the desired frequency, are: (a) the element reflection coefficient S_{xx} must be set and equal to a required value (i.e., S_{xx_t}) for each coil element, and (b) the element coupling between adjacent elements S_{xy} must be equal to a required value (i.e., S_{xy_t}) for each decoupled element pair. Hence

$$\mathrm{EF} = \sum_{\mathrm{Elem}} \left|S_{xx} - S_{xx\_t}\right|^2 + \sum_{\mathrm{all\_dec}} 0.5 \cdot \left|S_{xy} - S_{xy\_t}\right|^2$$

where Elem is the number of loops of the coil (namely 16) and all_dec is the number of decoupled element pairs (namely 32).

Both rows were excited in circular polarization (CP) mode with phase difference φ_{row} of 22.5° between rows. RF circuit and 3D EM co-simulation as detailed in [39] was used for calculations.

$SAR_{10\mathrm{g}}$ was calculated using an in-house procedure, which is consistent with the IEEE/IEC 62704-1 standard and validated by means of an IEEE TC 34 interlab comparison study [40].

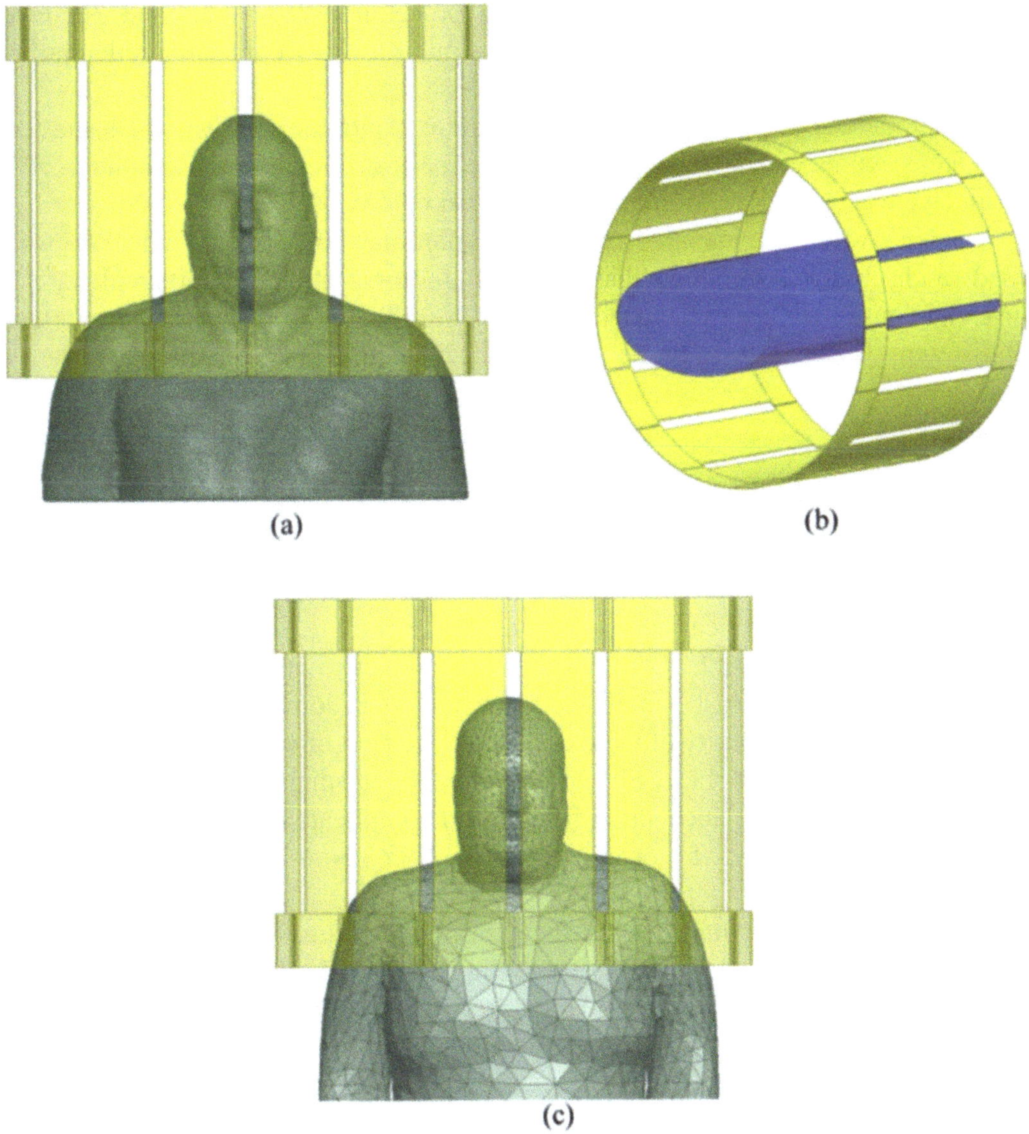

Fig. 13.14 3T birdcage coil geometry and loads: (**a**) AustinMan model, (**b**) an elliptical phantom, (**c**) NEVA Electromagnetics VHP high-resolution entire human model

13.2.7 3T MRI Application-Specific Case Study

The 3D EM model of the whole-body coil utilized a 123.2 MHz 16-rung high-pass birdcage of an equivalent design to those widely used in clinical 3T scanners (inner diameter 615 mm, total length 480 mm). The model head was positioned at the isocenter of the coil (Fig. 13.14a). The coil was shielded by a metal enclosure that mimicked a 1220 mm-long scanner bore. To mimic the scanner room, the coil was centered in an air box with the dimensions of $3 \times 2.25 \times 5$ m^3, surrounded by perfectly matched layer boundaries on all sides. The coil was tuned, matched, and decoupled using an elliptical phantom (length 700 mm, major radius 175 mm, minor radius 95 mm) positioned in the isocenter of the coil. The phantom material properties were: electrical conductivity σ = 0.52 S/m and relative permittivity

ε_r = 53.4 (Fig. 13.14b). The optimization procedure, RF circuit and 3D EM co-simulation for the 3T birdcage coil were similar to the 7T transmit coil simulations described in the previous section.

The amplitude of the two RF sources used to excite the coil was the same for both feeds, with a 90° phase shift between the feeds as in quadrature excitation. All results were calculated for a transmission power of 2 W.

The NEVA Electromagnetics VHP high-resolution whole human model [24] was used to check our assumption that the head and torso model is sufficient for 3T investigations of the head landmark position (Fig. 13.14c). The electrical properties of different types of human tissue were adopted from the IT'IS database [37].

13.2.8 RF Safety of Transcranial Direct Current Stimulation Equipment During MRI Case Studies

The impedance of electrical contact between an electrode and the skin should be low, e.g. during tDCS or EEG procedures. A conductive gel is used to minimize this impedance, which must be included in the numerical domain because it modifies the RF field in the proximity of the electrode. Placement of a gel patch between the electrode and the skin surface (Fig. 13.15a) resulted in small faces on the edges of the patch (Fig. 13.15b). Such small faces could complicate the generation of high quality numerical meshes, for example, in the ANSYS Non-Linear Thermal (NLT) platform.

Therefore, the triangular faces of the skin object in areas around the electrode were merged into a single NURBS face (Fig. 13.16a). Designed using the native geometrical capability of the ACIS kernel in ANSYS HFSS, the electrode and gel patch were located at the required positions in close proximity to the skin

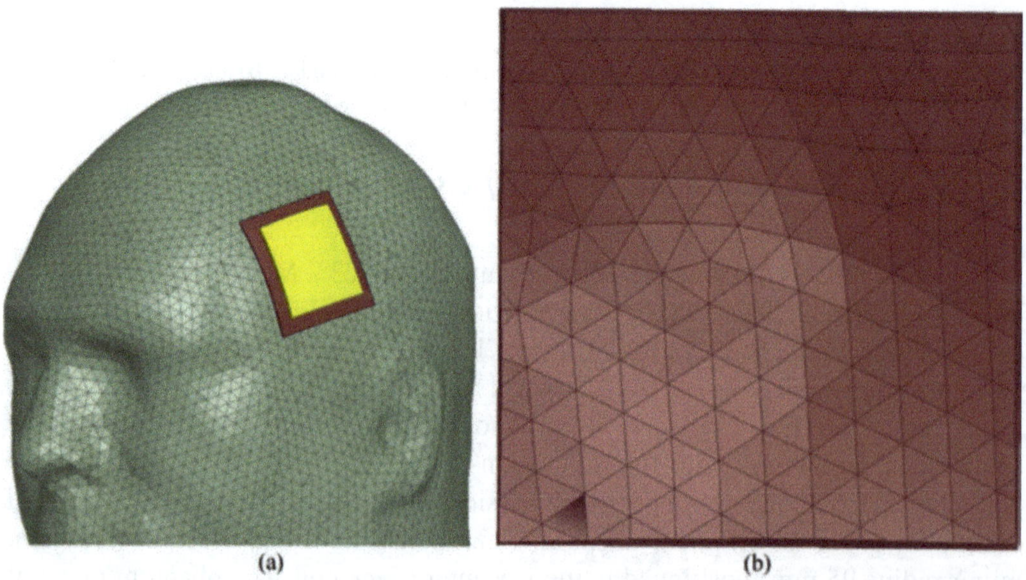

Fig. 13.15 (**a**) Skin object of AustinMan with an electrode and gel patch. (**b**) Surface of the gel patch

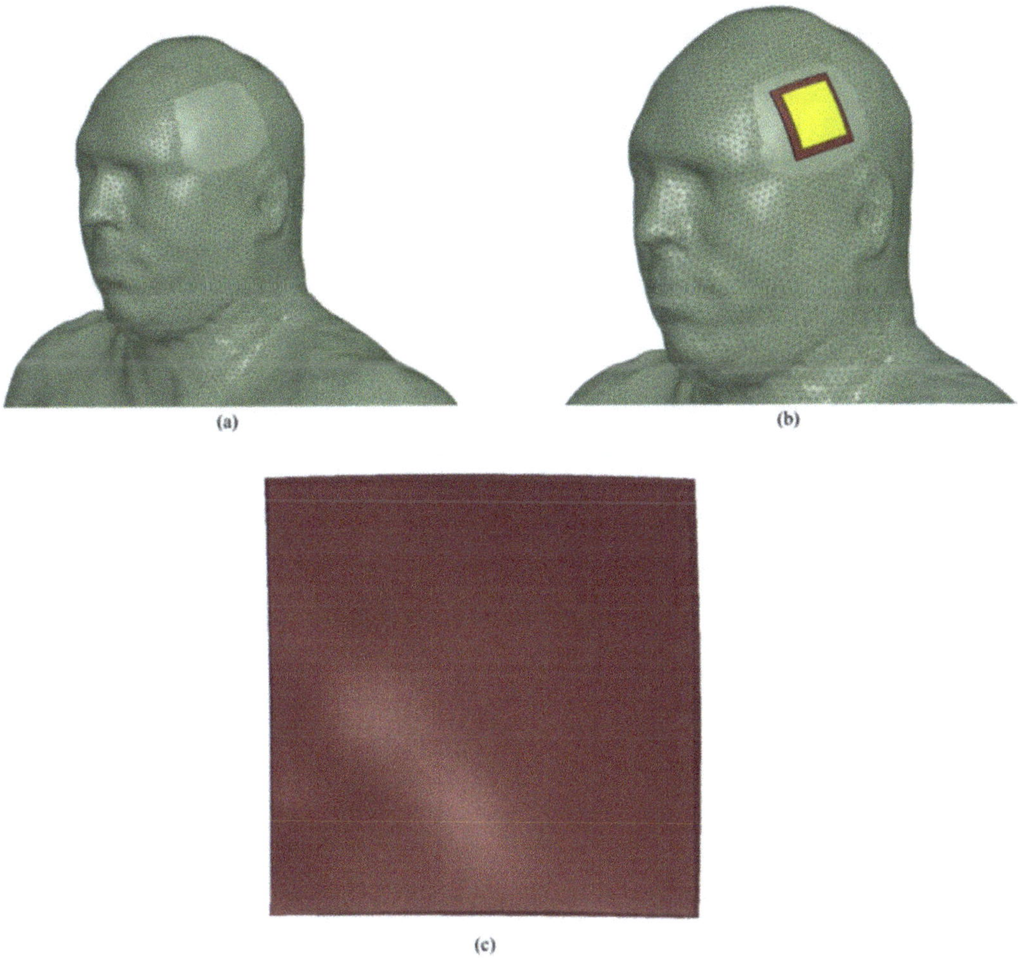

Fig. 13.16 (a) The enclosing exterior structure of AustinMan with an electrode and gel patch. (b) The enclosing exterior structure of AustinMan with an electrode and gel patch. (c) Surface of the gel patch

(Fig. 13.16b). After boolean subtraction of the gel object by the skin object, a correct single face contact between the gel object and skin object was obtained (Fig. 13.16c).

The tDCS setup consisted of two electrodes, two leads, and a metal connection box located 410 mm away from the coil enclosure. A composite-material quadratic tDCS electrode was simulated as a conductive medium with $\varepsilon_r = 3$ and $\sigma = 4$ S/m. The serial resistors integrated in the leads were located 100 mm away from the electrodes. Three resistor values were simulated: 1 mΩ, 5 kΩ, and 1 GΩ to simulate conditions of a short (potential manufacturing fault), normal operation, and an open connection (resistor failure after long-term operation), respectively. The tDCS lead included several straight segments and one helical segment. The lead copper wire was 1.2 mm in diameter with an insulation of 2.2 mm diameter. The helix pitch was 12.5 mm. Two tDCS lead trajectories were simulated: first on the axis of the scanner bore (recommended in the device manual) and then shifted towards the edge of the patient table.

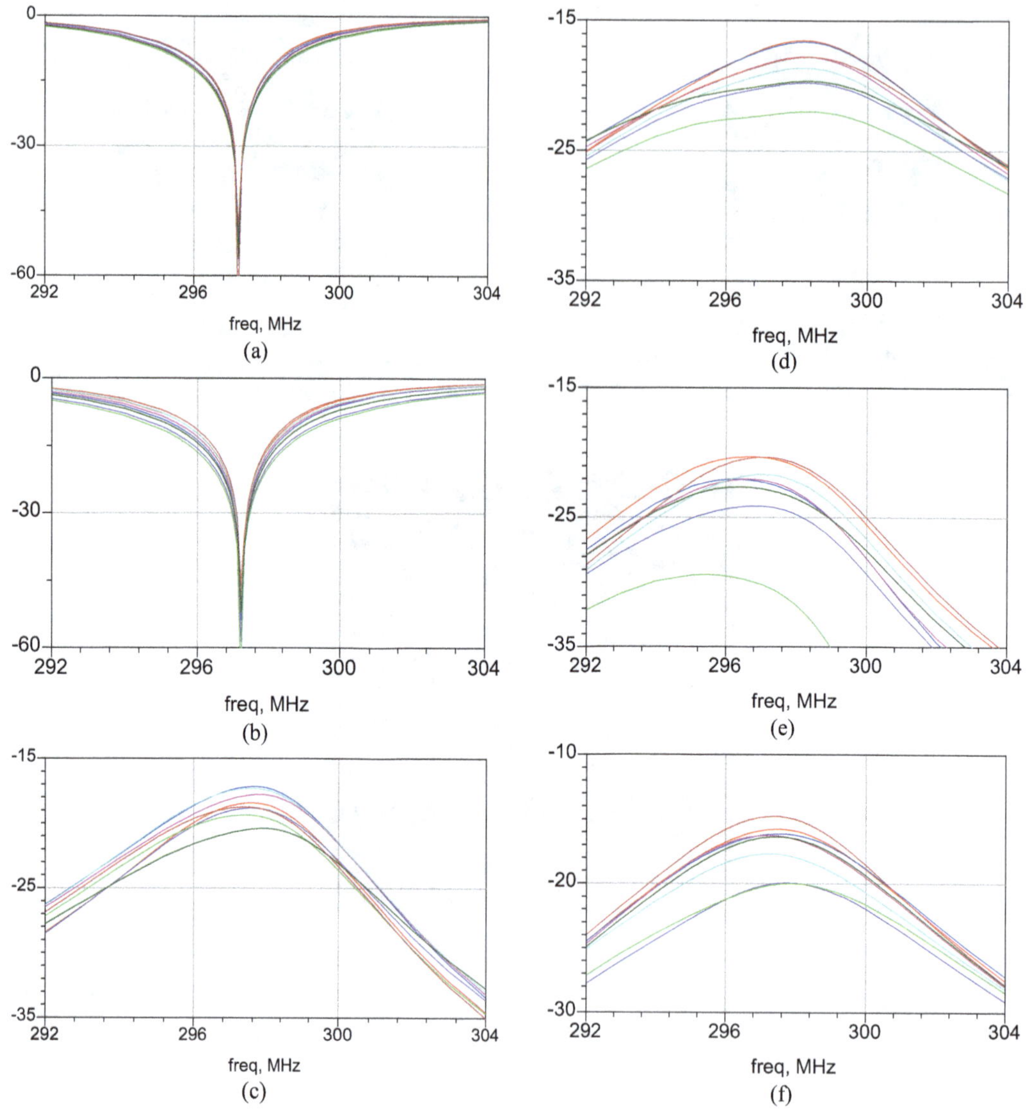

Fig. 13.17 Circuit level results: (**a**) S_{xx} for the top row, (**b**) S_{xx} for bottom row, (**c**) S_{xy} for the top row, (**d**) S_{xy} for the bottom row, (**e**) S_{xy} between the inductively decoupled adjacent elements between rows, (**f**) S_{xy} between the nearest non-adjacent elements between rows

13.3 Numerical Simulation Results

13.3.1 7T MRI Coil Simulation Results

The coil appeared to be correctly tuned for a given load with S_{xx} (the element reflection coefficient) values of less than −30 dB and S_{xy} (the element coupling between adjacent elements) values of less than −16 dB (Fig. 13.17).

CSF acted as a weak RF screen (Figs. 13.18 and 13.19), resulting in: *(i)* a decrease of B1+ at the skull/CSF boundary at the top part of the scalp, *(ii)* a substantial drop of the magnetic transmit field, B1+, in GM and WM, and *(iii)* a significant

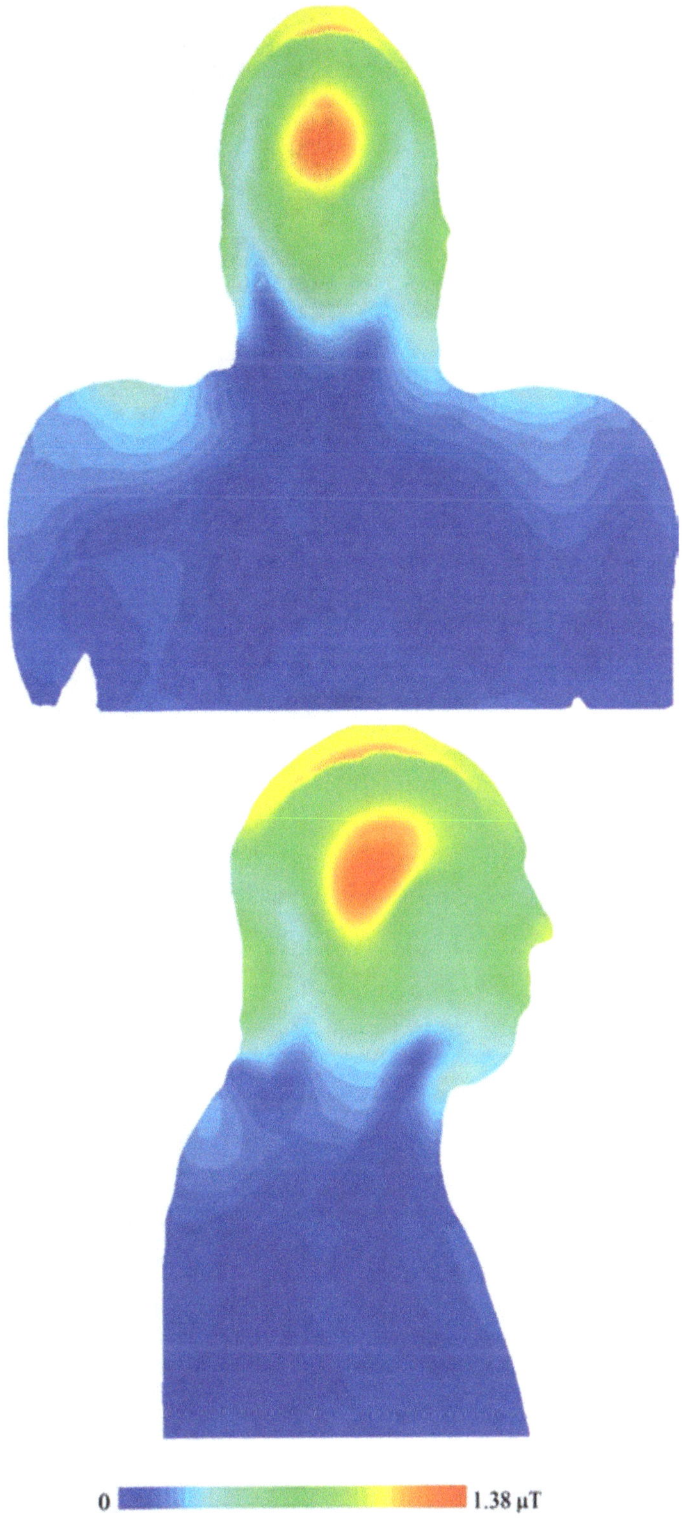

Fig. 13.18 B1+ maps for AustinMan model for the coil excited in CP mode

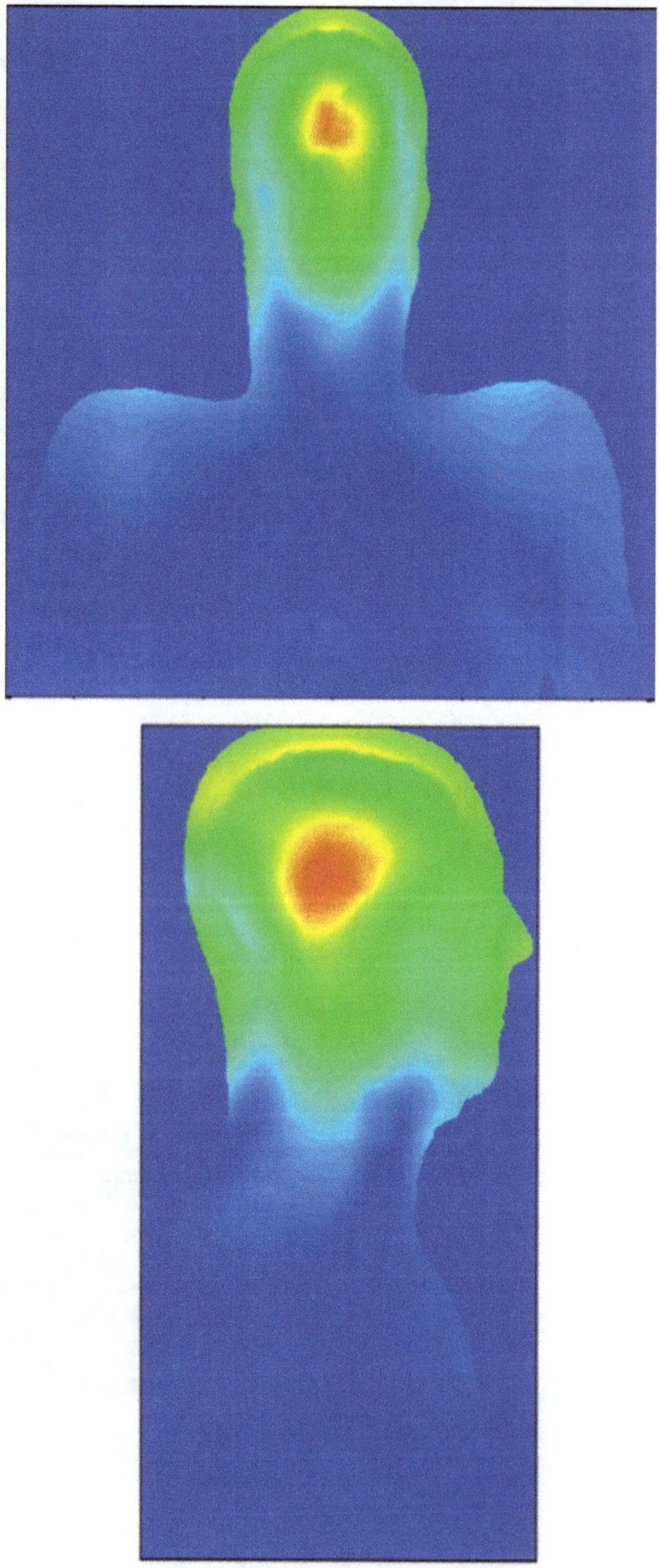

Fig. 13.19 B1+ maps for Model 1 for the coil excited in CP mode

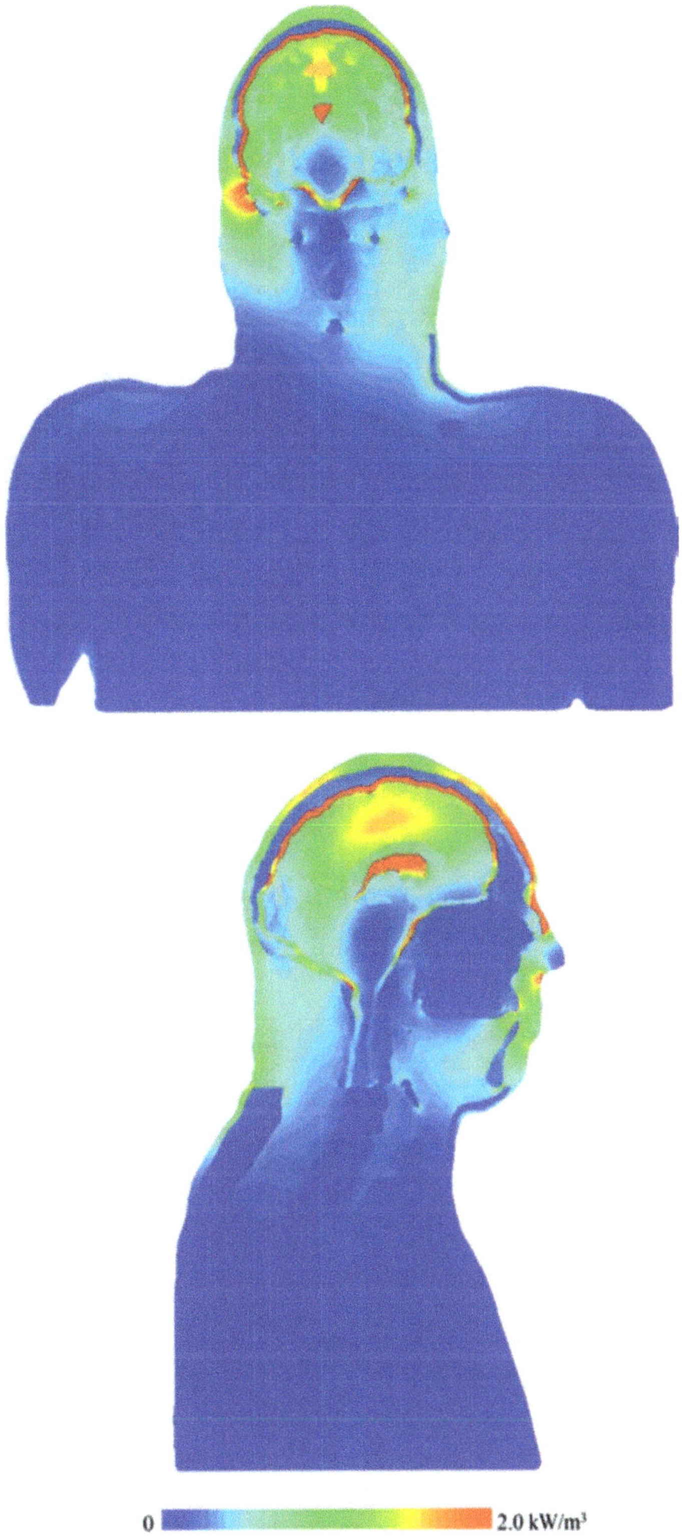

Fig. 13.20 Volume loss density maps for AustinMan for the coil excited in CP mode

Fig. 13.21 SAR_{10g} maps for Model 1 for the coil excited in CP mode

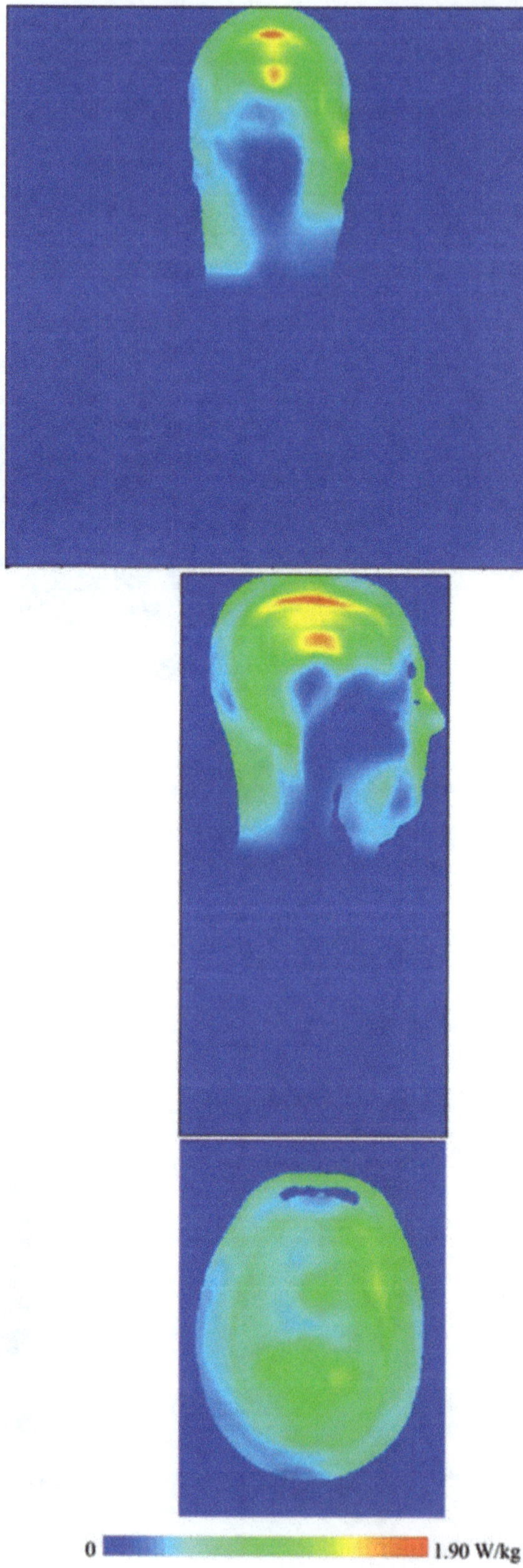

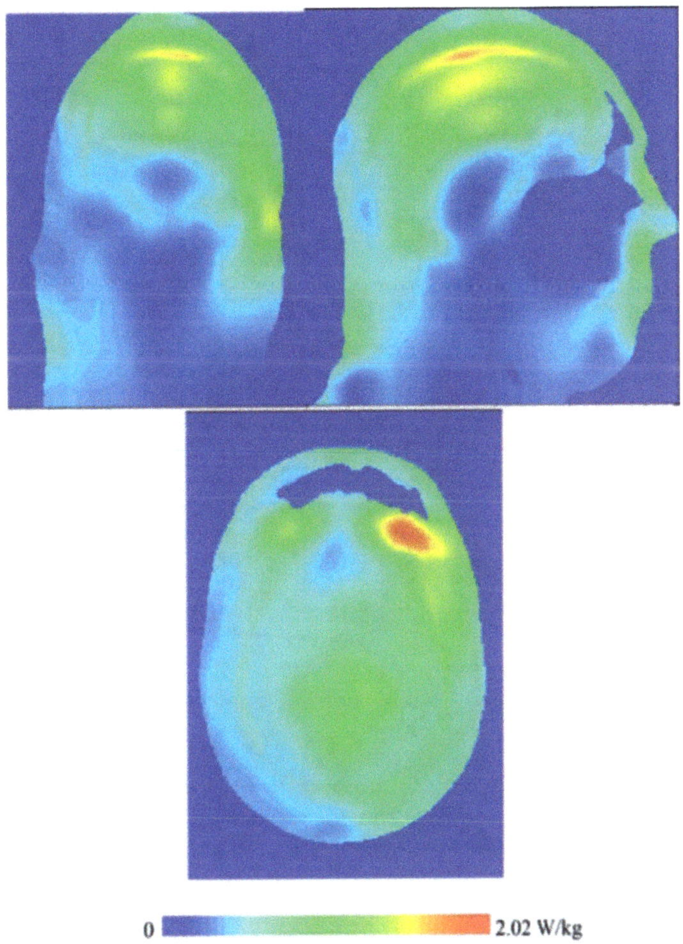

Fig. 13.22 SAR_{10g} maps for AustinMan model for the coil excited in CP mode

redistribution of volume loss density. Concomitantly, power deposition increased in the CSF space (Fig. 13.20). Visible variation of B1+ and volume loss density was observed for the investigated models.

Changing the human model resulted in some variation of B1+ and SAR_{10g} profiles (Figs. 13.21 and 13.22). Additionally, the transmission efficiency and the safety excitation efficiency were higher for Model 1.

13.3.2 3T MRI Coil Simulation Results

Circuit-level optimization resulted in an appropriately tuned birdcage coil with an elliptical phantom present in the bore (Fig. 13.23a). Unsurprisingly, the S parameters were visibly affected when the coil was loaded with human models at the head landmark position, which resulted in asymmetrical coil loading (Fig. 13.23b). No substantial difference in S parameters were observed for the head and torso of the AustinMan model or the NEVA Electromagnetics VHP entire human model.

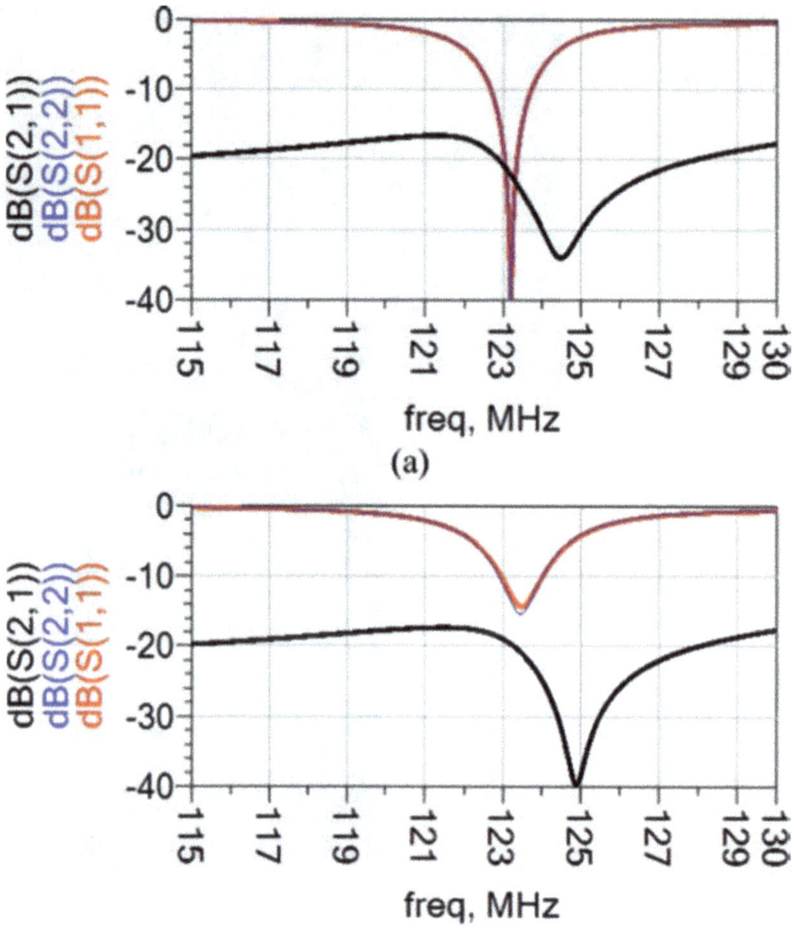

(b)

Fig. 13.23 S parameters of the birdcage coil loaded with (**a**) an oval phantom, and (**b**) a human model at the head landmark position

For both human models, 3D EM results were consistent with common observations in the literature, for example [41]: B1+ was rather homogeneously distributed across the head, and the maximum deposition of power occurred in the neck region (Figs. 13.24 and 13.25). As for the 7T coil simulation, if CSF was represented in the numerical domain as a non-separated, continuous segment within a space, it acted as a weak RF screen resulting in: *(i)* a decrease of B1+ at the skull/CSF boundary at the top part of the scalp and *(ii)* a significant volume loss density in CSF.

Truncation of the AustinMan model at the torso did not significantly affect the birdcage coil circuit level results or field distributions. Only a very weak scattered field was observed in the area located in close proximity to the torso cut plane.

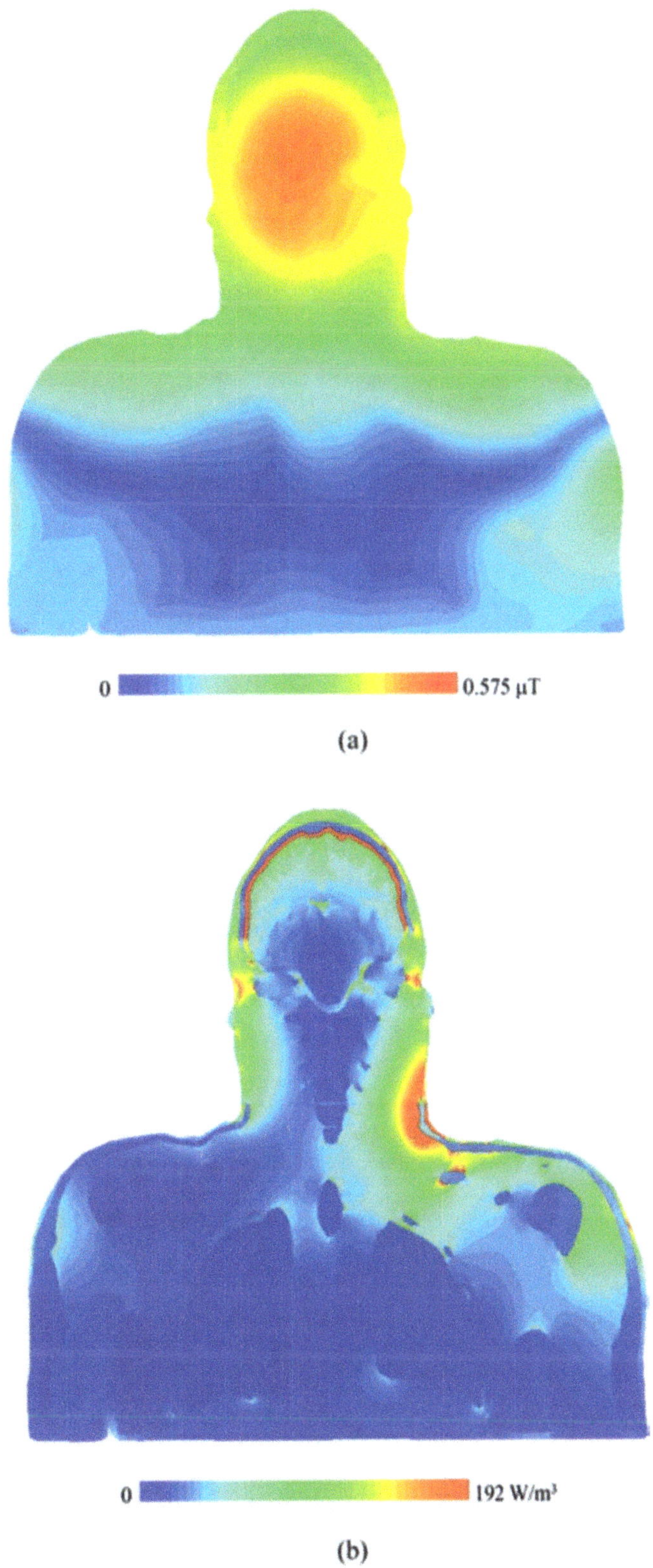

Fig. 13.24 AustinMan model in 3T birdcage coil. (**a**) B1+ map and (**b**) volume loss density map

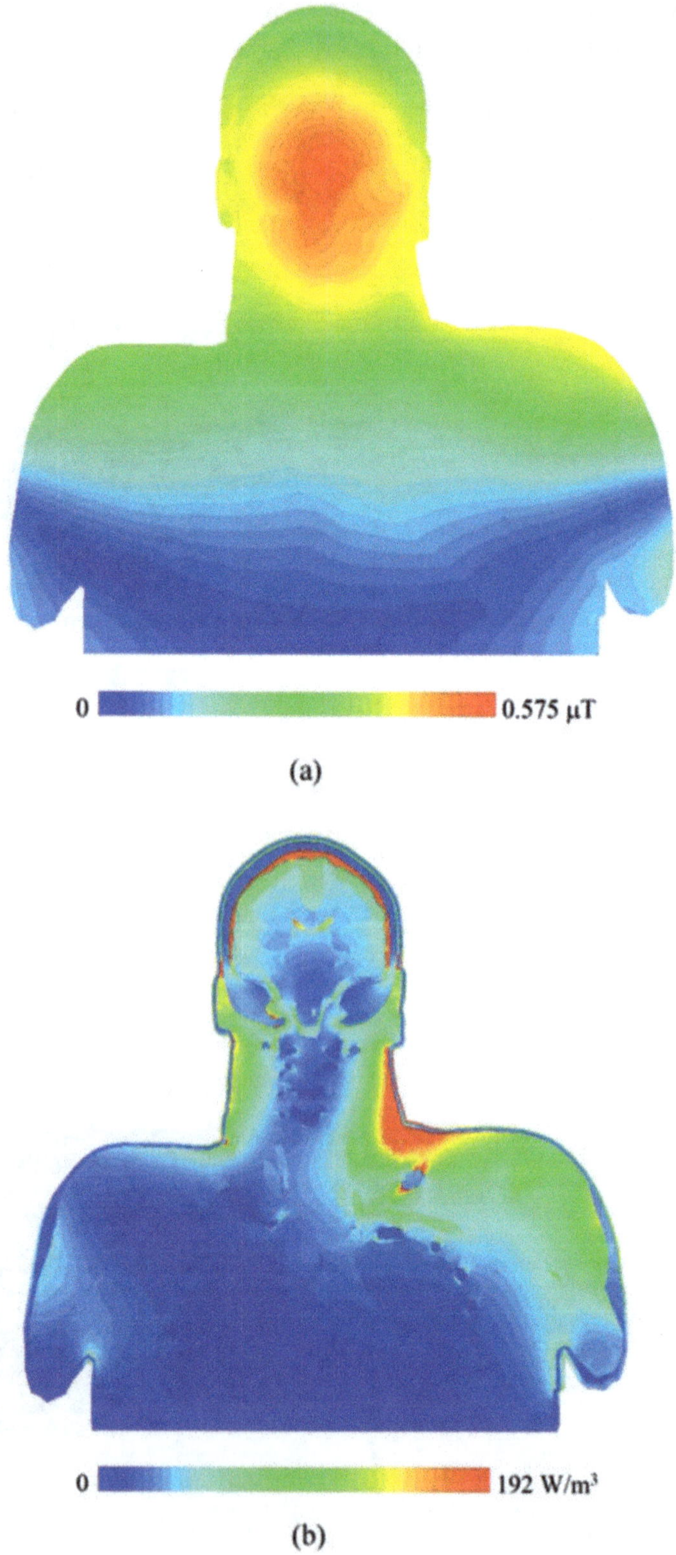

Fig. 13.25 The NEVA Electromagnetics VHP entire human model in 3T birdcage coil. (**a**) B1+ map and (**b**) volume loss density map

13.3.3 Transcranial Direct-Current Stimulation Results

After adding the tDCS setup with the lead directed along the magnet axis, substantial power deposition was observed in close proximity to the tDCS electrode edges for all values of the serial resistor. Unsurprisingly, the B1+ disturbance in close proximity to the electrode location was highest for $R = 1$ mΩ. For a normal tDCS setup operation with $R = 5$ kΩ, shifting the tDCS lead from the scanner axis toward the edge of the patient table resulted in a small variation of power deposition in the proximity of the electrode edges. Assuming a (pulsed) peak value of 30 kW of the total transmission power (which can be generated by the scanner's power amplifier) yielded voltages across the serial resistor up to 850 V for normal tDCS operation ($R = 5$kΩ) and up to 1.4 kV for an open connection ($R = 1$ GΩ). For a whole-body SAR level of 4 W/kg, average voltages across the serial resistor were 130 V and 225 V for $R = 5$ kΩ and $R = 1$ GΩ, respectively. The obtained range of voltages underscores how sufficient electrical strength (e.g., order of 1 kV) is required for the tDCS serial resistor. Due to the similarity of the power deposition in the proximity of the electrode edges for all investigated conditions, we conclude that the tDCS electrodes and the straight segments of the leads between them and the serial resistor predominantly determine the power deposition in human subjects.

13.4 Discussion

Our investigation explored the impact of patient-specific human models on MRI safety assessment from different perspectives. Future work should address how many different human models, head positions, and non-ideal tuning conditions need to be investigated and how many different excitation conditions need to be validated in order to demonstrate MRI RF transmit coil robustness, as well as MRI multimodal setup and implant RF safety.

Our mesh optimization procedure for the 3D EM simulation workflow is specifically tailored toward performing simulations with ANSYS HFSS and ANSYS NLT. Use of other simulation tools could require some modification of geometry preparation steps, for example, the generation of NURBS surfaces instead of faceted objects.

In our previous work, we have introduced a semi-automated processing pipeline to generate individualized surface-based models from MRI data of individual subjects. While this pipeline offers a high level of automation, especially concerning the segmentation of the MRI data and segmentation mask generation, so far it is limited to model a few relevant structures (i.e., the enclosing exterior structure, bone, air, GM, WM, and CSF). Limitations mainly arise from difficulties in segmenting certain inter-subject variable tissue types in MR images.

MRI data provides good contrast for different types of soft tissue, but additional effort is required to segment skin and bone, especially when this should be achieved in an algorithm-driven manner, without supervision of an expert. Fully unsupervised segmentation of highly variable structures, for example, muscle and fat tissue, from MRI data across subjects is challenging to achieve using our atlas-based approaches and is therefore still subject to further research. However, if corresponding segmentation images were available, our segmentation mask generation workflow could be extended to include these additional structures, as detailed in this work.

To prevent geometrical model errors in most simulation tools and to accelerate geometrical export and preprocessing, our segmentation mask generation process enforces the topological constraint that adjacent structures should not share a common boundary. The segmented structures were modified according to topological constraints for the human anatomy: *(i)* to being either strictly nested or *(ii)* not in contact with boundaries of neighboring structures. As a result, for example, the ventricles of the brain are entirely surrounded by WM, which again is fully surrounded by GM even at the brain stem, and there is a space between the rib cage and the lung object.

The more structures that are represented in the model, the more difficult it becomes to maintain this topological constraint. For example, the vascular system runs through a major subset of all the other structures, which made it impossible to fully nest it inside another single structure. Furthermore, introducing a space between its boundary and the boundaries of all the other structures would create holes in those structures.

Additionally, some tissue segments were too small to be represented, which, for example, was the case for the pieces of CSF in some narrow sulci in the brain. We therefore opted for an approach that eliminates the CSF in these sulci to ensure a continuous boundary for the subarachnoid CSF, resulting in trade-off of a less accurate representation of sulcal CSF.

As a consequence of both aforementioned problems, we did not include fat tissue in the head region. More specifically, fat tissue in the head is present in several types of tissue, for example, skin and muscle tissue. As a result, the fat exhibits common boundaries with several other structures, such as the skull, cartilage tissue, tendon tissue, and the eyes, which made it impossible to entirely nest it inside one structure. Additionally, the fat tissue was not segmented in a continuous way, larger gaps existed that could not be closed with morphological closing operations and some segments of fat tissue were as thin as only one voxel. A possible solution to address these obstacles might be to divide the class of fat tissue into subclasses for which compliance to the topological constraints can be achieved more easily. We are working on defining a set of rules on how to reasonably combine the mentioned classes of tissue in an informed anatomical way, and how to handle the discontinuous fat tissue and thereby ensure compliance with the necessary topological constraints.

In this work, we have elaborated on the necessary workflow using the AustinMan model. However, we expect our workflow to also work for other segmented data sets, such as NAOMI [42] and NORMAN [43] and voxel model databases of the average Japanese male and female [44]. If segmentation images are already available for a person from a previous investigation, our segmentation mask generation workflow can be applied to generate a surface-based head and torso model for this individual. Depending on the quality and continuity of the segmented structures, adaptations will only be necessary with regard to the integration of the available tissue classes into the desired set of structures, the order of structures for which the segmentation mask generation will be executed, from the innermost to the outermost structure, and the position of where to split the head section of the segmentation image from the torso section. These adjustments can be achieved in a time frame of approximately 1 day.

In addition, for these new models it is important to investigate whether certain structures need dedicated treatment, for example, as was observed in the narrow sulci of the GM, the vascular system, or the fat tissue in the head of the AustinMan model. Resolving these special cases may require adaptations as simple as adjusting the parameters for morphological operations (i.e., closing or filling), which was the case for the narrow sulci in the GM. Alternatively, they may require a dedicated sub-workflow to be developed, which was the case for the vascular system, and which would be the case for handling fat tissue in the head of the AustinMan. In the latter case, the necessary time frame of adapting the proposed workflow may easily increase to several days.

An extension of the presented workflow to create whole-body models will be the next step. We expect similar difficulties with body fat, as we discovered for the head and limbs, especially in the abdominal region where the intestines are located.

The time required for geometry modification, import, preprocessing, and mesh generation was ten times longer than the solver time of approximately 2 hours on an up-to-date Dell workstation. This is not compatible with real-time patient-specific safety assessment. However, it is reasonable for investigating more realistic distributions of human body shapes and sizes to explore the variation of SAR values between subjects, as well as SAR dependences on intracranial geometric variation (e.g., variation of CSF spaces with age).

Further development of ANSYS SpaceClaim and ANSYS HFSS capabilities: *(i)* to reduce the amount of facets in surface meshes without creating geometrical problems in ANSYS HFSS, and *(ii)* fast geometry import, preprocessing, and mesh generation for geometries with a large number of facets in ANSYS HFSS, could substantially decrease the time needed for 3D EM simulation of high-resolution human models.

Acknowledgments The authors acknowledge the Computational Electromagnetics Group at the University of Texas at Austin for developing and making the AustinMan human body model available at http://sites.utexas.edu/austinmanaustinwomanmodels/. The authors also acknowledge the ANSYS Partnership Program for providing the numerical simulation software for these investigations.

References

1. International Electrotechnical Commission (IEC). (2010). *Medical electrical equipment-part 2–33: Particular requirements for the basic safety and essential performance of magnetic resonance equipment for medical diagnosis*. Geneva, Switzerland: International Electrotechnical Commission, 60601-2-33 Ed. 3.
2. Shajan, G., Kozlov, M., Hoffmann, J., Turner, R., Scheffler, K., & Pohmann, R. (2014). A 16-channel dual-row transmit array in combination with a 31-element receive array for human brain imaging at 9.4 T. *Magnetic Resonance in Medicine, 71*(2), 870–879.
3. Oh, S., Webb, A. G., Neuberger, T., Park, B., & Collins, C. M. (2010). Experimental and numerical assessment of MRI-induced temperature change and SAR distributions in phantoms and in vivo. *Magnetic Resonance in Medicine, 63*, 218–223.
4. Murbach, M., Neufeld, E., Cabot, E., Zastrow, E., Córcoles, J., Kainz, W., & Kuster, N. (2016). Virtual population-based assessment of the impact of 3 Tesla radiofrequency shimming and thermoregulation on safety and B1+ uniformity. *Magnetic Resonance in Medicine, 76*(3), 986–997.
5. Murbach, M., Neufeld, E., Kainz, W., Pruessmann, K. P., & Kuster, N. (2014). Wholebody and local RF absorption in human models as a function of anatomy and position within 1.5T MR body coil. *Magnetic Resonance in Medicine, 71*, 839–845.
6. Voigt, T., Homann, H., Katscher, U., & Doessel, O. (2012). Patient-individual local SAR determination: In vivo measurements and numerical validation. *Magnetic Resonance in Medicine, 68*, 1117–1126.
7. Wu, X., Tian, J., Schmitter, S., Vaughan, J. T., Uğurbil, K., & Van De Moortele, P.-F. (2016). Distributing coil elements in three dimensions enhances parallel transmission multiband RF performance: A simulation study in the human brain at 7 Tesla. *Magnetic Resonance in Medicine, 75*(6), 2464–2472.
8. Ryan, K., Wawrzyn, K., Gati, J. S., Chronik, B. A., Wong, D., Duggal, N., & Bartha, R. (2018). 1H MR spectroscopy of the motor cortex immediately following transcranial direct current stimulation at 7 Tesla. *PLoS One, 13*(8). Article number e0198053.
9. Lee, M. B., Kim, H. J., Woo, E. J., & Kwon, O. I. (2018). Anisotropic conductivity tensor imaging for transcranial direct current stimulation (tDCS) using magnetic resonance diffusion tensor imaging (MR-DTI). *PLoS One, 13*(5). Article number e0197063.
10. Keinänen, T., Rytky, S., Korhonen, V., Huotari, N., Nikkinen, J., Tervonen, O., Palva, J. M., & Kiviniemi, V. (2018). Fluctuations of the EEG-fMRI correlation reflect intrinsic strength of functional connectivity in default mode network. *Journal of Neuroscience Research, 96*(10), 1689–1698.
11. Bailey, W., Mazur, A., McCotter, C., Woodard, P.K., Rosenthal, L., Johnson, W., & Mela, T. (2016). Clinical safety of the ProMRI pacemaker system in patients subjected to thoracic spine and cardiac 1.5-T magnetic resonance imaging scanning conditions. *Heart Rhythm, 13*(2), 464–471.
12. Bhusal, B., Bhattacharyya, P., Baig, T., Jones, S., & Martens, M. (2018). Measurements and simulation of RF heating of implanted stereo-electroencephalography electrodes during MR scans. *Magnetic Resonance in Medicine, 80*(4), 1676–1685.
13. Guerin, B., Serano, P., Iacono, M.I., Herrington, T.M., Widge, A.S., Dougherty, D.D., Bonmassar, G., Angelone, L.M., & Wald, L.L. (2018). Realistic modeling of deep brain stimulation implants for electromagnetic MRI safety studies", *Physics in Medicine and Biology, 63*(9), Article number 095015.
14. Atefi, S. R., Serano, P., Poulsen, C., Angelone, L. M., & Bonmassar, G. (2018). Numerical and experimental analysis of radiofrequency-induced heating versus lead conductivity during EEG-MRI at 3 T. *IEEE Transactions on Electromagnetic Compatibility*, (99). https://doi.org/10.1109/TEMC.2018.2840050.
15. Kozlov, M., & Kainz, W. (2018). Lead electromagnetic model to evaluate RF-induced heating of a coax lead: A numerical case study at 128 MHz. *IEEE Journal of Electromagnetics, RF and Microwaves in Medicine and Biology*. https://doi.org/10.1109/JERM.2018.2865459.

16. ASTM F2182-11a. (2011). *Standard test method for measurement of radio frequency induced heating on or near passive implants during magnetic resonance imaging*. West Conshohocken, PA: ASTM International, www.astm.org.
17. Xu, X. G. (2014). An exponential growth of computational phantom research in radiation protection, imaging, and radiotherapy: A review of the fifty-year history. *Physics in Medicine and Biology, 59*, R233–R302.
18. Gosselin, M.-C., Neufeld, E., Moser, H., Huber, E., Farcito, S., Gerber, L., Jedensjo, M., Hilber, I., Gennaro, F.D., Lloyd, B., Cherubini, E., Szczerba, D., Kainz, W., & Kuster, N. (2014). Development of a new generation of high-resolution anatomical models for medical device evaluation: The virtual population 3.0. *Phys. Med. Biol., 59*(18), 5287–5303.
19. Christ, A., Kainz, W., Hahn, E.G., Honegger, K., Zefferer, M., Neufeld, E., Rascher, W., Janka, R., Bautz, W., Chen, J., Kiefer, B., Schmitt, P., Hollenbach, H.-P., Shen, J., Oberle, M., Szczerba, D., Kam, A., Guag, J.W., & Kuster, N. (2010). The virtual family—Development of surface-based anatomical models of two adults and two children for dosimetric simulations. *Physics in Medicine and Biology, 55*(2), 23–38.
20. Yu, D., Wang, M., & Liu, Q. (2015). Development of Chinese reference man deformable surface phantom and its application to the influence of physique on electromagnetic dosimetry. *Physics in Medicine and Biology, 60*(17), 6833–6846.
21. Segars, W.P., Tsui, B.M.W., Cai, J., Yin, F.-F., Fung, G.S.K., & Samei, E. (2018). Application of the 4-D XCAT phantoms in biomedical imaging and beyond. *IEEE Transactions on Medical Imaging, 37*(3), 680–692.
22. Elemance: The sole distributor of the global human body models consortium family of virtual models of the human body, [online] Available: http://www.elemance.com/
23. Genc, K.O., Segars, P., Cockram, S., Thompson, D., Horner, M., Cotton, R., & Young, P. (2013). Workflow for creating a simulation ready virtual population for finite element modeling. *Journal of Medical Devices, 7*(4), 1–2. https://doi.org/10.1115/1.4025847.
24. Makarov, S. N., Noetscher, G. M., Yanamadala, J., Piazza, M. W., Louie, S., Prokop, A., Nazarian, A., & Nummenmaa, A. (2017). Virtual human models for electromagnetic studies and their applications. *IEEE Reviews in Biomedical Engineering, 10*, 95–121. http://ieeexplore.ieee.org/document/7964701/.
25. Spitzer, V., Ackerman, M. J., Scherzinger, A. L., & Whitlock, D. (1996). The visible human male: A technical report. *Journal of the American Medical Informatics Association: JAMIA, 3*, 118–130.
26. Homann, H., Börnert, P., Eggers, H., Nehrke, K., Dössel, O., & Graesslin, I. (2011). Toward individualized SAR models and in vivo validation. *Magnetic Resonance in Medicine, 66*, 1767–1776.
27. Kozlov, M., Bazin, P.-L., Möller, H. E., & Weiskopf, N. (2016). Influence of cerebrospinal fluid on specific absorption rate generated by 300 MHz MRI transmit array. *Proceedings of 10th European Conference on Antennas and Propagation (EuCAP)*. https://doi.org/10.1109/EuCAP.2016.7481666.
28. Massire, A., Cloos, M.A., Luong, M., Amadon, A., Vignaud, A., Wiggins, C.J., & Boulant, N. (2012). Thermal simulations in the human head for high field MRI using parallel transmission. *J. Magn. Reson. Imag., 35*(6), 1312–1321.
29. Kalloch, B., Bode, J., Kozlov, M., Pampel, A., Hlawitschka, M., Sehm, B., Villringer, A., Möller, H. E., & Bazin, P.-L. (2019). Semi-automated generation of individual computational models of the human head and torso from MR images. *Magnetic Resonance in Medicine, 81*(3), 2090–2105.
30. Kozlov M., Bode J., Bazin P.-L., Kalloch B., Weiskopf N., Moeller H.E. (2017). Building a high resolution surface-based human head and torso model for evaluation of specific absorption rates in MRI (pp. 1–6), *Proceedings of COMCAS 2017*. Tel-Aviv, Isreal.
31. Gjonaj, E., Bartsch, M., Clemens, M., Schupp, S., & Weiland, T. (2002). High-resolution human anatomy models for advanced electromagnetic field computations. *IEEE Transactions on Magnetics, 38*(2), 357–360.
32. Massey, J. W., & Yilmaz A. E. (2016). AustinMan and AustinWoman: high-fidelity, anatomical voxel models developed from the VHP color images. In *Proc. 38th Annual International*

Conference of the IEEE Engineering in Medicine and Biology Society (IEEE EMBC). Orlando, FL.

33. Mcauliffe, M., Lalonde, F., McGarry, D. P., Gandler, W., Csaky, K., Trus, B. (2001). Medical image processing, analysis & visualization in clinical research. In *Proceedings of the 14th IEEE Symposium on Computer-Based Medical Systems* (pp. 381–386).
34. Lucas, B. C., Bogovic, J. A., Carass, A., Bazin, P. L., Prince, J. L., Pham, D. L., & Landman, B. A. (2010). The Java Image Science Toolkit (JIST) for rapid prototyping and publishing of neuroimaging software. *Neuroinformatics, 8*, 5–17.
35. Moreland, K., & Geveci, B. (2014). *A pervasive parallel processing framework for data visualization and analysis at extreme scale*. Clifton Park, NY/Albuquerque, NM: Tech. rep. Kitware, Inc./Sandia National Lab.(SNL-NM).
36. Lorensen, W. E., Cline, H. E. (1987). Marching cubes: a high resolution 3D surface construction algorithm. In *Proceedings of the 14th annual conference on computer graphics and interactive techniques* (pp. 163–169).
37. Hasgall, P. A., Di Gennaro, F., Baumgartner, C., Neufeld, E., Gosselin, M. C., Payne, D., Klingenböck, A., Kuster, N. (2015). IT'IS database for thermal and electromagnetic parameters of biological tissues, Version 3.0, September 1. www.itis.ethz.ch/database
38. Kozlov, M., Turner, R., Weiskopf, N., Möller, H. E., Shajan, G. (2017). Investigation of 7T 16-channel dual-row transmit array coils: a case study of static RF shimming. In *Proceedings of COMCAS 2017* (pp. 1–5). Tel Aviv.
39. Kozlov, M., & Turner, R. (2009). Fast MRI coil analysis based on 3-D electromagnetic and RF circuit co-simulation. *Journal of Magnetic Resonance, 200*, 147–152.
40. Monebhurrun, V., Braux, Y., Devos, H., Kozlov, M., Simon, W., & Wittig, T. (2014). A benchmark CAD mobile phone model for specific absorption rate calculations. *IEEE Transactions on Magnetics, 50*(2). https://doi.org/10.1109/TMAG.2013.2282470.
41. Yeo, D., Wang, Z., Loew, W., Vogel, M., & Hancu, I. (2011). Local SAR in high pass birdcage and TEM body coils for multiple human body models in clinical landmark positions at 3T. *Journal of Magnetic Resonance Imaging, 33*(5), 1209–1217. https://doi.org/10.1002/jmri.22544.
42. Dimbylow, P. (2005). Development of the female voxel phantom, NAOMI, and its application to calculations of induced current densities and electric fields from applied low frequency magnetic and electric fields. *Physics in Medicine and Biology, 50*(6), 1047–1070. https://doi.org/10.1088/0031-9155/50/6/002.
43. Findlay, R. P., & Dimbylow, P. J. (2009). Spatial averaging of fields from half-wave dipole antennas and corresponding SAR calculations in the NORMAN human voxel model between 65 MHz and 2 GHz. *Physics in Medicine and Biology, 54*(8), 2437–2447. https://doi.org/10.1088/0031-9155/54/8/012.
44. Nagaoka, T., Watanabe, S., Sakurai, K., Kunieda, E., Watanabe, S., Taki, M., & Yamanaka, Y. (2004). Development of realistic high-resolution whole-body voxel models of Japanese adult male and female of average height and weight, and application of models to radio-frequency electromagnetic-field dosimetry. *Physics in Medicine and Biology, 49*, 1–15.

Chapter 14
Calculation of MRI RF-Induced Voltages for Implanted Medical Devices Using Computational Human Models

James E. Brown, Rui Qiang, Paul J. Stadnik, Larry J. Stotts, and Jeffrey A. Von Arx

14.1 Introduction

Magnetic resonance imaging (MRI), notwithstanding its status as the preferred imaging modality for soft tissue imaging and non-ionizing radiation, is generally contraindicated for patients with active implantable medical devices (AIMDs). However, it has been estimated that within 12 months of device implantation, 17% of pacemaker patients will need an MRI [1]. In order to assess the safety of devices, AIMD manufacturers work together with MR manufacturers, regulatory scientists, and academia to develop an international technical specification [2] which identifies the potential hazards for these patients in an MR environment. This test specification covers AIMDs such as deep brain stimulators, pacemakers, cochlear implants, etc.

These hazards are separated by the specific electromagnetic field component which causes that hazard. Then, the MR conditional safety of a device can be assessed in a laboratory by generating the worst case of each field component, each of which is unlikely to be observed in a clinical setting. Computational human models (CHMs) are used for three main hazards: radiofrequency (RF)-induced heating, RF-induced malfunction, and RF-induced rectification. RF-induced energy incident on the device can be rectified by non-linear electronics, which then can represent a safety issue if unintended tissue stimulation were to occur. RF-induced malfunction is a related hazard but represents the ability of the device to operate within its acceptable tolerances and free from damage during the MRI. RF-induced heating usually refers to heating of the tissue surrounding the lead tip, a phenomenon which has been shown to be linked with resonant behavior.

There are numerous advantages of using CHMs in concert with in vitro testing for this process, rather than more extensive clinical trials [4]. As the evaluation of

J. E. Brown (✉) · R. Qiang · P. J. Stadnik · L. J. Stotts · J. A. Von Arx
Micro Systems Engineering, Inc., Lake Oswego, OR, USA
e-mail: james.brown@biotronik.com

S. Makarov et al. (eds.), *Brain and Human Body Modeling*,
https://doi.org/10.1007/978-3-030-21293-3_14

MR conditional safety of AIMDs contains the interaction of three subsystems, the benefits of modeling techniques are here described in relation to these three subsystems: the patient, the MR system, and the AIMD.

Patient variability can be assessed using the Virtual Family [5], which includes 15 high-resolution CHPs. These models consist of males and females of all ages in various body weights and heights and are optimized for electromagnetic simulations. Numerical modeling provides a rapid and low cost way to extract quantities such as deposited power or temperature rise in vivo.

When a human body is present in the MRI bore, the uniform EM fields, especially the electric field, will be greatly disturbed. These accumulated field distortions produce multiple local specific absorption rate (SAR)/thermal hotspots inside the human bodies, while these local hotspots further contribute to the RF-induced voltages for AIMDs. However, since each human body is unique, the body/coil interaction of each CHM will be different. The simulated electric field map in Fig. 14.1 from a CHM inside the MRI bore illustrates this effect. As seen in the figure, the hazard area positions and strengths in each human body could be significantly different. To account for this effect, a library of multiple CHMs, which spans the population in terms of BMI, shall be selected in a safety assessment. Moreover, any tissue parameter variation of the human models can be easily addressed by the computational model.

In a clinical study, only a limited number of measurements can be performed and the size of any collected statistical extreme cases would be very limited. However, in numerical modeling, millions of simulations of any combination of patients and leads, device orientations, etc. can be easily achieved. A device-based risk analysis can then be completed utilizing such a large database.

There are various MRI coil manufacturers in the market and each manufacturer builds its own MR system, i.e., each of their coils has different geometries. To calculate the statistical extreme cases, multiple body positions inside the bore have to be investigated. This variability can be assessed through modeling in a manner beyond what is practical in a clinical setting.

In a 3T system, the RF field homogeneity is usually much worse than a 1.5T system. To improve the field homogeneity and picture signal-to-noise ratio (SNR), a technique called shimming has drawn extensive attention in the industry and has been implemented in commercial coils. Since the induced field distribution will be altered significantly after shimming is used, this technique has to be incorporated into the safety assessment. However, each manufacturer has its own shimming technique which is invisible to the public. Computational modeling has been shown as a very convenient way to rapidly evaluate any kind of shimming technique [7].

Through the use of CHMs, the impact of variability in device orientation, including lead pathway, can be thoroughly assessed. The use of CHMs in this process enables many more data points to be evaluated than would be practical in a clinical setting. For instance, it would be impossible (and likely unethical) to vary the lead pathway and study the resulting variability in heating near the lead tip in a single patient. Expanding a clinical study to account for all potential device variabilities would be undesirable.

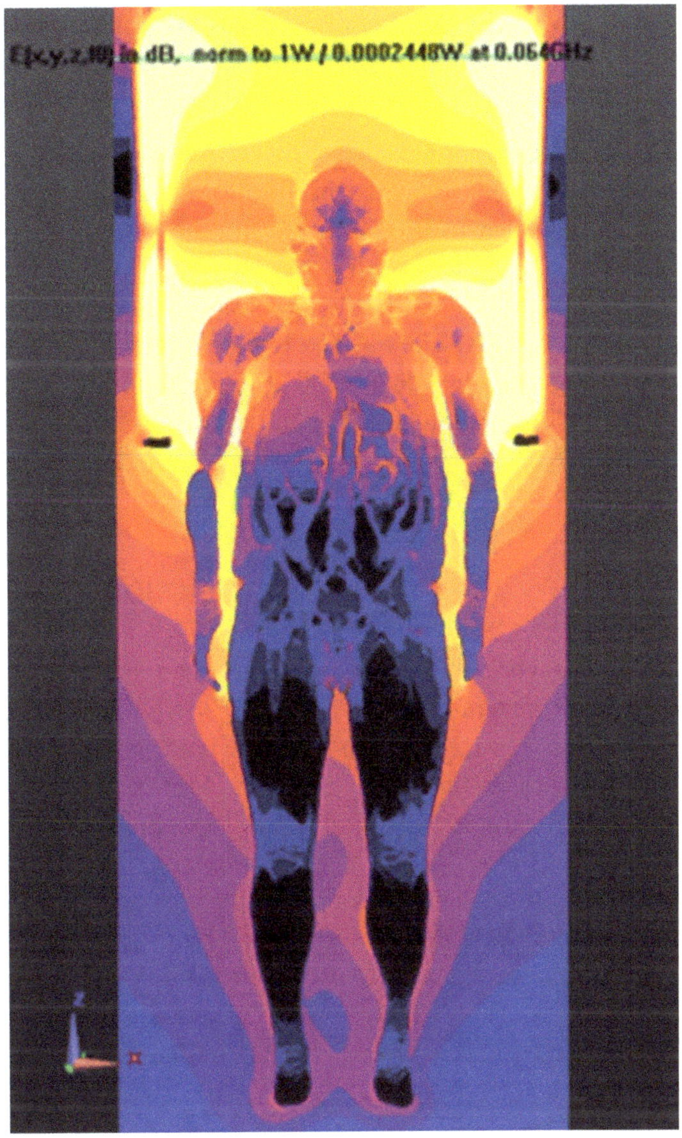

Fig. 14.1 Electric field distribution for obese male model at 64 MHz

In addition, comparative statements may easily be made between AIMDs that have been modeled within the same set of CHMs. In a clinical sense it would be impossible to gauge two devices under the exact same set of circumstances.

Finally, the use of computational techniques enables MR conditional safety to be evaluated for device prototypes. This is in contrast to clinical or experimental techniques, where the wait for a particular AIMD to be manufactured represents a delay in the timeline to evaluate the safety of a particular product. Thus, the use of CHMs in the assessment of MR conditional safety of AIMDs speeds up the development cycle and allows for new products to be developed specifically to meet MR conditional safety guidelines. Overall, this improves and accelerates the patient's access to both the benefits of MRI and to new therapies (provided by new AIMDs).

14.2 Evaluation of RF-Induced Malfunction Using Computational Human Models

This work focuses on the RF-induced energy incident on the RF antenna port, which may interfere with device operation. Manufacturers must perform a safety assessment of this hazard to determine the ability of the device to perform within its acceptable tolerances during and after the MRI. This assessment is performed via bench top testing, where the test conditions are conservatively computed through the use of CHMs. The voltage is induced at the antenna port simultaneously with any voltage which is developed at the lead port.

At the lead ports, AIMD manufacturers follow the well-known transfer function method [8, 9] to conservatively estimate these induced voltages according to the equation

$$V_{\mathrm{DUT}} = A\int_0^L S(\tau)\cdot E_{\mathrm{tan}}(\tau)\,d\tau. \tag{14.1}$$

A typical orientation of the implant for a dual-chamber pacemaker, with implantable pulse generator (IPG) location as well as atrial and ventricular pathways is shown in Fig. 14.2. A library of CHMs spanning the population in terms of height and BMI in different body positions, MR coils, and landmark positions is used to study the distribution of expected electromagnetic fields along the lead pathway. These fields (specifically, the component of the electric field tangent to the lead pathway) are then used to predict the response at the entry points to the cardiac implantable electronic device (CIED) by using a transfer function as a lead model.

Lead models are developed experimentally [10], in one or more tissue simulating media (TSM). The homogenous TSM should be chosen to accurately compute the RF-induced voltage once the transfer function is applied in the human body (via CHM).

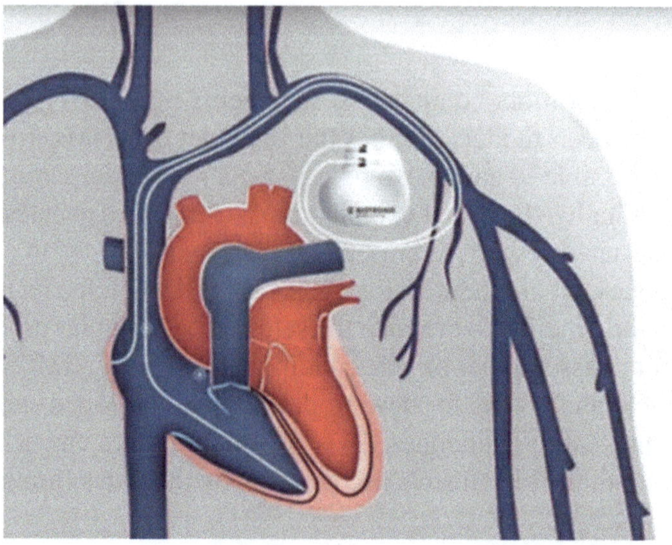

Fig. 14.2 Example orientation of dual-chamber pacemaker in the human body

The geometric accuracy of the CHM along the lead trajectory, including the continuity of organs through which the leads are placed, is paramount to the accuracy of the model. Variations in critical parameters such as surrounding anatomy and tissue properties must be included in the set of simulations used to generate the worst-case predicted RF-induced energy.

After deriving a probability distribution of the RF-induced energy at the AIMD (which can be expressed in terms of voltage, current, or power), the device is injected via bench-top test. The value from the distribution is chosen using a risk-based analysis and scaled by appropriate uncertainties to provide a conservative analysis of the device performance during MRI. The device is monitored to ensure it operates within its acceptable range during the injection and a post-test further examines the AIMD to ensure no damage has occurred to any of the device's subsystems.

The developing standard for leaded CIEDs establishes a test method for performing the RF-induced malfunction assessment of the antenna port [3]. Though this standard has been developed for these devices, it is extensible to devices of similar construction, such as spinal cord stimulators and deep brain stimulators. The standard method [3] consists of modeling the AIMD in two phantoms, high permittivity medium (HPM) and low permittivity medium (LPM), and exposing it to a uniform E-field oriented along each of the three axes, for a total of six models to derive a coupling coefficient (i.e., a scalar which gives the RF-induced voltage due to a 1 V/m incident field in the given direction). Separately, the expected peak E-field at the device location is derived by simulating a set of CHMs within a series of RF birdcage coils. Finally, the test condition for bench top RF injection testing is derived by multiplying this coupling coefficient with the predicted in vivo E-field according to the equation

$$V_{\text{ant}} = E_{\text{scanner}}\sqrt{A_x^2 + A_y^2 + A_z^2} + B_{\text{scanner}}\sqrt{C_x^2 + C_y^2 + C_z^2}. \tag{14.2}$$

This method involves separately simulating the CHM without the device present, simulating the device in a phantom and combining the results. A flowchart illustrating the process is shown in Fig. 14.3.

14.3 Approach

This work explores the accuracy of the method compared to a more rigorous model containing both the implanted device geometry and the birdcage coil. That method is more computationally intensive than performing the simulations separately, and is thus less desirable, as all of a large number of CHM-birdcage coil models would need to be resolved for every device. The method is explored for a pacemaker-type device but could be applied to other AIMDs of similar construction.

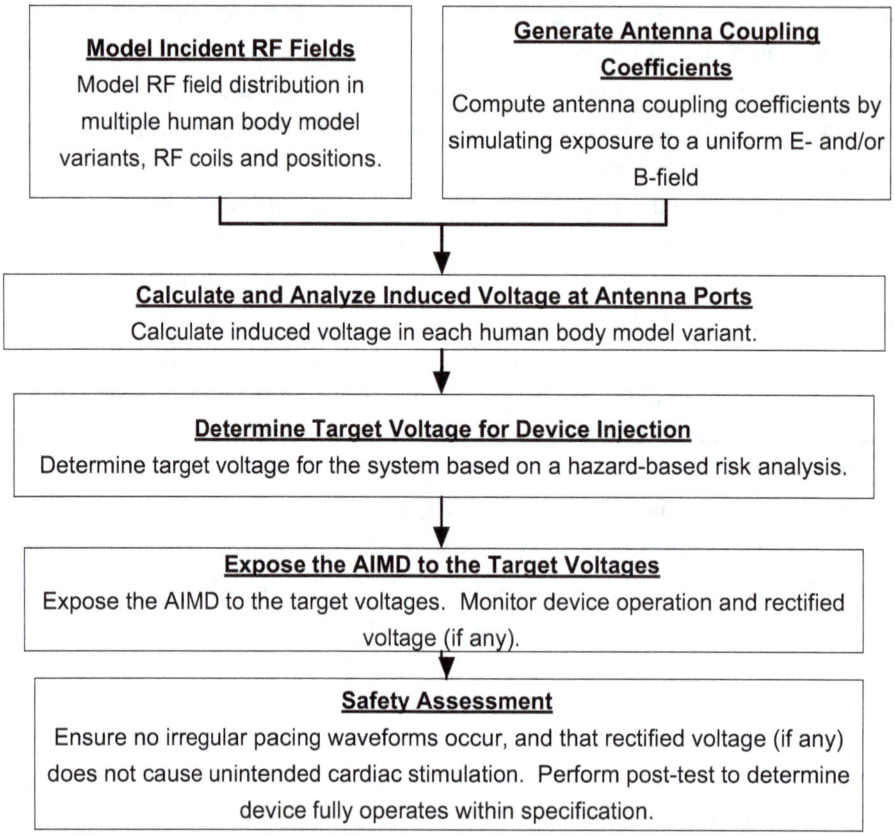

Fig. 14.3 Flow chart illustrating workflow for the assessment of protection from harm to the patient caused by RF-induced malfunction and rectification

14.3.1 Computational Human Model

A full device safety assessment would likely include a variety of human body models in a number of scan positions (e.g., supine, prone, etc.) and landmark locations (e.g., eyes, hips, or other anatomical region at isocenter) with respect to the MR coil geometry. The methods in this work are demonstrated by using one CHM in one scan position in one 3T coil and are extensible to a full range of conditions for a device safety assessment.

The CHM used in this study is the Visible Human Project (VHP)-Female v.3.0 [11]. The subject is 162 cm in height, with BMI of about 33.5. For electromagnetic simulation, this model has been shown to be computationally efficient and platform independent. In order to prevent CAD import errors, techniques such as using fully enclosed objects to avoid Boolean subtractions were used in the development of the model.

The CHM is shown in Fig. 14.4. The IPG is placed in the left pectoral region. Pacemaker implantation is usually in the area between fat and pectoral muscle on either the left or right side. The incident electric field would necessarily be very dif-

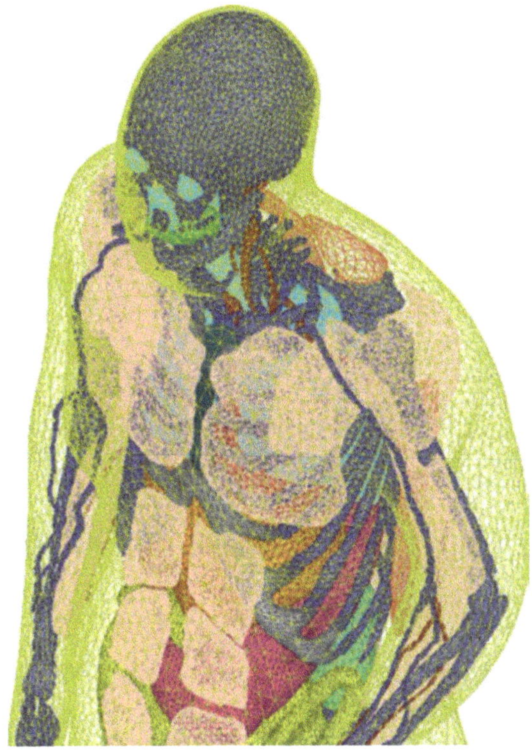

Fig. 14.4 The VHP-Female v. 3.0 model used in this study

ferent within these two tissues, due to the high contrast in permittivity. The device placement intersects these materials and a Boolean subtraction is performed with the device as the tool object. Objects in this example which require these subtractions include the pectoralis major and rib cartilage. This is due to the inability to actually deviate the anatomy around the implant and potentially represents a source of error between these computational techniques and the in vivo condition.

14.3.2 *Device Model*

The antenna used in this study is a loop antenna (classified as a B-field coupled antenna in [3]) in an epoxy header over a metallic enclosure. As this work does not consider optimization of the antenna, the model was constructed after the antenna used in [12]. The dimensions of the antenna are chosen for operation in the 2.4 GHz industrial, scientific, and medical (ISM) radio band, which is a higher frequency than the RF field of the MR system. The device is shown in Fig. 14.5. The can (device base) is a hollow (air-filled) PEC box. The header is modeled as epoxy and the feedthrough is ceramic. The antenna terminals are left open-circuited and the voltage drop is calculated between the two terminals inside the can.

Fig. 14.5 The device model used for this work, which is meant to represent a pacemaker geometry and includes a loop antenna

14.3.3 MRI Birdcage Coil Model

Two sets of simulations were performed using ANSYS Electronics Desktop. The 3T birdcage coil model was implemented using electromagnetic and circuit co-simulation [13], while the phantom models require only electromagnetic simulation. After solving the 3T birdcage coil model, the incident field was linked to a model including the CHM within the bore. Two simulations were then performed – one with and one without the device present. When the device was not present, some 2D and 3D regions were included to ease the computation of the incident field at the implant location. These included the rectangular solids shown in Fig. 14.5, for the can and header volume, as well as 2D sheets centered in these respective volumes.

14.3.4 Uniform Field Excitation

As the antenna in this study is a B-field coupled antenna, the test procedure outlined in [3] requires the exposure of the device to a uniform incident B-field. In order to expose the device to this B-field, a set of dual plane wave simulations were performed using ANSYS HFSS. By configuring each plane wave to have the same magnitude, a standing wave is generated in the modeled space which leads to cancellation of the E-field and uniform B-field in a particular direction. In this way, the RF-induced voltage at the antenna terminals may be extracted and normalized to a 1 μT incident field.

The phantom is a cube, 100 cm wide on each edge. The cube is assigned to be alternately the LPM ($\varepsilon_r = 11.5$, $\sigma = 0.045$) and the HPM ($\varepsilon_r = 78$, $\sigma = 0.47$) material properties as described in [3]. The model is extended by an enclosing vacuum object which is 300 cm on each side. The radiation boundary is assigned to the vacuum object. The phantom, with an example of one B-field orientation and the device model placement, is shown in Fig. 14.6.

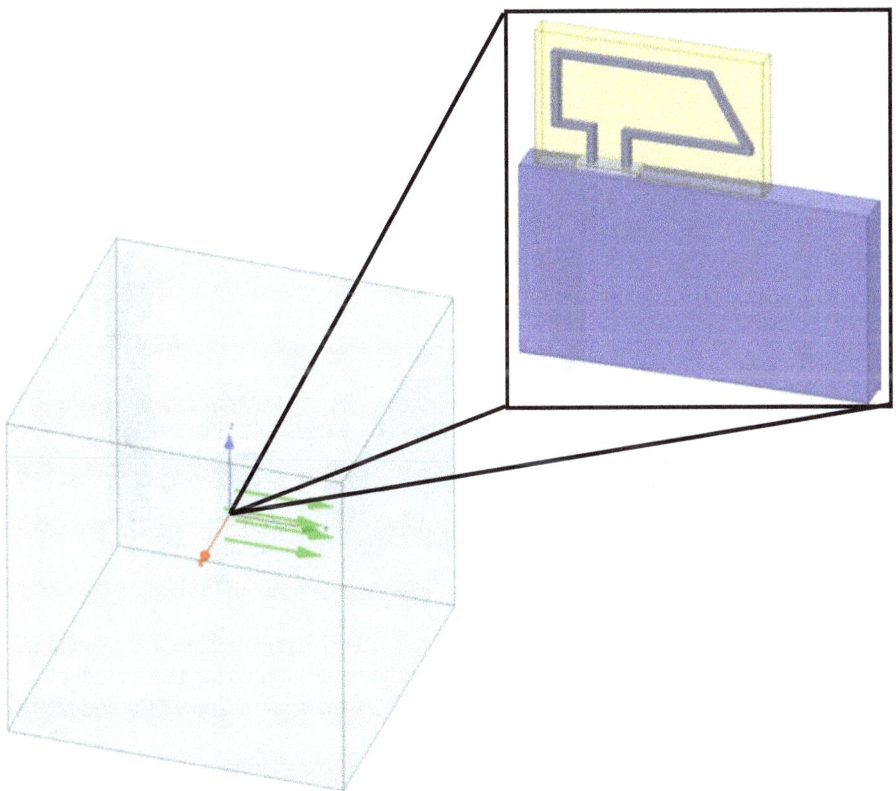

Fig. 14.6 The phantom model used for this study. The magnetic field is shown oriented in the y-direction at the device location. The device is not shown to scale with the phantom dimensions

14.4 Results

The results of the method in [3] and the more complete model are compared with each other. The field distribution of the CHM without the device present is then combined with these coefficients to give the predicted induced voltage at the antenna terminal. In order to quantify the field distribution at the CHM location, a number of methods are used. These are to take either the peak or the average fields values in a plane coinciding with the central axis of the device, the entire device volume, and the header volume. The non-model 2D and 3D objects described in the previous section are used to calculate volume and area averages for the relevant objects.

These six field distribution quantities are then combined with the results from either of the two phantoms, where the six coupling coefficients using Eq. 14.2 then give a voltage to be compared with the more rigorous and computationally expensive method of modeling the device in the CHM. These 12 results are compared to the calculated value in Fig. 14.7. In the figure, the term "device" means "can and header."

The results show that a number of these methods give a very conservative result. In fact, care should be taken to not overestimate the induced voltage, as this would lead to overly conservative test conditions.

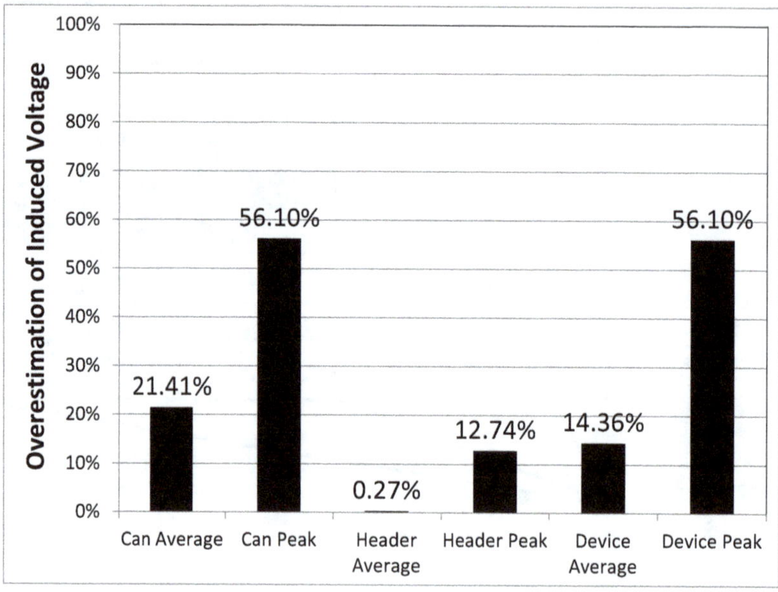

Fig. 14.7 The estimated RF-induced voltage from the HPM phantom model compared to the actual voltage from the CHP simulation. Different methods of quantifying the incident field distribution at the implant location result in varying degrees of overestimation of the induced voltage. "Device" refers to the combination of the can and header

A risk analysis for a device safety assessment following a process outlined in Fig. 14.1 would require many more simulations and a larger database of incident field distributions from the birdcage coil.

All investigated methods were conservative in estimating the induced voltage calculated via the more computationally expensive method for the case investigated. The method outlined in [3] and presented by Eq. (14.1) assumes worst-case alignment of the incident field vector and the antenna coupling coefficient, and thus is assumed to always be conservative. Additionally, the method requires taking the worst-case result from two phantoms, which further ensures the results to be conservative to the in vivo case.

14.5 Future Work

In the future, this work will be extended to 1.5T MR systems. While this work considered only a loop (B-field coupled) antenna, E-field coupled antennas, as well as more complex structures (e.g., helices), may be investigated. Additionally, the results presented here are for a single CHM, and a wider anatomical variety spanning the range of the patient population should be considered. Further, the impact of variability of the real situation such as the patient's position within the bore, variations in coil design, etc., can be considered as well. Finally, shimming techniques [7], which are used in 3T MR systems to improve field homogeneity, can be investigated. For this work, only one excitation vector for the coil ports was used.

There are many challenges related to the combination of modeling an MR coil, CHM, and device geometry in a single simulation. While some studies of these techniques exist [14, 15], the authors identified challenges related to computational resources for these efforts. These challenges arise from the fine details of the device model being solved in the large simulation volume of the CHP, MR coil, and surrounding space.

In the future, this work should be extended to realistic AIMD IPG geometries, as well as including leads. Lead geometries are sometimes coiled tightly (this is especially true for CIEDs) and include multiple conductors within a small volume. In addition, the impact of multiple leads on the induced voltages should be studied. Device operation can be incorporated through linking the electromagnetic simulation to a circuit model. The bulk of the literature in this area focuses on the analysis of RF-induced heating near leads for homogenous computational phantoms using computational techniques such as the method of moments [16] and the finite element method [17, 18]. Many of these techniques are extensible to calculating RF-induced energy at the proximal end of the lead as well. To date, in-depth device modeling has not been extensively performed in CHMs because of the challenges related to lead geometries. Future work could enable the evaluation of heat-reducing lead designs in realistic anatomies, instead of inhomogeneous phantoms, as is done in the literature today [19]. The further development of CHMs to provide optimal conditions for device modeling is an enabling technology.

Finally, the impact of any observed rectified voltage can be considered by incorporating a physiological model at the electrode interface to determine the probability of unintended stimulation, if any. Increased model development may further improve this area, as simulation could be used to speed the development of the physiological model. In particular, a Medical Device Development Tool could include co-simulation of electromagnetic and physiological phenomena [6]. Additionally, evaluation of, for example, unintended cardiac stimulation due to RF-induced rectified voltage, could be investigated via co-simulation rather than through developing probability models based on extensive lead testing.

References

1. Kalin, R., & Stanton, M. S. (2005). Current clinical issues for MRI scanning of pacemaker and defibrillator patients. *Pacing and Clinical Electrophysiology, 28*(4), 326–328.
2. ISO/TS 10974:2018, "Assessment of the safety of magnetic resonance imaging for patients with an active implantable medical device".
3. AAMI PC76 (Draft), "Requirements and Test Protocols for Safety of Patients with Pacemakers and ICDs Exposed to MRI", to be published.
4. Brown, J. E., et al. (2016). MR conditional safety assessment of implanted medical devices: Advantages of computational human phantoms. *Proceedings of the 38th Annual International Conference IEEE EMBC*, Orlando, pp. 6465–6468.
5. Christ, A., et al. (2010). The Virtual Family—development of surface-based anatomical models of two adults and two children for dosimetric simulations. *Physics in Medicine and Biology, 55*, N23–N38.

6. US FDA. (2013). Medical Device Development Tools – Draft Guidance for Industry, Tool Developers, and Food and Drug Administration Staff.
7. Ibrahim, T. S., et al. (2000). Application of finite difference time domain method for the design of birdcage RF head coils using multi-port excitations. *Magnetic Resonance Imaging, 10*, 733–742.
8. Park, S.-M., et al. (2007). Calculation of MRI-induced heating of an implanted medical lead wire with an electric field transfer function. *Journal of Magnetic Resonance Imaging, 26*, 1278–1285.
9. Feng, S., et al. (2015). A technique to evaluate MRI-induced electric fields at the ends of practical implanted lead. *IEEE Transactions on Microwave Theory and Techniques, 63*(1), 305–313.
10. Zastrow, E., Capstick, M., & Kuster, N. (2016). Experimental system for RF-heating characterization of medical implants during MRI. *Proceedings of the 24th Annual Meeting ISMRM*, Singapore.
11. Noetscher, G. M., et al. (2016). Computational human model VHP-female derived from datasets of the National Library of Medicine. *Proceedings of the 38th Annual International Conference IEEE EMBC*, Orlando, pp. 3350–3353.
12. ETSI TR 102 655 v. 1.1.1. (2008). Electromagnetic compatibility and Radio spectrum Matters (ERM); System reference document; Short Range Devices (SRD); Low Power Active Medical Implants (LP-AMI) operating in a 20 MHz band within 2 360 MHz to 3 400 MHz.
13. Bonmassar, G., Serano, P., & Angelone, L. M. (2013). Specific absorption rate in a standard phantom containing a deep brain stimulation lead at 3 Tesla MRI. *Proceedings of the 6th International IEEE/EMBS Conference NER*, San Diego, CA, pp. 747–750.
14. Zastrow, E., Cabot, E., & Kuster, N. (2014, August). Assessment of local RF-induced heating of AIMDs during MR exposure. *XXXIth URSI GASS*, Beijing, China.
15. Cabot, E., Zastrow, E., & Kuster, N. (2014, August). Safety assessment of AIMDs under MRI exposure: Tier3 vs. Tier4 evaluation of local RF-induced heating. *International Symposium EMC*, Tokyo.
16. Brown, J. E. (2012). Radiofrequency heating near medical devices in magnetic resonance imaging. Ph.D. dissertation, Bobby B. Lyle School of Engineering, Southern Methodist University, Dallas.
17. Brown, J. E., & Lee, C. S. (2013). Radiofrequency resonance heating near medical devices in magnetic resonance imaging. *Microwave and Optical Technology Letters, 55*(2), 299.
18. McCabe, S.O., & Scott, J.B. (2014, November). Cause and amelioration of MRI-induced heating through medical implant lead wires. *21st Elect New Zealand Conference*, Hamilton.
19. Nordbeck, P., et al. (2012). Reducing RF-related heating of cardiac pacemaker leads in MRI: Implementation and experimental verification of practical design changes. *Magnetic Resonance in Medicine, 68*, 1963–1972.

Chapter 15
Dose Coefficients for Use in Rapid Dose Estimation in Industrial Radiography Accidents

Haegin Han, Yeon Soo Yeom, Chansoo Choi, Hanjin Lee, Bangho Shin, Xujia Zhang, Rui Qiu, Nina Petoussi-Henss, and Chan Hyeong Kim

15.1 Introduction

Accidents by industrial radiography sources, which usually involve powerful gamma-emitting sources, could result in very high radiation doses to workers, leading to serious injuries or even death [1]. Even members of the public could be accidentally irradiated by such sources when the sources are not properly controlled or regulated. Moreover, industrial radiography accidents are one of the most frequently reported accidents among all reported cases in nuclear-related industries [2]. According to Lima et al. [3], a total of 80 industrial radiography accidents involving 120 workers and 110 members of the public, including 12 deaths, were recorded globally between 1896 and 2014.

Radiation accidents could lead to high radiation doses, which could induce acute radiation syndrome (ARS); these are primarily classified into hematopoietic (0.2–2 Gy), gastrointestinal (6–10 Gy), and cerebrovascular (>10 Gy) syndromes [4]. For the effective management of patients (i.e., exposed individuals) with ARS, initial medical triage should be performed accurately and rapidly, whereby those patients who could develop symptoms should be separately identified from those

H. Han · C. Choi · H. Lee · B. Shin · X. Zhang · C. H. Kim (✉)
Department of Nuclear Engineering, Hanyang University, Seoul, Republic of Korea
e-mail: chkim@hanyang.ac.kr

Y. S. Yeom
Division of Cancer Epidemiology and Genetics, National Cancer Institute, National Institutes of Health, Rockville, MD, USA

R. Qiu
Department of Engineering Physics, Tsinghua University, Beijing, People's Republic of China

N. Petoussi-Henss
Institute of Radiation Protection, Helmholtz Zentrum München, Deutsches Forschungszentrum für Gesundheit und Umwelt (GmbH), Neuherberg, Germany

S. Makarov et al. (eds.), *Brain and Human Body Modeling*,
https://doi.org/10.1007/978-3-030-21293-3_15

who do not require medical intervention [5]. Therapies for ARS, such as dose mitigators, should be promptly administered during the latent period in which the exposed person still feels well, i.e., before symptoms occur [6]. For this purpose, it is essential not only to carefully document clinical signs and symptoms but also to accurately and quickly estimate radiation doses to exposed individuals [7].

Individual radiation doses can be estimated using various dosimetric techniques based on biological, physical, and computational approaches. However, all the existing dosimetry techniques have limitations and none of them can be used as a stand-alone tool in a satisfactory manner for most radiation accident scenarios [8]. For instance, biological or physical dosimetry techniques generally require more than several days for sample collection and analysis, and these time-consuming processes make it difficult to perform timely medical triage. These techniques are therefore impractical, especially for use in a large-scale accident involving many individuals [5, 6, 9, 10]. Moreover, these techniques are generally limited to estimating the whole-body dose, without information on organ/tissue doses or dose distributions [8]. It should be noted that information about the whole-body absorbed dose is insufficient, particularly for partial-body or localized exposures. Organ/tissue absorbed doses and/or absorbed dose distributions can be estimated from computational dosimetry approaches (e.g., Monte Carlo simulations with computational human phantoms) if reliable information on the accident scenario is available, including accurate source geometry and duration of exposure [11], which are often unclear in accident situations [8, 12]. No single technique fully meets the criteria of an ideal dosimetry technique for use in accident situations; therefore, an integrated approach using multiple dosimetry techniques is considered the best strategy [8, 13, 14]. Absorbed doses calculated with computational human phantoms could be used as one of the dose estimators, particularly as an "initial, rapid estimator."

In the present study, a comprehensive set of organ/tissue absorbed dose coefficients (DCs) for industrial radiography sources was obtained by Monte Carlo simulations with the adult male and female mesh-type reference computational phantoms (MRCPs) recently developed by Task Group 103 of the International Commission on Radiological Protection (ICRP) [15]. In addition, adult male and female non-reference computational phantoms that represent the 10th and 90th percentiles of the Caucasian population were also used. These phantoms were constructed by deformation and modification of the MRCPs [16]. Note that for dose estimation of individuals in accident situations, consideration of the reference person would be insufficient, particularly when the body size of the individual involved in the accident is significantly different from that of the reference person. In this case, the dose estimation could be better approximated by using DCs calculated with a non-reference computational phantom whose body size is close to that of the actual individual.

The phantoms were implemented into the Geant4 Monte Carlo code [17] to calculate DCs by simulating the most commonly used industrial radiography sources, i.e., ^{192}Ir and ^{60}Co, for various source locations. The behavior of the calculated DCs was investigated, primarily by comparing the DCs among the different body size phantoms to study the influence of body size on the DCs.

15.2 Material and Methods

15.2.1 *Mesh-Type Computational Phantoms*

Figure 15.1 shows the adult male and female MRCPs and 10th and 90th percentile phantoms used in the DC calculations. The adult MRCPs were constructed by converting the adult voxel-type reference computational phantoms of ICRP

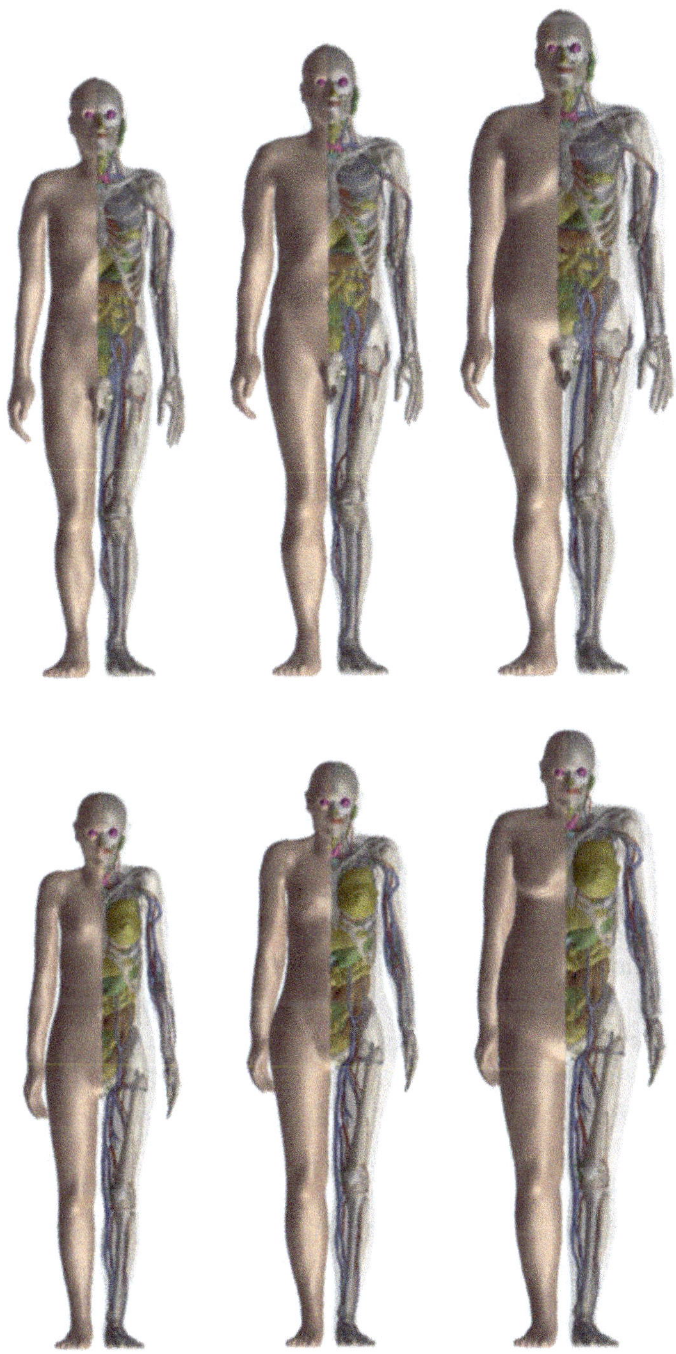

Fig. 15.1 10th percentile (left), mesh-type reference computational phantoms (MRCPs) (middle), and 90th percentile (right) phantoms for an adult male (upper figure) and adult female (lower figure)

Publication 110 [18] into a high-quality mesh format. This results in an improved representation of various organs/tissues and their ability to accurately estimate doses for very small organs/tissues such as the eye lens, the skin, and respiratory and alimentary tracts [15]. Note that recent developments enabled the direct use of mesh phantoms in Monte Carlo dose calculations, i.e., without going through a "voxelization" process, which is required for most existing surface phantoms (also called hybrid phantoms) [19, 20].

The MRCPs represent the adult reference male and female, as established in ICRP *Publication* 89 [21] with respect to the reference standing height, body mass (male: 1.76 m and 73 kg; female: 1.63 m and 60 kg) and individual organ and tissue masses. The 10th and 90th percentile phantoms were constructed by Lee et al. [16], by selectively deforming and modifying the MRCPs. The 10th percentile phantoms, which represent small statures, have the 10th percentile standing height and 10th percentile body mass (male: 1.67 m and 56 kg; female: 1.55 m and 44 kg), with organ masses ~25% lighter than those of MRCPs. The 90th percentile phantoms, which represent large statures, have the 90th percentile standing height and the 90th percentile body mass (male: 1.86 m and 108 kg; female: 1.72 m and 94 kg), and organ masses ~25% heavier than those of MRCPs.

15.2.2 Calculation of Organ Absorbed Dose Coefficients for Industrial Radiography Sources

The most commonly used industrial radiography sources of the ^{192}Ir and ^{60}Co radioisotopes were simulated as point sources placed near the phantoms. ^{192}Ir emits gamma rays up to 0.820 MeV with mean energy of 0.377 MeV, whereas ^{60}Co emits 1.33 and 1.17 MeV gamma rays. The point sources were assumed to be located at three different distances (0.005, 0.1, and 0.3 m) in four directions (anterior, posterior, right lateral, and left lateral) and at five locations along the height of the phantom (ground, middle thigh and lower, middle, and upper torso). In addition, for the lower torso level, three longer distances (1, 1.5, and 3 m) were considered for the four directions. The source distances used in the calculations are the distances from the phantom surface to the sources, except for the anterior and posterior directions at the ground and middle thigh levels, for which the source distances were measured from the center of an imaginary segment tangent to the surfaces of feet or left and right legs at the given level.

The doses of organs/tissues were calculated as organ/tissue-averaged absorbed dose per radioactive decay (Gy s^{-1} Bq^{-1}), by using the Geant4 Monte Carlo code [17]. For the simulations, we used a range cut of 5 μm for secondary electrons and photons, and the Livermore physics model (*G4EmLivermorePhysics*), which is provided by Geant4 for the transportation of electrons and photons with the energies ranging from 250 eV to 1 GeV. The statistical errors of the calculated values were less than 5% for all cases. The red bone marrow (RBM) DCs were calculated by using the fluence-to-absorbed dose response functions provided in Annex D of

ICRP *Publication* 116 [22] in order to consider the microscopic structure of bone geometry for different skeletal regions.

Note that the DCs in the present study (assuming point sources) were calculated without considering the exact source geometry. In order to allow users to consider the self-shielding effect of the sources, source self-shielding factors were additionally calculated by investigating the ratio of DCs for the spherical volume sources enclosed by capsule material to those for the bare point sources. For this purpose, combinations of four different thicknesses of radioactive material (1, 2, 3, and 4 mm) and two different thicknesses of capsule wall (1 and 2 mm) were considered, which cover the geometries of most radiography sources. The composition of the capsule material was assumed to correspond to that of 316 L stainless steel.

15.3 Results and Discussion

DCs for organ absorbed dose of the reference and 10th and 90th percentile phantoms were produced for two industrial radiography sources (^{192}Ir and ^{60}Co) placed at 72 different locations around the phantoms. The analysis of the calculated DCs for industrial radiography sources showed that the DCs generally tend to decrease with an increase in the source distance, while the opposite tendency was also observed for some cases. Figure 15.2, for example, shows the DCs for the lungs of the adult male MRCP for ^{60}Co. It can be seen that the DCs at the source levels of the upper, middle, and lower torso decrease with the increase of the source distance, whereas the opposite behavior is observed at the other source levels (i.e., the middle thigh and ground). This opposite trend is due to the different factors influencing the DCs of the lungs. For the first group of source levels (i.e., the upper, lower, and middle torso), the lung DCs are mainly influenced by the change of distance between the lungs and source (i.e., the inverse-square-law attenuation), rather than that of the shielding effect of the intervening tissue (i.e., the exponential-law attenuation). On the other hand, for the second group of source levels (i.e., the middle thigh and ground), which are much further away from the level of the lungs compared to the first group, the lung DCs are mainly influenced by the change of the shielding effect of the intervening tissue, which significantly decreases with the increase of the source distance. Similar trends were also observed for the DCs of the brain, which tend to increase with increasing source distance for the height levels of ground, middle thigh, and lower torso.

Our analysis also showed that the DC difference due to different body sizes tends to be larger when the source is closer to the body. When the source distance is shorter than 1 m, in general, the DCs of the 90th percentile phantoms were significantly smaller than those of the MRCPs; the maximum difference was a factor of ~8 for the female brain for the ^{192}Ir source located at 0.3 m from the phantom surface in the right-lateral direction at the ground level. The DCs of the 10th percentile phantoms were significantly larger than those of the MRCPs; the maximum difference was 2.3 times for the female brain for the ^{192}Ir source located at 0.1 m

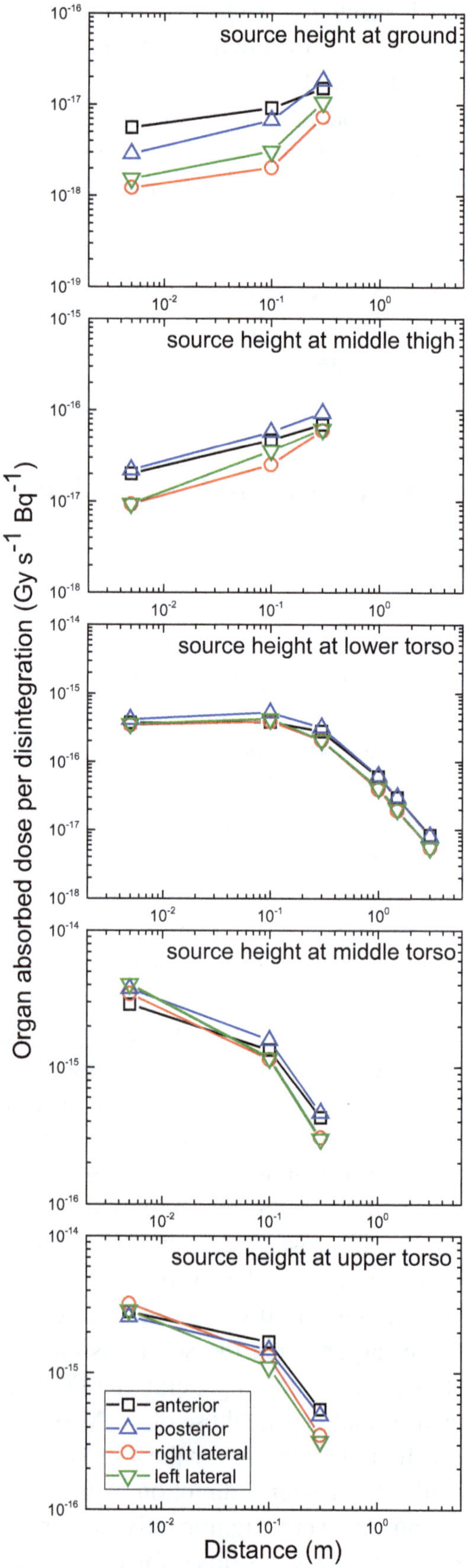

Fig. 15.2 Lungs absorbed dose coefficients for the adult male mesh-type reference computational phantom as a function of distance from the phantom surface, for a point source of ^{60}Co placed in four different directions (anterior, posterior, right lateral, and left lateral) and at five different heights (ground, middle thigh and lower, middle, and upper torso)

from the phantom surface in the left-lateral direction at the ground level. It should be noted that, for both cases, maximum differences were not found at the shortest source distance (0.005 m), because in these cases the source particles could directly reach the head only for the smaller phantoms, while this was not the case for the other phantoms.

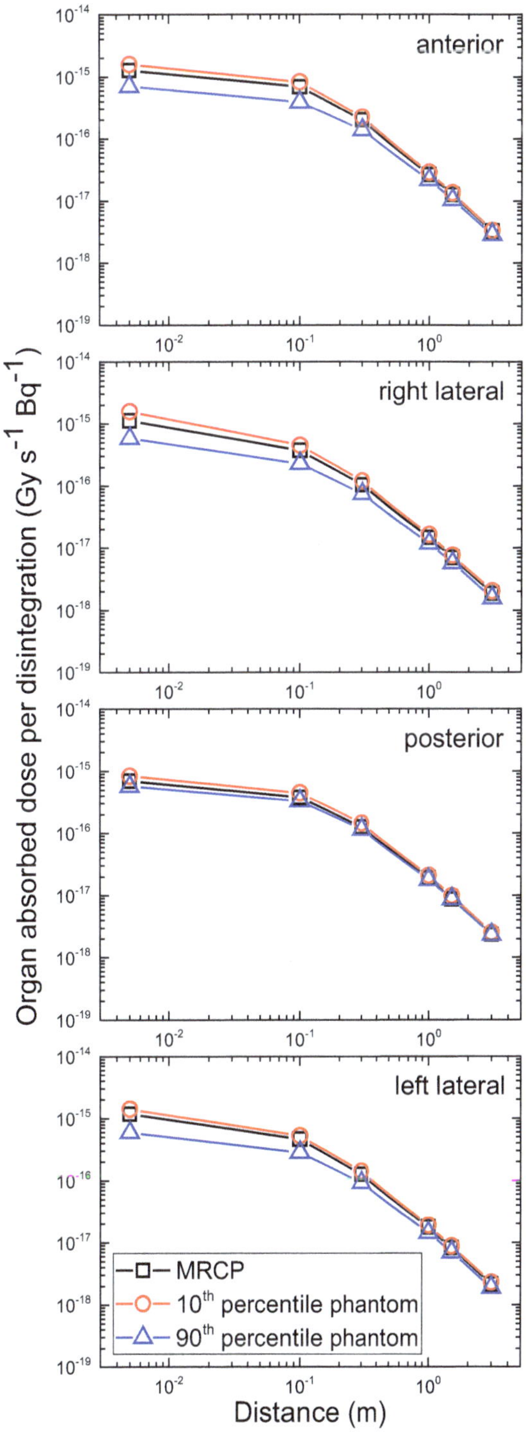

Fig. 15.3 Large intestine absorbed dose coefficients for the adult male mesh-type reference computational phantom (black squares) as a function of distance from the phantom surface, for a point source of ^{192}Ir in four different directions (anterior, posterior, right lateral, and left lateral) at the height of the lower torso level, along with those of the 10th percentile adult male phantom (red circles) and the 90th percentile adult male phantom (blue triangles)

Table 15.1 Source self-shielding factors for ^{192}Ir and ^{60}Co

	Radioactive material thickness							
	1 mm		2 mm		3 mm		4 mm	
Capsule-wall thickness	^{192}Ir	^{60}Co	^{192}Ir	^{60}Co	^{192}Ir	^{60}Co	^{192}Ir	^{60}Co
1 mm	0.840	0.972	0.717	0.965	0.627	0.958	0.556	0.949
2 mm	0.803	0.953	0.694	0.947	0.606	0.938	0.536	0.929

On the other hand, for the longer source distances, the DC differences due to the different body sizes were small, i.e., less than 20–30% for both genders and all organs/tissues, source directions, and isotopes considered in the present study. Figure 15.3 shows the DCs of the large intestine of the adult male MRCP for ^{192}Ir sources in the four directions at the lower torso level as a function of the source distance, along with those of the 10th and 90th percentile phantoms. It can be seen that the DC differences among the different body size phantoms decrease with an increase in the source distance. This trend indicates that the DCs among the different body sizes are influenced mainly by the change of the distance from the target organs/tissues to the source, rather than that of the shielding effect of the intervening tissue.

Finally, the DCs of MRCPs were generally found to be closer to the DCs of 10th percentile phantom than those of 90th percentile phantom. This is mainly because the MRCPs do not exactly correspond to the 50th percentile in body weight; the adult male and female MRCPs correspond to ~30th and ~40th percentile in body weight, respectively, which are closer to 10th percentile than 90th percentile. It can also be seen that the body size effect tends to be less significant for sources in the posterior direction as opposed to those in the other directions (anterior and left and right laterals). This can be explained by the fact that the change of the residual soft tissue (i.e., adipose tissue) among the different body size phantoms in the back is smaller than that in the other sides, especially in the abdominal region [23–25].

Table 15.1 shows the source self-shielding factors of ^{192}Ir and ^{60}Co, by which the DCs could be multiplied according to the different thicknesses of the radioactive material (1, 2, 3, and 4 mm) and capsule wall (1 and 2 mm). It can be seen that while the self-shielding effects of ^{60}Co sources are less than 10% in all cases, the self-shielding effects of ^{192}Ir sources are up to 56.4%, which is due to the extremely high density of iridium (22.5 g cm^{-3}). This implies that source self-shielding factors should be considered, especially for ^{192}Ir sources.

15.4 Conclusion

In the present study, a comprehensive data set of organ absorbed dose coefficients (DC) for industrial radiography sources of two radioisotopes (^{192}Ir and ^{60}Co), placed in 72 different source locations were obtained by performing Monte Carlo simulations with the mesh-type reference computational phantoms (MRCPs) and

10th and 90th percentile phantoms. In addition, the characteristics of the DC data were investigated, thereby finding that the body size indeed significantly influences the DCs, especially when the sources are closer than 1 m to the human body, which is often the case for accidents involving industrial radiography sources. The DC data are expected to be used as an initial tool for the rapid dose estimation of individuals accidentally exposed to industrial radiography sources. A full set of the DCs will appear in a forthcoming ICRP *Publication* prepared by the ICRP Task Group 103 of Committee 2.

Acknowledgments The authors would like to thank Dr. Wi-Ho Ha and Mr. Han Sung Kim of Korea Institute of Radiological & Medical Sciences (KIRAMS) for helpful discussions. This work was supported by the Nuclear Safety Research Development (NSR&D) Program through Korea Foundation of Nuclear Safety (KoFONS), funded by the Nuclear Safety and Security Commission (NSSC), and additionally by the National Research Foundation of Korea (NRF) funded by the Ministry of Science, ICT and Future Planning (Project No.: 1705006, 2016R1D1A1A09916337). Two of the authors (Haegin Han and Chansoo Choi) were supported by the Global Ph.D. Fellowship program (Project No.: NRF-2018H1A2A1059767 and NRF-2017H1A2A1046391).

References

1. IAEA. (2011). IAEA Safety Standards No. SSG-11, Radiation safety in industrial radiography.
2. IAEA. (1998). Safety Reports Series No. 7, Lessons learned from accidents in industrial radiography.
3. Lima, C. M. A., Lima, A. R., Degenhardt, A. L., et al. (2015). Reconstructive dosimetry for cutaneous radiation syndrome. *Brazilian Journal of Medical and Biological Research, 48*, 895.
4. Donnelly, E. H., Nemhauser, J. B., Smith, J. M., et al. (2010). Acute radiation syndrome: assessment and management. *Southern Medical Journal, 103*, 541.
5. Gougelet, R. M., Nicolalde, R. J., Rea, M., & Swartz, H. M. (2010). The View from the Tranches Part 1: Emergency Medical Response Plans and the Need for EPR Screening. *Health Physics, 98*, 118.
6. Rea, M. E., Gougelet, R. M., Nicolalde, R. J., et al. (2010). Proposed triage categories for large-scale radiation incidents using high-accuracy biodosimetry methods. *Health Physics, 98*, 136.
7. Cerezo, L. (2011). Radiation accidents and incidents. What do we know about the medical management of acute radiation syndrome? *Reports of Practical Oncology and Radiotherapy, 16*, 119.
8. Ainsbury, E. A., Bakhanova, E., Barquinero, J. F., et al. (2011). Review of retrospective dosimetry techniques for external ionising radiation exposures. *Radiation Protection Dosimetry, 147*, 573.
9. Swartz, H. M., Williams, B. B., & Flood, A. B. (2014). Overview of the principles and practice of biodosimetry. *Radiation and Environmental Biophysics, 53*, 221.
10. Kulka, U., Abend, M., Ainsbury, E., et al. (2017). RENEB-Running the European Network of biological dosimetry and physical retrospective dosimetry. *International Journal of Radiation Biology, 93*, 2.
11. Lu, W., Wu, Z., Qiu, R., et al. (2017). Physical dosimetric reconstruction of a radiological accident at Nanjing (China) for clinical treatment using THUdose. *Health Physics, 113*, 327.
12. Clairand, I., Trompier, F., & Bottollier-Depois, J. F. (2006). EX vivo ESR measurements assiciated with Monte Carlo calculations for accident dosimetry: application to the 2001 Georgian accident. *Radiation Protection Dosimetry, 119*, 500.

13. Sullivan, J. M., Prasanna, P. G. S., Grace, M. B., et al. (2013). Assessment of biodosimetry methods for a mass-casualty radiological incident: medical response and management considerations. *Health Physics, 105*, 540.
14. Ainsbury, E. A., Badie, C., Barnard, S., et al. (2017). Integration of new biological and physical retrospective dosimetry methods into EU emergency response plans-joint RENEB and EURADOS inter-laboratory comparisons. *International Journal of Radiation Biology, 93*, 99.
15. Kim, C. H., Yeom, Y. S., Nguyen, T. T., et al. (2018). New mesh-type phantoms and their dosimetric applications, including emergencies. *Annals of the ICRP, 47*, 45. https://doi.org/10.1177/0146645318756231.
16. Lee, H., Yeom, Y. S., Thang, T. N., et al. (2017). Transactions of the Korean Association for Radiation Protection Autumn Meeting.
17. Allison, J., Amako, K., Apostolakis, J., et al. (2016). Recent developments in Geant4. *Nuclear Instruments and Methods in Physics Research Section A, 835*, 186.
18. ICRP, ICRP Publication 110, *Adult Reference Computational Phantoms*, 2009.
19. Kim, C. H., Jeong, J. H., Bolch, W. E., et al. (2011). A polygon-surface reference Korean male phantom (PSRK-Man) and its direct implementation in Geant4 Monte Carlo simulation. *Physics in Medicine and Biology, 56*, 3137.
20. Yeom, Y. S., Jeong, J. H., Han, H. C., et al. (2014). Tetrahedral-mesh-based computational human phantom for fast Monte Carlo dose calculations. *Physics in Medicine and Biology, 59*, 3173.
21. ICRP, ICRP Publication 89, *Basic Anatomical and Physiological Data for Use in Radiological Protection: Reference Values*, 2002.
22. ICRP, ICRP Publication 116, *Conversion Coefficients for Radiological Protection Quantities for External Radiation Exposures*, 2010.
23. Borkan, G. A., Gerzof, S. G., Robbins, A. H., et al. (1982). Assessment of abdominal fat content by computed tomography. *The American Journal of Clinical Nutrition, 36*, 172.
24. Dixon, A. K. (1983). Abdominal fat assessed by computed tomography: sex difference in distribution. *Clinical Radiology, 34*, 189.
25. Grauer, W. O., Moss, A. A., Cann, C. E., et al. (1984). Quantification of body fat distribution in the abdomen using computed tomography. *The American Journal of Clinical Nutrition, 39*, 631.

Chapter 16
Effect of Non-parallel Applicator Insertion on 2.45 GHz Microwave Ablation Zone Size and Shape

Austin W. White, Dwight D. Day, and Punit Prakash

16.1 Introduction

Microwave ablation (MWA) is an established modality for minimally invasive treatment of tumors in the liver and other organs. During MWA, a microwave antenna (also referred to as an ablation applicator) is inserted into the target tissue and radiates microwave power, which is absorbed by the surrounding tissue and leads to heating. Compared to other energy modalities for thermal ablation, MWA offers the advantage of rapid heating of large tissue volumes via the simultaneous use of multiple applicators [1]. The electromagnetic power absorbed within tissue during multiple applicator MWA is a function of the system operating frequency, tissue biophysical properties, applied power levels, radiation pattern of each applicator, and the relative spacing between the applicators. Preclinical experiments in ex vivo tissue are often used to characterize the size and shape of ablation patterns for specific devices, and these data may be used to assist in treatment planning [2]. Experimental studies to characterize multiple applicator ablation patterns typically employ the applicators inserted into the tissue sample in a parallel manner [1]. In clinical practice however, parallel applicator insertion may not be feasible due to anatomical constraints. As illustrated in [3], the spacing between applicator tips may be up to twice as much as the applicator spacing at the insertion point on the skin surface.

Studies investigating the impact of non-parallel applicator insertion during interstitial hyperthermia with microwave antenna arrays determined that skewed applicator placement shifted the location of the region of maximum power absorption [4]. However, prior simulation studies to assess ablation zones created by MWA with non-parallel applicator configurations were limited due to the use of models

A. W. White (✉) · D. D. Day · P. Prakash
Department of Electrical and Computer Engineering, Kansas State University, Manhattan, KS, USA
e-mail: prakashp@ksu.edu

S. Makarov et al. (eds.), *Brain and Human Body Modeling*,
https://doi.org/10.1007/978-3-030-21293-3_16

which did not incorporate temperature-dependent tissue dielectric properties [3]. Recent studies have highlighted the importance of incorporating temperature-dependent tissue dielectric properties in computational models of MWA [5].

This chapter summarizes the use of multi-physics computational models and experiments in ex vivo tissue to compare the size and shape of ablation zones with parallel and non-parallel applicators.

16.2 Materials and Methods

We implemented finite element method (FEM) computational models of microwave tissue heating to comparatively assess ablation profiles achieved with a pair of 2.45 GHz interstitial dipole antennas spaced 10–20 mm apart. The antennas were inserted in parallel and non-parallel configurations. Experimental studies were conducted ex vivo using liver tissue for a subset of these configurations. Computational modeling results were compared against the experimental observations of ablation zone profiles.

16.2.1 MWA Applicator Configurations

We investigated nine different MWA applicator insertion configurations, including three cases with parallel applicator insertion and six cases with non-parallel applicator insertion. We considered inter-applicator spacings of 10–20 mm to capture the range of applicator spacing commonly encountered in clinical applications. As illustrated in Fig. 16.1, the parallel cases had antennas inserted into the liver with the center of the microwave applicators at separation distances denoted D_{ins}. The non-parallel cases had the tip of each applicator moved toward the centerline by 2.5 mm, 5 mm, or 7.5 mm each, resulting in tip spacings (D_{tip}) of 15 mm, 10 mm, or 5 mm, respectively.

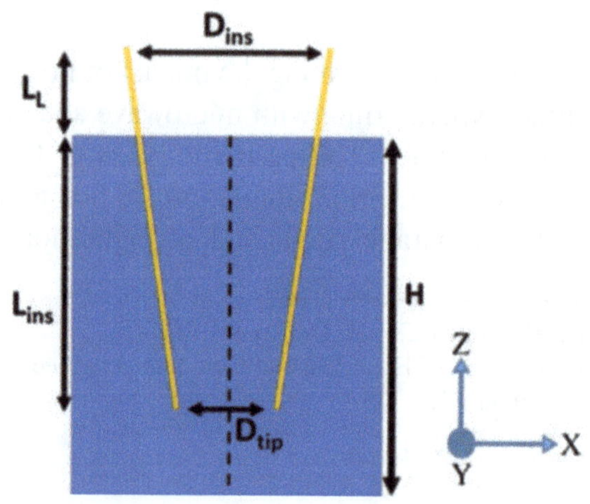

Fig. 16.1 Illustration of non-parallel MWA applicator insertion

In Fig. 16.1, D_{ins} is the nominal distance between the antennas and D_{tip} is the distance between the tips of the antennas. For the parallel cases, D_{ins} is equal to D_{tip}. For the non-parallel cases, D_{tip} was changed in multiples of 5 mm (each antenna moved 2.5 mm). The tip of each antenna was moved inward by half the total tip displacement, so that the setup was always symmetric with respect to the centerline. L_{ins} is the distance between the antenna tip and the upper surface of the liver tissue. L_{ins} corresponds to the length of the antenna inside the liver for a parallel case. The depth of the antenna tip from the top of the liver remained constant in all experiments. Therefore, the total length of the antenna in the liver slightly increases for a non-parallel case in order to maintain the same tip depth from the surface of the liver. L_L was chosen to act as a reference to define the insertion distance.

For this study, the MWA applicators were uncooled, interstitial, dipole antennas operating at 2.45 GHz. The antennas were created using UT-85 coaxial cable with the distal tip of the cable modified as described by Curto et al. [6]. Each antenna was energized with 15 W applied power (at the antenna input). We chose this power to be within the clinically relevant range, while yielding ablation zone sizes not impacted by the limited size of the tissue samples used in our experiments.

As illustrated in Fig. 16.1, ablation zone extents were assessed primarily in the XZ-plane.

16.2.2 Computational Model of Microwave Ablation

Three-dimensional multi-physics computational models of microwave tissue heating were implemented using the FEM solver within COMSOL Multiphysics (v5.2a). Models were employed to compute the electric fields radiated into tissue and the subsequent heat transfer due to absorbed microwave power. The Helmholtz electromagnetic wave equation, Eq. 16.1, was used to calculate the electric field at all mesh points.

$$\nabla \times \left(\mu_r^{-1} \nabla \times \mathbf{E} \right) - \frac{\omega^2}{c_0^2} \left(\varepsilon_r - \frac{i\sigma}{\omega \epsilon_0} \right) \mathbf{E} = 0 \tag{16.1}$$

Here μ_r is the relative permeability, ε_r is the relative permittivity and σ is the electrical conductivity of the material, c_0 is the speed of light in free space, ω is angular frequency, and $\mathbf{E}$ [V/m] is the electric field vector. From the electric field, we can calculate how much power is absorbed within the tissue using Eq. 16.2, which serves as the heat source for the transient heat equation, Eq. 16.3.

$$Q_e = \frac{1}{2} \sigma |\mathbf{E}|^2 \tag{16.2}$$

$$\rho c \frac{\partial T}{\partial t} = \nabla \cdot k \nabla T + Q_e \tag{16.3}$$

where T is temperature [K], ρ is tissue density [kg/m^3], and C is the specific heat capacity of tissue [J/(kg K)].

We did not include blood perfusion in our simulations as our ex vivo experimentation would also lack blood perfusion and the primary goal of this work was to compare simulation results with experimental measurements. The electrical properties of liver tissue at 2.45 GHz were dynamically adjusted as a function of the liver temperature as described in [7]. The nominal value of tissue relative permittivity was $\varepsilon_r = 57.9$ and effective conductivity $\sigma = 1.09$ S/m [8].

The applicators were inserted into a cuboidal volume of liver tissue (length and width of 60 mm and a height of 80 mm). A non-uniform tetrahedral mesh was employed in the simulations. In our case, the finest resolution was within the antenna and coarsest resolution was used at distances further away from the applicator. The conducting elements in the antennas were approximated as perfect electric conductors. Electrical scattering and thermally insulating boundary conditions were imposed on the exterior surfaces of the liver and the insulating catheter.

16.2.3 Experimental Assessment of Dual Applicator Microwave Ablation Zones

Instrumentation and Procedure

Two insulated dipole antennas were fabricated according to the dimensions given in the paper by Curto et al. [6]. The antennas had an active dipole length of 10.9 mm and a gap of 1 mm and were tuned to operate at 2.45 GHz. These antennas did not have any active cooling. Antennas were fabricated using Micro-Coax UT-85, with a thin heat shrink tubing (3M, FP-301 26-28 AWG) acting as an insulating sheath. An SMA connector (Huber+Suhner) was attached to the end of the coax to form the connector. Antenna reflection coefficient measurements using a Vector Network Analyzer (VNA) verified that the antennas were properly matched (S_{11} of at least −10 dB at the operating frequency) to the feeding transmission line.

For the experimental setup, a Flash Forge Creator Pro 3D printer was used to create a base from polylactic acid (PLA). The base was designed to hold the antennas and liver tissue during experimental testing. The internal measurements of the base were 60 mm × 60 mm × 85 mm. The template blocks held the antennas in the same configurations as the simulations. The following antenna configurations were investigated experimentally: 20 mm parallel, 20 mm to 15 mm (converging), 15 mm parallel, and 15 mm to 10 mm (converging).

Ex vivo ablations were conducted in fresh bovine liver. Liver samples were procured from a local meat-processing plant and cut to fit into the base. The liver samples were placed in plastic bags and then warmed in a water bath (Polyscience 7306) until they reached a temperature of ~30 °C. Five experiments were performed for each configuration. For each experiment, one liver piece was removed from the water bath and placed into the base and the template block was then snapped into place. The antennas were inserted through the template block to a depth of 6 cm into the liver.

We applied a 30 W forward power signal at 2.45 GHz (SAIREM microwave generator) and used an equal power splitter (Pasternack PE-T1000) to divide the power into two separate, phase-matched signals fed to each antenna. We then used two power meters (Bird Technologies 7022-1-02020 and Bird Technologies 5012D) to track the power level delivered to each antenna. We adjusted the total average forward power to remain close to 30 W during each experimental ablation.

The liver was sectioned in half after the ablation, revealing the ablation zone in the XZ plane. This was done by removing the template block along with the red connector piece and slicing the liver with a knife down the slot in the base as shown in Fig. 16.2. Both halves of the liver were then placed on a flatbed scanner (Epson V550 Photo Color Scanner) which was used to acquire a color (RGB) image of the ablation zone at 400 dpi. The scanner was used to prevent skew from photos that might appear in pictures taken with a hand-held camera. The scanner has the added benefit of keeping the lighting and resolution constant for all pictures, which is desirable for automated processing of acquired images.

Image-Based Analysis of Experimental Ablation Zones

The resulting pictures were processed using the basic image processing techniques available in Python 3.6 to measure the size and compare the shape of the resulting ablation zones. A basic thresholding operation in both the RGB color space as well as the Hematoxylin-Eosin-DAB (HED) color space (HED used often in detecting blood composition) was the primary method used to detect the ablated area of the

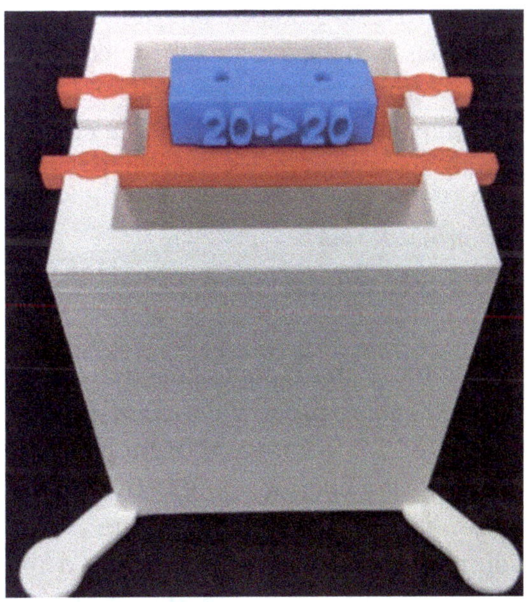

Fig. 16.2 3D printed base with template block in place to hold the antennas

liver tissue. We used *skimage* to convert the RGB image into the HED color space. Then *scipy* was used to perform a small Gaussian blur (σ = .19) on the resulting non-binary image to remove noise, and then converted back to a binary image. A series of morphological operators (erosion and then dilation using a disk of size 3 pixels) further reduced noise and smoothed the edges of the ablation region. After the morphological operators, the *scipy* function – binary fill holes – was used to fill in the holes in the middle of the ablated region. *Skimage* region properties were used to single out the ablation region, which corresponded to the largest region in the image after noise removal. An edge detection operation was used on the largest area to determine the size of the ablation zone. Further details of the image analysis procedure are provided in [9].

Once the shape of the ablation region was segmented in all ablations for a given scenario, they were overlaid for aggregate analysis. Each segmented ablation zone was centered and aligned to the same orientation as part of the overlay process. Region properties of the ablation zone were used to find the center of mass for each ablation area. We also applied a rotation on an image-by-image basis depending on the placement of the liver sample on the scanner bed. The angle of rotation was determined qualitatively for each of the ablation areas. After aligning the ablation shapes, we added the binary score (0 or 1) for each pixel from multiple repetitions of each ablation scenario together to find the composite ablation image. Since we performed $n = 5$ experiments for each experimental scenario, this composite image contained only integers with values that ranged between zero and five. The value of five occurred in regions where the tissue was ablated in each of the five experiments; likewise, the areas that had a value of one are locations where the tissue was ablated in only one of the five experiments. The darkest mass in the center corresponded to an ablation area in all experiments. From this, we found the maximum and minimum ablation shape for each scenario, as well as the 75% contour (i.e., the region that we would expect to see ablated in 75% of ablations). To find the maximum, we converted the image to a binary image format where only the highest number (five) would be a one, all else would be zero. A canny edge detection algorithm was then used to find the maximum shape.

We compared the shape of the ablation zone extents from experiments using the Dice Similarity Coefficient (DSC) metric. This metric is commonly used in image processing applications to compare shape similarity. The DSC is calculated using Eq. 16.4,

$$\text{DSC} = \frac{2|A \cap B|}{|A| + |B|} \tag{16.4}$$

For the experiments, A and B are binary images that have been detected through image processing. Specifically, the areas inside some contours (typically the 75% contour). For the simulations, we use the XZ plane that bisects the liver and antennas and considered only points on that plane that are at or above 55 °C as an estimate of the simulated ablation zone. We also employed a volumetric approach that compared the entire volume of the ablation region that is above 55 °C to the 3D simulation results.

16.3 Results and Discussion

Figure 16.3 illustrates experimental results for the ablation zone segmentation, alignment, aggregation, and averaging process. Table 16.1 lists Dice Similarity Coefficients between microwave ablation zones for antennas spaced 10–20 mm apart in parallel and non-parallel configurations.

Although both experiments and simulations predicted similar trends, experimentally measured DSCs were ~15% lower. This discrepancy is possibly due to the jagged edges of experimental ablation zones, which confound image similarity metrics.

Figure 16.4 illustrates simulated and experimental ablation zones for 15 mm parallel and non-parallel (2.5 mm displacement of antenna tips) configurations. Both simulations and experiments indicated small changes in ablation zone shape. For the 20 mm parallel vs. non-parallel configurations illustrated in Fig. 16.5, the difference between ablation zone shapes is slightly more pronounced, as evidenced by the smaller DSC and observed ablation zone shapes. The presented method therefore provides a means for comparing ablation zone shapes for various configurations, in contrast to the widely used approach of comparing parameterized measurements of ablation zone geometries. This approach is especially helpful for analysis of shapes that are not close approximations of spheres/ellipses.

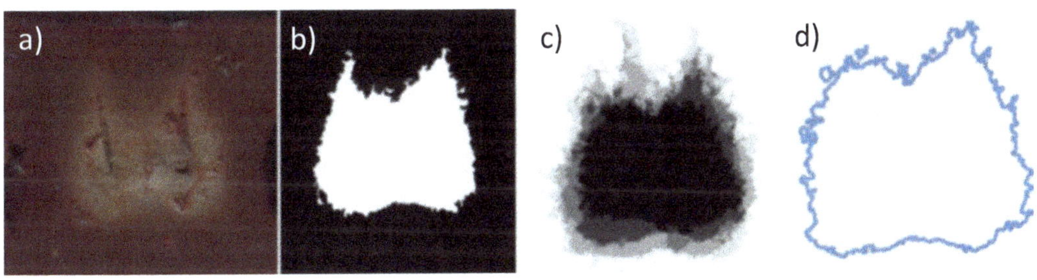

Fig. 16.3 Image processing at various stages- from left to right: (**a**) scanned image of ablation zone in ex vivo liver, (**b**) segmented ablation region, (**c**) overlay of segmented regions, and (**d**) the resulting 75% contour

Table 16.1 Dice Similarity Coefficients between microwave ablation zones

Antenna spacing at insertion	Antenna tip spacing	DSC simulated	DSC experimental
10	5	0.95	
15	10	0.93	0.81 + 0.003
15	5	0.89	
20	15	0.85	0.73 + 0.06
20	10	0.78	

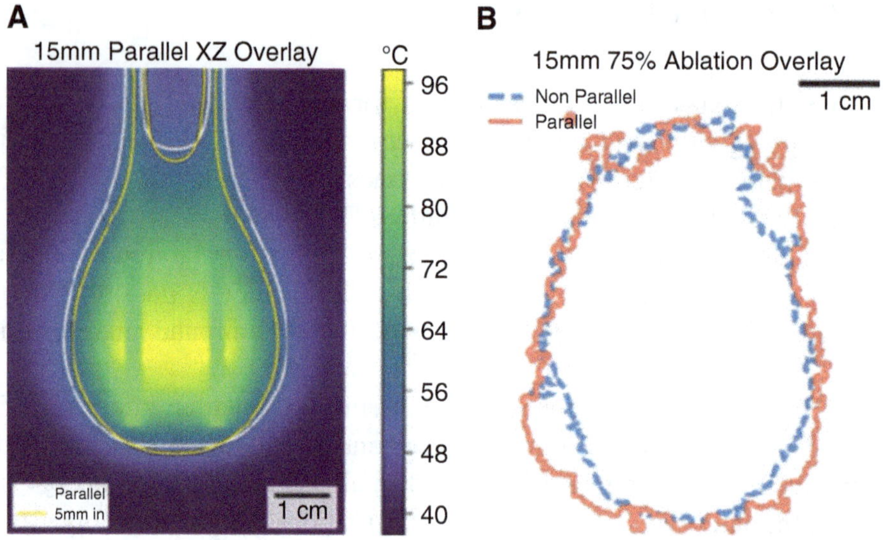

Fig. 16.4 (**a**) Illustration of simulated ablation zones for 15 mm parallel vs. non-parallel configurations; (**b**) 75% ablation overlay from experiments

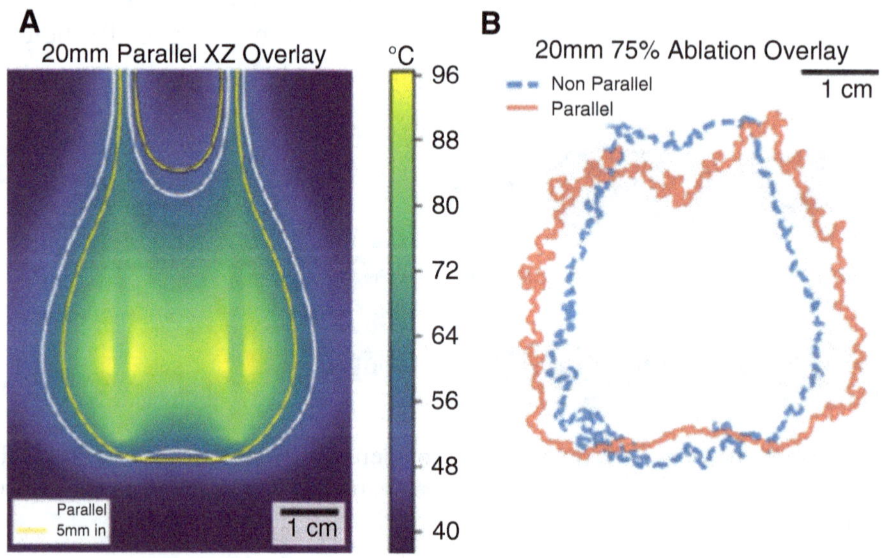

Fig. 16.5 (**a**) Illustration of simulated ablation zones for 20 mm parallel vs. non-parallel configurations; (**b**) 75% ablation overlay from experiments

16.4 Conclusion

The use of microwave ablation for treating tumors in the liver and other organs is growing. Clinicians who deploy multiple antennas for treating hepatocellular carcinoma (HCC) may not always be able to insert the antennas parallel to each other, and it is not yet known what impact this has on the ablation outcome. This research was an investigation to assess the potential change in ablation volume for non-parallel insertion. These results may serve as a guide to practitioners to determine the potential impact of non-parallel antenna insertion when conducting clinical procedures.

We found that non-parallel insertion could have a measurable change on the ablation size and volume for our setup. There was up to a 30% variability for some of the more severe cases of non-parallel insertions studied. We also found that the rate of change for ablation size does vary considerably and did not vary linearly with the distance of the tip displacement from the original parallel position.

The model we used for simulating ablation was relatively accurate at predicting the shape of the ablation zone for our experiment and was also capable of showing how the ablation shape would change given a small tip offset distance. The amount of similarity between the parallel and non-parallel insertions did vary when comparing the simulation and the experimentation, possibly due to unaccounted tissue shrinkage in the simulations.

Acknowledgments This research was funded in part through NIH grant R01 CA 218357.

References

1. Brace, C. L., Laeseke, P. F., Sampson, L. A., Frey, T. M., van der Weide, D. W., & Lee, F. T. (2007). Microwave ablation with multiple simultaneously powered small-gauge triaxial antennas: Results from an in vivo swine liver model. *Radiology, 244*(1), 151–156.
2. Amabile, C., et al. (Feb. 2017). Microwave ablation of primary and secondary liver tumours: Ex vivo, in vivo, and clinical characterisation. *International Journal of Hyperthermia: the Official Journal of European Society for Hyperthermic Oncology, North American Hyperthermia Group, 33*(1), 34–42.
3. Mukherjee, S., Curto, S., Albin, N., Natarajan, B., & Prakash, P. (2015). *Multiple-antenna microwave ablation: Analysis of non-parallel antenna implants*. Presented at the Energy-based Treatment of Tissue and Assessment VIII, vol. 9326, p. 93260U.
4. Clibbon, K. L., & McCowen, A. (1993). Thermal modelling of nonideal interstitial microwave antenna array hyperthermia for the treatment of cancer. In *1993 IEEE MTT-S International Microwave Symposium Digest*, Atlanta, pp. 1147–1150.
5. Deshazer, G., Hagmann, M., Merck, D., Sebek, J., Moore, K. B., & Prakash, P. (2017). Computational modeling of 915 MHz microwave ablation: Comparative assessment of temperature-dependent tissue dielectric models. *Medical Physics, 44*, 4859.
6. Curto, S., Taj-Eldin, M., Fairchild, D., & Prakash, P. (2015). Microwave ablation at 915 MHz vs 2.45 GHz: A theoretical and experimental investigation: Microwave ablation at 915 MHz vs 2.45 GHz. *Medical Physics, 42*(11), 6152–6161.
7. Ji, Z., & Brace, C. L. (2011). Expanded modeling of temperature-dependent dielectric properties for microwave thermal ablation. *Physics in Medicine and Biology, 56*(16), 5249–5264.
8. IT'IS Foundation. (2015). "Tissue Properties Database V3.0." IT'IS Foundation.
9. White, A. (2018). Effect of non-parallel applicator insertion on microwave ablation zone size and shape.

Part IV
Mesh Construction, Manipulation and Material Augmentation

Chapter 17
A Robust Algorithm for Voxel-to-Polygon Mesh Phantom Conversion

Justin L. Brown, Takuya Furuta, and Wesley E. Bolch

17.1 Introduction

Since their early development in the late 1950s, general-purpose Monte Carlo (MC) radiation transport codes have utilized primitive geometric structures to define material interfaces in their transport geometry, e.g., planes, spheres, ellipsoids, and truncated cones. These structures were used from the 1960s to mid-1980s to geometrically represent the human body in both its outer body contour and internal organ structure. Their geometric simplicity was ideal for the limited computer technology at the time, and addressed the need for computational efficiency in particle tracking. These "stylized" phantoms, while at the time fit for purpose, did not provide an anatomically realistic representation of the human body, particularly in regard to organ shape and inter-organ tissue separation.

Beginning in the late 1980s, the need for improved anatomical accuracy, along with concurrent advances in computational memory and processor speed, led to the subsequent development and use of voxel-based human computational phantoms. Those phantoms were defined by a collection of rectangular parallelepipeds (voxels) of equal or non-equal size defining each tissue material. Voxel phantoms originate from the segmentation of CT or MR image data sets. Consequently, all tissue elements within a voxel phantom are generally of uniform size and shape (x,y,z dimen-

J. L. Brown
Medical Physics Graduate Program, University of Florida, Gainesville, FL, USA

T. Furuta
Nuclear Science and Engineering Centre, Japan Atomic Energy Agency, Tokai, Ibaraki, Japan

W. E. Bolch (✉)
Medical Physics Graduate Program, University of Florida, Gainesville, FL, USA

J. Crayton Pruitt Family Department of Biomedical Engineering, University of Florida, Gainesville, FL, USA
e-mail: wbolch@ufl.edu

S. Makarov et al. (eds.), *Brain and Human Body Modeling*,
https://doi.org/10.1007/978-3-030-21293-3_17

sions). The transition from stylized to voxel phantoms necessitated an increase in computational steps during radiation transport, as boundary crossing checks shifted from those associated with entering or leaving an organ or body region to those associated with entering or leaving each voxel defining that organ or body region.

Beginning in the early 2000s, a third generation of human computational phantom – mesh phantoms – was advanced, in which body regions and internal organ structures were once again represented, not by a collection of voxels, but by surfaces defined by 3D control points (non-uniform rational B-splines or NURBS) or arrays of polygons. These mesh-type phantoms allowed for the scalability and deformability provided by stylized phantoms, yet they retained the anatomical realism of voxel phantoms. In the coupling of mesh phantoms to radiation transport codes, however, a final step of voxelization had to be performed as the particle tracking algorithms employed at that time did not recognize NURBS or polygon mesh surfaces. Mesh phantom voxelization thus entailed filling these surfaces with an array of voxels of user-defined dimensions. A second advantage of mesh phantom voxelization was a resolution of potential surface overlaps and intersections introduced during phantom construction, rescaling, and/or deformation. The voxelization processes, by definition, eliminated these tissue incongruencies. Within the past few years, however, significant advances have been made in particle tracking algorithms so as to now enable the direct use of meshed geometries during MC radiation transport simulation. These developments were initially introduced into the MCNP code in 2009 [1], into the GEANT4 code in 2013 [2], and into the PHITS code in 2015 [3, 4]. Thus, there are a tremendous number of existing voxel-based computational phantoms that would now benefit from a conversion to mesh-type format.

This chapter reviews a computational algorithm developed to convert voxel phantoms to polygon mesh phantoms suitable for MC transport and importable into modern CAD software. The method eliminates geometric redundancies, allowing for a minimal and optimized geometric representation of the meshed structures. This feature is beneficial for computational human phantoms as voxel size is typically governed by the smallest anatomical structure to be represented, while a mesh phantom is not limited in this respect. The resulting algorithm allows users to continue to use the significant number of existing voxel phantoms that have been developed over the past 20 years without the need for labor-intensive manual modification. Additionally, the algorithm can be used with segmented image data to form mesh geometries free of intersections and incongruences so as to be used in simulation or CAD software.

17.2 Materials and Methods

17.2.1 Voxel to Mesh Conversion Procedure

The voxel-to-mesh conversion procedure is divided into six main steps: (1) data preparation, (2) gridded surface generation, (3) surface simplification, (4) line simplification, (5) polygon detection, and (6) polygon correction. The details of each step are briefly described in this section, which also includes a discussion of the benchmarking procedures used to evaluate the conversion process.

Data Preparation

First, the phantom voxels are read into a three-dimensional array of specified size $<n_x, n_y, n_z>$. Next, two additional four-dimensional arrays are created of dimensions $<n_x, n_y, 3, 12>$ and $<n_x, n_y, 3, 8>$ which represent sliding windows of temporary data used to ensure the uniqueness of every facet, vertex, and line that is generated in the newly created mesh phantom. The guarantee of element uniqueness is important to minimizing subsequent memory requirements during the handling of arbitrarily large arrays. It is important to note that the z-axis is chosen to be 3 units wide as this is typically the dimension along the phantom's cranial-caudal (and longest) direction this is chosen to minimize memory requirements. The array is 3 units wide as only adjacent z-slices of voxels can possibly contain information relevant to the current voxel. Several other arrays are also generated:

- The vertex array – an array of 3D points
- The line array – an array containing two integers representing two connected vertices within the vertex array
- The facet array – an array containing arbitrary numbers of integers representing connected lines within the line array
- The facet tag array – an array containing the ordered materials which separate the facets.

Gridded Surface Generation

The voxel data is parsed after the data is initialized. Each voxel is checked to determine if neighboring voxels are of a different material from the current voxel. If the neighboring voxels are of the same material, nothing is generated. If a neighboring voxel is found to be of a different material, the next step is to determine the facets to be produced. At this step, a facet is simply a rectangle between two voxels of different materials. As shown in Fig. 17.1, there is a possibility of 6 facets, 8 vertices, and 12 lines that could be produced for each voxel. Facets, vertices, and lines are produced depending on which adjacent voxels are of different materials. Given which neighbors are different materials, the required lines and vertices are determined. Once the required lines and vertices are determined, the sliding window of vertices and lines is checked to determine if this information already exists. If the data has already been generated, it is added to the current voxel position within the sliding window. If the data are not present, the data are generated and stored appropriately. The position of the vertices in 3D space is given by the required facets. At this point, a facet is composed of only four lines forming a rectangle. This process is repeated throughout the phantom array as the window is shifted along the longest axis through which it iterates. At this step, a surface mesh phantom has been

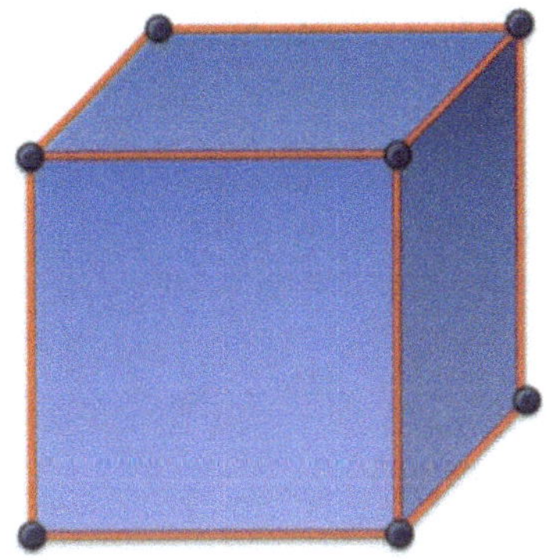

Fig. 17.1 Example of a single voxel and its potential 6 facets (blue), 8 vertices (black), and 12 lines (red)

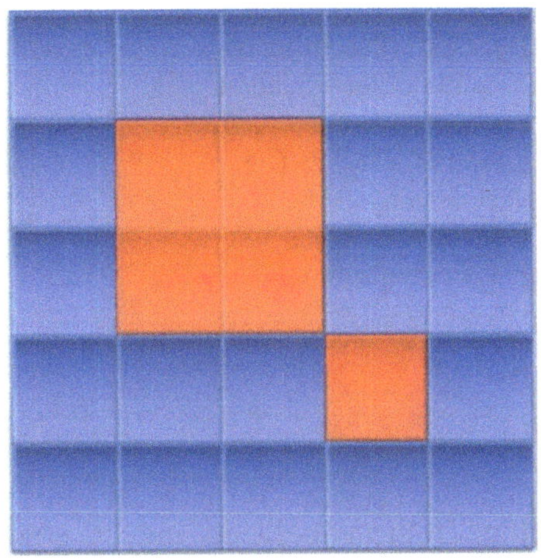

Fig. 17.2 A two-dimensional example of a voxelized surface after it has been converted to a gridded mesh. One material is depicted in blue and the other in orange

generated whose boundaries only differ between different materials (e.g., organs and tissue material of a given elemental composition and mass density). These boundaries are represented by a gridded surface which is further simplified and optimized as shown in Fig. 17.2.

Surface Simplification

The surface simplification process can begin once all necessary facets, vertices, and lines have been generated. Surfaces are first grouped by three values: (1) separated material, (2) whether or not x, y, or z is constant, and (3) the value of this constant. This grouping results in sets of surfaces which all separate the same material and are co-planar to one another (see Fig. 17.3). The purpose of this grouping is twofold. First, the grouping reduces the required computation time for the surface simplification step as comparisons only need to be made between grouped facets rather than across the entire list. Second, this grouping allows the surface simplification step to be performed in parallel.

The facets are then merged after grouping facets of the same material. A Boolean union operation is performed for every facet within each group. To determine if co-planar facets can be merged, the facets are checked to see if they share a common

Fig. 17.3 Illustration of the facet grouping procedure depicting separation of facets into coplanar groups of the same separated material

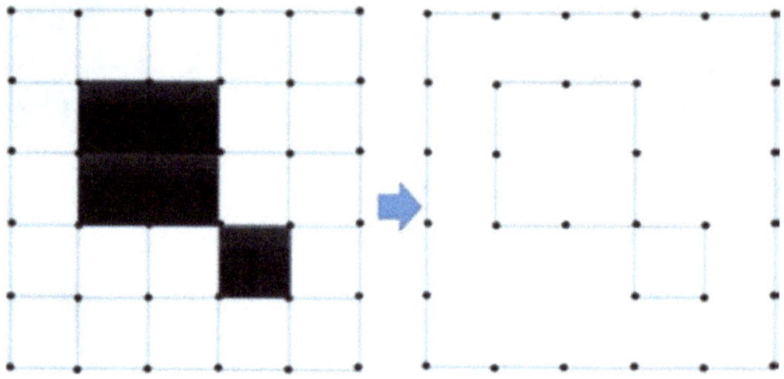

Fig. 17.4 Illustration of the surface simplification process for the blue surface group in Fig. 17.3. The black area on the left image depicts space no longer occupied by polygons

line. If the two facets indeed share one line, then they may be combined. The Boolean union process involves three sub-steps:

(i) The common line is determined and removed from both facet 1 and facet 2.
(ii) The remaining lines of facet 2 are added to facet 1.
(iii) Facet 2 is marked for removal.

This process is repeated until no more facets can be incorporated within each facet group. The facets marked for removal are then removed from the array. After this process is completed, unused lines and vertices are removed and facets and lines are updated to reflect new vertex and line positions within their respective arrays. At this point in the conversion process, the phantom surface mesh has been reduced to a minimal number of polygons, as shown in Fig. 17.4.

Line Simplification

After a minimum number of polygon surface representations have been generated, these polygons contain more lines than are necessary to enclose the required volume (e.g., organ or body region). Prior to simplifying the lines, they are grouped in a similar manner to the facets. First, lines are scanned iteratively to determine for every vertex how many lines use that vertex. Next, lines are subdivided into co-linear groups. This subdivision allows for the line simplification process to be performed in parallel and thus minimizes the required processing time needed since fewer comparisons need to be made.

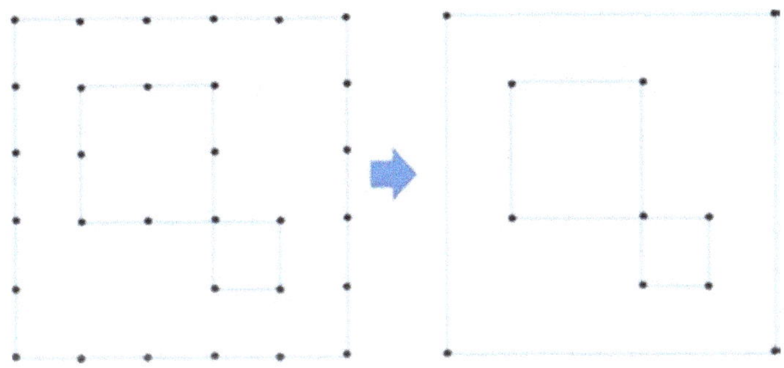

Fig. 17.5 Illustration of the line simplification process for the simplified surface group in Fig. 17.4

To simplify lines, each group of co-linear lines is scanned iteratively. Line pairs are flagged if both lines share a common vertex. If the lines share a vertex, a Boolean union operation can be performed if the vertex in common is only used by two lines globally within the phantom. If this is the case, the two lines and the vertex are not necessary to properly represent a given surface, and thus they can be removed without inducing a mesh overlap. The Boolean union process is performed for these two lines in a manner similar to that used for the facets:

(i) The common vertex shared by only line 1 and line 2 is determined.
(ii) The shared vertex in line 1 is replaced by the unshared vertex in line 2.
(iii) Line 2 is marked for removal.

This process is repeated until no additional lines can be incorporated within each group of lines. The lines marked for removal and all unused vertices are then removed. The line and facet arrays are then updated to reflect the new position of the vertices and lines in their respective arrays. At this stage, the mesh phantom is represented by the minimum possible number of surfaces and these surfaces are represented by the minimum possible number of lines as demonstrated in Fig. 17.5.

Polygon Detection

After these two simplification processes, the facets are now composed of an unordered set of lines and polygons. By construction, the lines contained within each facet must form at least one closed loop (i.e., a polygon). Within each facet, polygons are formed by simply end matching lines until all lines are used. If multiple polygons are formed by construction within one facet, one of these polygons must be interior to the other, thus forming a hole within the outer polygon. This is easily determined by a bounding box as demonstrated in Fig. 17.6. This process is repeated for each facet.

Polygon Correction and Hole Detection

Even though the technique described herein is computationally efficient, using an end-matching method to construct polygons can possibly create self-intersections within each polygon. These may occur because vertex repetition is not checked as each line is added to the polygon as it would result in a significant decrease in

Fig. 17.6 Illustration of the polygon detection process for the simplified set of lines in Fig. 17.5

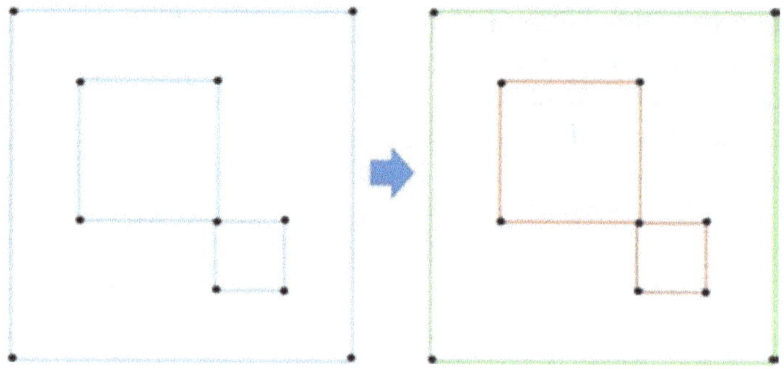

Fig. 17.7 Illustration of the polygon correction and hole detection process for the simplified set of lines in Fig. 17.6. Each color represents a polygon formed in the facet, with the black color illustrating the presence of a hole in the facet

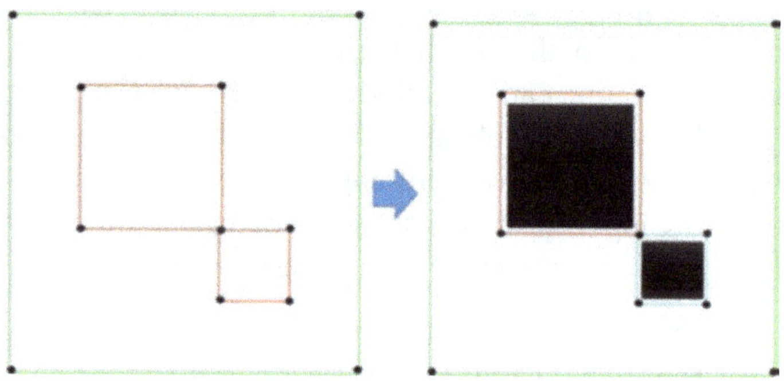

computational efficiency. Instead, after polygons have been created, they are checked to see if any vertices other than the start/end vertex have been used multiple times. An "ear-clipping" method is employed in this situation. To ear-clip a polygon, one creates a new polygon from the lines between the vertex that is used multiple times. These lines are then removed from the larger polygon and a new polygon is added to the facet as demonstrated in Fig. 17.7. At this point, the mesh is now an intersection-free and redundancy-free (all vertices, lines, and facets are unique) mesh that is represented by the least number of surfaces. If the surface-mesh phantom is to be converted to a tetrahedral-mesh phantom, as required by the PHITS radiation transport code, the open-source conversion code TETGEN [5] may be utilized. The mesh can also be triangularized and exported in a file format acceptable to most modern CAD software codes.

17.2.2 Conversion Process Benchmarking

In testing the performance of the voxel-to-mesh conversion algorithm, two benchmarking tasks were performed. First, it was important to test that the algorithm is robust and can handle arbitrary datasets correctly. Thus, a series of random square binary voxel arrays were generated and then meshed to contain between 10^3 and 10^8 elements. One example is shown in Fig. 17.8. Second, it was important that the conversion algorithm performed efficiently in a practical setting. Thus, mesh conversions were applied to the UF/NCI reference adult male phantom [6] at voxel resolutions ranging from 1 cm^3 to 1 mm^3 as shown in Fig. 17.9. Finally, it was

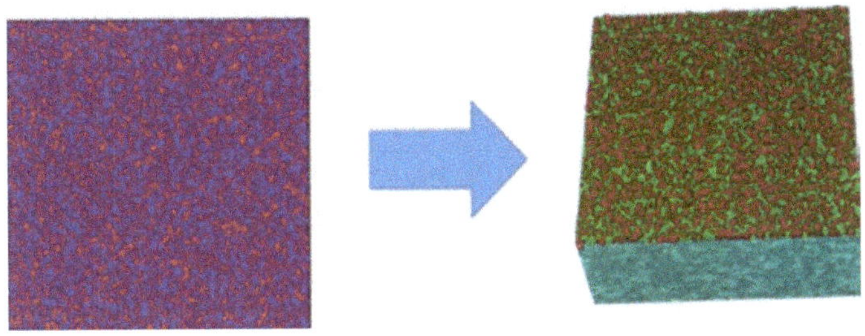

Fig. 17.8 Example of a 10^6 random binary voxel array (left) and its converted meshed format (right)

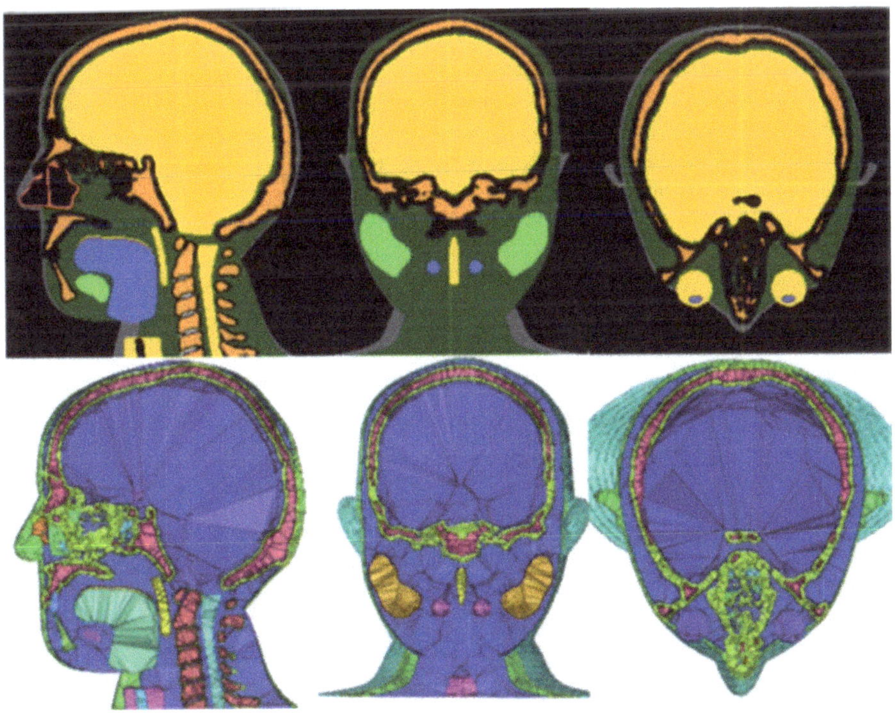

Fig. 17.9 Example of a voxel phantom (resolution of 1 mm^3) (top) converted into a mesh format (bottom)

important to assess how this conversion algorithm scales across multiple processors. Thus, the previous two benchmarking studies were performed using 1, 2, 4, 8, and 16 cores, respectively. All benchmarking tasks were run on the UF HiPerGator cluster using Intel E5-2698 v3 (2.3 GHz) processors. The code was compiled using Intel's C++ compiler with the –qopenmp and –O3 compiler flags.

17.3 Results

For the random array meshing benchmarks on a single core, the time to mesh for the highest resolution dataset (250 × 250 × 250) was 2.5 × 10^4 seconds, while the conversion time for the highest resolution head phantom was approximately 350 seconds.

Looking at the time breakdown for each step in the meshing algorithm, the majority of the compute time is devoted to the surface simplification step (see Sect. 2.1.3), which is expected as this step iteratively compares facets to one another causing this portion of the algorithm to have an order of n^2 performance (see Fig. 17.10). For the voxel-to-mesh phantom conversion, a more linear performance is seen, but this is

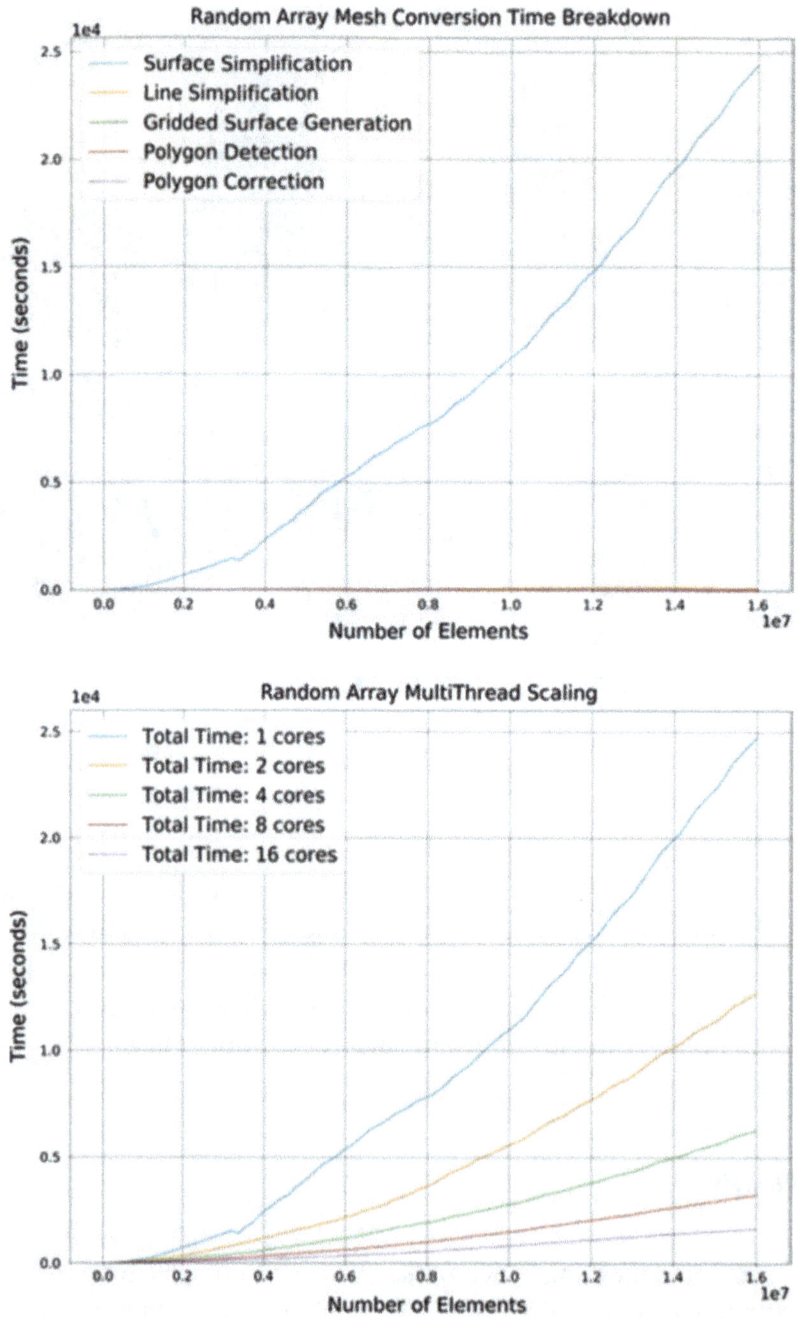

Fig. 17.10 Random array meshing time (units: 10^4 s) results per step (units: 10^7 steps) (top) and per multiprocessor scaling (bottom)

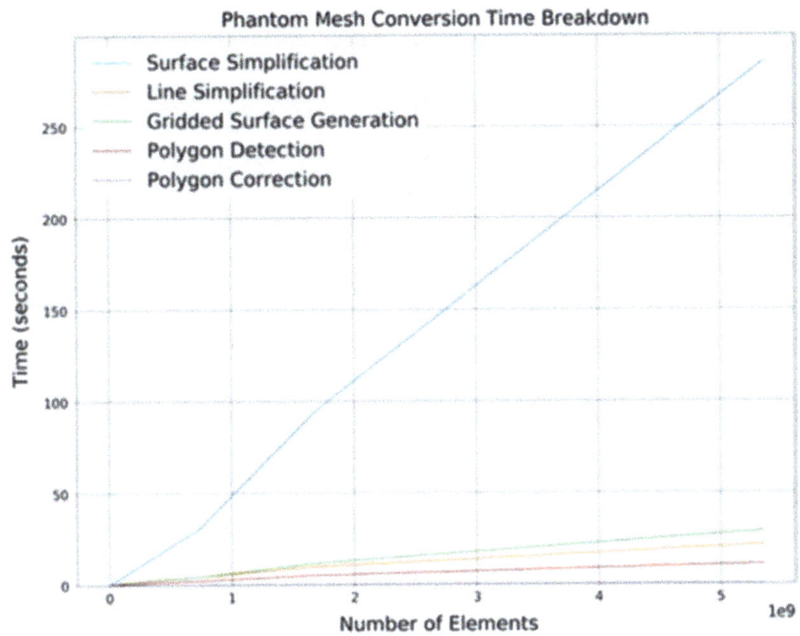

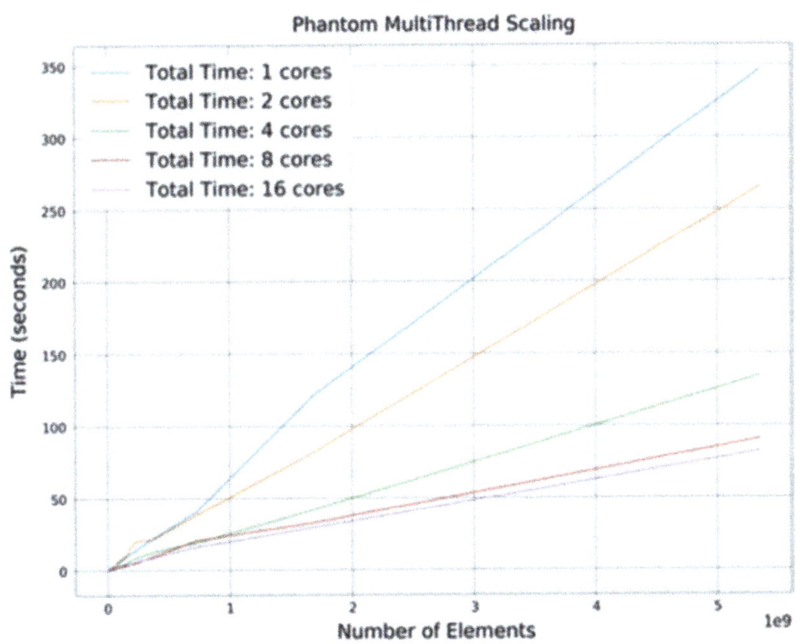

Fig. 17.11 Voxel phantom meshing time (units: s) results per step (units: 10^9 steps) (top) and per multiprocessor scaling (bottom)

likely due to the less randomized nature of the problem (see Fig. 17.11). Across multiple processors, both benchmarks saw performance gains although, as expected, they are not linear. The voxel-to-mesh phantom conversion speedup for 16 cores was approximately a factor of 4.1, whereas for 8 cores it was only a factor of 3.7. The diminishing returns are likely due to the implementation of OpenMP scheduling. The process can be better optimized in future development of this algorithm.

17.4 Conclusions

The presented methodology provides a fast and efficient method to convert voxel data to a polygon mesh format, containing no degenerate facets and no self-intersections, thus making it useful for input to Monte Carlo sampling codes and CAD programs. The algorithm can convert any segmented set of voxelized data to an optimized meshed surface suitable for a variety of applications such as Monte Carlo radiation transport or finite element simulations of the interactions between electromagnetic fields and the human body, e.g., during MRI.

Acknowledgments This work was supported in part by Contracts T72472 and T73057 with Wyle Laboratories and NASA Johnson Space Center, and grant R01 EB013558 with the National Cancer Institute.

References

1. Werner, C. J., et al. (2018). Title: MCNP Version 6.2 Release Notes. In *LA-UR-18-20808* (pp. 1–39).
2. Geant4 Collaboration 2017 Introduction to GEANT4, Release 10.4. https://geant4.web.cern.ch/support/user_documentation.
3. Sato, T., et al. (2018). Features of Particle and Heavy Ion Transport code System (PHITS) version 3.02. *Journal of Nuclear Science and Technology, 55*(6), 684–690.
4. Furuta, T., et al. (2017). Implementation of tetrahedral-mesh geometry in Monte Carlo radiation transport code PHITS. *Physics in Medicine and Biology, 62*(12), 4798–4810.
5. Si, H. (2015). TetGen, a quality tetrahedral mesh generator. *ACM Transactions on Mathematical Software, 41*(2), 11.
6. Lee, C., Lodwick, D., Hurtado, J., Pafundi, D., Williams, J. L., & Bolch, W. E. (2010). The UF family of reference hybrid phantoms for computational radiation dosimetry. *Physics in Medicine and Biology, 55*(2), 339–363.

Chapter 18
FEM Human Body Model with Embedded Respiratory Cycles for Antenna and E&M Simulations

Anh Le Tran, Gregory Noetscher, Sara Louie, Alexander Prokop, Ara Nazarian, and Sergey Makarov

18.1 Background

Human respiration is the exchange of air between the lungs and the ambient atmosphere. Below, we briefly summarize some major facts pertinent to our study.

Mechanics. Respiratory mechanics represent a complex multi-object deformation process. It predominantly involves the non-rigid motion of the (i) diaphragm; (ii) thoracic cage including ribs, cartilage, and sternum; (iii) lungs; (iv) heart; (v) liver; (vi) kidneys; and (vii) intestine. For inhalation, the diaphragm contracts and pushes the contents of the abdomen in the inferior direction as shown in Figs. 18.1 [1] and 18.2 [2]. Simultaneously, the external intercostal muscles expand the rib cage and slightly raise it. For exhalation, the diaphragm and the external intercostal muscles relax.

A. Le Tran
ECE Department, Worcester Polytechnic Institute, Worcester, MA, USA

G. Noetscher (✉)
Worcester Polytechnic Institute, Worcester, MA, USA
e-mail: gregn@wpi.edu

S. Louie
ANSYS, Inc., Canonsburg, PA, USA

A. Prokop
CST-Computer Simulation Technology AG, Darmstadt, Germany

A. Nazarian
Center for Advanced Orthopaedic Studies at Beth Israel Deaconess Medical Center, Harvard Medicine School, Boston, MA, USA

S. Makarov
Massachusetts General Hospital, Boston, MA, USA

Worcester Polytechnic Institute, Worcester, MA, USA

S. Makarov et al. (eds.), *Brain and Human Body Modeling*,
https://doi.org/10.1007/978-3-030-21293-3_18

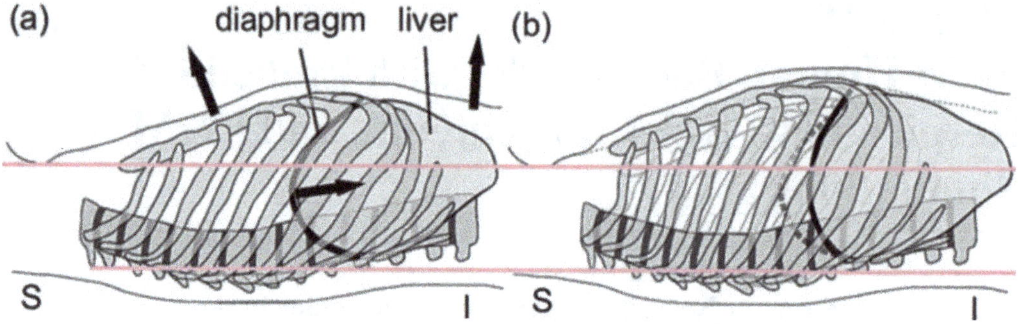

Fig. 18.1 (**a**) Maximum exhalation position, (**b**) maximum inhalation position, after [1]

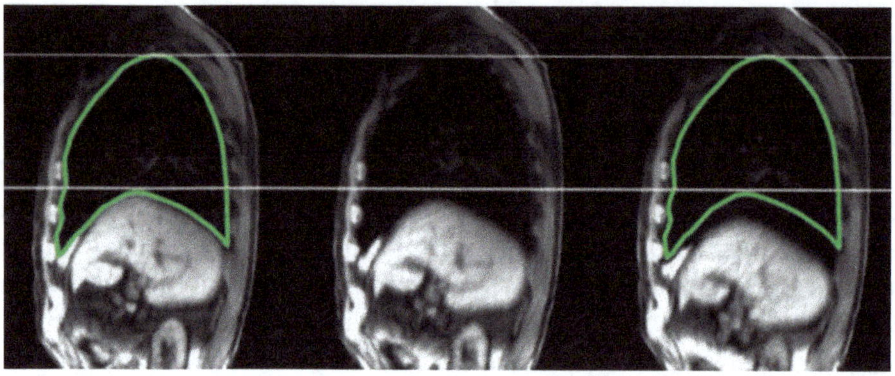

Fig. 18.2 Respiratory motion captured via MRI retrospective gating and averaging over multiple cycles, after [2]. The green contour indicates lung volume at maximum exhalation

Diaphragm motion. Respiration is chiefly driven by the diaphragm with primary motion in the superior-inferior direction; total travel is estimated as 10–30 mm during quiet breathing [1]. Other studies report 20 ± 7.0 mm average [2]. A simplified 1D diaphragm motion, $x(t)$, is non-harmonic, and the exhalation portion dominates the inhalation. Given the exhalation at origin, one has

$$x(t) = -A\cos^4 \omega t \tag{18.1}$$

where A is the corresponding amplitude [3, 4]. Furthermore, the respiratory motion often exhibits hysteresis in space, with an amplitude on the order of 2–4 mm [1].

Adjacent tissues. Closely adjacent structures (i.e., liver, etc.) show comparable motion amplitudes. Furthermore, the following motion amplitudes have been observed (cf. a review in Ref. [1]):

- Motion with an average amplitude of 12 mm in the lung for targets not attached to rigid structures
- 1–25 mm superior-inferior motion of the kidneys, 13 mm superior-inferior motion of the spleen, 2–8 mm motion of the heart (the heart motion is mostly a simple rigid-body translation [5, 6]), and 1–7 mm motion of the trachea
- 13 mm superior-inferior motion of the spleen

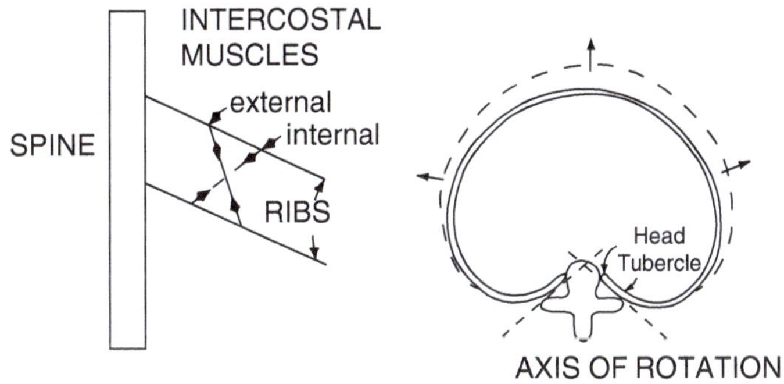

Fig. 18.3 Motion of the ribs during respiration, after [5, 7]. The ribs rotate about an axis through their costal neck

Thoracic cage kinematics. During respiration, the ribs rotate about an axis through their costal necks to affect the anteroposterior and transverse diameters of the thoracic cavity as shown in Fig. 18.3 [5, 7].

CAD B-Spline modeling. Modeling of the breathing cycle to date has been mostly performed via deformable NURBS surfaces (B-splines) for the lungs and surrounding tissues. The changes the phantoms undergo are then typically splined over time to create time continuous 4D respiratory models [5, 8, 9], which indeed utilize free-form deformations.

Challenges of FEM CAD Modeling. Commercial FEM codes do not operate with B-spline surfaces but rather with triangulated surfaces and tetrahedral/hexahedral volumes. This is in particular valid for most frequency-domain electromagnetics solvers such as ANSYS EM Suite/Maxell 3D and CST Microwave Studio. Therefore, a free-form breathing sequence has to be ultimately converted to a (large) discrete series of separate (full-body) triangulated CAD models, even if the original data were in the form of parametric B-splines. Generally, a conversion from NURBS surfaces to FEM triangular surfaces requires very significant additional meshing times.

The size of one detailed FEM full-body model is quite large (about 200–1000 Mbytes in ANSYS) and a computation with 20–30 such models would be a significant challenge from several points of view. For example, a user will need to create, run, and then post-process a number of large distinct project files, each of which must replicate his/her own excitation setup (e.g., a coil, an antenna, or a radar) and employ a new human model. Furthermore, manual repositioning is necessary for any and all on-body and in-body devices at every step, which would potentially create errors.

18.2 Methods

Built-in affine transformations. A commercial CEM package typically includes a set of nine affine transformations:

Three translations (in the x, y, z directions)
Three rotations (about the x, y, z axes)
Three directional scaling transformations (along the x, y, z axes)

applicable to any object (including a triangular tissue mesh) or to a group of objects and in the form of a parametric sweep. These transformations can be performed in either global or local coordinate systems. The user can initialize a discrete generic global variable, x_n, $n = 0,\ldots,N$, define object geometry parameters as certain unique functions of x_n, and then move, rotate, or deform every object of a multi-object structure independently within the framework of the same project file.

Our approach. We apply built-in parameterized affine transformations to construct breathing cycles (quiet, deep, shallow) using only one base full-body human model [10] source not found and using only one project file. Along with the base static human CAD model, this project file includes a parametric sweep or sweeps modeling deformations of involved tissues. Such an approach is not exact, but it may have sufficient accuracy when the parametric sweep is carefully designed. It will allow us to employ any temporal resolution, which is impossible with discrete models. To construct an anatomically relevant breathing cycle, we will try to follow the anatomical data collected from Refs. [1–9] as close as possible.

Challenges. To design an FEM-compatible and anatomically justified multi-tissue affine parametric sweep, a very extensive preprocessing of the static human CAD model is necessary, which is a significant undertaking.

18.2.1 Selecting a Sweeping Variable

The natural sweeping variable x_n is proportional to *the diaphragm motion*. Since the breathing cycle is periodic, only one period T must be considered. According to Eq. (18.1), physical time, t, is expressed through a sweep variable by

$$t = T\left\{\frac{1}{2} - \frac{1}{\pi} a\cos\left(\sqrt[4]{\frac{x_n}{N}}\right)\right\} \text{ when } 0 \le t \le T/2 \tag{18.2}$$

This result can be programmed in MATLAB as

```
E = 11; t_=0:E; T = 1; t = T*(pi/2-acos((t_/E).^0.25))/pi; plot(t_, t, '-*'); grid on.
```

Table 18.1 gives the corresponding numerical time values. Sweeping variable x_n runs from zero to $N = 11$ in 12 uniform steps. Its zero value corresponds to maximum exhalation; its maximum value of 11 corresponds to maximum inhalation. Higher N values can be considered for a better accuracy.

18.2.2 Static CAD Model

As a base human model at maximum exhalation, we will choose the VHP-Female v.3.1 CAD model (http://www.nevaem.com/) shown in Fig. 18.4. The source data for this model was provided by the National Library of Medicine's Visible Human Project in the form of full color cryosection images. These images were hand segmented and registered in a global coordinate frame. The model has 26 individual tissues, 270 individual tissue parts, major blood vessels and peripheral nerves, and

Table 18.1 Time values in terms of period, T, corresponding to the sweeping variable x_n, $n = 0, \ldots, N$ for $N = 11$

x_n	t/T
0	0.0000
1	0.1850
2	0.2265
3	0.2571
4	0.2830
5	0.3066
6	0.3292
7	0.3515
8	0.3747
9	0.4000
10	0.4308
11	0.5000
10	0.5692
9	0.6000
8	0.6253
7	0.6484
6	0.6708
5	0.6934
4	0.7170
3	0.7429
2	0.7735
1	0.8150
0	1.0000

a superior resolution in the spinal cord/cranium. All tissue structures are manifold shells and no shell intersects with any neighboring shell. The sweep for the respiratory motion will be implemented for both BASE and SMOOTH sub-models. Only the results for the BASE sub-model will be reported here.

The subject is a ~60-year-old white female with a height h of 162 cm measured from top of the scalp to the average center of both heels. The body mass M, computed using standard tissue densities [11] and assigning the average body shell, which includes internal tissues, the density of muscle, is ~88 kg. The computed BMI is ~33.5 (moderately obese). The subject has a heart pathology.

18.2.3 Respiratory Cycle and CAD Tissues Affected by Respiration Motion

The overall change in lung volume is set at 0.32 L, which is close to a normal-to-shallow breathing sequence for this subject. Default temporal resolution includes *12 discrete uniform steps* from 0 to 11 in steps of 1 from maximum exhalation to

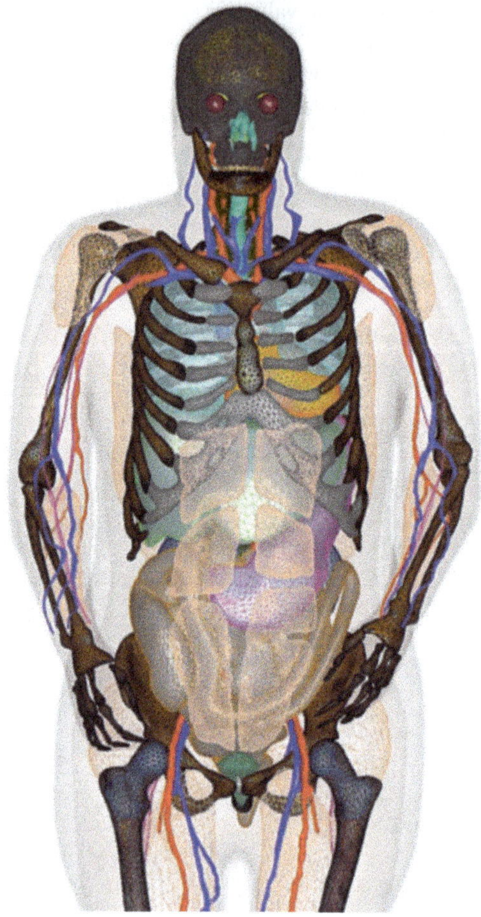

Fig. 18.4 Static VHP-Female v.3.1 CAD model at maximum exhalation (http://www.nevaem.com/)

maximum inhalation. The default full cycle includes *23 discrete steps*. Breathing cycles with finer resolution may be trivially constructed.

We choose the following major set of tissue parts (35 in total) to be affected by the respiratory motion:

- Lungs
- Ribcage with 24 ribs (every rib is moved *independently*)
- Thoracic cage cartilage
- Sternum
- Pectoralis muscles (major/minor)
- Abdominal muscles
- Erector spinae muscles
- Heart (muscle)
- Liver
- Stomach
- Outer shell – average body
- Outer shell – skin

These objects are transformed so that there are no intersections between any of them at any time moment, *with the minimum deformation factors*. These transformations are to be performed in global or local coordinate systems.

18.2.4 Required Accuracy: Total Body Mass

Since the respiratory motion modeled with multiple deformed CAD objects is an approximation, a requirement should be made with regard to the total mass error. We will require that the maximum relative body mass variation shall not exceed 0.1% during the entire respiratory cycle.

18.2.5 Algorithm

Below, we briefly review suggested kinematics and dynamics for the individual tissues. All quantitative approximations and the final formulas are thoroughly described in Appendix A.

Lung dynamics This is the first deformation step described in detail in Appendix A. In a local coordinate system associated with the top of the lung, the lung is deformed in all three directions and is moved in one direction in order to guarantee the expected diaphragm movement of 20 mm and simultaneously the volume change of 0.32 L, while maintaining anatomically sound overall deformations.

Thoracic cage kinematics This is the second deformation step. Since the rotation axes in Fig. 18.3 are very loosely defined for the actual anatomical data, we have rotated each rib pair about a fixed axis passing through the heads of two ribs (the end parts closest to the spine). We have also rotated slightly the rib pairs about the vertical axis. Thus, every rib pair is subject to rotation about two axes. All permissible variations of rotation angles have been tested, for every rib pair, in order to satisfy two criteria: (i) avoid intersections with the lung and (ii) stay as close to the lung as possible.

Sternum/cartilage dynamics This is the next deformation step. The sternum is subject to a translation motion, without rotation. Fixed *control points* on its surface are introduced. Those control points, along with the rib tips, form lines, along which the corresponding cartilage parts will further be deformed (moved and expanded).

Muscles dynamics In this case, we apply rotations, movements, and slight deformations. The goal is to minimize overall movement while avoiding intersections with the thoracic cage.

Heart kinematics The heart is moved in two respective directions without rotations and deformations. The cardiac cycle is not considered.

Liver/stomach kinematics Liver and stomach are moved in two respective directions and are slightly deformed; see Appendix A.

Outer full-body shells This is the only case where we cannot apply affine transformations. However, we may apply Boolean operations with the tissue CAD objects. A number of deformed chests objects are created internally, and then they are united with the otherwise static full-body shells. This operation requires greater care since we have two very closely spaced (1 mm) body shells.

18.2.6 Polynomial Interpolation

After a discrete set of affine transformations has been established, this set was converted to polynomials applicable to any temporal resolution and reported in Appendix A. The polynomial approximations have been independently tested with a fine grid. As an example, Table 18.2 reports affine polynomial approximations for several muscles. Note that the dynamic variable t in Table 18.1 is not the actual time, but is proportional to the diaphragm motion $x(t)$ in Eq. (18.1).

18.3 Results

The corresponding full-body VHP-Female model with the embedded respiratory motion in the form of a parametric sweep described in Appendix A has been independently realized in

- ANSYS Electronics Desktop software package
- CST Studio Suite software package
- MATLAB

The maximum body mass variation during the entire respiratory cycle is 80 g, *which is less than 0.1% of the total body mass*. The parametric sweep may be adjusted/modified at any time in response to further anatomical evaluations and customer needs.

18.3.1 RF Test at 300 MHz

The problem geometry is shown in Fig. 18.5. An incident plane wave at 300 MHz has a horizontal polarization. The simulations have been performed in ANSYS HFSS with three adaptive mesh refinement passes and with the final meshes approaching 1 M tetrahedra.

Near field Figure 18.5 shows the near-field results at three observation points given a 1 V/m incident wave. The scattered field is plotted. In the illuminated zone, the co-pol near field data may vary by about 3% due to the respiratory motion. In the

Table 18.2 Affine transformations of some muscles (inhalation only) of the VHP-Female model

		Muscles	Polynomials of deformation factors (angles recorded in degrees)
Pectoralis minor (in local CS)	*Left*	Rot z	$-7.149e-5*t^6+0.00252*t^5-0.03393*t^4+0.2181*t^3-0.681*t^2$ $+1.406*t+0.005579$
		Move y	$0.0002042*t^6-0.007194*t^5+0.09695*t^4-0.6231*t^3+1.946*t^2$ $-4.016*t\ -0.01594$
		Scale y	$2.042e-7*t^6-7.194e-6*t^5+9.695e-5*t^4-0.0006*t^3+0.001946*t^2$ $-0.004*t+0.999984$
	Right	Rot z	$1.083e-05*t^6+8.348e-05*t^5-0.00957*t^4+0.1344*t^3-0.7021*t^2$ $+1.774*t+0.01398$
		Move y	$-3.095e-5*t^6-0.0002385*t^5+0.02734*t^4-0.3841*t^3+2.006*t^2$ $-5.067*t-0.03994$
		Scale y	$3.09e-8*t^6-2.385e-7*t^5+2.73e-5*t^4-0.00038*t^3+0.002*t^2-$ $0.005067*t+0.99996$
Pectoralis major (in local CS)	*Left*	Move z	$-0.00013*t^6+0.004709*t^5-0.0656*t^4+0.4366*t^3-1.388*t^2$ $+2.397*t-0.005142$
		Move y	$-0.05+0.0002525*t^6-0.00914*t^5+0.1272*t^4-0.8445*t^3+2.673*t^2$ $-4.721*t+0.01132$
		Scale y	$2.6e-7*t^6-9.418e-6*t^5+0.0001312*t^4-0.00087*t^3+0.0028*t^2$ $-0.0048*t+1.00001028$
	Right	Move z	$-4.642e-05*t^6+0.001553*t^5-0.02174*t^4+0.1655*t^3-0.6858*t^2$ $+1.801*t+0.02082$
		Move y	$9.284e-05*t^6-0.003107*t^5+0.04348*t^4-0.331*t^3+1.372*t^2$ $-3.602*t-0.04165$
		Scale y	$9.284e-8*t^6-3.107e-6*t^5+4.348e-5*t^4-0.00033*t^3+0.00137*t^2$ $-0.0036*t+0.99995835$

(continued)

Table 18.2 (continued)

		Muscles	Polynomials of deformation factors (angles recorded in degrees)
Erector Spinae (in local CS)	*Left*	Scale y	$8.987e-06*t^6-0.0003339*t^5+0.004839*t^4-0.03445*t^3+0.1242*t^2$ $-0.2474*t+0.998756$
		Scale x	$4.493e-06*t^6-0.0001669*t^5+0.00242*t^4-0.01723*t^3+0.06208*t^2$ $-0.1237*t+0.9993778$
		Move y	$-1.123e-05*t^6-0.0004174*t^5+0.006049*t^4-0.04307*t^3+0.1552*t^2$ $-0.3093*t-0.3093$
	Right	Scale y	$-0.02*t+1$
		Scale x	$-0.02*t+1$
		Move y	$-0.05*t$
Abdominal (in global CS)		Move z	$0.09091*t*10^{-3}$
		Move y	$-0.35*t*10^{-3}$

shadow zone, the corresponding variation is negligibly small. *Cross-polarization components may exhibit considerably larger relative near-field variations.*

RCS Figure 18.6 shows the monostatic radar cross section (RCS) of the heterogeneous breathing VHP-Female model during the respiratory cycle. The RCS variations are about 1%. More data may be acquired from the website www.nevaem.com.

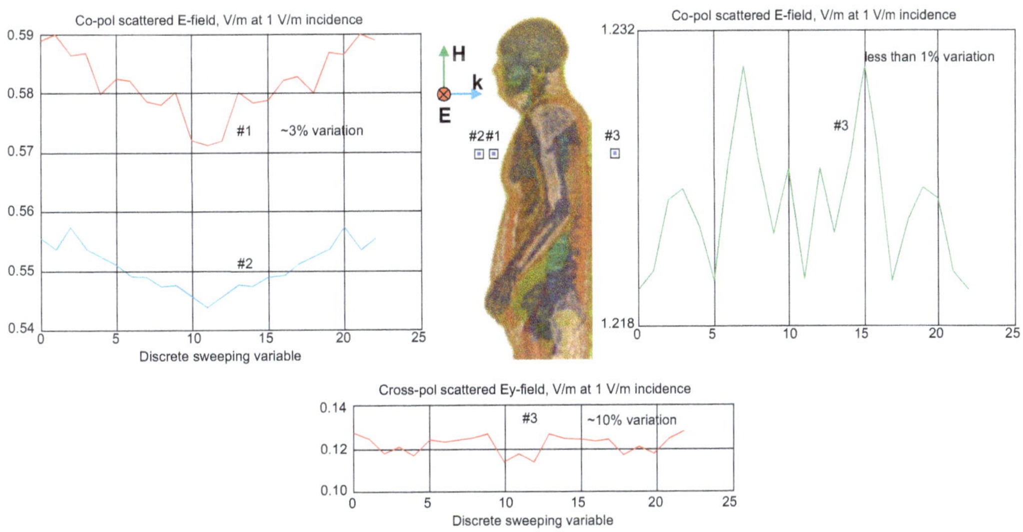

Fig. 18.5 Scattered field in the Fresnel region at 300 MHz

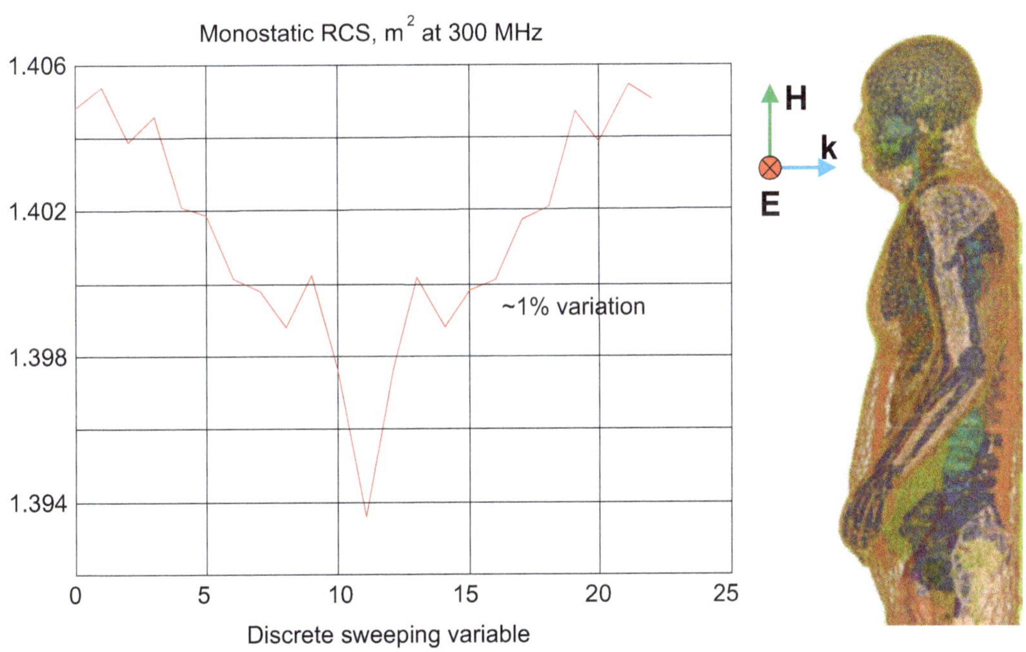

Fig. 18.6 Monostatic RCS during the respiratory cycle

Appendix A: Realization of the Respiratory Cycle for the VHP-Female CAD Model

Lung Deformation Sequence

New global coordinate system: Lung_CS. The origin is located at (0, max (Py), max (Pz)) with P being the point cloud of the lungs. The origin coordinates are given by

$$X = 0,\ Y = 122.8347,\ Z = -131.3727 \tag{18.A1}$$

Scaling in Lung_CS over N (N = 11) iterations total: Resulting Parametric Sweep in ANSYS

- 10% size increase in the z-direction: lung_scalez = $\left(1+(0.1/N)\right)^{t}$
- 1% size increase in the x-direction: lung_scalex = $\left(1+(0.01/N)\right)^{t}$
- 1% size increase in the y-direction: lung_scaley = $\left(1+(0.01/N)\right)^{t}$

Variable t (sweeping variable, not time!) is running from 0 to N. This will result in the overall volume change from 2.22 L to 2.54 L, i.e., 0.32 L. Other sequences may be constructed in a similar fashion.

Translation in **Lung _ CS** over N (N = 11) iterations total: Resulting parametric sweep in ANSYS

- 3 mm overall in the y-direction: lung_movey $= -\frac{3}{N} * t * 10^{-3}$ (m)

Rotation: None

Ribs Deformation Sequence

New global coordinate system: None

Scaling: None

Translation: None

Rotation: Every rib is rotated individually for a particular lung deformation so that there are no intersections between ribs and lungs given the minimum separation distance. Two rotation angles are used:

- Rotation about a rib axis, which is created by connecting two control points of two adjacent ribs closest to the vertebral column
- Rotation about the z-axis, in a new local CS, which is obtained by translation of the origin of the global CS to the rib control point(s) (individually for every rib)

Control points: Closest points to the vertebral column

Definition of rotation angles:

θ – Rotation angle about the rib axis, which results in an upward motion of the rib pair

φ – Rotation angle about the local z-axis, which results in an outward motion of the rib pair

See Tables 18.A1, 18.A2, and 18.A3

Sternum Deformation Sequence

New global coordinate system: **Sternum _ CS**. The origin is located at ((3∗(min (P*x*) + max (P*x*)/5), max (P*y*), (3∗max (P*z*)/5)) with P being the point cloud of the sternum. The origin coordinates are given by

$$X = 20.78,\ Y = -28.86,\ Z = -290.3$$

Scaling: None

Translation: None

Rotation in **Sternum _ CS**: one degree about the new global y-axis over $N\ (N = 11)$ iterations total.

Resulting parametric sweep in ANSYS

$$\text{sternum\_rot} = 0.09091 * \boldsymbol{t}.$$

Cartilage Deformation Sequence (Implemented in MATLAB)

New global coordinate system: None

Scaling: Two movement vectors are determined for every cartilage component at each iteration which will decide its scaling factor as follows:

$$New\ Movement\ Vector = \vec{m}_n$$

$$Old\ Movement\ Vector = \vec{m}_0$$

$$Scaling\ Factor = {|m_n|}\big/{|m_0|}$$

$$Scaling\ Vector = {-m_n}\big/{|m_n|}$$

Table 18.A1 Rib deformation sequence: table of computed control points (mm)

Coord./Rib Par#	1	2	3	4	5	6	7	8	9	10	11	12
X	44.690	43.810	43.42	44.11	39.57	32.07	30.61	34.9	34.31	35.91	46.5	54.13
	−27.49	−19.62	−17.15	−18.25	−13.45	−4.1	−1.296	−0.242	−1.062	−2.352	−10.33	−21.09
Y	58.37	80.38	91.95	104.3	107.7	104.5	106.3	103.3	102.7	106.3	112.5	109.3
	61.14	76.38	93.32	100.7	106.7	102.6	106.3	106.9	102.9	103	108	115.1
Z	−117.2	−131.4	−149.7	−167.3	−187.8	−215.3	−233.8	−260.5	−281.3	−306.4	−334	−367.3
	−121.7	−134	−156.6	−170	−192.2	−217.8	−232.7	−257	−279.9	−304.8	−334	−365.4

Table 18.A2 Rib deformation sequence: table of rotation angles (deg.) extracted from MATLAB

Iter. #/Rib pair #		1	2	3	4	5	6	7	8	9	10	11
$n_R = 1$	Θ	0.1667	0.3333	0.6667	0.8333	1.1667	1.5000	2.3333	3.0000	3.6667	4.3333	4.8333
	φ	0.0800	0.1600	0.3200	0.4000	0.5600	0.7200	1.1200	1.4400	1.7600	2.0800	2.3200
$n_R = 2$	Θ	0.6667	1.0000	1.3333	1.6667	2.0000	2.3333	2.6667	3.0000	3.3333	3.6667	4.0000
	φ	0.1600	0.2400	0.3200	0.4000	0.4800	0.5600	0.6400	0.7200	0.8000	0.8800	0.9600
$n_R = 3$	Θ	1.5000	2.0000	2.5000	3.0000	3.5000	4.0000	4.5000	5.0000	5.5000	6.0000	6.5000
	φ	0.2400	0.3200	0.4000	0.4800	0.5600	0.6400	0.7200	0.8000	0.8800	0.9600	1.0400
$n_R = 4$	Θ	0.6667	1.3333	2.0000	2.6667	3.3333	4.0000	4.6667	5.3333	6.0000	6.6667	7.3333
	φ	0.0800	0.1600	0.2400	0.3200	0.4000	0.4800	0.5600	0.6400	0.7200	0.8000	0.8800
$n_R = 5$	Θ	0.4167	0.6250	0.8333	1.0417	1.4583	1.8750	2.0833	2.5000	2.9167	3.1250	3.3333
	φ	0.8000	1.2000	1.6000	2.0000	2.8000	3.6000	4.0000	4.8000	5.6000	6.0000	6.4000
$n_R = 6$	Θ	0.2500	0.9583	1.6667	2.3750	3.0833	3.7917	4.5000	5.0000	5.2500	5.5000	5.7500
	φ	0.4000	1.5333	2.6667	3.8000	4.9333	6.0667	7.2000	8.0000	8.4000	8.8000	9.2000
$n_R = 7$	Θ	0.5833	0.8750	1.1667	1.4583	1.7500	2.0417	2.3333	2.6250	2.9167	3.2083	3.5000
	φ	0.8000	1.2000	1.6000	2.0000	2.4000	2.8000	3.2000	3.6000	4.0000	4.4000	4.8000
$n_R = 8$	Θ	0.6667	1.0000	1.3333	1.6667	2.0000	2.3333	2.6667	3.0000	3.3333	3.6667	4.0000
	φ	0.8000	1.2000	1.6000	2.0000	2.4000	2.8000	3.2000	3.6000	4.0000	4.4000	4.8000
$n_R = 9$	Θ	0.3750	0.7500	1.1250	1.5000	1.8750	2.2500	2.6250	3.0000	3.3750	3.7500	4.1250
	φ	0.4000	0.8000	1.2000	1.6000	2.0000	2.4000	2.8000	3.2000	3.6000	4.0000	4.4000
$n_R = 10$	Θ	0.4167	0.8333	1.2500	1.6667	2.0833	2.5000	2.9167	3.3333	3.7500	4.1667	4.5833
	φ	0.4000	0.8000	1.2000	1.6000	2.0000	2.4000	2.8000	3.2000	3.6000	4.0000	4.4000
$n_R = 11$	Θ	0.2292	0.4583	0.6875	0.9167	1.1458	1.3750	1.6042	1.8333	2.0625	3.6667	6.1875
	φ	0.2000	0.4000	0.6000	0.8000	1.0000	1.2000	1.4000	1.6000	1.8000	3.2000	5.4000
$n_R = 12$	Θ	0.5000	1.0000	1.5000	2.0000	2.5000	3.0000	3.5000	4.0000	4.5000	5.0000	5.5000
	φ	0.4000	0.8000	1.2000	1.6000	2.0000	2.4000	2.8000	3.2000	3.6000	4.0000	4.4000

Table 18.A3 Rib deformation sequence: table of rotation angles (deg.) extracted from MATLAB

Iter. #/Rib pair #		Polynomials of rotation angle (deg)
$n_R = 1$	Θ	$5.901e-05 * t^6 - 0.002019 * t^5 + 0.02502 * t^4 - 0.1346 * t^3 + 0.3259 * t^2 - 0.08703 * t + 0.006159$
	φ	$0.0001035 * t^6 - 0.00367 * t^5 + 0.0489 * t^4 - 0.3052 * t^3 + 0.9405 * t^2 - 1.114 * t + 0.5939$
$n_R = 2$	Θ	$0.3333*t$
	φ	$0.08*t$
$n_R = 3$	Θ	$0.5*t$
	φ	$0.08*t$
$n_R = 4$	Θ	$0.6667*t$
	φ	$0.08*t$
$n_R = 5$	Θ	$-1.844e-05 * t^6 + 0.0004494 * t^5 - 0.003782 * t^4 + 0.01514 * t^3 - 0.03083 * t^2 + 0.4792 * t - 0.001689$
	φ	$-3.54e-05 * t^6 + 0.0008629 * t^5 - 0.007261 * t^4 + 0.02908 * t^3 - 0.0592 * t^2 + 0.4488 * t - 0.003243$
$n_R = 6$	Θ	$0.45*t$
	φ	$0.4*t$
$n_R = 7$	Θ	$0.2917*t$
	φ	$0.4*t$
$n_R = 8$	Θ	$0.3333*t$
	φ	$0.4*t$
$n_R = 9$	Θ	$0.375*t$
	φ	$0.4*t$
$n_R = 10$	Θ	$0.4*t$
	φ	$0.4167*t$
$n_R = 11$	Θ	$2.986e-05 * t^6 - 0.00111 * t^5 + 0.01608 * t^4 - 0.1145 * t^3 + 0.4126 * t^2 - 0.09056 * t - 0.004135$
	φ	$2.156e-05 * t^6 - 0.000801 * t^5 + 0.01161 * t^4 - 0.08265 * t^3 + 0.2979 * t^2 + 0.01943 * t - 0.002985$
$n_R = 12$	Θ	$0.25*t$
	φ	$0.5 * t$

Translation: A translation vector determines the movement of the cartilage for every iteration, given by:

$$Translation\ Vector = \vec{m}_n - \vec{m}_0$$

Rotation: The rotation axis and the rotation degree are given by:

$$Rotation\ Vector - \cos^{-1}\left(\vec{m}_n \cdot \vec{m}_0\right) / \left(\left|\vec{m}_n\right| \times \left|\vec{m}_0\right|\right)$$

$$Rotation\ Axis = \vec{m}_0 \times \vec{m}_n$$

See Tables 18.A4 and 18.A5

Muscle Deformation Sequence (Pectoralis Major, Pectoralis Minor, Abdominal Muscles, and Erector Spinae)

New local coordinate systems: A local coordinate system is defined for each muscle using a simple translation. The origins are located at (min(Px), max (Py), max (Pz)) with P being the point cloud of each left muscle and (max(Px), max (Py), max (Pz)) of each right muscle. Abdominal muscles are only transformed with respect to the global coordinate system: the origin at (0, 0, 0)

Scaling in local **CS**s: See the following tables for individual muscles

Translation in local **CSs**: See the following tables for individual muscles

Rotation in local **CS**s: See the following tables for individual muscles

See Tables 18.A6, 18.A7, 18.A8, 18.A9, 18.A10, 18.A11, 18.A12, 18.A13, and 18.A14

Heart Deformation Sequence

New local coordinate systems: According to literature, the pumping motion of the heart is independent of breathing. As a result, the heart object will only be transformed to avoid intersection with lungs in breathing sequence, with respect to the origin of the global coordinate system (0, 0, 0). See Tables 18.A15 and 18.A16

Table 18.A4 Cartilage deformation sequence: table of rotation angles (radians)

Iter. #/Cartilage pair #	1	2	3	4	5	6	7	8	9	10	11
Left1	0	0	0.01445	0.01388	0.01334	0.03706	0.03272	0.07020	0.01471	0.00713	0.00692
Right1	0	0	0.01445	0.01388	0.01334	0.03706	0.03272	0.07020	0.01471	0.00713	0.00692
Left2	0	0.02217	0.02194	0.02169	0.02143	0.02114	0.02084	0.02053	0.02021	0.01988	0.01954
Right2	0	0.02287	0.02281	0.02272	0.02260	0.02246	0.02230	0.02211	0.02190	0.02167	0.02142
Left3	0	0.03000	0.02975	0.02945	0.02910	0.02870	0.02826	0.02779	0.02728	0.02674	0.02618
Right3	0	0.02900	0.02858	0.02813	0.02763	0.02710	0.02655	0.02597	0.02537	0.02476	0.02413
Left4	0	0.03671	0.03598	0.03517	0.03430	0.03338	0.03242	0.03142	0.03041	0.02939	0.02837
Right4	0	0.03461	0.03470	0.03470	0.03461	0.03443	0.03416	0.03381	0.03338	0.03288	0.03231
Left5	0	0.01392	0.01361	0.01331	0.02113	0.02019	0.01195	0.01883	0.01801	0.01073	0.01049
Right5	0	0.01605	0.01575	0.01544	0.02521	0.02417	0.01399	0.02263	0.02167	0.01262	0.01234
Left6	0	0.00617	0.00610	0.00603	0.00596	0.00590	0.00583	0.00577	0.00571	0.00565	0.00559
Right6	0	0.01445	0.01388	0.01334	0.03706	0.03272	0.07020	0.01471	0.00713	0.00692	0.00672

Table 18.A5 Cartilage deformation sequence: table of expansion factors

Iter. #/Cartilage pair #	1	2	3	4	5	6	7	8	9	10	11
Left1	0	0.002864	0.004915	0.003289	0.005533	0.005889	0.013425	0.012064	0.012891	0.013652	0.011283
Right1	0	0.003292	0.005705	0.003689	0.006288	0.006622	0.015209	0.013425	0.014189	0.014884	0.012137
Left2	0	0.005041	0.005525	0.005993	0.006444	0.006876	0.007290	0.007685	0.008061	0.008417	0.008753
Right2	0	0.001314	0.001875	0.002430	0.002978	0.003518	0.004046	0.004562	0.005064	0.005552	0.006023
Left3	0	0.003696	0.004616	0.005512	0.006378	0.007211	0.008008	0.008765	0.009481	0.010153	0.010781
Right3	0	0.006657	0.007496	0.008297	0.009057	0.009774	0.010446	0.011073	0.011653	0.012188	0.012677
Left4	0	0.009183	0.010484	0.011701	0.012829	0.013864	0.014804	0.015651	0.016405	0.017069	0.017648
Right4	0	0.001770	0.000510	0.000766	0.002033	0.003287	0.004517	0.005714	0.006871	0.007980	0.009035
Left5	0	0.010654	0.010767	0.010868	0.023283	0.023299	0.011074	0.023284	0.023185	0.011103	0.011113
Right5	0	0.009079	0.009300	0.009506	0.020776	0.021132	0.010164	0.021585	0.021758	0.010557	0.010633
Left6	0	0.007867	0.007894	0.007918	0.007940	0.007959	0.007977	0.007992	0.008006	0.008017	0.008027
Right6	0	0.019481	0.019443	0.019386	0.062856	0.061347	0.165538	0.035730	0.016683	0.016521	0.016358

Table 18.A6 Origin coordinates for the local coordinate systems

Muscle	Local *X* (mm)	Local *Y* (mm)	Local *Z* (mm)	0.008753
Pectoralis minor left	89.03	27.64	−201.93	0.006023
Pectoralis minor right	−56.28	29.22	−189.24	0.010781
Pectoralis major left	18.83	76.19	−192.42	0.012677
Pectoralis major right	14.82	−31.92	−197.71	0.017648
Erector spinae left	0	146.73	−450.22	0.009035
Erector spinae right	0	146.79	−453.65	0.011113

Table 18.A7 Deformation factors for pectoralis minor left muscle. All angles are recorded in degrees

Configuration number	Rotation about *Z* axis (*Θ*)	Movement in *Y* direction	Scaling in *Y* direction
1	±0.9545	−2.7273	−0.0027
2	±1.2727	−3.6364	−0.0036
3	±1.9091	−5.4545	−0.0055
4	±2.2273	−6.3636	−0.0064
5	±2.8636	−8.1818	−0.0082
6	±3.1818	−9.0909	−0.0091
7	±3.8182	−10.9091	−0.0109
8	±4.1364	−11.8182	−0.0118
9	±4.4545	−12.7273	−0.0127
10	±5.0909	−14.5455	−0.0145
11	±5.4091	−15.4545	−0.0155

Table 18.A8 Deformation factors for pectoralis minor right muscle. All angles are recorded in degrees

Configuration number	Rotation about *Z* axis (*Θ*)	Movement in *Y* direction	Scaling in *Y* direction
1	± 1.2727	−3.6364	−0.0036
2	± 1.5909	−4.5455	−0.0045
3	±1.9091	−5.4545	−0.0055
4	±2.2273	−6.3636	−0.0064
5	± 2.5455	−7.2727	−0.0073
6	±3.1818	−9.0909	−0.0091
7	±3.8182	−10.9091	−0.0109
8	± 4.4545	−12.7273	−0.0127
9	± 5.0909	−14.5455	−0.0145
10	± 5.4091	−15.4545	−0.0155
11	± 6.0455	−17.2727	−0.0173

Table 18.A9 Deformation factors for pectoralis major left muscle. All angles are recorded in degrees

Configuration number	Movement in Z direction	Movement in Y direction	Scaling in Y direction
1	1.3636	−2.7273	−0.0027
2	1.8182	−3.6364	−0.0036
3	2.2727	−4.5455	−0.0045
4	2.7273	−5.4545	−0.0055
5	3.6364	−7.2727	−0.0073
6	4.0909	−8.1818	−0.0082
7	5.0000	−10.0000	−0.0100
8	5.4545	−10.9091	−0.0109
9	5.9091	−11.8182	−0.0118
10	6.3636	−12.7273	−0.0127
11	7.2727	−14.5455	−0.0145

Table 18.A10 Deformation factors for pectoralis major right muscle. All angles are recorded in degrees

Configuration number	Movement in Z direction	Movement in Y direction	Scaling in Y direction
1	1.3636	−2.7273	−0.0027
2	1.8182	−3.6364	−0.0036
3	2.2727	−4.5455	−0.0045
4	2.7273	−5.4545	−0.0055
5	3.1818	−6.3636	−0.0064
6	3.6364	−7.2727	−0.0073
7	4.0909	−8.1818	−0.0082
8	5.0000	−10.0000	−0.0100
9	5.4545	−10.9091	−0.0109
10	5.9091	−11.8182	−0.0118
11	6.8182	−13.6364	−0.0136

Table 18.A11 Deformation factors for erector spinea left muscles. All angles are recorded in degrees

Configuration number	Scaling in Y direction	Scaling in X direction	Movement in Y direction
1	−0.1600	−0.0800	0.2000
2	−0.2000	−0.1000	0.2500
3	−0.2400	−0.1200	0.3000
4	−0.2800	−0.1400	0.3500
5	−0.3200	−0.1600	0.4000
6	−0.3600	−0.1800	0.4500
7	−0.4000	−0.2000	0.5000
8	−0.4400	−0.2200	0.5500
9	−0.4800	−0.2400	0.6000
10	−0.5200	−0.2600	0.6500
11	−0.5600	−0.2800	0.7000

Table 18.A12 Deformation factors for erector spinea right muscles. All angles are recorded in degrees

Configuration number	Scaling in *Y* direction	Scaling in *X* direction	Movement in *Y* direction
1	−0.0200	−0.0200	0.0500
2	−0.0400	−0.0400	0.1000
3	−0.0600	−0.0600	0.1500
4	−0.0800	−0.0800	0.2000
5	−0.1000	−0.1000	0.2500
6	−0.1200	−0.1200	0.3000
7	−0.1400	−0.1400	0.3500
8	−0.1600	−0.1600	0.4000
9	−0.1800	−0.1800	0.4500
10	−0.2000	−0.2000	0.5000
11	−0.2200	−0.2200	0.5500

Table 18.A13 Deformation factors for abdominal muscles in the global coordinate system. All angles are recorded in degrees

Configuration number	Movement in *Z* direction	Movement in *Y* direction
1	−0.0909	−0.0909
2	−0.1818	−0.1818
3	−0.2727	−0.2727
4	−0.3636	−0.3636
5	−0.4545	−0.4545
6	−0.5455	−0.5454
7	−0.6364	−0.6363
8	−0.7273	−0.7272
9	−0.8182	−0.8181
10	−0.9091	−0.9090
11	−1.0000	−1.0000

Liver Deformation Sequence

New local coordinate systems: The liver object is deformed to avoid intersection with lungs in breathing sequence, with respect to the origin of a local coordinate system: (0, max (Py), max (Pz)). See Tables 18.A17, 18.A18, and 18.A19

Table 18.A14 Muscle deformations: Polynomials of deformation factors

		Muscles	Polynomials of deformation factors (angles recorded in degrees)
Pectoralis minor (in local CS)	*Left*	Rot z	$-7.149e-5*t^6+0.00252*t^5-0.03393*t^4+0.2181*t^3-0.681*t^2$ $+1.406*t+0.005579$
		Move y	$0.0002042*t^6-0.007194*t^5+0.09695*t^4-0.6231*t^3+1.946*t^2$ $-4.016*t-0.01594$
		Scale y	$2.042e-7*t^6-7.194e-6*t^5+9.695e-5*t^4-0.0006*t^3+0.001946*t^2$ $-0.004*t+0.999984$
	Right	Rot z	$1.083e-05*t^6+8.348e-05*t^5-0.00957*t^4+0.1344*t^3-0.7021*t^2$ $+1.774*t+0.01398$
		Move y	$-3.095e-5*t^6-0.0002385*t^5+0.02734*t^4-0.3841*t^3+2.006*t^2$ $-5.067*t-0.03994$
		Scale y	$3.09e-8*t^6-2.385e-7*t^5+2.73e-5*t^4-0.00038*t^3+0.002*t^2-$ $0.005067*t+0.99996$
Pectoralis major (in local CS)	*Left*	Move z	$-0.00013*t^6+0.004709*t^5-0.0656*t^4+0.4366*t^3-1.388*t^2$ $+2.397*t-0.005142$
		Move y	$-0.05+0.0002525*t^6-0.00914*t^5+0.1272*t^4-0.8445*t^3+2.673*t^2$ $-4.721*t+0.01132$
		Scale y	$2.6e-7*t^6-9.418e-6*t^5+0.0001312*t^4-0.00087*t^3+0.0028*t^2$ $-0.0048*t+1.00001028$
	Right	Move z	$-4.642e-05*t^6+0.001553*t^5-0.02174*t^4+0.1655*t^3-0.6858*t^2$ $+1.801*t+0.02082$
		Move y	$9.284e-05*t^6-0.003107*t^5+0.04348*t^4-0.331*t^3+1.372*t^2$ $-3.602*t-0.04165$
		Scale y	$9.284e-8*t^6-3.107e-6*t^5+4.348e-5*t^4-0.00033*t^3+0.00137*t^2$ $-0.0036*t+0.99995835$

(continued)

Table 18.A14 (continued)

		Muscles	Polynomials of deformation factors (angles recorded in degrees)
Erector spinae (in local CS)	*Left*	Scale y	$8.987e-06*t^6-0.0003339*t^5+0.004839*t^4-0.03445*t^3+0.1242*t^2$ $-0.2474*t+0.998756$
		Scale x	$4.493e-06*t^6-0.0001669*t^5+0.00242*t^4-0.01723*t^3+0.06208*t^2$ $-0.1237*t+0.9993778$
		Move y	$-1.123e-05*t^6-0.0004174*t^5+0.006049*t^4-0.04307*t^3+0.1552*t^2$ $-0.3093*t-0.3093$
	Right	Scale y	$-0.02*t+1$
		Scale x	$-0.02*t+1$
		Move y	$-0.05*t$
Abdominal (in global CS)		Move z	$0.09091*t*10^{-3}$
		Move y	$-0.35*t*10^{-3}$

Table 18.A15 Deformation factors for the heart

Configuration number	Movement in *Z* direction	Movement in *Y* direction
1	−0.15	−0.05
2	−0.3	−0.1
3	−0.45	−0.15
4	−1.5	−0.5
5	−2.85	−0.95
6	−4.35	−1.45
7	−5.7	−1.9
8	−7.2	−2.4
9	−8.7	−2.9
10	−10.05	−3.35
11	−11.55	−3.85

Table 18.A16 Heart deformations: polynomials of deformation

Heart	Polynomials of deformation factors (angles recorded in degrees)
Move z	$-6.672e-06 * t^6 + 0.0008203 * t^5 - 0.02038 * t^4 + 0.2019 * t^3 - 0.917 * t^2 + 0.7346 * t - 0.03539$
Move y	$2.451e-06 * t^6 - 8.201e-05 * t^5 + 0.0009 * t^4 - 0.0015 * t^3 - 0.04611 * t^2 + 0.03447 * t - 0.0046$

Table 18.A17 Local coordinate system: liver

	Local *X* (mm)	Local *Y* (mm)	Local *Z* (mm)
Liver	0	120.136	−373.331

Table 18.A18 Deformation factors for the liver

Configuration number	Movement in Z direction	Movement in Y direction	Scale in Z direction
1	−0.18	−0.04	−0.001
2	−0.36	−0.8	−0.002
3	−1.44	−0.32	−0.008
4	−2.52	−0.56	−0.014
5	−3.6	−0.8	−0.020
6	−5.04	−1.12	−0.028
7	−6.48	−1.44	−0.036
8	−7.92	−1.76	−0.044
9	−9.36	−2.08	−0.052
10	−10.8	−2.4	−0.060
11	−12.24	−2.72	−0.068

Table 18.A19 Liver deformations: polynomials of deformation

Heart		Polynomials of deformation factors (angles recorded in degrees)
Move	Z	$-6.672e-06*t^6+0.0008203*t^5-0.02038*t^4+0.2019*t^3-0.917*t^2+0.7346*t-0.03539$
Move	Y	$2.451e-06*t^6-8.2e-05*t^5+0.0009106*t^4-0.001488*t^3-0.0461*t^2+0.0344*t-0.004638$
Scale	Z	$6.1e-8*t^6-2.05e-6*t^5+2.7e-5*t^4-3.7e-5*t^3-0.00115*t^2+0.00086*t+0.99989$

Skin Shell Deformation

First, the skin shell deformation starts with a generation of N chest objects for each step via non-rigid transformations. This process was accomplished in MATLAB. A deformed chest region is defined as

$$141.3\,\text{mm} < P(:,1) < 173.4\,\text{mm}$$

$$P(:,2) < 64\,\text{mm}$$

$$-330.6\,\text{mm} < P(:,3) < -150.6\,\text{mm}$$

All nodes in the chest region of the skin shell are selected and transformed in the y-direction using the following equation:

$$P(:,2) = Pbase(:,2) - \frac{10}{N} * t * \sin\left(P(:,3) - \min\left(P(:,3)\right)\right)$$

We chose nodes in the chest region so that $P(:,3) - \min(P(:,3))$ goes from 180 to 0. Therefore, nodes that are closer to the upper and lower boundaries of the region will move less than the nodes that are closer to the center. With maximum inhalation, the center node of the chest region will move by 10 mm in the Y direction. Thus, only coordinates of nodes belonging to the chest area are changed. Also, the connectivity matrix, t, of the entire skin shell still remains the same. As a result, 11 skin shell objects with different chest regions will be generated.

Second, these new skin shells are subtracted from the original skin shell in HFSS, which results in N smaller deformed chest objects. These chest objects are spaced evenly (400 mm in Y direction) in front of the original shell and then united. A moving box is carefully designed so that it covers only one chest object at any time instant *t*. Then, the intersection is performed. The process is illustrated in Fig. 18.A1.

Box original position is given by: −300mm, $(200 - t * 400) * 10^{-3}$, − 350mm.

An intersection operation is performed with the box and the chest array object, which results in one chest object for a particular time t. Finally, the chest object is moved and a unite operation is performed with the original skin shell (shown in Fig. 18.A2).

Fig. 18.A1 A box is carefully designed so that each iteration covers only one chest object at a time

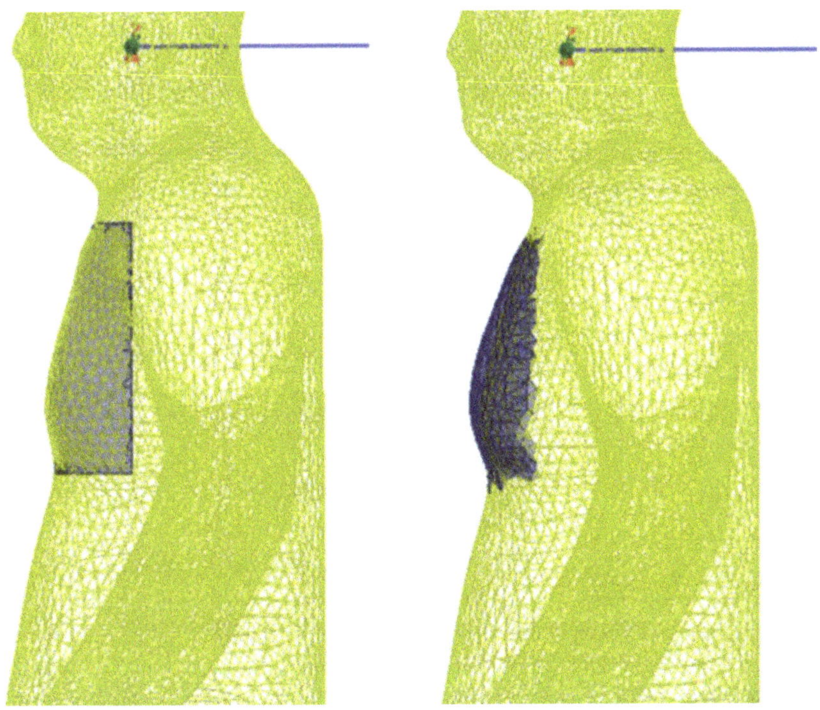

Fig. 18.A2 (**a**) (left). Skin Shell with $t = 0$. (**b**) (right). Skin shell with $t = 11$

References

1. Siebenthal, M. V. (2008). *Analysis and modelling of respiratory liver motion using 4DMRI*. Ph.D. dissertation, Elect. Eng. and Inform. Technology Dept., ETH Zurich, Switzerland.
2. Grimm, R., et al. (2015). Self-gated MRI motion modeling for respiratory motion compensation in integrated PET/MRI. *Medical Image Analysis, 19*, 110–120.
3. Lujan, A. E., et al. (1999). A method for incorporating organ motion due to breathing into 3D dose calculations. *Medical Physics, 26*(5), 715–720.
4. Lujan, A. E., Balter, J. M., & Ten Haken, R. K. (2003). A method for incorporating organ motion due to breathing into 3D dose calculations in the liver: Sensitivity to variations in motion. *Medical Physics, 30*(10), 2643–2649.
5. Segars, W. P., Lalush, D. S., & Tsui, B. M. W. (1999). Modelling respiration mechanics in the MCAT and spline-based MCAT phantom. *Nuclear Science Symposium*, Seattle, WA,, *2*, 985–989.
6. Wang, Y., Riederer, S., & Ehman, R. (1995). Respiratory motion of the heart: Kinematics and the implications for the spatial resolution in coronary imaging. *Magnetic Resonance in Medicine, 33*(5), 713–719.
7. West, J. (1995). *Respiratory physiology* (5th ed.). Baltimore: Williams and Wilkins.
8. Zeng, R.(2007). *Estimating respiratory motion from CT images via deformable models and priors*. Ph.D. dissertation, Elect. Eng. Dept., University of Michigan, Ann Arbor, Michigan.
9. Eom, J., et al. (2010). Predictive modeling of lung motion over the entire respiratory cycle using measured pressure-volume data, 4DCT images, and finite element analysis. *Medical Physics, 37*(8), 4389–4400.
10. Yanamadala, J., Noetscher, G. M., Louie, S., Prokop, A., Kozlov, M., Nazarian, A., & Makarov, S. N. (April, 2016). *Multi-purpose VHP-female version 3.0 cross-platform computational human model, 10th European conference on antennas and propagation 2016 (EuCAP16)*, Davos, Switzerland.
11. Hasgall, P. A., Di Gennaro, F., Baumgartner, C., et al. (13 Jan, 2015). *IT'IS database for thermal and electromagnetic parameters of biological tissues, version 2.6*.

Chapter 19
Radio Frequency Propagation Close to the Human Ear and Accurate Ear Canal Models

Louis Chen, Gerry Eaton, Sergey Makarov, and Gregory Noetscher

19.1 Introduction

When contemplating on- or near-body wireless networks, many factors contribute to creation of an optimal communications link. Transmission signal interface with the human body in the near-field can contribute to significant antenna mismatch and disruption of the transfer of power between the transmitter and receiver(s). For this reason, realistic and anatomically correct human body models are required when conducting simulations aimed at optimizing a transceiver antenna and characterizing the corresponding link budget. This chapter provides detail on the construction of a CAD-based computational human phantom with highly detail outer ear and ear canal structures and compares the simulated power transmission coefficient to models that do not possess this level of anatomic detail.

L. Chen · G. Eaton
Bose Corporation, Framingham, MA, USA

S. Makarov
Massachusetts General Hospital, Boston, MA, USA

Worcester Polytechnic Institute, Worcester, MA, USA

G. Noetscher (✉)
Worcester Polytechnic Institute, Worcester, MA, USA
e-mail: gregn@nevaem.com

S. Makarov et al. (eds.), *Brain and Human Body Modeling*,
https://doi.org/10.1007/978-3-030-21293-3_19

19.2 Model Construction

19.2.1 Base Model

Numerous full-body computational phantoms are in use today [1], each with a corresponding resolution and level of anatomical precision; each user must decide which model is most appropriate for the simulation methodology and end application. A non-homogenous, multi-structure model, known as the NELLY model, was the basis for the modifications described below. The NELLY model is CST's implementation of the Visible Human Project Female (VHP-Female) model shown in Fig. 19.1 [2]. This model was constructed by hand segmenting cryosection images provided as part of the National Library of Medicine's multi-modal Visible Human Project (VHP) executed in the 1990s and utilized by thousands of users worldwide for a number of eclectic purposes since its completion. The VHP-Female model is composed of approximately 270 individual components, all manifold surfaces that have been registered via a co-registration process. This model was specifically created to enable cross-platform compatibility from an open source data set. While triangular surface meshes fitting the finite element method are the primary means of construction, these meshes may be voxelized to also allow for time domain based methods of simulation.

While this model is perfectly acceptable for a number of simulation problems, including calculation of specific absorption rate (SAR) [3, 4] and antenna development [5, 6], its anatomical accuracy was lacking for specific applications involving transmission in or around the ear; this can be seen in Fig. 19.2 where fine features around the ear have been removed. The original model was optimized for fast runtimes on applications that required this level of detail. For this reason, the source material was revisited and further segmented to augment the existing model with accurate ear and ear canal features.

19.2.2 Accurate Ear Canal Segmentation

Since the VHP cryosection data was used in the creation of the initial model, this data was examined again using the same workflow that enabled creation of the VHP-Female model. This consisted of reviewing 450 full color images, each with a pixel resolution of 0.33 mm × 0.33 mm. Each image was a representation of an axial slice of the original patient, depicting all internal organs and structures. Identification of the ear canal was particularly easy in this format and can be seen clearly in Fig. 19.3, which shows a number of axial slices progressing along the vertical (z) axis.

The VHP model (and consequently, the NELLY model) is akin to the layers of an onion; the outermost skin layer completely surrounds a layer of fat, which completely surrounds a layer of muscle (or "average body") tissue. Contained within this

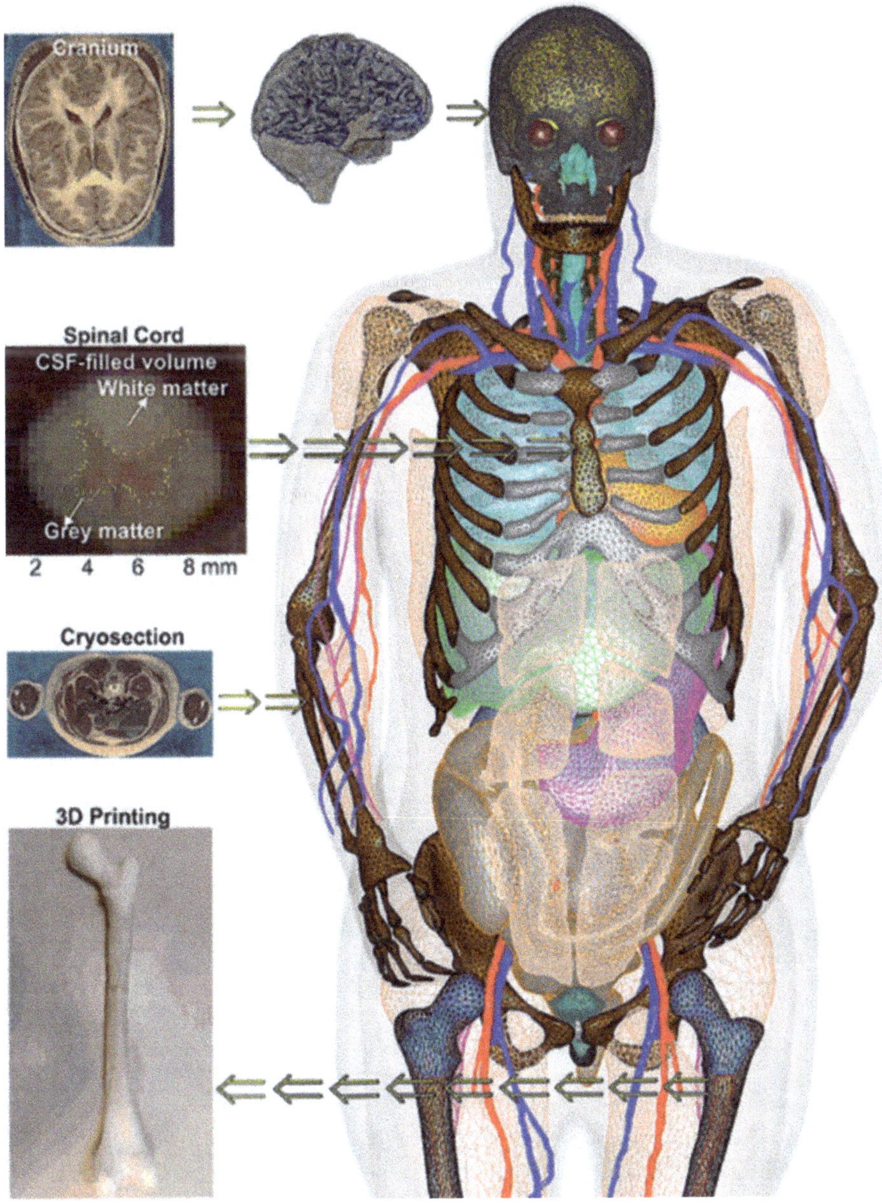

Fig. 19.1 The Visible Human Female model, used as the basis for the NELLY model implemented in CST

muscle layer are all subsequent internal organs and tissues. In order to create a realistic ear and ear canal, the three outermost layers (skin, fat, and average body) required modification to enable integration with the refined features. As shown in Fig. 19.3, the thicknesses between the various layers (skin to fat and fat to muscle) were not homogeneous. This makes sense from an intuitive perspective given that different areas of the body have different material content. Given this fact, the thickness of the skin was tapered from 1 mm around the subject's scalp to 0.3 mm within the outer ear. Likewise, the thickness of the fat layer was tapered from 2 mm to 0.3 mm.

Fig. 19.2 The original skin shell of the NELLY model, defeatured and optimized for fast simulations

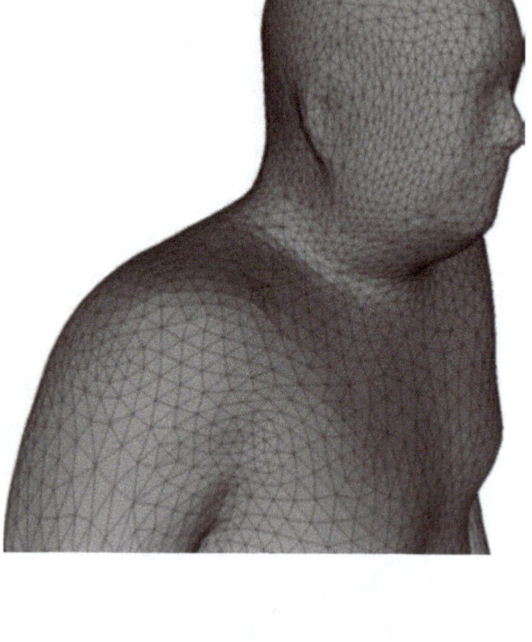

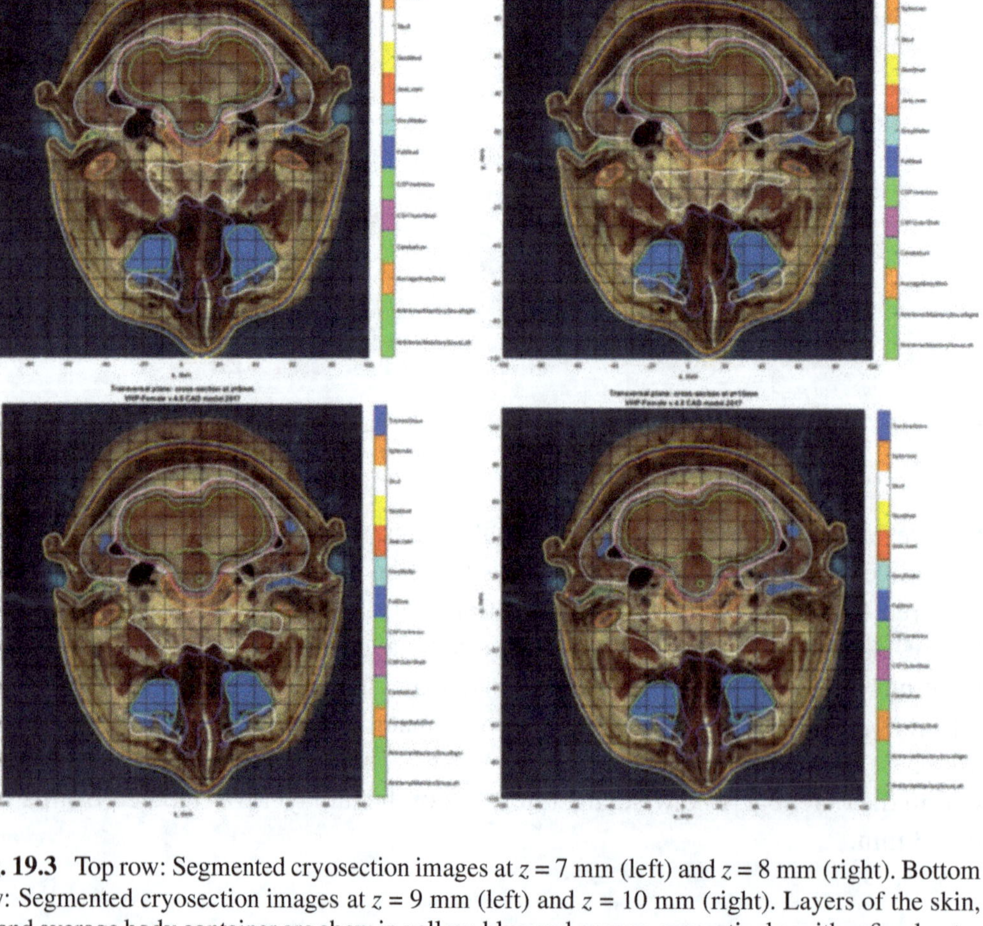

Fig. 19.3 Top row: Segmented cryosection images at $z = 7$ mm (left) and $z = 8$ mm (right). Bottom row: Segmented cryosection images at $z = 9$ mm (left) and $z = 10$ mm (right). Layers of the skin, fat and average body container are show in yellow, blue and orange, respectively, with refined outer ear and ear canal structures

19.2.3 Levels of Segmentation

During the segmentation process, the left and right ears were segmented using several levels of refinement. The first segmentation level was highly refined and allowed for the construction of a highly accurate model with over 10 k triangles per ear, shown in Fig. 19.4. Each ear canal was approximately 30 mm long, stretching into the head and enabling the modeling of this air-filled cavity to assess its impact on antennas placed in or near to the ear.

Following this time-consuming operation, the highly accurate models were decimated via Quadric Edge Decimation to produce left and right components with

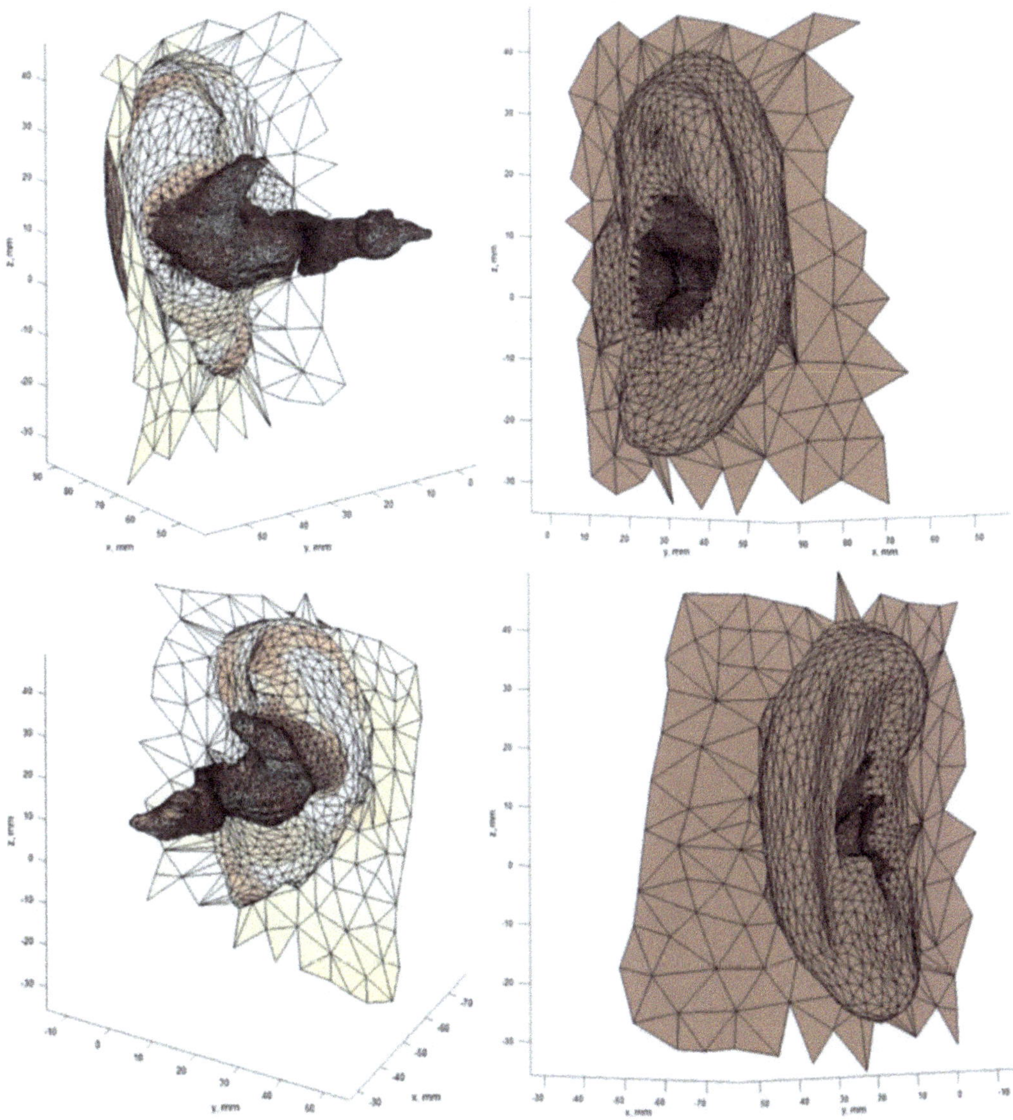

Fig. 19.4 Top row: Highly refined right outer ear (left) and ear canal (right). Bottom row: Highly refined right outer ear (left) and ear canal (right). Not the very high triangle count within the ear canal and stretching approximately 30 mm into the head

several hundred triangles each. This decimation was accomplished to enable very fast simulation times when a highly refined model was not necessary. These reduced models are shown in Fig. 19.5.

19.2.4 Integration of Ear and Ear Canal

After segmentation and mesh construction of the outer ear and ear canal were completed, these components were stitched together with the skin, fat, and average body shells using the mesh manipulation tool meshmixer. This freeware enables advanced mesh processing and allows for straightforward combination of multiple mesh structures. This operation was validated by overlaying the resulting meshes on top of the cryosection images, as shown in Fig. 19.3. In this way, any user of this product can independently verify that the required accuracy is present in the model.

19.2.5 Simulation Setup

The augmented model was read into CST Microwave Studio as a series of stereolithographic model files and complemented with the standard set of human body material properties inherent in the software. A simple simulation designed to establish a wireless link between two antennas located on the body was set up:

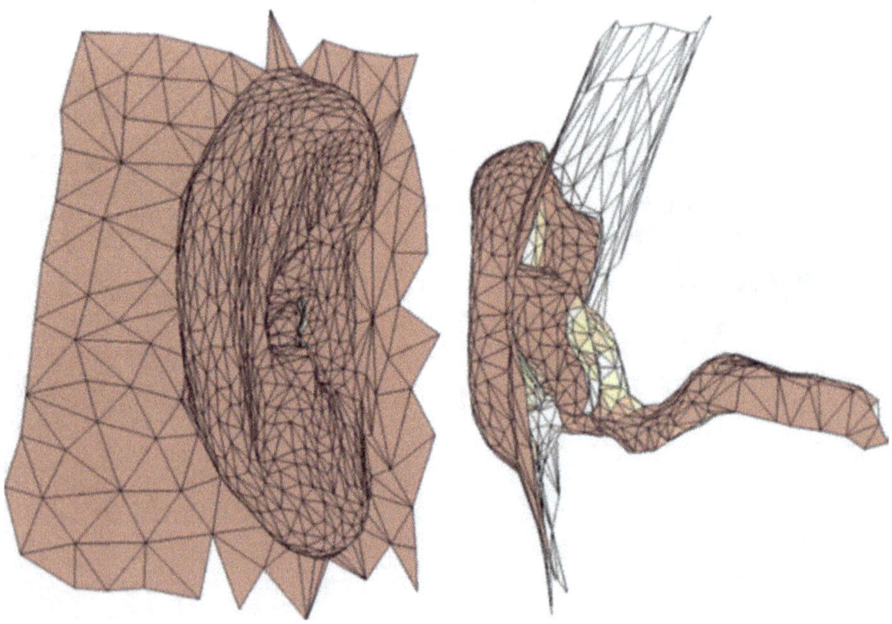

Fig. 19.5 Decimated mesh of the right ear. The number of triangles is drastically reduced from the examples shown in Fig. 19.4

- Antenna 1 was constructed as a Planar Inverted F Antenna (PIFA). This antenna was located in the right ear and enclosed in a plastic container. This antenna was mirrored in the left ear to obtain transmission coefficient information.
- Antenna 2 was also a PIFA, modeled after a popular cellular phone antenna design. This antenna was placed at the left hip to simulate carrying a phone at this body location.

Table 19.1 below describes the CST simulation settings used for this work.

19.3 Results

As outlined in Table 19.1, both the GPU memory usage and the total simulation time were larger when using the augmented model. This was to be expected since a higher global mesh density of six cells per wavelength near the model was used on the augmented model versus four cells per wavelength on the original NELLY. This step was taken due to hardware resource limitations.

The results of the simulations with both models are presented in Table 19.2. It is clear that antenna reflection coefficient is relatively independent of model construction due to similar loading. However, the transmission between the left and right ear is significantly different when incorporating true ear geometry into the model. The transmission from the ear to the pocket is also considerably different with the newly augmented model very closely matching what was experimentally measured.

19.4 Discussion

Based on the results shown in Table 19.2, it is clear that including the true anatomic structure into the model will substantially improve the accuracy of the simulation. While this model does require more time and memory to properly simulate, the results are much closer to experiment than those obtained with the previous model.

Table 19.1 Simulation parameters used for the models under discussion

	Voxel Nelly	Bose-Neva VHP
Mesh cells	896,691,180	566,555,542
Solver	TLM	TLM
GPU	4	4
Mesh cell reduction	94.3%	90.4%
GPU memory usage	6.7 GB	11.3 GB
Total simulation time	39 h 3 min 22 s	49 h 13 min 17 s

Table 19.2 Comparison of simulated results between both models and with experimental measurements

Model comparisons with experiment in dB			
	Voxel Nelly	Bose-Neva VHP	Measurement
Antenna (in ear) reflection coefficient	−22.4	−21.3	–
S21 (ear-to-pocket)	−99	−89	−88
S21 (ear-to-ear)	−72	−84	–

One additional source of difference is likely the fact that the surfaces of the Voxel NELLY model are not smooth. Due to its voxel nature, any curved surface on this model is actually a staircase approximation of the true smooth geometry. This may likely play a part in the transmission coefficient calculation and could add further justification for use of a triangular surface based mesh.

19.5 Conclusions

This work describes the construction of a finite element compatible model with highly refined outer ear and ear canal structures. This model was created to address the specific need of simulating advanced wireless wearable technologies and to efficiently design in virtual space wearable antennas for on-body propagation. Simulation results were shown to closely match experimental measurements.

Future improvements may consist of augmenting this model with internal ear cartilage structures and assessing the impact of ear morphology on antenna performance. This latter item could include examining the size and spacing of the ear relative to the scalp of the model.

References

1. Makarov, S. N., et al. (2017). Virtual human models for electromagnetic studies and their applications. *IEEE Reviews in Biomedical Engineering, 10*, 95–121.
2. Tankaria, H., et al. (2016). VHP-female full-body human CAD model for cross-platform FEM simulations — Recent development and validations. In *2016 38th Annual International Conference of the IEEE Engineering in Medicine and Biology Society* (pp. 2232–2235). Orlando, FL.
3. Kozlov, M., Tankaria, H., Noetscher, G., & Makarov, S. (2017). Comparative analysis of different versions of a human model located inside a 1.5T MRI whole body RF coil. In *39th Annual International Conference of the IEEE Engineering in Medicine and Biology Society (EMBC 2017)*. Jeju Island, Korea.
4. Kozlov, M., Bazin, P. -L., Möller, H. E., & Weiskopf, N. (2016). Influence of cerebrospinal fluid on specific absorption rate generated by 300 MHz MRI transmit array. In *2016 10th European conference on antennas and propagation (EuCAP)* (pp. 1–5). Davos.

5. Noetscher, G. M., et al. (2017). Virtual humans for antenna/implant modeling. In *2017 IEEE international symposium on antennas and propagation & USNC/URSI National Radio Science Meeting* (pp. 223–224). San Diego: CA.
6. Venkatasubramanian, A., & Blair, C. (2015). Modeling and design of antennas for implantable telemetry applications. In *2015 IEEE 16th annual wireless and microwave technology conference (WAMICON)* (pp. 1–5). Cocoa Beach, FL. https://doi.org/10.1109/WAMICON.2015.7120355.

Chapter 20
Water-Content Electrical Property Tomography (wEPT) for Mapping Brain Tissue Conductivity in the 200–1000 kHz Range: Results of an Animal Study

Cornelia Wenger, Hadas Sara Hershkovich, Catherine Tempel-Brami, Moshe Giladi, and Ze'ev Bomzon

20.1 Introduction and Motivation

Methods for non-invasively imaging the electrical properties (EPs) of tissues in vivo have been a recent topic of significant interest. Specifically, many research groups have attempted to create volumetric images of electrical conductivity (σ) and relative permittivity (ε_r) at specific frequencies. These studies have been motivated by a number of reasons, the primary one being that the interaction between applied electromagnetic fields and biological tissue is determined by their EPs [1]. Thus, non-invasively imaging EPs within the body offers important insight for a variety of clinical applications.

One example is the safety studies concerned with understanding tissue heating that occurs during high-field MRI applications [2–5]. The calculation of the local specific absorption rate (SAR) of tissue requires knowledge of the electric field induced by radiofrequency excitation and the local EPs of tissues [4]. And as summarized in [3], the actual local SAR pattern varies significantly with patient size and position, as well as patient-specific EPs. Antenna design for wearable and implantable medical devices is another high-frequency application that would benefit from knowledge of the patient-specific EP distribution [6]. And virtual human models are often used to estimate induced electric fields, which would benefit from individual and frequency-dependent EP maps [6]. Major low frequency medical applications which are being studied with computational models include neurostimulation techniques (such as TMS [7] or tDCS [8]) or MEG/EEF source detection studies. In all of these studies, knowledge of the spatial distribution of the induced electric field is necessary to interpret experimental results and to optimize field delivery [9].

C. Wenger (✉)
Novocure GmbH, Lucerne, Switzerland
e-mail: cwenger@novocure.com

H. S. Hershkovich · C. Tempel-Brami · M. Giladi · Z. Bomzon
Novocure Ltd, Haifa, Israel

S. Makarov et al. (eds.), *Brain and Human Body Modeling*,
https://doi.org/10.1007/978-3-030-21293-3_20

Therefore, a method for rapidly creating accurate and patient-specific EP maps would be important for future improvement of these medical applications.

Furthermore, experimental measurements have revealed that the EPs of pathological tissues often differ from those of healthy tissues, with generally higher σ and ε_r values in cancerous tissue at both low and high frequencies [10–16]. The authors associate the EP difference with the changes in water content that occur in neoplastic tissue, which are due to varying protein hydration, vascularization, alterations in membrane permeability and amount of extracellular fluid, as well as packing density, and cancer cell orientation. Thus, mapping EPs may help to detect malignant regions with higher sensitivity than current imaging techniques. This knowledge contributes to further development and treatment planning of medical applications such as radiofrequency tissue ablation, electro-chemotherapy and gene therapy with reversible electroporation, nanoscale pulsing, and irreversible electroporation, as described and summarized in [16, 17].

Another example, and the motivation for this study, is a cancer treatment modality called Tumor Treating Fields (TTFields). TTFields are an antimitotic cancer treatment using alternating electric fields with intermediate frequencies (between 100 and 300 kHz) and low intensities (between 1 and 3 V/cm) to disrupt cell division in tumors [18, 19]. TTFields are FDA approved as a monotherapy for recurrent glioblastoma [20] and as a combination therapy with chemotherapy for newly diagnosed glioblastoma patients [21]. TTFields treatment of glioblastoma is performed with the Optune™ device, which delivers alternating electric fields at a frequency of 200 kHz via two pairs of transducer arrays [22, 23]. Since TTFields treatment efficacy depends on delivered field intensity at the target, patient-specific treatment planning [24] provides array layouts with personalized locations of transducers to optimize induced field intensity in the tumor. Several computational modeling studies have used realistic computational head models to study the induced electric field distribution during Optune™ treatment [25–27]. These studies all showed that the field intensity delivered to the tumor is dependent on the dielectric properties, specifically the electric conductivity σ of the healthy and pathological tissues of the patient's brain [26, 28, 29]. These computational studies usually assume homogenous and isotropic EPs in different tissue types, which is a reasonable assumption for well-studied healthy brain tissues like the gray matter (GM), white matter (WM), and cerebrospinal fluid (CSF) [30, 31]. Yet, the glioblastoma tumor area is typically a very heterogeneous region composed of enhancing or non-enhancing parts, and cystic or necrotic compartments. Furthermore, EP measurements of cancerous tissue also predict higher patient variability compared to healthy tissue (e.g [32]). Thus, a patient-specific EP map of the brain would be optimal to accurately evaluate the induced electric field distribution and intensities at the tumor bed. For a detailed review on preclinical, clinical, and modeling studies related to TTFields, see [33].

A variety of different approaches have been proposed to non-invasively image EPs of biological tissue at different frequencies over the past two decades. The oldest methods for mapping EPs in a volume include electrical impedance tomography [34] and magnetic induction tomography [35]. These techniques suffer from the inherently ill-posedness of the inverse problem, leading to low resolution, low sensitivity, and limited applicability in the clinic. Other attempts to obtain EP maps with high spatial resolution include magnetic resonance electrical impedance tomography (MREIT) [36], as well as other (magneto-) acoustic methods. MREIT was proposed to perform conductivity imaging at DC using the MR current density imaging approach [37]. According to [38], it remains a technical problem to reduce current injection down to a level for routine clinical use while maintaining the spatial resolution of the resulting conductivity image without long imaging sequences. More recently, in an attempt to overcome the limitations of these techniques, MR-based electrical properties tomography (EPT) was introduced by Haacke et al. [39]. EPT is based on B_1 mapping, i.e., the EPs are reconstructed from measurable RF-coil-induced magnetic fields (B_1 fields) by employing Maxwell's equation. The EPT approach has been well studied and a comprehensive review can be found in [40]. However, a recent study [41] points out certain challenges associated with EPT, including boundary reconstruction errors due to the fact that the Helmholtz equation does not hold at regions where tissue types coexist and the high degree of symmetry that is needed to obtain higher signal-to-noise ratio.

The authors of this study propose a new, alternative approach termed water-content-based electrical properties tomography (wEPT) [41]. wEPT utilizes two T1-MR images to map σ and ε_r within tissue. The foundation for wEPT is Maxwell's mixture theory, which suggests that tissue conductivity is highly correlated with water content, which in turn can be estimated from the ratio between two T1-weighted MR images with different repetition times (TR). wEPT has been used to map brain conductivity at 128 MHz without concern about inhomogeneity among tissues. Because wEPT mapping is based on relatively standard and rapid MRI imaging sequences, it is a highly attractive approach for implementation in research and in the clinic.

Having TTFields treatment of glioblastoma patients in mind, the aims of this study were

(a) To test whether wEPT can be applied to the frequency range of 200–1000 kHz
(b) To establish that wEPT can be used to map σ of healthy brain tissue in this frequency range
(c) To investigate whether or not wEPT can be used to map EPs in brain tumors

To achieve these goals, a series of experiments was performed utilizing BSA phantoms, samples of calf brain and in vivo experiments in tumor bearing-rats. A detailed description of the experiments and their results follow.

20.2 Methods and Results

This section starts with a summary of the theoretical basis behind the wEPT approach (Sect. 20.2.1).

In order to address our above-stated aims, the study was performed in three stages. Materials and methods as well as obtained results will be presented in chronological order of experiments: First the feasibility of applying wEPT at 200 kHz was established by performing wEPT measurements on phantoms (Sect. 20.2.2). Then, wEPT was applied to estimate the EPs of healthy brain tissue derived from calf brains (Sect. 20.2.3). Finally, we tested the applicability of wEPT to map σ in tumor tissue derived from tumor bearing-rats (Sect. 20.2.4).

All imaging was performed on an in-house Bruker ICON 1.0 T MRI machine. Animal experiments were approved by the Israeli Animal Care.

20.2.1 Theory of wEPT

The specific EPs of a tissue are determined by relative intra- and extracellular volumes, membrane permeability, ion concentration, and mobility [15]. During studies of the dielectric properties of various types of tumor and normal tissues with varying water content, it has been demonstrated that their EPs conform to the Maxwell mixture theory for high and low water-content tissues [42–44]. Since cell membranes have low impedance at RF frequencies, the EPs of a tissue become highly correlated with the water fraction. It has been reported that the tissues' conductivity at these frequencies is mostly determined by the ionic conductivity of cytoplasm and that the permittivity is likely determined by the water fraction [45]. For ultra-high frequencies, the EPs are almost entirely determined by the water content due to the polarization of polar media-like water [15, 42].

Thus, the wEPT approach assumes that at an imaging frequency of 128 MHz, the electrical conductivity σ and the relative permittivity ε_r can be modeled as a monotonic function of tissue water content (WC) under the principle of Maxwell's mixture theory. The wEPT approach seeks to circumvent the need for lengthy imaging processes by computing the WC maps using a transfer function derived from the image ratio (IR) of two T1-weighted images with different repetition times (TR) via

$$\mathrm{WC} = w_1 e^{-w_2 \mathrm{IR}} \tag{20.1}$$

where w_1 and w_2 are determined through curve fitting.

The image ratio assumption is based on the fact that both properties, the WC and the IR, can be expressed as functions of the T_1 relaxation values (T1). On the one hand, early works had already demonstrated that the WC can be directly related to the T1 value in living tissue [46]. On the other hand, the IR can be expressed by

$$\mathrm{IR} = \kappa \left(\frac{1 - e^{-\mathrm{TR_s}/\mathrm{T_1}}}{1 - \cos\theta_1 \cos 2\theta_1 e^{-\mathrm{TR_s}/\mathrm{T_1}}} \right) / \left(\frac{1 - e^{-\mathrm{TR_l}/\mathrm{T_1}}}{1 - \cos\theta_1 \cos 2\theta_1 e^{-\mathrm{TR_l}/\mathrm{T_1}}} \right),$$

where θ_1 and $\theta_2$are the nominal angles of the excitation and refocusing pulses, κ is a factor compensating signal gain fluctuations, and $\mathrm{TR_s}$ and $\mathrm{TR_l}$ are the short and long repetition times of the two images, respectively.

In order to connect WC and the EPs of brain tissues, the authors in [41] used landmark points for the three primary brain tissues (GM, WM, and CSF) and gathered corresponding values for WC, σ, and ε from an extensive literature search. Two separate curve fitting procedures revealed the best fit for a monotonically increasing function between WC and the EPs at 128 MHz:

$$\sigma = c_1 + c_2 e^{c_3 \mathrm{WC}} \tag{20.2a}$$

$$\varepsilon = p_1 \mathrm{WC}^2 + p_2 \mathrm{WC} + p_3 \tag{20.2b}$$

In order to find appropriate values for short and long repetition times, the authors in [41] plotted IR as a function of T1 with different combinations of $\mathrm{TR_s}$ and $\mathrm{TR_l}$ and assuming $\theta_1 = 90^\circ$, $\theta_2 = 180^\circ$, and $\kappa = 1$. The preferred combination was determined by the largest IR signal and greatest difference between the WM with the highest T1 and the CSF with the lowest T1. In the original study, $\mathrm{TR_s}$=700 ms and $\mathrm{TR_l}$=3000 ms were considered.

20.2.2 BSA Phantom

We initially conducted a phantom study to test the feasibility of performing wEPT studies for a lower frequency of 200 kHz. The phantom was created by placing different solutions of BSA (bovine serum albumin) in DPBS (Dulbecco's phosphate-buffered saline) into a plastic plate with cylindrical chambers. Each chamber was filled with solutions of varying concentrations of albumin (20%, 25%, 30%, and 40%). In order to obtain the coefficients of the wEPT model equations, we measured four parameters of the BSA solutions at 200 kHz: T1, WC, σ, and ε_r (Table 20.1). The T1 value of the solutions was estimated using the variable TR method, in which the T1 value is estimated from exponential curve fitting of the signal intensity in a voxel for a selection of spin-echo images with varying $\mathrm{TR_s}$ (compare Fig. 20.1b). The WC of the solutions can be deduced from the known

Table 20.1 BSA experiment

	BSA20	BSA25	BSA30	BSA40
T1 *measured*	1119.4	928.0	741.0	483.2
WC *measured*	79%	75%	70%	60%
WC *wEPT*	80%	75%	70%	60%
σ *measured*	0.90	0.79	0.69	0.49
σ *wEPT*	0.88	0.79	0.69	0.49
ε *measured*	315.60	284.89	277.17	262.51
ε *wEPT*	300.44	284.94	277.20	261.92

Measured and wEPT-estimated values of T1 (ms), WC (%), σ (S/m), and ε_r at 200 kHz for four BSA solutions

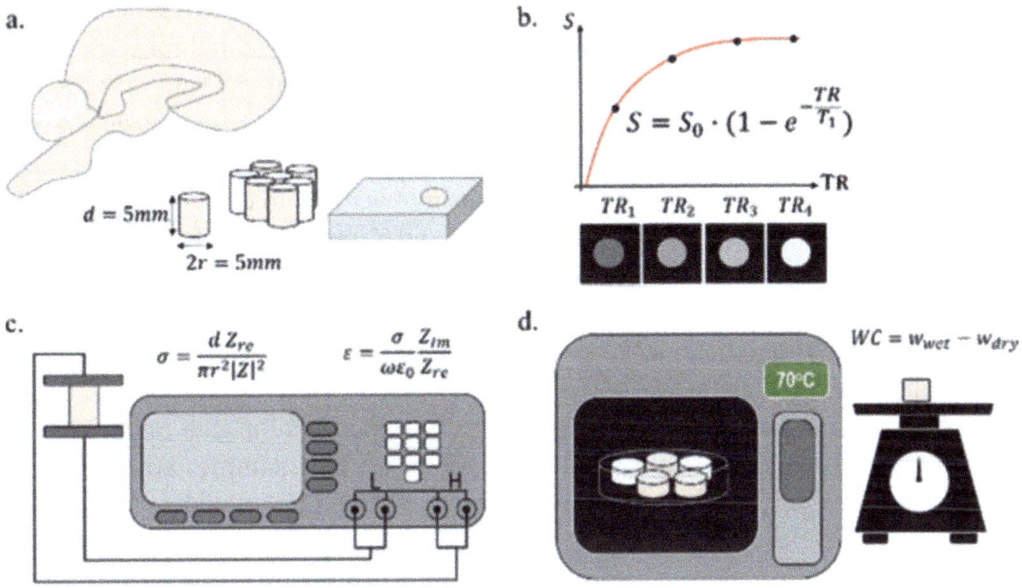

Fig. 20.1 Experimental setup of brain tissue measurements. (**a**) GM and WM sample preparation (**b**) T1 values were estimated with the variable TR method. (**c**) An LCR meter was used to estimate σ and ε for frequencies between 20 Hz and 1 MHz. (**d**) WC is estimated as the difference between the weight of wet and dry tissue samples

concentration of albumin in each solution. The EPs of the solutions were measured utilizing the parallel-plates method with a cylindrical chamber (10 mm diameter, 9 mm thickness) sandwiched between two Ag/AgCl electrodes, which were connected to an LCR meter (Keysight E4980AL). Sample impedance was measured in the range of 20 Hz to 1 MHz, and σ and ε_r were derived from these measurements using a well-established procedure. By testing different combinations of 13 spin-echo images with TR ranging from 50 ms to 5 s, we found that a combination of TR_s/TR_l=700/4000 yields optimal distinction of IR values between the BSA solutions tested. Furthermore, a few different curve-fitting methods were tested for mapping EPs at 200 kHz from WC (including polynomial and exponential fitting), which all yielded similar wEPT estimations. Table 20.1 summarizes experimentally measured values of T1, WC, σ, and ε at 200 kHz, as well as the estimated values with wEPT, which are the mean values of the properties in the central ROIs. Obtained results predict that wEPT estimations of the WC and EPs in all solutions match very well with corresponding measurements. Only minor discrepancies are found for the BSA 20% solution with the highest WC.

20.2.3 Calf Brain Samples

In order to establish a wEPT model for mapping brain tissue EPs at 200 kHz, appropriate coefficients have to be found via curve fitting experimental measurements to Eqs. 20.1, 20.2a, and 20.2b. Thus, measurements of T1, WC, σ, and ε were performed for three tissue types: GM, WM, and CSF. The first experiments with brain tissue were conducted with excised samples of three calf brains received 48–72 hours postmortem. Although there was some uncertainty about storage and handling of the first bovine brains prior to delivery, we wanted to establish a workflow for experimental procedures and refine our planned methodology.

For sample preparation, the brain was cut using tweezers to expose the thalamus where cylindrical samples of GM were punched out. Cylindrical WM samples were punched out of thick coronal slices. Different punches with diameters ranging from 5 to 7 mm were used, depending on the operator and the brain region. The samples were placed in a 5 mm thick polycarbonate plate with holes of 5 mm diameter, and were sealed with cellophane (Fig. 20.1a).

To estimate T1, plates containing up to eight samples were placed in the MRI to acquire spin-echo images with variable TR_s (Fig. 20.1b). Samples were initially kept about 2 mm thicker than the plates to ensure good image quality. The plates were moved to the LCR measurement station after MRI image acquisition (Fig. 20.1c). The cover of each sample was removed just before measurement, when it was cut to the plate thickness of 5 mm. While assuring full contact, the Ag/AgCl electrodes were placed on the top and bottom of each tissue sample. The LCR meter was used to record the impedance of each sample for a frequency sweep between 20 Hz and 1 MHz, at a lab temperature of about 21 °C. σ and ε values were derived from the impedance measurements assuming a parallel plate measurement system geometry.

The WC of the samples was defined as the weight difference between wet and dry tissue samples. After the LCR measurement, the samples were transferred to a pre-weighted weighing paper. The gross weight of sample and paper was documented with a scale (ME104 Metler Toledo SN004) prior to and then after placing them in an oven at 70 °C for 48 h of drying (Fig. 20.1d). A preliminary experiment assured that the weight of the paper is not affected by the drying process.

Using this process, measurements of T1, WC, σ, ε of 32 different calf brain samples (16 WM, 13 GM, 3 mixed samples) were obtained. For model completion, two porcine CSF samples were also measured. In order to implement wEPT on these samples, T_1-weighted images with $TR_s = 700$ ms and $TR_l = 3000$ ms were chosen for calculating IR. IR was plotted against the measured WC of all 34 tissue samples, and coefficients w_1 and w_2 in Eq. 20.1 were found via curve fitting (Fig. 20.2a). Subsequently, the same procedure was performed for equations that relate the WC to the EPs (Eqs. 20.2a and 20.2b). We created two sets of σ coefficients c_1, c_2, c_3 and ε_r coefficients p_1, p_2, p_3. One set of coefficients was for a frequency of 200 kHz (Fig. 20.2b, c); the second set was for the highest measured frequency of 1 MHz (data not shown).

The following procedure was used to analyze the wEPT model of the bovine brain samples and estimate EPs at 200 kHz and 1 MHz: IR was calculated for each voxel in the sample, corresponding WC values are calculated via Eq. 20.1 and EP values with Eqs. 20.2a and 20.2b. The estimated properties were averaged over the sample volume to yield the average IR, WC, and EPs of each sample, which were compared to the measured values (compare Table 20.2). Figure 20.3 illustrates the

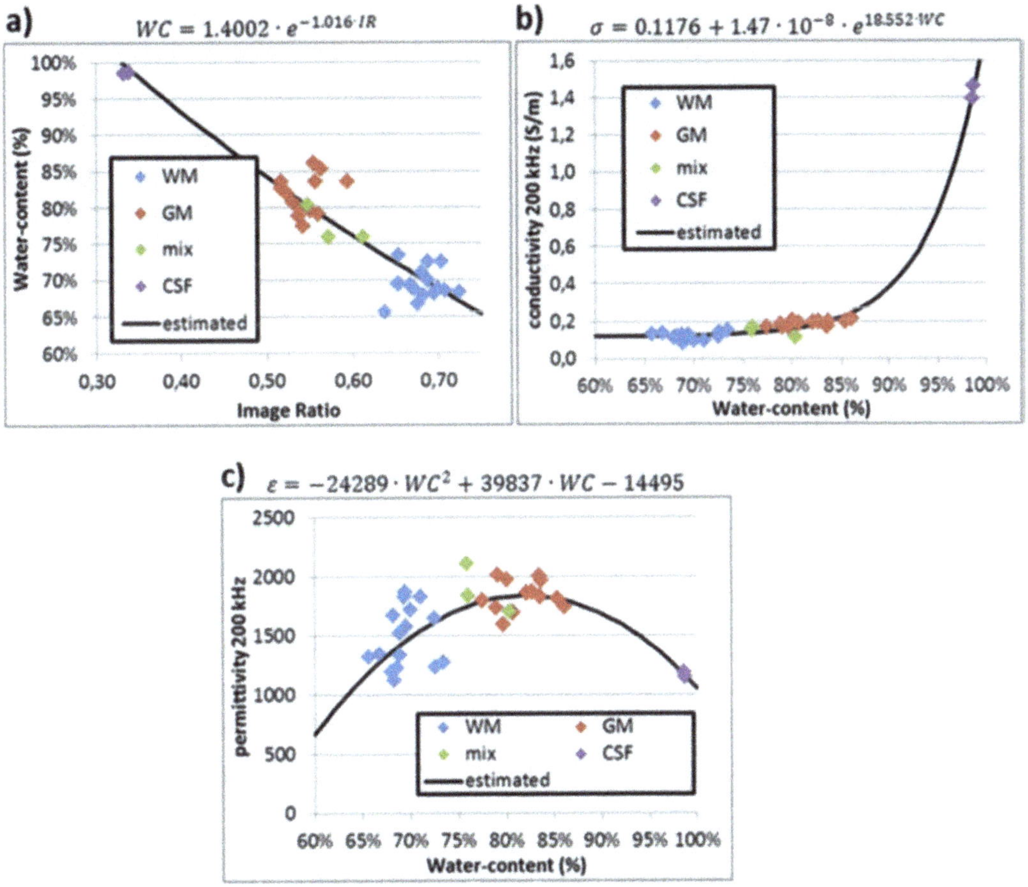

Fig. 20.2 wEPT model creation for brain tissue of calves1–3, for WM (blue), GM (red), mixed samples (green), and pig CSF (purple). Model coefficients were obtained by curve fitting for mapping WC from IR with Eq. 20.1 (**a**), for σ mapping with Eq. 20.2a (**b**), and ε_r mapping with Eq. 20.2b (**c**). The equations with their coefficients are listed above each panel

Table 20.2 Average relative error of wEPT estimates in all brain samples of calves 1–3 compared to measured values

	IR (%)	WC (%)	σ^{200kHz} (%)	σ^{1MHz} (%)	ε^{200kHz} (%)	ε^{1MHz} (%)
mod_a1	1.3	3.8	14.2	13.3	10.3	11.1
mod_a2	0.9	3.5	13.2	12.3	10.4	11.7

results of this procedure for two GM and two WM samples. The WC map and EP maps at 200 kHz and 1 MHz clearly show the different tissue types. The estimated values corresponding to the color scale match measured mean values of these properties in all GM and WM samples (displayed in each panel).

It is important to note that the dimensionless factor $\kappa = k_{short}/k_{long} \approx 1$ in the equation for calculating IR compensates for the signal gain fluctuations of the scanner at different TR_s. Preliminary analysis predicted that $\kappa = 0.9894$ should be used for our studies. Indeed, introducing this factor into the calculation of IR minimized the difference between measured and wEPT estimated EPs (Table 20.2). Thus, this factor

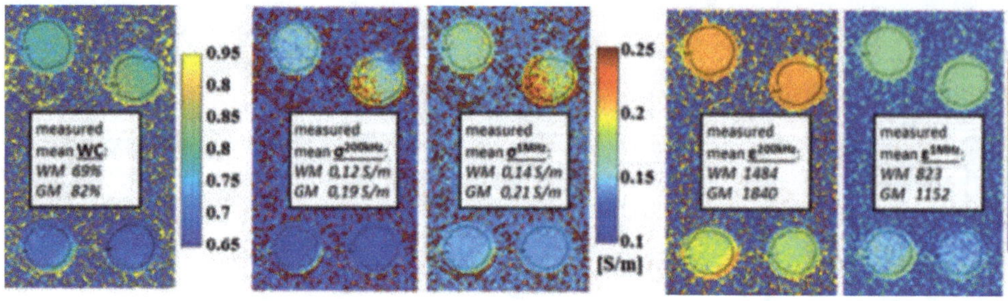

Fig. 20.3 wEPT-estimated maps of two GM (id's = {2,3} top samples) and two WM (id's = {1,4} bottom samples) excised tissue samples of calf1. Model equations used correspond to Fig. 20.4. The WC map (left), σ at 200 kHz and 1 MHz (middle), and ε_r at 200 kHz and 1 MHz (right) are presented with fixed color scales for the EPs to show their frequency dependence. In each panel, the corresponding mean values of the measurements are displayed for the two tissues. These values compare well with obtained maps

was introduced into all subsequent wEPT calculations presented in this study. Errors in wEPT estimation of IR, WC, and EPs compared to measured values were averaged among all calf tissue samples. Results are presented in Table 20.2 for wEPT estimations before κ corrections (*mod_a1*) and after (*mod_a2*). It should further be mentioned that two more adaptions have been analyzed for mapping σ at 200 kHz. When a fourth-order polynomial is used instead of the exponential form of Eq. 20.2a, σ^{200kHz} error further reduces to 11.2%.

This first study established the feasibility of using wEPT to measure the EPs of healthy brain tissue samples. However, the study had several shortcomings:

(a) Tissue samples were measured 48–72 hours after death. Thus, we wanted to test how time from death might influence the measured EPs.
(b) Tissues samples were measured at 20–22 °C. Thus, we wanted to test how temperature might influence the measured EPs.
(c) Samples were excised prior to MRI imaging. Thus, we wanted to test if removing tissue from a bulk might influence the measured and wEPT-estimated EPs. This last point is important to clarify when establishing the use of wEPT for in vivo models.

To address these issues, a set of measurements was performed on three freshly excised calf brains (calves 4–6). Previous measurements in calves 1–3 showed that the EPs measured in different hemispheres for the same tissue type are essentially equal. Thus, the left and right hemispheres were separated to test the influence of time from death on EPs. The left hemisphere was used for measurements on the same day ("fresh" samples). The right hemisphere was preserved at 4 °C for 48 h before measurements were collected ("48 h" samples). The temperature of the brain was measured with a thermocouple to establish the effect of temperature on measurements. To account for differences in wEPT estimates that may occur when small samples are excised from the bulk, wEPT imaging was first performed on a large bulk section of cow brain. After imaging, a rectangular piece of cortex containing the thalamus was removed from the hemisphere (Fig. 20.4a) for sample

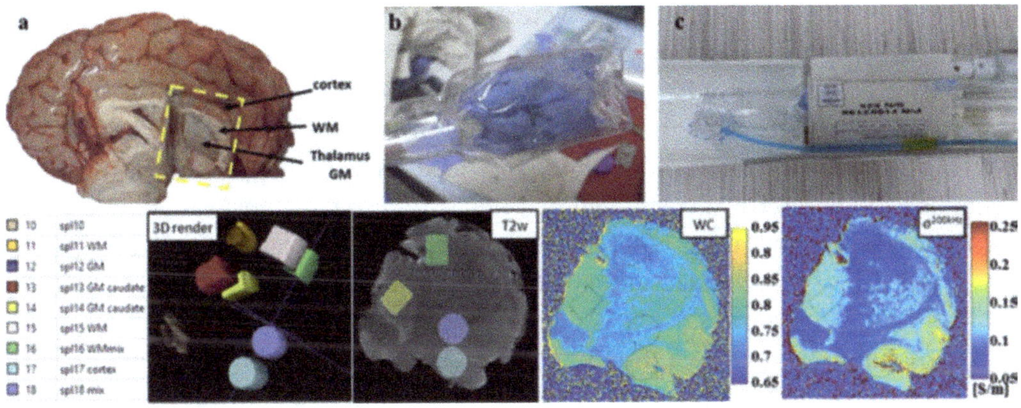

Fig. 20.4 Experimental procedures (top) and wEPT analysis (bottom) of the calf4 brain. The excised brain is separated into two hemispheres and a rectangular piece of cortex containing the thalamus is removed (**a**). Prior to sample punching, the whole brain piece is covered and wrapped for assuring tissue hydration (**b**), image taken prior to placing the sample in the coil for MR image acquisition (**c**). All samples that will be measured for their WC and EPs are segmented in the T2w image of the calf piece (bottom left). A 3D rendering of all samples is represented and one slice through samples with id's = {14, 16, 17, 18}. Corresponding T1w images with TR_s/ TR_l = 700/4000 ms are loaded into MATLAB and their IR is calculated. The WC and EP maps are evaluated with the wEPT model equations; WC and σ^{200kHz} maps are displayed for the same slice in the bottom right. The outlines of the segmented tissue samples are also indicated, which can be used for evaluating mean values of estimated properties in order to be compared to their measured values

punching. To prevent dehydration, the piece was covered with a glove and a sealed-by-heat plastic bag (Fig. 20.4b) before placing it into the coil for imaging (Fig. 20.4c). Each MRI session included T1-weighted (T1w) and T2-weighted (T2w) axial scans. T1w scans were used for wEPT mapping and T2w axial scans were used for identifying brain anatomy and segmentation of excised samples (Fig. 20.4 bottom left). Details of the imaging sequences in each brain piece are as follows:

T1w axial: RARE protocol, TR = 700 ms or TR = 4000 ms, TE = 11 ms, RARE factor = 1, 4 averages, 10 slices of 1.25 mm thickness and gap of 1.25 mm between slices, FOV = 3 cm, matrix 140, in-plane resolution 214 μm, acquisition time = 6.32 min or 37.20 min. Two intercalated scans were acquired to avoid slice cross-talk artifacts.

T2w axial: RARE protocol, TR = 4400 ms, TE = 85 ms, RARE factor = 12, 12 averages, 20 slices of 1.25 mm thickness, FOV = 3 cm, matrix 140, in-plane resolution 214 μm, acquisition time = 9.40 min.

Calf brain sections were prepared for EPs and WC measurements immediately after the MRI. The orientation was marked and kept clear throughout the entire procedure. At all times, the brain sections were covered either with a glove or with a cellophane film to minimize dehydration. A minimum of nine samples per brain section was collected with 5 or 6 mm punches including 2–3 WM samples, 2 GM samples from the thalamus, and 1 additional GM sample and 2 mixed tissue samples from the cortex. Samples were segmented from the T2w images using ITK-SNAP (Fig. 20.4 bottom left). The impedance measurements with the LCR system, the calculation of Eps, and the weighing method for WC estimation have already been described previously. At the end of the first day, two more samples were measured from the first calf (calf 4) to compare changes in the first few hours post excision. Results showed that measurements 2 h post excision are similar to those collected from samples analyzed 12 h post excision. On the following day, the EP and WC measurement procedure was repeated for the three right hemisphere brain sections.

In summary, we excised a total of 57 samples from three different calves, 30 of which were analyzed immediately after excision ("fresh") and the other 27 samples with a delay of a total of 48 h. After discarding uncertain measurements, we decided on 56 reliable WC measurements and 39 reliable EP measurements originating from calves 4–6. Results showed no significant difference between measurements in fresh calf samples compared to 48 h samples. However, certain discrepancies in comparison to the measurements of calves 1–3 can be observed (Fig. 20.5). The WC measurements were similar in calves 1–3 and calves 4–6, with a higher mean value of 73% in the WM of calves 4–6 in comparison to a mean of 69% in calves 1–3. The mean WC value in the GM remained at 82%. Higher differences have been found for EP measurements, whereas at 200 kHz calves 4–6 show lower σ values and higher ε values in both tissues, i.e., the mean σ^{200kHz} in the WM decreased from 0.12 S/m to 0.10 S/m, in the GM from 0.19 S/m to 0.13 S/m for calves 1–3 and

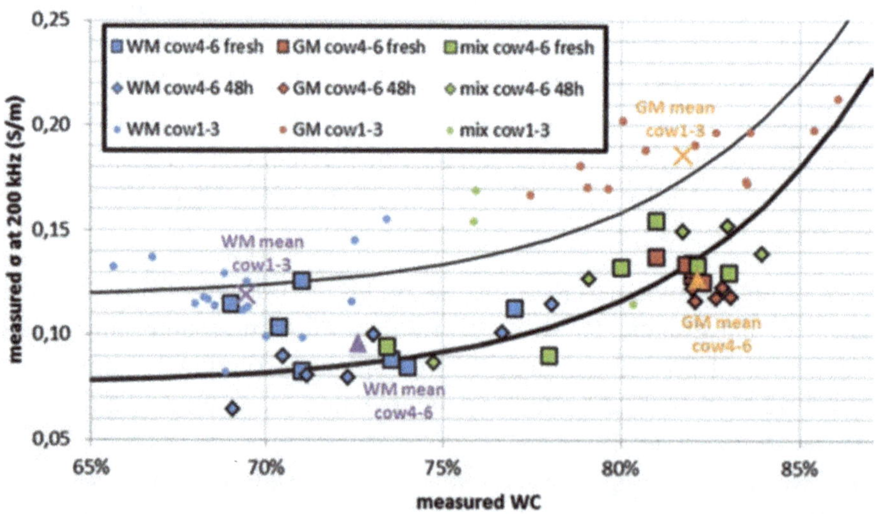

Fig. 20.5 Measurements of WC and σ^{200kHz} in all WM (blue), GM (red), and mixed (green) samples of calves 1–3 (dots) compared to "fresh" samples of calves 4–6 (squares) and "48 h" samples of calves 4–6 (diamonds). The black lines show the best fit according to Eq. 20.2a for all calves 1–3 samples (thin line) and all calves 4–6 samples (thick line). Mean values are indicated for WM (purple) and GM (orange) for calves 1–3 samples (crosses) and calves 4–6 samples (triangle). Note the discrepancies between the first and second set of measurements in bovine brain samples, which were a result of the different storage condition before experimental procedures were established, i.e., the temperature in the refrigerator of calves 1–3 was too low

calves 4–6, respectively. For ε^{200kHz}, mean values are 1484 vs 1630 in the WM and 1840 vs 2487 in the GM of calves 1–3 and calves 4–6, respectively.

All calf WC and σ^{200kHz} measurements are presented in Fig. 20.5. Colors correspond to the tissue type; marker symbols show the differences between calves 1–3 (dots) and fresh (squares) and 48 h (diamonds) samples of calves 4–6. The pairs of mean values of the WC and σ^{200kHz} are also indicated as crosses for calves 1–3 and triangles for all calves 4–6 samples, purple markers correspond to WM samples and orange markers to GM samples. Furthermore, the lines indicate optimal fits to Eq. 20.2a for mapping σ from WC. It is obvious that the same coefficients used in Eq. 20.2a for the calves 1–3 measurements (thin black line) will not best represent the new measurements from calves 4–6 (thick black line).

Furthermore, a slightly different combination of TR_s was used in the two T1w images of the wEPT imaging sequence for calves 4–6, i.e., TR_s = 700 ms and TR_l = 4000 ms were used instead of TR_l being 3000 ms as was the case for calves 1–3 studies. Therefore, the coefficients in Eq. 20.1 had to be refitted for altered values of IR. This model will be designated as *mod_b*, with coefficients $w_1 = 1.3028$ and $w_2 = 0.9134$. In a next step, two new sets of coefficients for Eq. 20.2a were created from the measurements of calves 4–6, one for mapping from WC to σ^{200kHz} (thick black line in Fig. 20.5) and one for σ^{1MHz}. The new coefficients were found via curve fitting, $c_1 = 0.0759$, $c_2 = 1.2351E\text{-}08$, and $c_3 = 18.7575$ for 200 kHz and $c_1 = 0.0952$, $c_2 = 2.1919e\text{-}08$, and $c_3 = 18.1696$ for 1 MHz. This model is called *mod_c*.

Typical wEPT model analysis, including WC and σ map creation, is illustrated in Fig. 20.4 at the bottom right. The presented slice corresponds to the T2w image that was used for segmentation. Therefore, it is possible to use the segmentation data and calculate mean values of WC and σ in a 3D representation of each tissue sample. The mean values can be compared to the corresponding measurement as presented in Table 20.3 for both tested models. Although an average error for WC estimation of 2.5% with *mod_b* is very low, the errors in the σ estimates are high. As expected, Eq. 20.2a derived from the data for calves 1–3 (*mod_b*, thin line in Fig. 20.5) does not optimally fit the measured WC and σ values of calves 4–6. The adequate wEPT model *mod_c* performs well for σ estimates with average errors as low as 13.6% for 200 kHz and 13.3% for 1 MHz.

We also conducted experiments to estimate the temperature dependence of the properties. Another calf brain was measured 24 hour post excision for this experiment. Four samples were extracted for these tests: two GM samples from the thalamus and two WM samples. Each sample was 5 mm thick and 5 mm in diameter. The base temperature of the samples was 20 °C. Samples were placed in the plate to be

Table 20.3 Average relative error of wEPT estimates in all brain samples of calves 4–6 compared to measured values

	WC (%)	σ^{200kHz} (%)	σ^{1MHz} (%)
mod_b	2.5	41.1	30.7
mod_c	3.3	13.6	13.3

measured with the Ag/AgCl electrodes. In order to gradually warm the tissue samples in the plate, lab gloves filled with hot water (~70 °C) were positioned around the setup and fixated to keep it heated. A thermocouple was placed on the plate next to the electrode to monitor the temperature. When the temperature increased sufficiently, the electrodes were removed and the thermocouple was inserted into the sample to measure the actual tissue temperature. When the temperature increased a few degrees, LCR measurements of the impedance were manually logged in the notebook at a single frequency of 200 kHz.

Measurements of σ^{200kHz} for increasing sample temperature are shown in Fig. 20.6. The exponential trend lines for GM (blue) and WM (red) are displayed. Two conclusions can be drawn from this study. When the body temperature of 37 °C is assumed, this would result in GM conductivity of 0.194 S/m and for the WM 0.102 S/m, which are in good agreement with values reported in the literature, e.g., $\sigma^{GM} = 0.25$ S/m and $\sigma^{GM} = 0.12$ S/m at 200 kHz were assumed in previous simulation studies [47]. Second, our measurements of excised samples were conducted at a room temperature of about 21 °C. According to these results, the multiplication factor for increasing σ from 21 °C to 37 °C is very similar. Note that this will be important when conducting in vivo wEPT studies because the temperature of tissues imaged in live animals are higher than in excised samples.

20.2.4 In Vivo wEPT Study: Healthy and Tumor-Bearing Rats

One of the main objectives of this study was to investigate if the wEPT approach can be used to create in vivo maps of WC and σ in rat brain tumor models. In order to establish optimal experimental procedures and validate the wEPT model, a study was performed in one healthy rat (Fisher F334 rat, male, 10.2 weeks old from Envigo, USA). Similar to the bovine brain experiments, 16 tissue samples were taken from the excised brain and T1, WC, and EPs were measured. But prior to euthanizing the animal (Pentane 1 ml injected into the thorax cage after strong

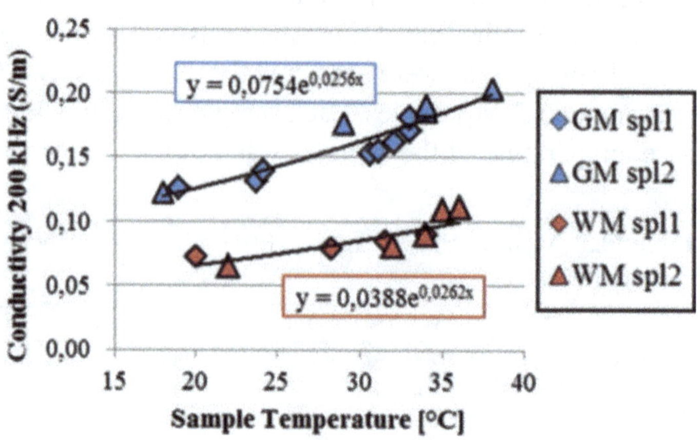

Fig. 20.6 Temperature-dependence of conductivity at 200 kHz for sample temperatures between ~18 and 38 °C. σ was measured for two GM (blue) and two WM (red) samples of one calf brain. Exponential trend lines are displayed for both tissues

anesthetization with ketamine/xylasine), we performed in vivo MRI imaging of the rat with the inclusion of the wEPT sequence (two T1w images with different TR_s).

In more detail, the live animal was anesthetized with isoflurane (2%) and placed in the rat head coil of the MRI in the prone position. The brain was first localized in the axial, sagittal, and coronal position. Twelve slices were chosen orthogonally to the brain covering the olfactory bulb (front part of brain) to the cerebellum (back of the brain) (Fig. 20.7 top). A T2w image and corresponding stacks of T1w images for T1 value estimation and wEPT analysis were acquired. General image sequence parameters remain the same as previously listed (compare Sect. 20.2.3) The whole brain was excised and placed in a rat brain slicer (Zivic labs) to subsequently extract cylindrical samples (Fig. 20.7 middle). Two blades with a measured thickness of 0.2–0.3 mm were inserted in the slicer. We used a 3.5 mm punch and obtained three samples from the front of the brain, six samples from the midbrain (Fig. 20.7 middle), three samples from the hindbrain, and four more samples of the remainder of the back brain, which were not imaged in vivo because it was outside the FOV.

After sample preparation, impedance measurements were taken with the LCR meter and their EPs estimated using the known sample dimensions. The T1 value of the samples was evaluated with MR imaging using the variable TR method, which includes images for further wEPT analysis (T1w images with TR_s/TR_l = 700/4000 ms). In order to monitor tissue dehydration in the MRI, the samples have been weighed pre- and post imaging. The sample WC was again estimated as the difference between wet and dry weights.

Experimental results predicted lower WC values in the rat samples than expected, which could be a result of the very small sample size possibly leading to fast drying of tissue. Furthermore, tissue samples of the rat are far more heterogeneous than in the calf. Thus, σ measurements of rat samples only showed a very minor increase in average values of GM compared to the WM samples. In conclusion, it is not recommended to use the 16 measurements of the rat samples to create a new wEPT model (finding coefficients that best match Eqs. 20.1 and 20.2a). Instead, we employed the wEPT models originating from calf measurements.

The ex vivo MRIs were analyzed in the same manner as calves 1–3 and calves 4–6, i.e., the two T1w images were loaded, maps were created, and mean values of estimated properties were calculated according to the segmentation masks of each sample. We used the previously described *mod_b* and *mod_1* settings for wEPT analysis. Table 20.4 summarizes average errors of wEPT estimates of WC, $\sigma^{200\text{kHz}}$, $\sigma^{1\text{MHz}}$ compared to measurements in all 16 rat1 samples. Higher deviations of measured values were observed for the rat versus calf data. The higher WC errors might be related to the measurement problems of WC encountered in the small rat samples.

Surprisingly, the model that performed the best for σ estimation with wEPT was *mod_b* (calves 1–3 data) with the lowest average relative errors for both frequencies, 200 kHz and 1 MHz. However, the worst case of average error is 23.9% for σ estimations at 200 kHz, which is still reasonable given the high heterogeneity in the

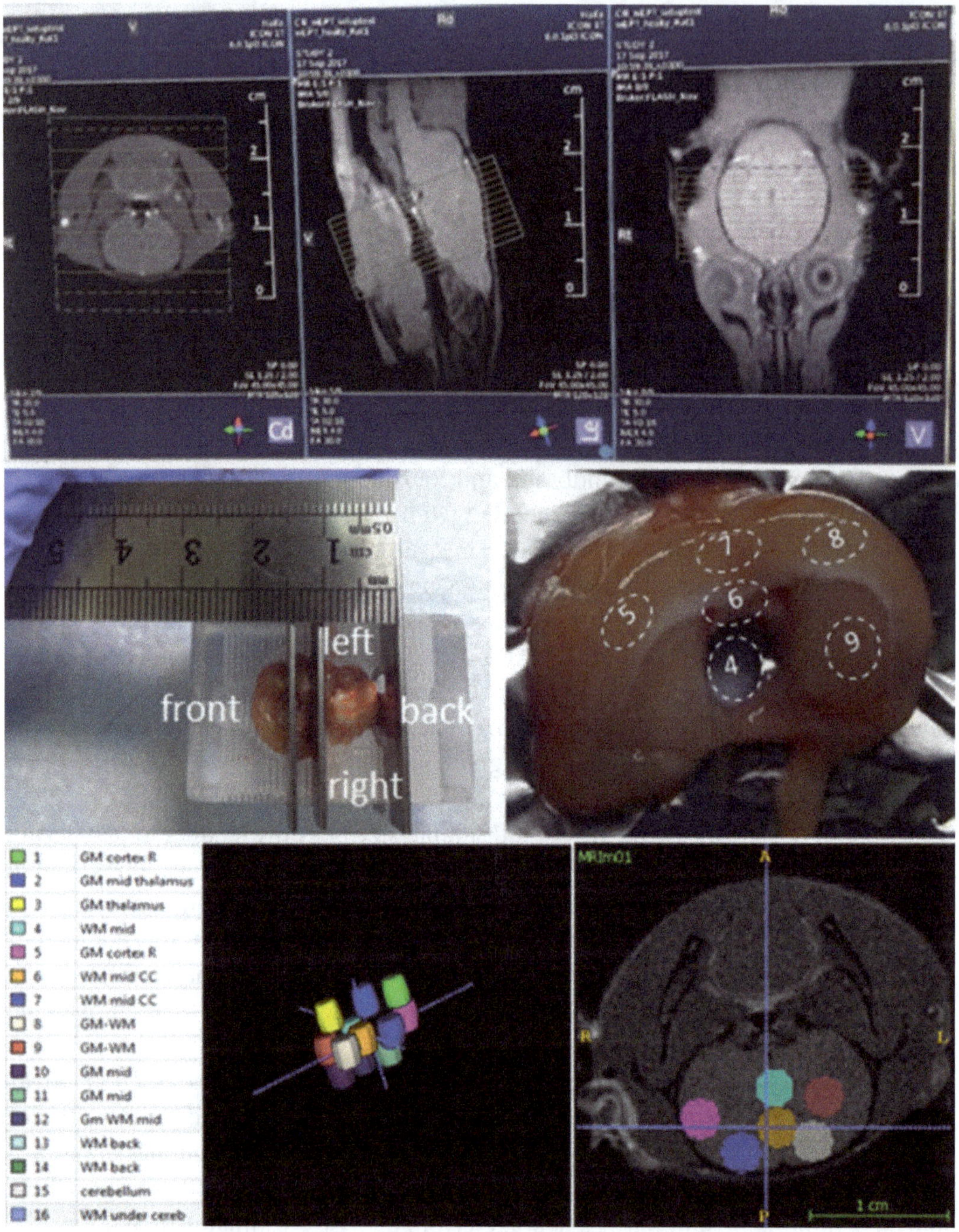

Fig. 20.7 Experimental procedures in healthy rat1. (top) The in vivo MRI localization of the brain in axial, coronal, and sagittal views. (middle) Excised brain in the brain slicer and slice of mid-brain, where one sample with id = 4 is already punched out. The locations of five additionally planned punch sites are indicated. (bottom) Corresponding sample segmentation in ITK-Snap along with a 3D rendering of 12 tissue samples with their labels and one corresponding slice through samples with the same id = {4–9}

Table 20.4 Average error of wEPT estimates in all brain samples of rat1 compared to measured values

	WC (%)	σ^{200kHz} (%)	σ^{1MHz} (%)
mod_b	7.4	21.7	19.1
mod_c	6.6	23.9	20.2

samples as compared to calf experiments. Also note that given the small sample size, many pixels in the MRI image corresponded to more than one tissue sample which might also affect the results. In general, reported mean values of wEPT estimations depend on the quality of the segmentation.

Subsequently, the in vivo MRI data of healthy rat1 was analyzed. This was the first time an imaging dataset of a live animal was used with a more realistic 3D representation of the excised samples (Fig. 20.7 bottom). However, recall that only 12 of the 16 samples were visible in the in vivo wEPT studies. The wEPT sequence consisted of two T1w images (TR_s/TR_l = 700/4000 ms) each with 12 slices through the rat's brain (Fig. 20.7). wEPT analysis again yielded maps of IR, WC, σ^{200kHz}, and σ^{1MHz}, which were created with MATLAB (data not shown, but compare Fig. 20.9 of a tumor-bearing rat) and the corresponding segmentation consisting of 3D masks for each tissue sample can be used to evaluate mean values of the estimations.

In addition to *mod_b* and *mod_c*, another wEPT model termed *mod_d* was created that used that accounts for the temperature dependency of T1 and σ based on the calves 4–6 dataset (Fig. 20.6). The T_1 relaxation value of tissue is not only dependent on the imaging frequency but also on the temperature. In order to account for these changes, the estimated T1 values in the ex vivo samples of rat1 at 20–22 °C were compared to the corresponding T1 values in the in vivo images at 35–37 °C. Results predicted a multiplication factor of 1.3003 of T1 values for increasing the temperature. Since the WC of tissue is independent of temperature, it follows that the calculated IR has to be multiplied by a factor of $IR^{TempCor}$ = 0.8587 to obtain adapted IR values that are valid for in vivo MRI data. The T1 plateau for the high CSF value at 37 °C could not be reached; therefore, no reliable results could be obtained for the CSF with our scanner. Thus, for temperature adjustment of the CSF values, the same factor of 0.8587 was used to reduce the IR. After correcting IR values, curve fitting was applied to find the new coefficients of Eq. 20.1 in *mod_d*, which were found to be w_1 = 1.2598 and w_2 = 1.0384. Subsequently, the conductivity values of all brain tissue samples of calves 4–6 were multiplied by a factor of $\sigma^{TempCor}$ = 1.49 (at 200 kHz and 1 MHz) to account for increased conductivity at an in vivo temperature of 37 °C. A previous experiment (Fig. 20.6) revealed that the increase of σ with temperature is only slightly different for GM and WM; therefore, the average of the two evaluated factors was chosen for $\sigma^{TempCor}$. The same factor was used for conductivity adjustments of all brain samples at 1 MHz, although the preliminary study was only performed for 200 kHz. The conductivity of CSF was assumed to be 1.67 S/m at both frequencies. For the completion of *mod_d*, curve fitting was applied to find the coefficients in Eq. 20.2a, c_1 = 0.1214,

c_2 = 2.4469e-8, c_3 = 18.2031 for 200 kHz and c_1 = 0.1481, c_2 = 6.9562e-8, c_3 = 17.1286 for 1 MHz.

It should be mentioned that curve fitting was also repeated for σ mapping for adjusting sample values with a tissue-type specific factor (one for GM, one for WM). The results for the coefficients c_1,c_2,c_3 were the same for 200 kHz and only slightly different at 1 MHz. Therefore, any other temperature-dependent adaptations of the conductivity were neglected.

Average values for WC, $\sigma^{200\text{kHz}}$, and $\sigma^{1\text{MHz}}$ estimations of the three wEPT models (*mod_b*, *mod_c*, *mod_d*) were evaluated in the 12 tissue samples visible in the in vivo study. The average error to corresponding measurements are presented in Table 20.5. For comparing measured σ values to wEPT estimates with *mod_d*, the recorded mean values in the samples which correspond to in vivo temperature have to be adjusted by multiplying with 0.67 (the reciprocal of σ^{TempCor}). As was the case for the ex vivo studies, errors of WC estimates are relatively high, which might be due to already mentioned measurement errors. Surprisingly, conductivity estimations were best for *mod_b* and higher average errors were found for *mod_c*, as expected. Model *mod_d*, which was specifically derived for in vivo studies, shows average errors for the σ estimations of 18.6% and 16.8% for 200 kHz and 1 MHz, respectively.

It should be noted that the difference in average errors in the in vivo wEPT model might appear lower compared to the ex vivo errors (Table 20.4) because the segmentation of the in vivo images represents a larger portion of the actual tissue sample. The brain samples of the rat brain are far more heterogeneous than those of the calf, making it difficult to estimate EP values from the impedance measurements. Furthermore, the small brain samples are only partially visible in the ex vivo MRIs, which may explain the variation in reported values. Therefore, to improve the measurement and imaging data for the tumor-bearing rats, the protocol was adjusted such that larger (but fewer) tissue samples were collected.

Four tumor-bearing F334 rats (rats 2–5) have been investigated during this study. The main goals were to create in vivo maps of WC and σ and to obtain measurements of excised tumor samples and compare them to measurements of their contralateral counterparts. Two tumor types, RG2 cells in rat2 and F98 cells in rats 3–5, were considered.

Under general anesthesia (isoflurane for induction and then i.p. injection of xylasine 10 mg/kg/ketamine 75 mg/kg), the rat head was shaved and the animal immobilized on the stereotactic unit. After disinfection and incision of the skin of the

Table 20.5 Average error of *in vivo* wEPT estimates from the 12 imaged brain samples of rat1 compared to measured values

	WC (%)	$\sigma^{200\text{kHz}}$ (%)	$\sigma^{1\text{MHz}}$ (%)
mod_b	5.2	12.5	11.1
mod_c	7.2	41.8	32.9
mod_d	5.5	18.6	16.8

head, a hole was drilled through the skull at coordinates corresponding to the right thalamus (2.5 mm lateral to midline, 1 mm anterior to the coronal fissure). A 10 μl mixture of cells (5000 cells in rats 2–3 and 10,000 cells in rats 4–5) and matrigel matrix ref. 354,263 (1:1) was slowly injected with a 0.5 cc insulin syringe 30 G, at a depth of 3 mm under the dura. The syringe was placed vertically. The muscles and skin were sutured 2 minutes after the injection. The analgesic Calmogine (Vitoquenol) was injected subcutaneously 10 mg, and another analgesic (Buprenorphine, 2 mg/ml) was delivered via the drinking water throughout the experiment.

MRI images were acquired at selected time points: rat2 was imaged 7, 13, an 15 days after cell injection; rat3 at 7,13, 16, and 19 days after injection; and rats 4–5 at 10 and 14 days. Before each imaging session, rats were anesthetized using 2% isoflurane in air (SomnoSuite, Kent Instruments) and placed in the prone position into the dedicated head coil. The MRI sequence protocol was identical to that of healthy rat1. However, the T2w axial scan now also contributes to tumor detection and to the brain slicing and sampling plans as well as the segmentation of the ex vivo samples and tumor. Figure 20.8 plots the estimated tumor volume in the four tumor-bearing rats according to the T2w MRIs that have been acquired at different days after tumor cell injection.

After the final in vivo MRI, each rat was sacrificed by overdose of pentobarbital and the whole brain was carefully removed in one piece as depicted in the top left of Fig. 20.9. The excised brain was placed in the brain slicer and three slices per animal were created. Samples were punched out using 5 mm punches whenever possible. One tumor sample and the corresponding contralateral sample were taken from the middle slice, as shown at the top of Fig. 20.9. The brain, the slices, and the tissue samples were covered with a glove or cellophane film when-

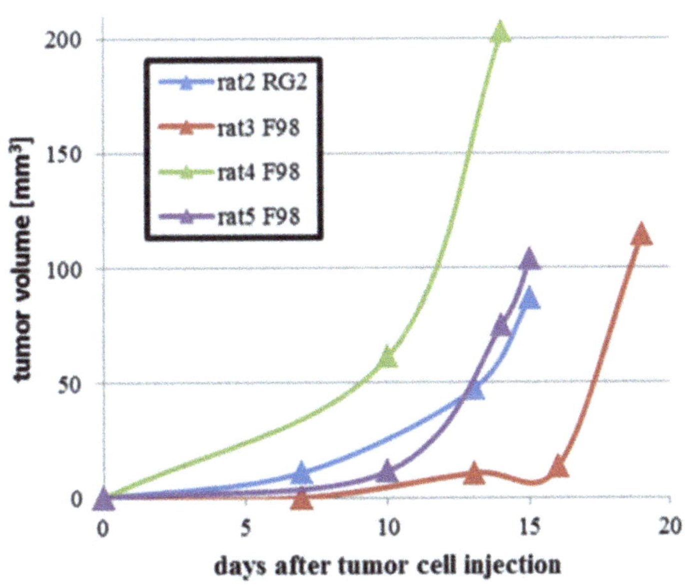

Fig. 20.8 Tumor volume estimated by T2w imaging segmentation over time for four tumor-bearing rats

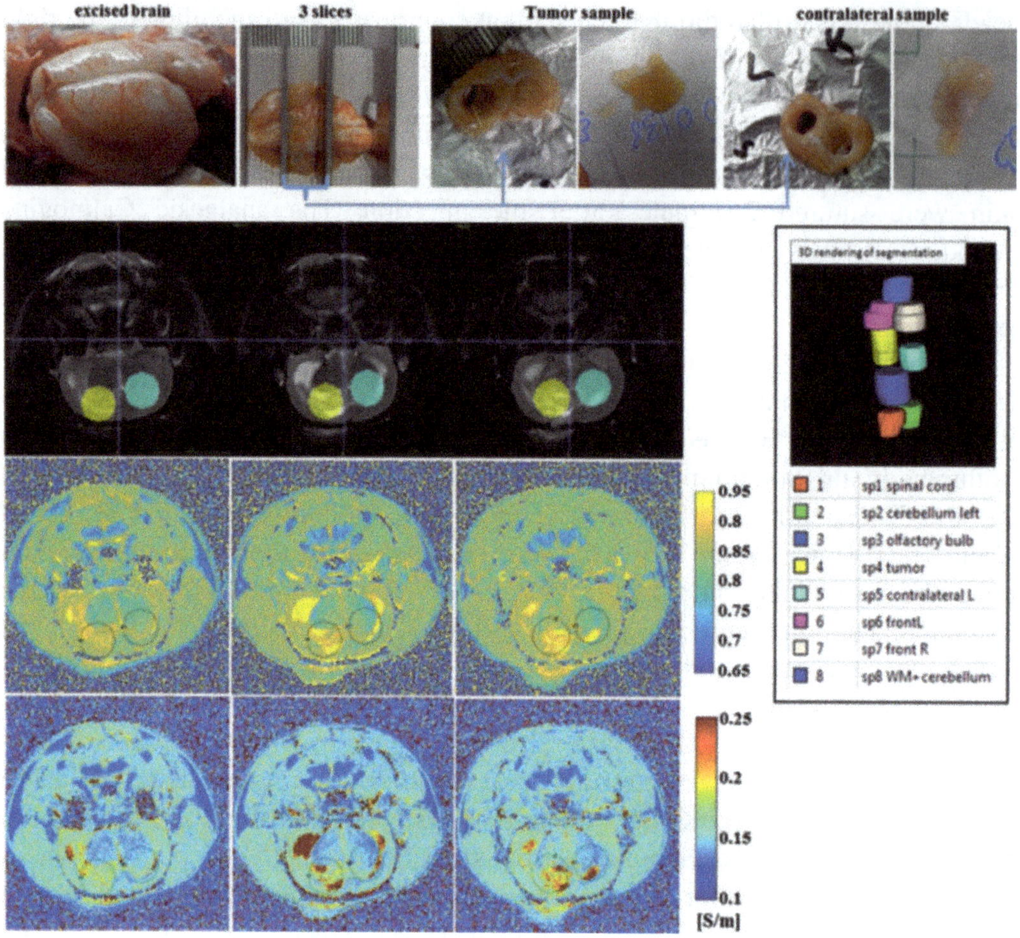

Fig. 20.9 Experimental procedures in F98 tumor-bearing rat5 for the final in vivo MRI images, which were acquired 14 days post-injection of tumor cells. (top) The excised brain was cut into three slices, the middle one containing the tumor and contralateral samples. The three slices corresponding to the middle brain slice containing the tumor (id = 4, yellow) and contralateral sample (id = 5, cyan). These three slices are reproduced in the T2w image with the overlaid segmentation (middle). A 3D rendering with the label description is provided to the right. (bottom) Corresponding WC and σ^{200kHz} maps are presented. Outlines of the segmentation are superimposed on these maps

ever possible during the experimental procedures to minimize dehydration. WC and EP measurements were obtained using the previously described LCR meter and the weighing method. Different numbers of samples were punched out from the four rat brains, the middle slice contains the tumor sample and the contralateral sample (Fig. 20.9). We did not measure EPs in all of the samples. However, we had combined measurements of WC and EPs in six samples of rats 2–4 and five samples of rat5 (id's = {4-8}), including tumor and contralateral samples in all animals.

It was expected that the tumor sample would have the highest WC and consequently also the highest σ values. This was almost always the case, except in rat2, where one sample from the right part of the front of the brain had an unrealistically high value of 94%. The measured WC values of the samples for each rat are shown in the top plot of Fig. 20.10. The red crosses indicate the tumor sample, which indeed have the highest measured WC in rats 3–5. The contralateral sample is indicated by a yellow cross. Also note, that rat2 not only had one very high measurement of a brain tissue sample but also the measured tumor WC is lower than the WC of the contralateral sample.

According to the σ measurements at 200 kHz (Fig. 20.10 bottom), the tumor only has the highest conductivity in rat3 and rat5, it is however always higher than the conductivity of the contralateral sample. This increase is even more pronounced for σ measurements at 1 MHz (data not shown).

Subsequently, we performed wEPT analysis with the final in vivo MRI data from the four tumor-bearing rats. As an example, WC and $\sigma^{200\text{kHz}}$ maps are reproduced for rat5 in the bottom part of Fig. 20.9. The tumor and contralateral samples were visible in three slices of the MRI dataset. Anatomical structures are well represented in both maps. The higher WC fraction and the higher conductivity in the tumor compared to the contralateral sample are clearly visible. The mean values of wEPT estimations of WC and σ were calculated according to the sample segmentation. Similar to the studies in healthy rat1, we considered *mod_b*, *mod_c*, and *mod_d*. Average errors obtained in all 23 samples from rats 2–5, including the tumor sample are presented in Table 20.6. A particularly low average error of WC estimations of 2.2% are predicted for *mod_b* and *mod_d*. Generally, the average errors for WC estimation are lower in the tumor-bearing rats than in healthy rat1, which might again indicate measurement errors in the WC of rat1. For wEPT esti-

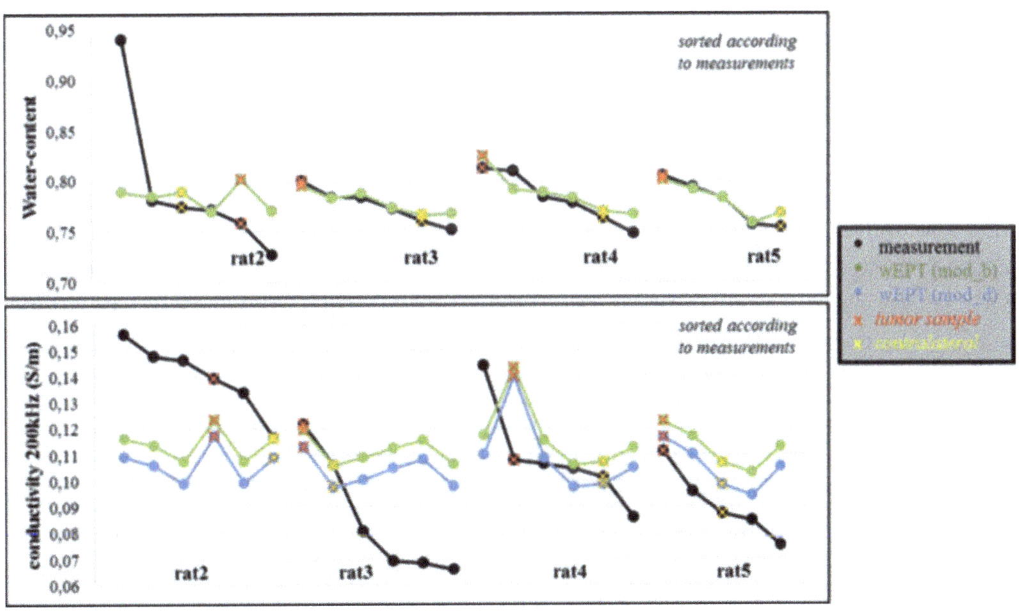

Fig. 20.10 Measured and wEPT-estimated values of WC (top) and $\sigma^{200\text{kHz}}$ (bottom) in six samples of rats 2–4 and five samples of rat5. The data is sorted according to the measurements (black). The mean values in the corresponding sample according to wEPT estimations are displayed, *mod_b* (green), *mod_d* (blue). Tumor samples are indicated with a red cross and contralateral samples in yellow

Table 20.6 Average error of *in vivo* wEPT estimates compared to measured values in all tissue samples (including the tumor sample) of rats 2–5

	WC (%)	σ^{200kHz} (%)	σ^{1MHz} (%)
mod_b	2.2	22.1	20.9
mod_c	6.7	110.2	80.2
mod_d	2.2	23.3	21.6

mations of σ, the best performing model is surprisingly *mod_b*. However, the average error for σ^{200kHz} in wEPT estimations with the in vivo model *mod_d* is only slightly higher with 23.3%. And as expected, *mod_c* does not produce accurate estimations of WC or σ.

The wEPT estimations of WC and σ^{200kHz} of the two best performing models are also displayed in Fig. 20.10 for each samples, *mod_b* in green and *mod_d* in blue. The top part of Fig. 20.10 indicates that WC estimations with the wEPT approach matches the measurements quite well. It is also likely that the wEPT estimations in rat2 are more realistic than the actual measurements, given the high outlier and also the low tumor WC. The trend for σ estimations is not the same when compared to the measurements in the rat samples (Fig. 20.10 bottom). When σ is estimated with the wEPT approach, the tumor always has the highest value, i.e., the sample with the red cross has the highest conductivity. Thus, the contralateral sample always shows lower conductivity than the tumor in each animal for measurements and wEPT-estimated values contrary to the measurements.

20.3 Summary and Discussion

In this study, we investigated if the wEPT approach can be adapted for mapping EPs, specifically the conductivity of brain tissue, between 200 kHz and 1 MHz. We conducted experimental measurements and wEPT analysis in a phantom study with BSA solutions, in tissue samples excised from calf and rat brains, and in tumor-bearing rat models. The wEPT analysis included model creation and estimation analysis, i.e., first the optimal coefficients of mapping functions for WC and EP have to be found, then the maps can be created and values in different areas can be estimated and compared to corresponding measurements.

The feasibility study with four BSA solutions of varying albumin concentrations predicted good agreement between experimentally measured values of IR, WC, σ, and ε at 200 kHz and 1 MHz and wEPT estimations (Table 20.1). We then performed extensive studies with excised calf brain samples. We measured properties (T1, WC, and EP between 20 Hz and 1 MHz) of excised samples of WM and GM (Fig. 20.1). New wEPT models have been developed based on this data (Figs. 20.2 and 20.5), i.e., optimal coefficients in mapping functions (IR→ WC, WC→ σ, WC→ ε) have been estimated by curve fitting. Maps of WC and EPs at 200 kHz and 1 MHz have been created for excised calf samples and pieces of calf brain with the wEPT imaging approach (Fig. 20.3 and Fig. 20.4). Tissue types can be distinguished in the maps and average errors compared to measurements are low for WC (<3.5%) and reasonable for σ estimations at 200 kHz (<13.6%), compare Tables 20.2 and 20.3.

The next set of studies was performed in rat models, using one healthy rat and four tumor-bearing rats. Experimental measurements revealed WC and EPs of brain tissue samples and tumor samples excised postmortem (Figs. 20.7 and 20.9). Previously derived wEPT model equations were used to study in vivo images of the rats. Furthermore, we created an adapted wEPT model that incorporates adjustments considering the temperature dependence of T1 values and σ of brain tissue (Fig. 20.6). WC and σ^{200kHz} maps were created for the animals and analyzed (Fig. 20.9). The brain anatomy, distribution of different tissue types, and also the tumor are clearly visible in the created maps. Average errors between measurements and wEPT estimations were calculated (Tables 20.5 and 20.6). WC estimations show low average errors and correspond very well to the measurements in each sample (Fig. 20.10). Average errors of σ^{200kHz} estimations in all samples including the tumor samples were reasonable (22–23%). Yet, wEPT estimations predict highest WC and σ in the tumor sample, which although expected was not that case for our measurements in all animals. For example rat2 and rat4 measurements predict higher σ values in healthy brain tissue.

However, at this point it should be noted that we estimated measurement errors for WC of ±1% and measurement errors of σ of ±10%. Measurement quality is dependent on tissue sample preparation and handling. For example, the differences between measurements of calves 1–3 and calves 4–6 may come from different storage conditions in the office vs. in the butcher refrigerator. Too cold temperature in the butcher refrigerator could generate the higher conductivity we observed in calves 1–3. Furthermore, we saw some discrepancies in the WC measurements of healthy rat1, possibly because of fast drying of samples of small sample size. Furthermore, the samples of rat brain are more heterogeneous and it is difficult to obtain samples that only represent one type of tissue. Figure 20.11 summarizes the measurements of WC and σ^{200kH}z in all tissue samples.

In the first studies, we also concentrated on estimations of the relative permittivity ε_r, which were neglected in later investigations. This is due to the fact that we are most interested in applying the wEPT approach for modeling induced electric fields during Optune treatment of patients with GBM. Computational studies have shown that at a frequency of 200 kHz, the field is almost completely shaped by the tissue conductivity [25]. Furthermore, relative errors from calves 1–3 data revealed better agreement than the conductivity mapping results. This is somewhat surprising given the fact that the permittivity is not strictly increasing for increasing water content of the tissue at our frequency range because the CSF with the highest water content has the lowest permittivity.

However, on a general note, all reported wEPT-estimated mean values are dependent on the underlying sample segmentation. For higher image resolution, a bigger section of the samples can be imaged and used for the estimations. For ex vivo wEPT studies, we usually had only one MRI slice through the sample with a lower

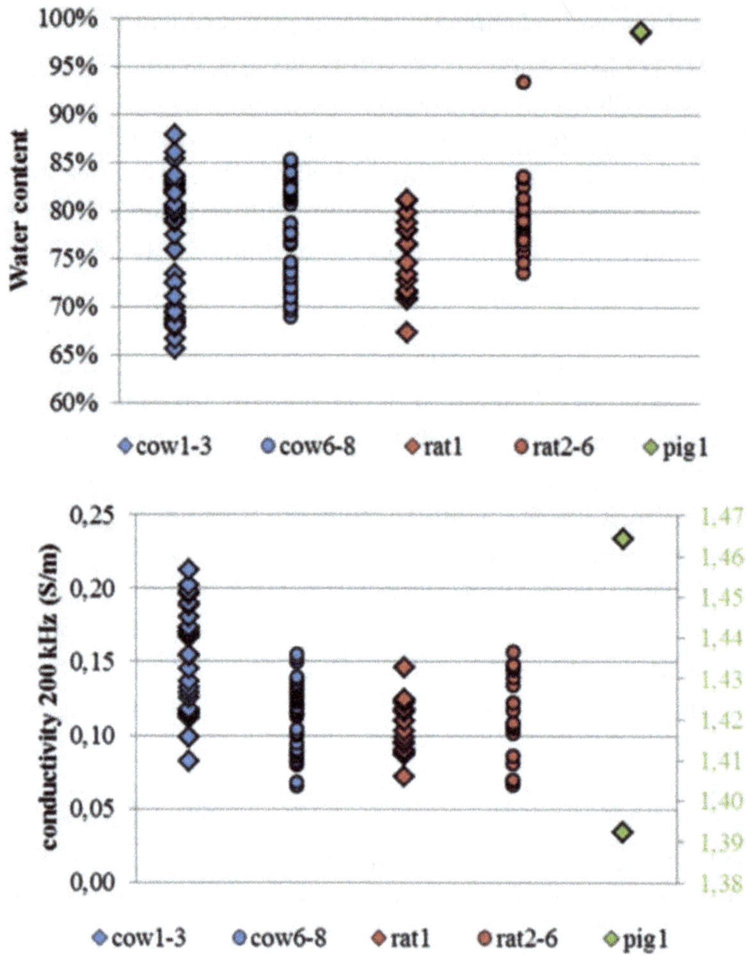

Fig. 20.11 Measurements of WC (top) and $\sigma^{200\mathrm{kHz}}$ (bottom) all tissue samples originating from different animal samples

number of pixels contributing to the tissue, compared to in vivo wEPT studies with a more realistic 3D representation of samples. Yet, additional experiments might be necessary for validation of the adapted in vivo wEPT model. Specifically, the temperature dependence of EPs of brain tissue might be investigated in more detail.

Also, wEPT is an imaging technique for estimating WC and EPs. Thus, the results are dependent on imaging quality and general procedures for reprocessing (such as denoising algorithms) might be studied in more detail. The authors of the original wEPT article further investigated the effect of B_1 inhomogeneities and found that errors may reach ±20% for σ and ± 11% for ε_r estimations. Yet their approach was described for a frequency of 128 MHz and it was conducted with a 3 T MRI scanner.

In conclusion, with our adapted wEPT approach, the frequency independent WC maps produce reliable results. It remains questionable if the relationship between WC and EPs of healthy and pathological tissues for frequencies in the 100–1000 kHz range is sufficient. At frequencies lower than 10 kHz, cell membranes act as capacitive elements because of their low conductivity. As the frequency increases, this membrane capacitance effect disappears, allowing electric currents to flow more freely according to the local ionic conductivity. As a result, the EPs of the tissues become highly correlated with the tissue water fraction. According to our results, adapting wEPT to frequencies between 100 and 1000 kHz is likely at the verge of producing reliable results. In future studies, the approach might be enhanced by including adaptations according to additional radiological features derived from other imaging modalities, such as T2-weighted imaging or diffusion imaging.

Improving and refining an approach for non-inversely mapping the EPs within the brain would be of high interest for planning and adapting the Optune treatment. Some studies have already been conducted with optimized EPT sequences for reconstructing EPs in brain tumors [48–50]. Results predict higher conductivity in the tumor compared to healthy tissue and a greater variability of EPs in individual tumor areas among patients. However, the wEPT approach would be of particular interest because it only requires two T1w images, which resemble the conventional T1 and proton density (PD) images that are acquired routinely. A preliminary study has already been conducted and WC and σ maps have been created for three GBM patients. Results will be presented in a future publication.

References

1. Miklavčič, D., Pavšelj, N., & Hart, F. X. (2006). Electric properties of tissues. In M. Akay (Ed.), *Wiley encyclopedia of biomedical engineering*. https://doi.org/10.1002/9780471740360.ebs0403.
2. Collins, C. M., & Wang, Z. (2011). Calculation of radiofrequency electromagnetic fields and their effects in MRI of human subjects. *Magnetic Resonance in Medicine, 65*(5), 1470–1482.
3. Voigt, T., Homann, H., Katscher, U., & Doessel, O. (2012). Patient-individual local SAR determination: In vivo measurements and numerical validation. *Magnetic Resonance in Medicine, 68*(4), 1117–1126.
4. Zhang, X., Schmitter, S., Van de Moortel, P.-F., Liu, J., & He, B. (2013). From complex B1 mapping to local SAP estimation for human brain MR imaging using multi-channel transceiver coil at 7T. *IEEE Transactions on Medical Imaging, 32*(6), 1058–1067.
5. Restivo, M. C., et al. (2015). Local specific absorption rate in brain tumors at 7 tesla. *Magnetic Resonance in Medicine, 00*, n/a-n/a.
6. Makarov, S., et al. (2017). Virtual human models for electromagnetic studies and their applications. *IEEE Reviews in Biomedical Engineering, 10*, 95–121.
7. Barker, A. T., Jalinous, R., & Freeston, I. L. (1985). Non-invasive magnetic stimulation of human motor cortex. *Lancet, 1*, 1106–1107.
8. Nitsche, M. A., & Paulus, W. (2001). Sustained excitability elevations induced by transcranial DC motor cortex stimulations in humans. *Neurology, 57*, 1899–1901.
9. Miranda, P. C., Salvador, R., Wenger, C., & Fernandes, S. R. (2016). Computational models of non-invasive brain and spinal cord stimulation. In *Engineering in medicine and biology society (EMBC), 2016 38th annual international conference of the IEEE.*
10. Joines, W. T., Zhang, Y., & Jirtle, R. L. (1994). The measured electrical properties of normal and malignant human tissues from 50 to 900 MHz. *Medical Physics, 21*(4), 547–550.

11. Haemmerich, D., Staelin, S. T., Tsai, J. Z., Tungjitkusolmun, S., Mahvi, D. M., & Webster, J. G. (2003). In vivo electrical conductivity of hepatic tumours. *Physiological Measurement, 24*(2), 251–260.
12. Haemmerich, D., Schutt, D. J., Wright, A. W., Webster, J. G., & David, M. (2009). Electric conductivity measurements of excised human metastatic liver tumours before and after thermal ablation. *Physiological Measurement, 30*(5), 459–466.
13. Surowiec, A. J., Stuchly, S. S., Barr, J. R., & Swarup, A. (1988). Dielectric properties of breast carcinoma and the surrounding tissues. *IEEE Transactions on Biomedical Engineering, 35*(4), 257–263.
14. Lu, Y., Li, B., Xu, J., & Yu, J. (1992). Dielectric properties of human glioma and surrounding tissue. *International Journal of Hyperthermia, 8*(6), 755–760.
15. Foster, K. R., & Schwan, H. P. (1989). Dielectric properties of tissues and biological materials: A critical review. *Critical Reviews in Biomedical Engineering, 17*(1), 25–104.
16. Laufer, S., Ivorra, A., Reuter, V. E., Rubinsky, B., & Solomon, S. B. (2010). Electrical impedance characterization of normal and cancerous human hepatic tissue. *Physiological Measurement, 31*(7), 995–1009.
17. Ivorra, A., Al-Sakere, B., Rubinsky, B., & Mir, L. M. (2009). In vivo electrical conductivity measurements during and after tumor electroporation: Conductivity changes reflect the treatment outcome. *Physics in Medicine and Biology, 54*(19), 5949–5963.
18. Kirson, E. D., et al. (2004). Disruption of cancer cell replication by alternating electric fields. *Cancer Research, 64*(9), 3288–3295.
19. Kirson, E. D., et al. (2007). Alternating electric fields arrest cell proliferation in animal tumor models and human brain tumors. *Proceedings of the National Academy of Sciences of the United States of America, 104*(24), 10152–10157.
20. Stupp, R., et al. (2012). NovoTTF-100A versus physician's choice chemotherapy in recurrent glioblastoma: A randomised phase III trial of a novel treatment modality. *European Journal of Cancer, 48*(14), 2192–2202.
21. Stupp, R., et al. (2015). Maintenance therapy with tumor-treating fields plus temozolomide vs temozolomide alone for glioblastoma. A randomized clinical trial. *JAMA, 314*(23), 2535–2543.
22. Fonkem, E., & Wong, E. T. (2012). NovoTTF-100A: A new treatment modality for recurrent glioblastoma. *Expert Review of Neurotherapeutics, 12*, 895–899.
23. Mrugala, M. M., et al. (2014). Clinical practice experience with NovoTTF-100A™ system for glioblastoma: The patient registry dataset (PRiDe). *Seminars in Oncology, 41 Suppl 6*(5), S4–S13.
24. Chaudhry, A., et al. (2015). NovoTTF™-100A system (tumor treating fields) transducer array layout planning for glioblastoma: A NovoTAL™ system user study. *World Journal of Surgical Oncology, 13*, 316.
25. Miranda, P. C., Mekonnen, A., Salvador, R., & Basser, P. J. (2014). Predicting the electric field distribution in the brain for the treatment of glioblastoma. *Physics in Medicine and Biology, 59*(15), 4137–4147.
26. Wenger, C., Salvador, R., Basser, P. J., & Miranda, P. C. (2015). The electric field distribution in the brain during TTFields therapy and its dependence on tissue dielectric properties and anatomy: A computational study. *Physics in Medicine and Biology, 60*(18), 7339.
27. Wenger, C., Salvador, R., Basser, P. J., & Miranda, P. C. (2016). Improving tumor treating fields treatment efficacy in patients with glioblastoma using personalized array layouts. *International Journal of Radiation Oncology, Biology, Physics, 94*(5), 1137.
28. Korshoej, A. R., Hansen, F. L., Thielscher, A., Von Oettingen, G. B., Christian, J., & Hedemann, S. (2017). Impact of tumor position, conductivity distribution and tissue homogeneity on the distribution of tumor treating fields in a human brain: A computer modeling study. *PLoS One, 12*(6), e0179214.
29. Lok, E., San, P., Hua, V., Phung, M., & Wong, E. T. (2017). Analysis of physical characteristics of tumor treating fields for human glioblastoma. *Cancer Medicine, 6*(6), 1286–1300.

30. Gabriel, C., Gabriel, S., & Corthout, E. (1996). The dielectric properties of biological tissues: I. Literature survey. *Physics in Medicine and Biology, 41*(11), 2231–2249.
31. Gabriel, C., Peyman, A., & Grant, E. H. (2009). Electrical conductivity of tissue at frequencies below 1 MHz. *Physics in Medicine and Biology, 54*(16), 4863–4878.
32. Lu, Y., Li, B., Xu, J., & Yu, J. (1992). Dielectric properties of human glioma and surrounding. *International Journal of Hyperthermia, 8*(6), 755–760.
33. Wenger, C., et al. (2018). A review on tumor treating fields (TTFields): Clinical implications inferred from computational modeling. *IEEE Reviews in Biomedical Engineering, 11*, 195–207.
34. Metherall, P., Barber, D. C., Smallwood, R. H., & Brown, B. H. (1996). Three-dimensional electrical impedance tomography. *Nature, 380*(6574), 509–512.
35. Griffiths, H., Stewart, W. R., & Gough, W. (1999). Magnetic induction tomography. A measuring system for biological tissues. *Annals of the New York Academy of Sciences, 873*, 335–345.
36. Woo, E. J., & Seo, J. K. (2008). Magnetic resonance electrical impedance tomography (MREIT) for high-resolution conductivity imaging. *Physiological Measurement, 29*, R1–R26.
37. Joy, M. L., Scott, G. C., & Henkelman, R. M. (1989). In vivo detection of applied electric currents by magnetic resonance imaging. *Magnetic Resonance Imaging, 7*(1), 89–94.
38. Katscher, U., Kim, D.-H., & Seo, J. K. (2013). Recent progress and future challenges in MR electric properties tomography. *Computational and Mathematical Methods in Medicine, 2013*, 546562.
39. Haacke, E. M., Brown, R. W., Thompson, M. R., & Venkatesan, R. (1999). *Magnetic resonance imaging physical principles and sequence design*. New York: Wiley.
40. Zhang, X., Liu, J., & He, B. (2014). Magnetic resonance based electrical properties tomography: A review. *IEEE Reviews in Biomedical Engineering, 7*, 87–96.
41. Michel, E., Hernandez, D., & Lee, S. Y. (2017). Electrical conductivity and permittivity maps of brain tissues derived from water content based on T 1 -weighted acquisition. *Magnetic Resonance in Medicine, 77*, 0094–1103.
42. Schwan, H. P. (1957). Electrical properties of tissue and cell suspensions. *Advances in Biological and Medical Physics, 5*, 147–209.
43. Stoy, R. D., Foster, K. R., & Schwan, H. P. (1982). Dielectric properties of mammalian tissues from 0.1 to 100 MHz; a summary of recent data. *Physics in Medicine and Biology, 27*(4), 501–513.
44. Foster, K. R., Schepps, J. L., Stoy, R. D., & Schwan, H. P. (1979). Dielectric properties of brain tissues between 0.01 and 10 GHz. *Physics in Medicine and Biology, 24*, 1177–1187.
45. Smith, S. R., & Foster, K. R. (1985). Dielectric properties of low-water-content tissues. *Physics in Medicine and Biology, 30*, 965–973.
46. Kiricuta, I.-C., & Simplăceanu, V. (1975). Tissue water content and nuclear magnetic resonance in normal and tumor tissues. *Cancer Research, 35*(5), 1164–1167.
47. Wenger, C., Salvador, R., Basser, P. J., & Miranda, P. C. (2015). The electric field distribution in the brain during TTFields therapy and its dependence on tissue dielectric properties and anatomy : A computational study. *Physics in Medicine and Biology, 60*(18), 7339–7357.
48. van Lier, A. L., et al. (2011). Electrical conductivity imaging of brain tumours. In *Proceedings of the 19th Annual Meeting of ISMRM* (Vol. 19, p. 4464). Montréal.
49. Huhndorf, M., Stehning, C., Rohr, A., Helle, M., Katscher, U., & Jansen, O. (2013). Systematic brain tumor conductivity study with optimized EPT sequence and reconstruction algorithm. *Proceedings of the 21th Annual Meeting of ISMRM, Salt Lake City, 66*(2011), 3626.
50. Voigt, T. (2011). Imaging conductivity using electric properties tomography – initial clinical results in glioma patients. In *2011 XXXth URSI general assembly and scientific symposium* (pp. 1–4).

Index

S. Makarov et al. (eds.), *Brain and Human Body Modeling*,
https://doi.org/10.1007/978-3-030-21293-3

www.ingramcontent.com/pod-product-compliance
Lightning Source LLC
LaVergne TN
LVHW082110200926
848484LV00008B/1937